Seat Leon
Owners Workshop Manual

Mark Storey

Models covered

(6408 - 464)

Leon 'Mk 2' Hatchback, inc. Cupra & Cupra R
Petrol: 1.6 litre (1595cc) & 2.0 litre (1984cc)
Turbo-diesel: 1.6 litre (1598cc), 1.9 litre (1896cc) & 2.0 litre (1968cc)

Includes coverage of models with DSG transmission
Does NOT cover 1.2, 1.4 or 1.8 litre petrol models, or 2.0 litre TSI petrol models with engine code CCZB
Does NOT cover fully-automatic transmission, 'Flex-fuel' models or new Leon 'Mk 3' range introduced October 2012

© Haynes Publishing 2018

ABCDE
FGHIJ
KLMNO
PQRST

A book in the **Haynes Owners Workshop Manual Series**

ISBN **978 1 78521 408 0**

British Library Cataloguing in Publication Data
A catalogue record for this book is available from the British Library.

Printed in Malaysia

Haynes Publishing
Sparkford, Yeovil, Somerset BA22 7JJ, England

Haynes North America, Inc
859 Lawrence Drive, Newbury Park, California 91320, USA

Printed using NORBRITE BOOK 48.8gsm (CODE: 40N6533) from NORPAC; procurement system certified under Sustainable Forestry Initiative standard. Paper produced is certified to the SFI Certified Fiber Sourcing Standard (CERT - 0094271)

Contents

Contents

The Leon models covered by this manual date from September 2005 to September 2012. These are commonly known as Mk2 (or 1P) models. Only one body style is available.

Models are available with a range of petrol or diesel engines. Petrol engines range in size from 1.2 litres to 2.0 litres, but note that only the popular 1.6 litre and 2.0 litre petrol engines are covered by this manual. Diesel engines range in size from 1.6 litre to 2.0 litre. Most models use 1.9 or 2.0 litre 'Pumpe Duse' (PD) engines. These are unit injector engines where the injector pressurises and injects the fuel. On later models (around 2009) 'Common Rail' (CR) engines were introduced to the range. On CR engines the injectors are supplied with high pressure fuel from a reservoir (the 'common rail'). Note that both CR and PD engines were both available until the end of production. All the engines are of a well-proven design and, provided regular maintenance is carried out, are unlikely to give trouble.

Fully-independent front and rear suspension is fitted, with the components attached to a subframe assembly; the rear suspension uses trailing arms together with multi-link transverse arms.

A five or six-speed manual gearbox is fitted, with a six-speed DSG semi-automatic transmission also available.

A wide range of standard and optional equipment is available within the model range to suit most tastes, including air conditioning (with manual or fully automatic climate control).

For the home mechanic, Leon models are straightforward vehicles to maintain, and most of the items requiring frequent attention are easily accessible.

Your Leon Manual

The aim of this manual is to help you get the best value from your vehicle. It can do so in several ways. It can help you decide what work must be done (even should you choose to get it done by a garage). It will also provide information on routine maintenance and servicing, and give a logical course of action and diagnosis when random faults occur. However, it is hoped that you will use the manual by tackling the work yourself. On simpler jobs it may even be quicker than booking the car into a garage and going there twice, to leave and collect it. Perhaps most important, a lot of money can be saved by avoiding the costs a garage must charge to cover its labour and overheads.

The manual has drawings and descriptions to show the function of the various components so that their layout can be understood. Tasks are described and photographed in a clear step-by-step sequence.

References to the 'left' and 'right' of the vehicle are in the sense of a person in the driver's seat facing forward.

Acknowledgements

Thanks are due to Draper Tools Limited, who provided some of the workshop tools, and to all those people at Sparkford who helped in the production of this manual.

This manual is not a direct reproduction of the vehicle manufacturer's data, and its publication should not be taken as implying any technical approval by the vehicle manufacturers or importers.

We take great pride in the accuracy of information given in this manual, but vehicle manufacturers make alterations and design changes during the production run of a particular vehicle of which they do not inform us. No liability can be accepted by the authors or publishers for loss, damage or injury caused by any errors in, or omissions from, the information given.

Seat Leon

Working on your car can be dangerous. This page shows just some of the potential risks and hazards, with the aim of creating a safety-conscious attitude.

General hazards

Scalding

• Don't remove the radiator or expansion tank cap while the engine is hot.
• Engine oil, transmission fluid or power steering fluid may also be dangerously hot if the engine has recently been running.

Burning

• Beware of burns from the exhaust system and from any part of the engine. Brake discs and drums can also be extremely hot immediately after use.

Crushing

• When working under or near a raised vehicle, always supplement the jack with axle stands, or use drive-on ramps.
Never venture under a car which is only supported by a jack.

• Take care if loosening or tightening high-torque nuts when the vehicle is on stands. Initial loosening and final tightening should be done with the wheels on the ground.

Fire

• Fuel is highly flammable; fuel vapour is explosive.
• Don't let fuel spill onto a hot engine.
• Do not smoke or allow naked lights (including pilot lights) anywhere near a vehicle being worked on. Also beware of creating sparks (electrically or by use of tools).
• Fuel vapour is heavier than air, so don't work on the fuel system with the vehicle over an inspection pit.
• Another cause of fire is an electrical overload or short-circuit. Take care when repairing or modifying the vehicle wiring.
• Keep a fire extinguisher handy, of a type suitable for use on fuel and electrical fires.

Electric shock

• Ignition HT and Xenon headlight voltages can be dangerous, especially to people with heart problems or a pacemaker. Don't work on or near these systems with the engine running or the ignition switched on.

• Mains voltage is also dangerous. Make sure that any mains-operated equipment is correctly earthed. Mains power points should be protected by a residual current device (RCD) circuit breaker.

Fume or gas intoxication

• Exhaust fumes are poisonous; they can contain carbon monoxide, which is rapidly fatal if inhaled. Never run the engine in a confined space such as a garage with the doors shut.
• Fuel vapour is also poisonous, as are the vapours from some cleaning solvents and paint thinners.

Poisonous or irritant substances

• Avoid skin contact with battery acid and with any fuel, fluid or lubricant, especially antifreeze, brake hydraulic fluid and Diesel fuel. Don't syphon them by mouth. If such a substance is swallowed or gets into the eyes, seek medical advice.
• Prolonged contact with used engine oil can cause skin cancer. Wear gloves or use a barrier cream if necessary. Change out of oil-soaked clothes and do not keep oily rags in your pocket.
• Air conditioning refrigerant forms a poisonous gas if exposed to a naked flame (including a cigarette). It can also cause skin burns on contact.

Asbestos

• Asbestos dust can cause cancer if inhaled or swallowed. Asbestos may be found in gaskets and in brake and clutch linings. When dealing with such components it is safest to assume that they contain asbestos.

Special hazards

Hydrofluoric acid

• This extremely corrosive acid is formed when certain types of synthetic rubber, found in some O-rings, oil seals, fuel hoses etc, are exposed to temperatures above 4000C. The rubber changes into a charred or sticky substance containing the acid. *Once formed, the acid remains dangerous for years. If it gets onto the skin, it may be necessary to amputate the limb concerned.*
• When dealing with a vehicle which has suffered a fire, or with components salvaged from such a vehicle, wear protective gloves and discard them after use.

The battery

• Batteries contain sulphuric acid, which attacks clothing, eyes and skin. Take care when topping-up or carrying the battery.
• The hydrogen gas given off by the battery is highly explosive. Never cause a spark or allow a naked light nearby. Be careful when connecting and disconnecting battery chargers or jump leads.

Air bags

• Air bags can cause injury if they go off accidentally. Take care when removing the steering wheel and trim panels. Special storage instructions may apply.

Diesel injection equipment

• Diesel injection pumps supply fuel at very high pressure. Take care when working on the fuel injectors and fuel pipes.

⚠ *Warning: Never expose the hands, face or any other part of the body to injector spray; the fuel can penetrate the skin with potentially fatal results.*

Remember...

DO

• Do use eye protection when using power tools, and when working under the vehicle.

• Do wear gloves or use barrier cream to protect your hands when necessary.

• Do get someone to check periodically that all is well when working alone on the vehicle.

• Do keep loose clothing and long hair well out of the way of moving mechanical parts.

• Do remove rings, wristwatch etc, before working on the vehicle – especially the electrical system.

• Do ensure that any lifting or jacking equipment has a safe working load rating adequate for the job.

DON'T

• Don't attempt to lift a heavy component which may be beyond your capability – get assistance.

• Don't rush to finish a job, or take unverified short cuts.

• Don't use ill-fitting tools which may slip and cause injury.

• Don't leave tools or parts lying around where someone can trip over them. Mop up oil and fuel spills at once.

• Don't allow children or pets to play in or near a vehicle being worked on.

The following pages are intended to help in dealing with common roadside emergencies and breakdowns. You will find more detailed fault finding information at the back of the manual, and repair information in the main chapters.

If your car won't start and the starter motor doesn't turn

- ☐ If it's a model with automatic transmission, make sure the selector is in the P or N position.
- ☐ Open the bonnet and make sure that the battery terminals are clean and tight.
- ☐ Switch on the headlights and try to start the engine. If the headlights go very dim when you're trying to start, the battery is probably flat. Try jump starting using another car.

If your car won't start even though the starter motor turns as normal

- ☐ Is there fuel in the tank?
- ☐ Is there moisture on electrical components under the bonnet? Switch off the ignition, and then wipe off any obvious dampness with a dry cloth. Spray a water-repellent aerosol product (WD-40 or equivalent) on the fuel system electrical connectors like those shown in the photos. Note that diesel engines do not normally suffer from damp.

A Check the condition and security of the battery connections (under cover).

B Check the fuses in the fusebox located on the left-hand side of the engine compartment.

C Check the security of all the accessible electrical connectors

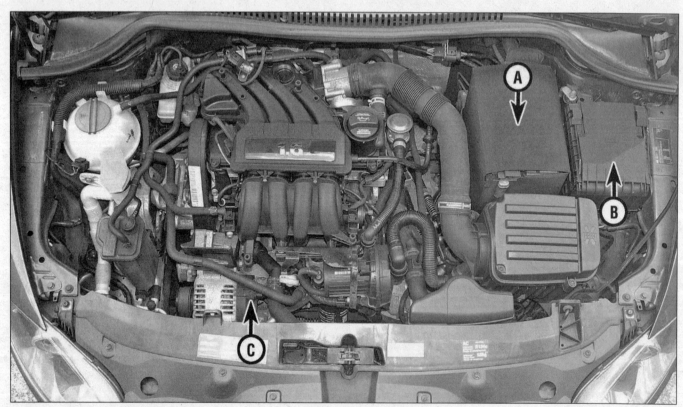

Check that electrical connections are secure (with the ignition switched off) and spray them with a water-dispersant spray like WD-40 if you suspect a problem due to damp.

Jump starting

1 The battery has been drained by repeated attempts to start, or by leaving the lights on.

2 The charging system is not working properly (alternator drivebelt slack or broken, alternator wiring fault or alternator itself faulty).

3 The battery itself is at fault (electrolyte low, or battery worn out).

When jump-starting a car, observe the following precautions:

Caution: Remove the key in case the central locking engages when the jump leads are connected.

✓ Before connecting the booster battery, make sure that the ignition is switched off.
✓ Ensure that all electrical equipment (lights, heater, wipers, etc) is switched off.
✓ Take note of any special precautions printed on the battery case.
✓ Make sure that the booster battery is the same voltage as the discharged one in the vehicle.

✓ If the battery is being jump-started from the battery in another vehicle, the two vehicles MUST NOT TOUCH each other.
✓ Make sure that the transmission is in neutral (or PARK, in the case of automatic transmission).

 Budget jump leads can be a false economy, as they often do not pass enough current to start large capacity or diesel engines. They can also get hot.

1 Connect one end of the red jump lead to the positive (+) terminal of the flat battery

2 Connect the other end of the red lead to the positive (+) terminal of the booster battery.

3 Connect one end of the black jump lead to the negative (-) terminal of the booster battery

4 Connect the other end of the black jump lead to a bolt or bracket on the engine block, well away from the battery, on the vehicle to be started.

5 Make sure that the jump leads will not come into contact with the fan, drive-belts or other moving parts of the engine.

6 Start the engine using the booster battery and run it at idle speed. Switch on the lights, rear window demister and heater blower motor, then disconnect the jump leads in the reverse order of connection. Turn off the lights etc.

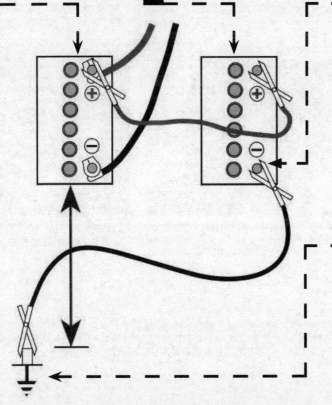

Wheel changing

⚠️ *Warning: Do not change a wheel in a situation where you risk being hit by other traffic. On busy roads, try to stop in a lay-by or a gateway. Be wary of passing traffic while changing the wheel – it is easy to become distracted by the job in hand.*

Note: *Some models do not have a spare wheel supplied as standard, instead a compressor and bottle of tyre sealant is supplied. It is possible to obtain and use a spare wheel.*

Preparation

☐ When a puncture occurs, stop as soon as it is safe to do so.

☐ Park on firm level ground, if possible, and well out of the way of other traffic.

☐ Use hazard warning lights if necessary.

☐ If you have one, use a warning triangle to alert other drivers of your presence.

☐ Apply the handbrake and engage first or reverse gear (or Park on models with automatic transmission).

☐ Chock the wheel diagonally opposite the one being removed – a couple of large stones will do for this.

☐ If the ground is soft, use a flat piece of wood to spread the load under the jack.

☐ Where fitted the spare wheel is located in the floor beneath the load area. Lift up the cover to access the wheel and tool kit.

1 Use the wire hook to remove the wheel bolt caps (some models) or wheel trim. Use the wheel brace to slacken each of the wheel bolts by half a turn.

2 Use the special adapter when slackening the locking wheel bolts.

3 On models with sill covers, remove the filler panel

4 Make sure the jack is located on firm ground, and engage the jack head correctly with the sill. Then raise the jack until the wheel is raised clear of the ground.

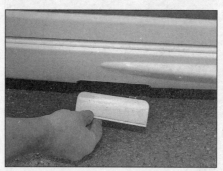

5 Unscrew the wheel bolts (using the adapter provided) and remove the wheel. Place the wheel under the vehicle sill in case the jack fails.

6 Fit the spare wheel and screw in the bolts. Lightly tighten the bolts with the wheel brace then lower the car to the ground.

7 Securely tighten the wheel bolts in a diagonal sequence.

Note: *The spare wheel is a temporary 'space-saver' spare wheel. Special conditions apply to its use. This type of spare wheel is only intended for use in an emergency, and should not remain fitted any longer than it takes to get the punctured tyre repaired. While the temporary wheel is in use, ensure it is inflated to the correct pressure, do not exceed 50mph, and avoid harsh acceleration, braking or cornering.*

Using the puncture repair kit

⚠️ *Warning: Do not use the puncture repair kit in a situation where you risk being hit by other traffic. On busy roads, try to stop in a lay-by or a gateway. Be wary of passing traffic while repairing the puncture – it is easy to become distracted by the job in hand.*

Preparation

- ☐ When a puncture occurs, stop as soon as it is safe to do so.
- ☐ Park on firm level ground, if possible, and well out of the way of other traffic.
- ☐ Use hazard warning lights if necessary.
- ☐ If you have one, use a warning triangle to alert other drivers of your presence.
- ☐ Apply the handbrake and engage first or reverse gear (or Park on models with automatic transmission).
- ☐ Where supplied the sealant and compressor are fitted in the spare wheel well beneath the load area.

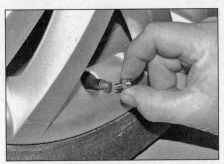

8 With the valve at the lowest point, remove the valve with the provided tool

9 Shake the sealant bottle and empty the entire contents into the tyre

10 Fit the compressor and inflate the tyre to a maximum of 2.5 Bar. If the pressure is low, drive the vehicle a metre to distribute the fluid and try again. If the pressure will not hold the tyre is beyond repair. If all is well drive for 10 minutes and check the pressure again. Replace the tyre at the earliest opportunity DO NOT exceed 50 mph.

Finally . . .

- ☐ Remove the wheel chock.
- ☐ Check the tyre pressure on the wheel just fitted. If it is low, or if you don't have a pressure gauge with you, drive slowly to the next garage and inflate the tyre to the correct pressure.
- ☐ The wheel bolts should be slackened and tightened to the specified torque at the earliest possible opportunity (see Chapter 10 Specifications).
- ☐ Have the damaged tyre or wheel repaired as soon as possible, or another puncture will leave you stranded.

Towing

☐ When all else fails, you may find yourself having to get a tow home – or of course you may be helping somebody else. Long-distance recovery should only be done by a garage or breakdown service. For shorter distances, DIY towing using another car is easy enough, but observe the following points:

☐ Use a proper tow-rope – they are not expensive. The vehicle being towed must display an ON TOW sign in its rear window.

☐ Always turn the ignition key to the 'on' position when the vehicle is being towed, so that the steering lock is released, and the direction indicator and brake lights work.

☐ The towing eye is kept inside the spare wheel (see Wheel changing). To fit the eye, unclip the access cover from the relevant bumper and screw the eye firmly into position **(see illustration)**.

☐ Before being towed, release the handbrake and select neutral on the transmission.

☐ Note that greater-than-usual pedal pressure will be required to operate the brakes, since the vacuum servo unit is only operational with the engine running.

☐ The driver of the car being towed must keep the tow-rope taut at all times to avoid snatching.

☐ Make sure that both drivers know the route before setting off.

☐ Only drive at moderate speeds and keep the distance towed to a minimum. Drive smoothly and allow plenty of time for slowing down at junctions.

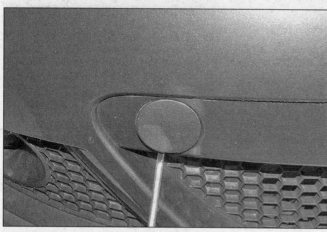

Flip open the cover.

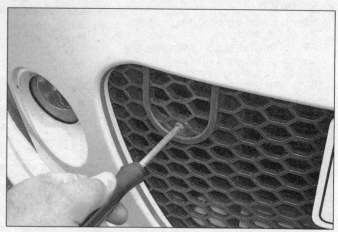

On some models, unscrew the cover...

... and remove it

Fit the towing eye (left-hand thread) and fully tighten with the wheel brace

Identifying leaks

Puddles on the garage floor or drive, or obvious wetness under the bonnet or underneath the car, suggest a leak that needs investigating. It can sometimes be difficult to decide where the leak is coming from, especially if an engine undershield is fitted. Leaking oil or fluid can also be blown rearwards by the passage of air under the car, giving a false impression of where the problem lies.

⚠ *Warning: Most automotive oils and fluids are poisonous. Wash them off skin, and change out of contaminated clothing, without delay.*

 The smell of a fluid leaking from the car may provide a clue to what's leaking. Some fluids are distinctively coloured. It may help to remove the engine undershield, clean the car carefully and to park it over some clean paper overnight as an aid to locating the source of the leak. Remember that some leaks may only occur while the engine is running.

Sump oil

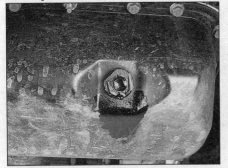

Engine oil may leak from the drain plug...

Oil from filter

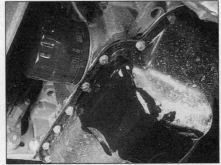

...or from the base of the oil filter.

Gearbox oil

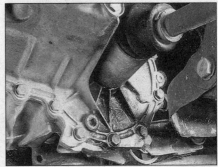

Gearbox oil can leak from the seals at the inboard ends of the driveshafts.

Antifreeze

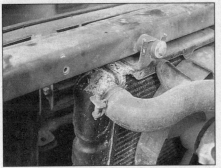

Leaking antifreeze often leaves a crystalline deposit like this.

Brake fluid

A leak occurring at a wheel is almost certainly brake fluid.

Introduction

There are some very simple checks which need only take a few minutes to carry out, but which could save you a lot of inconvenience and expense.

☐ These checks require no great skill or special tools, and the small amount of time they take to perform could prove to be very well spent, for example:

☐ Keeping an eye on tyre condition and pressures, will not only help to stop them wearing out prematurely, but could also save your life.

☐ Many breakdowns are caused by electrical problems. Battery-related faults are particularly common, and a quick check on a regular basis will often prevent the majority of these.

☐ If your car develops a brake fluid leak, the first time you might know about it is when your brakes don't work properly. Checking the level regularly will give advance warning of this kind of problem.

☐ If the oil or coolant levels run low, the cost of repairing any engine damage will be far greater than fixing the leak, for example.

Underbonnet check points

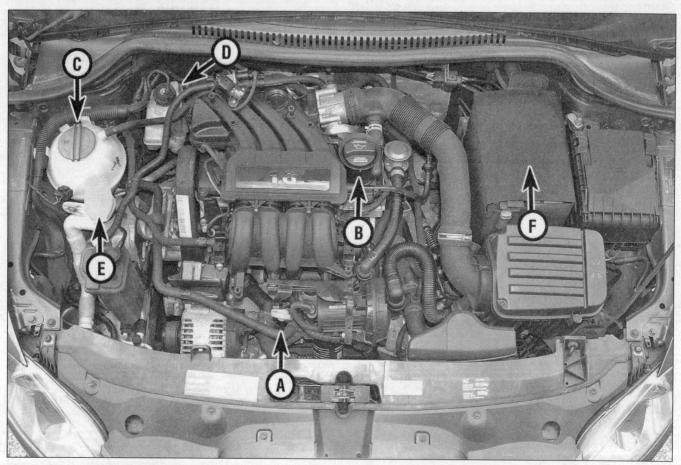

▲ **1.6 petrol engine**

A *Engine oil level dipstick*
B *Engine oil filler cap*

C *Coolant expansion tank*
D *Brake (and clutch) fluid reservoir*

E *Screen washer fluid reservoir*
F *Battery*

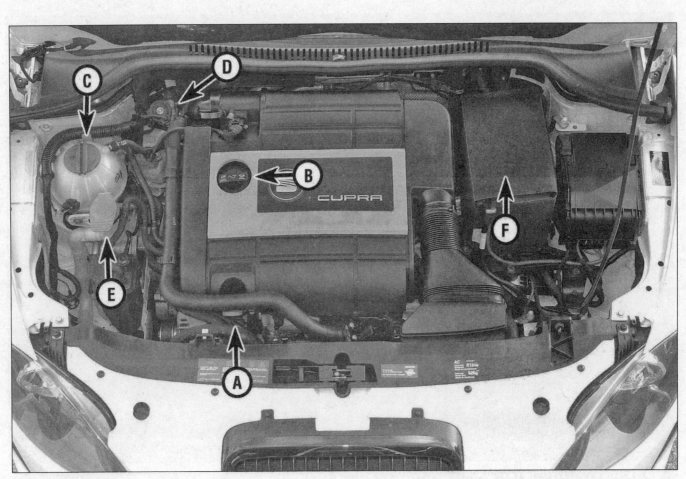

▲ **2.0 litre petrol (Cupra)**

A *Engine oil level dipstick*

B *Engine oil filler cap*

C *Coolant expansion tank*

D *Brake (and clutch) fluid reservoir*

E *Screen washer fluid reservoir*

F *Battery*

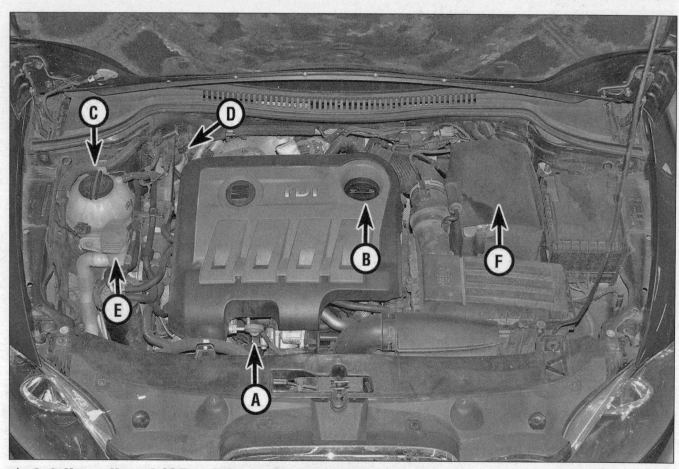

▲ **2.0 litre diesel (CR – PD engines similar)**

A *Engine oil level dipstick*

B *Engine oil filler cap*

C *Coolant expansion tank*

D *Brake (and clutch) fluid reservoir*

E *Screen washer fluid reservoir*

F *Battery*

Engine oil level

If the oil is checked immediately after driving the vehicle, some of the oil will remain in the upper engine components, resulting in an inaccurate reading on the dipstick.

Before you start

✔ Make sure that the car is on level ground.
✔ Check the oil level before the car is driven, or at least 5 minutes after the engine has been switched off.

The correct oil

Modern engines place great demands on their oil. It is very important that the correct oil for your car is used (see *Lubricants and fluids*).

Car care

● If you have to add oil frequently, you should check whether you have any oil leaks. Place some clean paper under the car overnight, and check for stains in the morning. If there are no leaks, then the engine may be burning oil.
● Always maintain the level between the upper and lower dipstick marks (see photo 2). If the level is too low, severe engine damage may occur. Oil seal failure may result if the engine is overfilled by adding too much oil.

1 The dipstick is located at the front of the engine (see *Underbonnet check points*); the dipstick is often brightly coloured or has a picture of an oil-can on the top for identification. Withdraw the dipstick.

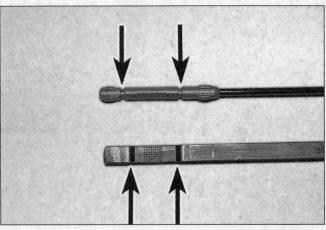

2 The oil level on the end of the dipstick, which should be between the upper (MAX) mark and lower (MIN) mark (both types of dipstick shown).

3 Oil is added through the filler cap aperture, unscrew the cap.

4 Top-up the level; a funnel may help to reduce spillage. Add the oil slowly, checking the level on the dipstick often. Don't overfill (see *Car care*).

Brake (and clutch) fluid level

 Warning: Brake fluid can harm your eyes and damage painted surfaces, so use extreme caution when handling and pouring it.

 Warning: Do not use fluid that has been standing open for some time, as it absorbs moisture from the air, which can cause a dangerous loss of braking effectiveness.

Before you start
✔ Make sure that your car is on level ground.

Safety first!
● If the reservoir requires repeated topping-up this is an indication of a fluid leak somewhere in the system, which should be investigated immediately.
● If a leak is suspected, the car should not be driven until the braking system has been checked. Never take any risks where brakes are concerned.

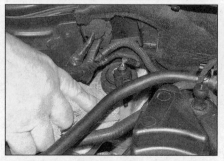

1 The upper (MAX) fluid level marking is on the side of the reservoir, which is located at the rear of the engine compartment. Remove the air intake pipe to improve access. First wipe clean the area around the filler cap with a clean cloth.

2 If topping-up is necessary, unscrew the cap and remove it along with the rubber seal.

3 Carefully add fluid, avoiding spilling it on the surrounding paintwork. Use only the specified hydraulic fluid. After filling to the correct level, refit the cap and diaphragm and tighten it securely. Wipe off any spilt fluid.

Coolant level

 Warning: Do not attempt to remove the expansion tank pressure cap when the engine is hot, as there is a very great risk of scalding. Do not leave open containers of coolant about, as it is poisonous.

Car care
● With a sealed-type cooling system, adding coolant should not be necessary on a regular basis. If frequent topping-up is required, it is likely there is a leak. Check the radiator, all hoses and joint faces for signs of staining or wetness, and rectify as necessary.

● It is important that antifreeze is used in the cooling system all year round, not just during the winter months. Don't top up with water alone, as the antifreeze will become diluted.

1 The coolant level must be checked with the engine cold; the coolant level should be between the MAX and MIN marks on the expansion tank.

2 If topping-up is necessary, remove the pressure cap (see Warning) from the expansion tank. Slowly unscrew the cap to release any pressure present in the cooling system, and remove the cap.

3 Add a mixture of water and antifreeze to the expansion tank until the coolant level is between the level marks. Once the level is correct, securely refit the cap.

Electrical systems

✔ Check all external lights and the horn. Refer to Chapter 12 Section 2 for details if any of the circuits are found to be inoperative.

✔ Visually check all accessible wiring connectors, harnesses and retaining clips for security, and for signs of chafing or damage.

 If you need to check your brake lights and indicators unaided, back up to a wall or garage door and operate the lights. The reflected light should show if they are working properly.

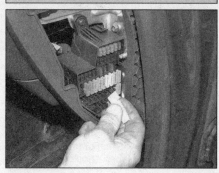

1 If a single indicator light, brake light, sidelight or headlight has failed, it is likely that a bulb has blown, and will need to be renewed. Refer to Chapter 12, Section 5 for details. If both brake lights have failed, it is possible that the switch has failed (see Chapter 9, Section 18).

2 If more than one indicator or tail light has failed, it is likely that either a fuse has blown or that there is a fault in the circuit. The main fuses are located behind the facia right-hand end panel. Additional fuses and relays are located in a fusebox on the left-hand side of the engine compartment.

3 To renew a blown fuse, simply pull it out and fit a new fuse of the correct rating (see **Wiring diagrams**). If the fuse blows again, it is important that you find out why – a complete checking procedure is given in Chapter 12, Section 2.

Battery

Caution: Before carrying out any work on the vehicle battery, read the precautions given in 'Safety first!' at the start of this manual.

✔ Make sure that the battery tray is in good condition, and that the clamp is tight. Corrosion on the tray, retaining clamp and the battery itself can be removed with a solution of water and baking soda. Thoroughly rinse all cleaned areas with water. Any metal parts damaged by corrosion should be covered with a zinc-based primer, then painted.

✔ Periodically (approximately every three months), check the charge condition of the battery, as described in Chapter 5A Section 2.

✔ If the battery is flat, and you need to jump start your vehicle, see *Jump starting*.

 Battery corrosion can be kept to a minimum by applying a layer of petroleum jelly to the clamps and terminals after they are reconnected.

1 The battery is located on the left-hand side of the engine compartment. Open the cover to gain access to the battery terminals. The exterior of the battery should be inspected periodically for damage such as a cracked case or cover.

2 Check the battery lead clamps for tightness to ensure good electrical connections, and check the leads for signs of damage.

3 If corrosion (white, fluffy deposits) are evident, remove the cables from the battery terminals, clean them with a small wire brush, then refit them. Automotive stores sell a tool for cleaning the battery post …

4 … as well as the battery cable clamps.

Wiper blades

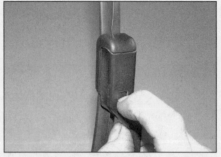

1 Check the condition of the wiper blades: if they are cracked or show signs of deterioration, or if the glass swept area is smeared, renew them. For maximum clarity of vision, wiper blades should be renewed annually.

2 To remove a windscreen wiper blade, turn the ignition on, then turn the ignition off, and press the wiper switch stalk down once. This places the arms in the 'service' position. Lift the wiper from the screen and press the securing clip …

3 … slide the blade upwards from the end of the wiper arm, taking care not to allow the wiper arm to spring back and damage the windscreen. When completed, return the blades to the park position by pressing the wiper switch stalk again.

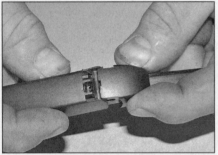

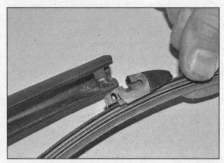

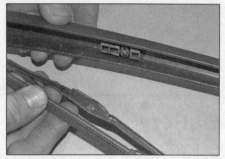

4 To remove the rear wiper blade, tilt the wiper blade away from the screen, depress the side tabs…

5 …and unhook the blade

6 Some rear wiper blades simply clip in place

Screen washer fluid level

● Screenwash additives not only keep the windscreen clean during bad weather, they also prevent the washer system freezing in cold weather – which is when you are likely to need it most. Don't top-up using plain water, as the screenwash will become diluted, and will freeze in cold weather.

 Warning: On no account use engine coolant antifreeze in the screen washer system – this may damage the paintwork.

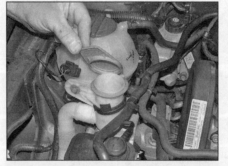

1 The washer fluid reservoir is located in the right-hand side of the engine compartment. To check the fluid level, open the cap and then look down the clear filler neck.

2 If topping-up is necessary, add water and a screenwash additive in the quantities recommended on the bottle.

Tyre condition and pressure

It is very important that tyres are in good condition, and at the correct pressure – having a tyre failure at any speed is highly dangerous.

Tyre wear is influenced by driving style – harsh braking and acceleration, or fast cornering, will all produce more rapid tyre wear. As a general rule, the front tyres wear out faster than the rears. Interchanging the tyres from front to rear ("rotating" the tyres) may result in more even wear. However, if this is completely effective, you may have the expense of replacing all four tyres at once!

Remove any nails or stones embedded in the tread before they penetrate the tyre to cause deflation. If removal of a nail does reveal that the tyre has been punctured, refit the nail so that its point of penetration is marked. Then immediately change the wheel, and have the tyre repaired by a tyre dealer.

Regularly check the tyres for damage in the form of cuts or bulges, especially in the sidewalls. Periodically remove the wheels, and clean any dirt or mud from the inside and outside surfaces. Examine the wheel rims for signs of rusting, corrosion or other damage. Light alloy wheels are easily damaged by "kerbing" whilst parking; steel wheels may also become dented or buckled. A new wheel is very often the only way to overcome severe damage.

New tyres should be balanced when they are fitted, but it may become necessary to re-balance them as they wear, or if the balance weights fitted to the wheel rim should fall off. Unbalanced tyres will wear more quickly, as will the steering and suspension components. Wheel imbalance is normally signified by vibration, particularly at a certain speed (typically around 50 mph). If this vibration is felt only through the steering, then it is likely that just the front wheels need balancing. If, however, the vibration is felt through the whole car, the rear wheels could be out of balance. Wheel balancing should be carried out by a tyre dealer or garage.

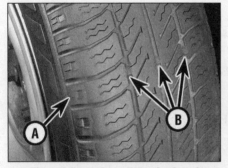

1 *Tread Depth - visual check*
The original tyres have tread wear safety bands (B), which will appear when the tread depth reaches approximately 1.6 mm. The band positions are indicated by a triangular mark on the tyre sidewall (A).

2 *Tread Depth - manual check*
Alternatively, tread wear can be monitored with a simple, inexpensive device known as a tread depth indicator gauge.

3 *Tyre Pressure Check*
Check the tyre pressures regularly with the tyres cold. Do not adjust the tyre pressures immediately after the vehicle has been used, or an inaccurate setting will result.

Tyre tread wear patterns

Shoulder Wear

Underinflation (wear on both sides)
Under-inflation will cause overheating of the tyre, because the tyre will flex too much, and the tread will not sit correctly on the road surface. This will cause a loss of grip and excessive wear, not to mention the danger of sudden tyre failure due to heat build-up.
Check and adjust pressures
Incorrect wheel camber (wear on one side)
Repair or renew suspension parts
Hard cornering
Reduce speed!

Centre Wear

Overinflation
Over-inflation will cause rapid wear of the centre part of the tyre tread, coupled with reduced grip, harsher ride, and the danger of shock damage occurring in the tyre casing.
Check and adjust pressures

If you sometimes have to inflate your car's tyres to the higher pressures specified for maximum load or sustained high speed, don't forget to reduce the pressures to normal afterwards.

Uneven Wear

Front tyres may wear unevenly as a result of wheel misalignment. Most tyre dealers and garages can check and adjust the wheel alignment (or "tracking") for a modest charge.
Incorrect camber or castor
Repair or renew suspension parts
Malfunctioning suspension
Repair or renew suspension parts
Unbalanced wheel
Balance tyres
Incorrect toe setting
Adjust front wheel alignment
Note: *The feathered edge of the tread which typifies toe wear is best checked by feel.*

Lubricants and fluids

Note: *Using lubricants and fluids that do not meet the VW standard may invalidate the warranty*

Standard (distance/time) service interval:	Multigrade (high-lubricity) engine oil, viscosity SAE 5W/40. Eg. Castrol Edge.

Petrol engines:

Fixed service intervals.	VW 502 00
LongLife service intervals	VW 504 00

Diesel engines:

Engines without particulate filter.	VW 505 01, 507 00 or better
Engines with particulate filter	VW 507 00 or better
LongLife (variable) service interval:	VW LongLife engine oil
Engines without particulate filter.	VW 506 01, 507 00 or better
Engines with particulate filter	VW 507 00 or better

Cooling system	VW additive G12 plus-plus (TL-VW 774 G) or G12 plus (TL-VW 774 F) – antifreeze and corrosion protection
Manual transmission	Synthetic gear oil, viscosity SAE 75W VW G52
DSG semi-automatic transmissions.	G 052 145
Braking system.	Hydraulic fluid to SAE J1703F or DOT 4

Tyre pressures

Note: *The recommended tyre pressures for each vehicle are given on a sticker attached to the inside of the fuel filler flap or the drivers side B-pillar. The pressures given are for the original equipment tyres – the recommended pressures may vary if any other make or type of tyre is fitted; check with the tyre manufacturer or supplier for latest recommendations. The following pressures are typical.*

All models	Front	Rear
Normal load	2.0 bars (29 psi)	2.0 bars (29 psi)
Full load	2.3 bars (33 psi)	2.8 bars (41 psi)
Emergency spare wheel.	4.2 bars (61 psi)	

Chapter 1 Part A
Routine maintenance and servicing – petrol engines

Contents

Degrees of difficulty

| **Easy,** suitable for novice with little experience | | **Fairly easy,** suitable for beginner with some experience | | **Fairly difficult,** suitable for competent DIY mechanic | | **Difficult,** suitable for experienced DIY mechanic | | **Very difficult,** suitable for expert DIY or professional | |

1 Servicing specifications – petrol models

Lubricants and fluids
Refer to Lubricants and fluids

Engine codes*
1.6 litre petrol engine: . . . . BGU, BSE and BSF (all SOHC)
2.0 litre petrol engine:
   Non-turbo. . . . . BLR, BVY, BLY and BVZ
   Turbo . . . . BWA, BWJ, CDLD and CDLA
See 'Vehicle identification' at the end of this manual for the location of engine code markings.

Capacities

Engine oil – including filter (approximate)
1.6 litre engines . . . . 4.5 litres
2.0 litre engines . . . . 4.6 litres

Cooling system (approximate)
1.6 litre engines . . . . 8.0 litres
2.0 litre engines . . . . 8.0 litres

Transmission
Manual transmission*:
   Type 0AF (5 speed). . . . . 2.0 litres
   Type 02S (6 speed). . . . . 1.9 litres
   Type 02Q (6 speed). . . . . 2.3 litres
DSG transmission (02E) . . . . 5.2 litres
See Chapter for application details

Fuel tank (approximate)
All models. . . . . 50 litres

Washer reservoirs
Models with headlight washers . . . . 5.5 litres
Models without headlight washers. . . . . 3.0 litres

Cooling system
Antifreeze mixture:
   40% antifreeze . . . . Protection down to -25°C
   50% antifreeze . . . . Protection down to -35°C
Note: *Refer to antifreeze manufacturer for latest recommendations.*

Ignition system

Spark plugs:	Type	Electrode gap
1.6 litre engine (BGU, BSE and BSF)	VW 101000033AA/NGK BKUR6ET-10	0.9 to 1.1 mm
2.0 litre engine codes:		
BLR, BLY, BLX, BVX and BVY	VW 101905620/NGK PZFR5N-11T	1.0 to 1.1 mm
BWA, BWJ, CDLA and CDLD.	VW 101905631B/NGK PFR7S8EG	0.7 to 0.8 mm

Brakes
Brake pad lining minimum thickness:
   Front . . . . 2.0 mm
   Rear . . . . 2.0 mm

Torque wrench settings

	Nm	lbf ft
Manual gearbox drain/filler plug plug (0AF transmission only):		
Spline head type bolt	25	18
Hex head bolt.	30	22
Manual gearbox drain plug (02S transmission only)	35	25
Manual gearbox filler/level plug	30	22
Oil filter (models with a canister type filter)	20	15
Oil filter cap (models with a paper filter and plastic cap))	25	18
Pivot pin bolt (02S transmission only)	25	18
Reversing light switch	20	15
Roadwheel bolts.	120	89
Spark plugs	30	22
Sump drain plug*	30	22

* Do not re-use

2 Maintenance schedule – petrol models

1 The maintenance intervals in this manual are provided with the assumption that you, not the dealer, will be carrying out the work. These are the minimum intervals recommended by us for vehicles driven daily. If you wish to keep your vehicle in peak condition at all times, you may wish to perform some of these procedures more often. We encourage frequent maintenance, since it enhances the efficiency, performance and resale value of your vehicle.

2 When the vehicle is new, it should be serviced by a dealer service department (or other workshop recognised by the vehicle manufacturer as providing the same standard of service), in order to preserve the warranty.

The vehicle manufacturer may reject warranty claims if you are unable to prove that servicing has been carried out as and when specified, using only original equipment parts or parts certified to be of equivalent quality.

3 Depending on the model specification service intervals will either be at standard times (or distance) or set to 'LongLife' service intervals. On models set to the LongLife display, the service interval is variable according to the number of starts, length of journeys, vehicle speeds, brake pad wear, bonnet opening frequency, fuel consumption, oil level and oil temperature, however the vehicle must be serviced at least every two years. Note that if the variable (LongLife) service interval is being used, the engine must only be filled with the recommended long-life engine oil (see Lubricants and fluids).

4 All Seat Leon models are equipped with a Service Interval Display (SID) indicator in the instrument panel. When a service date is approaching a message will appear in the message centre or a spanner symbol will appear in the instrument panel. The message or spanner symbol will also show the distance (or time) to the next service and will count down as the due date approaches. A minus symbol or a 'Service – - days ago' (on models with a message centre) will be shown when the due date (or mileage) has passed. Note that on models on Longlife service intervals that have the battery disconnected for a long time the service interval will not be correctly calculated. Where this is the case the vehicle should be serviced at the standard intervals.

5 After completing a service, Seat technicians use a special instrument to reset the service display to the next service interval, and a print-out is put in the vehicle service record. The display can be reset by the owner as described in Section

Every 250 miles

☐ Refer to *Weekly checks*

Every 10 000 miles or 12 months, whichever comes first

☐ Renew the engine oil and filter (Section 6)
Note: *Frequent oil and filter changes are good for the engine. We recommend changing the oil at least once a year.*
☐ Check the front and rear brake pad thickness (Section 7)
☐ Reset the service interval display (Section 8).

Every 20 000 miles or 2 years, whichever comes first

In addition to the items listed above, carry out the following:
☐ Check the condition of the exhaust system and its mountings (Section 9)
☐ Check all underbonnet components and hoses for fluid and oil leaks (Section 10)
☐ Check the condition of the auxiliary drivebelt (Section 11)
☐ Check the coolant antifreeze concentration (Section 12)
☐ Check the brake hydraulic circuit for leaks and damage (Section 13)
☐ Check the headlight beam adjustment (Section 14)
☐ Renew the pollen filter element (Section 15)
☐ Check the underbody protection for damage (Section 18)
☐ Check the condition of the driveshaft gaiters (Section 19)
☐ Check the steering and suspension components for condition and security (Section 20)
☐ Check the battery condition and security (Section 21)
☐ Lubricate all hinges and locks (Section 22)
☐ Check the condition of the airbag unit(s) (Section 23)
☐ Check the operation of the windscreen/tailgate/headlight washer system(s) (as applicable) (Section 24)
☐ Check the engine management self-diagnosis memory for faults (Section 25)
☐ Carry out a road test and check exhaust emissions (Section 27)

Every 40 000 miles or 4 years, whichever comes first

Note: *Many dealers perform these tasks at every second service.*
☐ Renew the air filter element (Section 28)
☐ Check the condition of the auxiliary drivebelt (Section 11)
☐ Renew the DSG transmission oil (where applicable) (Section 17).

Every 60 000 miles or 4 years

☐ Renew the timing belt and tensioner roller as described in Chapter 2A Section 7 or Chapter 2B Section 4.
Note: *The interval recommended by Seat is approximately 60 000 miles for 1.6 litre engines, and 100 000 miles for 2.0 litre engines (or 4 years whichever occurs sooner). However, our recommendation is that the belt is changed every 60 000 miles or 4 years on all engine. It is strongly recommended that the interval be further reduced on vehicles that are subjected to intensive use, ie, mainly short journeys or a lot of stop-start driving. The actual belt renewal interval is very much up to the individual owner, but bear in mind that severe engine damage will result if the belt breaks.*
Note: *It is increasingly considered best practice to always replace the idlers, tensioner and coolant pump whenever the timing belt is replaced. Note also that most manufacturers will not guarantee a belt unless the tensioners and rollers have been replaced at the same time.*

Every 2 years

☐ Renew the brake (and clutch) fluid (Section 32)
☐ Renew the coolant* (Section 33)
* Note: *This work is not included in the Seat schedule and should not be required if the recommended G12 Plus (Purple colour) coolant antifreeze/inhibitor is used. G12 Plus can be mixed with older versions (G11 and G12).*

3 Component location – petrol models

Underbonnet view of a 1.6 litre (BSE)

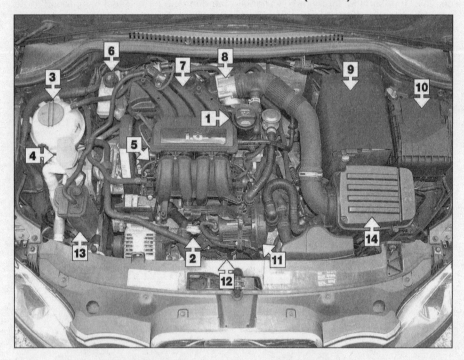

1 Engine oil filler cap
2 Engine oil dipstick
3 Coolant expansion tank
4 Windscreen/headlight washer fluid reservoir
5 Spark plugs (hidden under manifold)
6 Brake master cylinder fluid reservoir
7 Inlet manifold
8 Throttle body
9 Battery (under cover)
10 Fusebox
11 Secondary air pump
12 Oil filter
13 Evaporative emission charcoal canister
14 Air filter housing

Front underbody view of a 1.6 litre

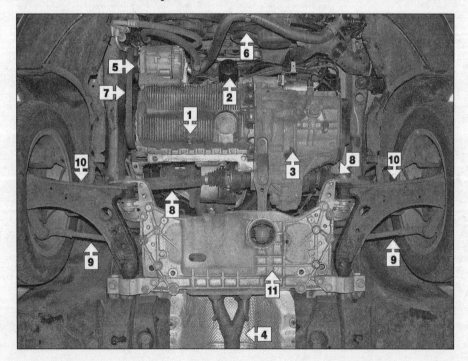

1 Sump drain plug
2 Oil filter
3 Manual transmission
4 Exhaust front pipe
5 Air conditioning compressor
6 Radiator and electric cooling fans
7 Auxiliary drivebelt
8 Driveshafts
9 Steering track rods
10 Front suspension lower arms
11 Front subframe

Rear underbody view of a 1.6 litre

1 Fuel tank
2 Fuel filter
3 Track control arms
4 Rear subframe
5 Rear anti-roll bar
6 Trailing arms
7 Lower transverse links
8 Handbrake cables
9 Exhaust rear silencers and tail pipe

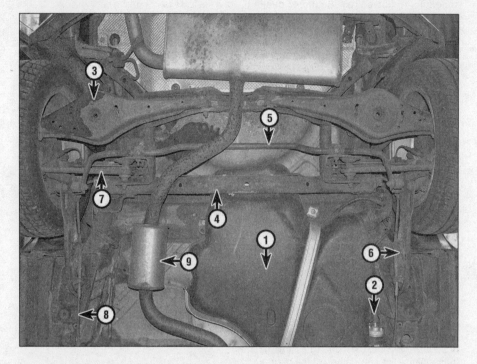

Underbonnet view of a 2.0 litre

1 Oil filler cap
2 Oil dipstick
3 Coolant expansion reservoir
4 Windscreen/headlight washer fluid reservoir
5 Brake master cylinder fluid reservoir
6 Evaporative emission charcoal canister
7 Battery (under cover)
8 Fusebox
9 Air filter (part of the engine cover)

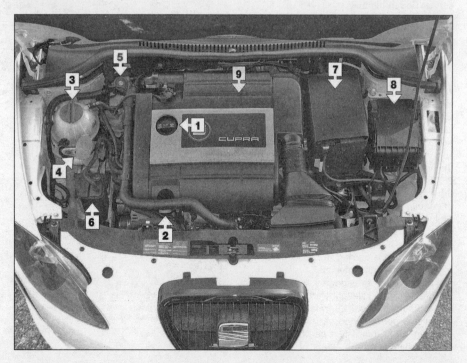

4 Introduction

1 This Chapter is designed to help the home mechanic maintain his/her vehicle for safety, economy, long life and peak performance.

2 The Chapter contains a master maintenance schedule, followed by Sections dealing specifically with each task in the schedule. Visual checks, adjustments, component renewal and other helpful items are included. Refer to the accompanying illustrations of the engine compartment and the underside of the vehicle for the locations of the various components.

3 Servicing your vehicle will provide a planned maintenance programme, which should result in a long and reliable service life. This is a comprehensive plan, so maintaining some items but not others will not produce the same results.

4 As you service your vehicle, you will discover that many of the procedures can – and should – be grouped together, because of the particular procedure being performed, or because of the proximity of two otherwise unrelated components to one another. For example, if the vehicle is raised for any reason, the exhaust can be inspected at the same time as the suspension and steering components.

5 The first step in this maintenance programme is to prepare yourself before the actual work begins. Read through all the Sections relevant to the work to be carried out, then make a list and gather all the parts and tools required. If a problem is encountered, seek advice from a parts specialist, or a dealer service department.

5 Regular maintenance

1 If, from the time the vehicle is new, the routine maintenance schedule is followed closely, and frequent checks are made of fluid levels and high-wear items, as suggested throughout this manual, the engine will be kept in relatively good running condition, and the need for additional work will be minimised.

2 It is possible that there will be times when the engine is running poorly due to the lack of regular maintenance. This is even more likely if a used vehicle, which has not received regular and frequent maintenance checks, is purchased. In such cases, additional work may need to be carried out, outside of the regular maintenance intervals.

3 If engine wear is suspected, a compression test (refer to the relevant Part of Chapter 2) will provide valuable information regarding the overall performance of the main internal components. Such a test can be used as a basis to decide on the extent of the work to be carried out. If, for example, a compression test indicates serious internal engine wear, conventional maintenance as described in this Chapter will not greatly improve the performance of the engine, and may prove a waste of time and money, unless extensive overhaul work is carried out first.

4 The following series of operations are those most often required to improve the performance of a generally poor-running engine:

Primary operations

a) *Clean, inspect and test the battery (See 'Weekly checks').*
b) *Check all the engine-related fluids (See 'Weekly checks').*
c) *Check the condition and tension of the auxiliary drivebelt (Section 11).*
d) *Renew the spark plugs (Section 29).*
e) *Check the condition of the air filter, and renew if necessary (Section 28).*
f) *Check the condition of all hoses, and check for fluid leaks (Section 10).*

5 If the above operations do not prove fully effective, carry out the following secondary operations:

Secondary operations

6 All items listed under Primary operations, plus the following:

a) *Check the charging system (see Chapter 5A Section 4).*
b) *Check the ignition system (see Chapter 5B).*
c) *Check the fuel system (see Chapter 4A).*
d) *Renew the ignition HT leads (where applicable).*

6 Engine oil and filter renewal – petrol models

1 Frequent oil and filter changes are the most important maintenance procedures which can be undertaken by the DIY owner. As engine oil ages, it becomes diluted and contaminated, which leads to premature engine wear.

2 Before starting this procedure, gather all the necessary tools and materials. Also make sure that you have plenty of clean rags and newspapers handy, to mop-up any spills. Ideally, the engine oil should be warm, as it will drain better, and more built-up sludge will be removed with it. Take care, however, not to touch the exhaust or any other hot parts of the engine when working under the vehicle. To avoid any possibility of scalding, and to protect yourself from possible skin irritants and other harmful contaminants in used engine oils, it is advisable to wear gloves when carrying out this work. Access to the underside of the vehicle will be greatly improved if it can be raised on a lift, driven onto ramps, or jacked up and supported on axle stands (see *Jacking and vehicle support*). Whichever method is chosen, make sure that the vehicle remains level, or if it is at an angle, that the drain plug is at the lowest point. Undo the retaining screws and remove the engine undertray(s), then also remove the engine top cover where applicable **(see illustration)**.

3 Using a socket and wrench or a ring spanner, slacken the drain plug about half a turn **(see**

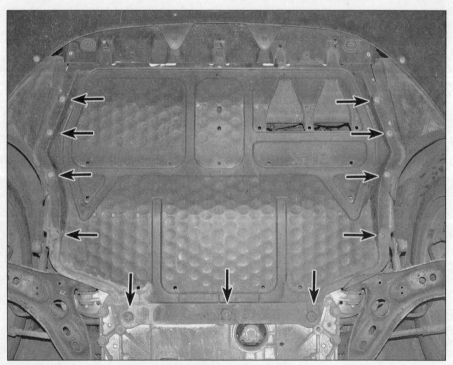

6.2 The engine undershield

6.3a The engine oil drain plug location on the sump (1.6 litre engine)

6.3b The seal is integral with the plug

Keep the drain plug pressed into the sump while unscrewing it by hand the last couple of turns. As the plug releases, move it away sharply so the stream of oil issuing from the sump runs into the container, not up your sleeve.

illustrations). Position the draining container under the drain plug, then remove the plug completely (see Haynes Hint). The seal is integral with the drain plug. Consequently, the drain plug must be renewed.

4 Allow some time for the old oil to drain, noting that it may be necessary to reposition the container as the oil flow slows to a trickle.

5 After all the oil has drained, clean the area around the drain plug opening, and fit the new plug. Tighten the plug to the specified torque.

6 If the filter is also to be renewed, move the container into position under the oil filter **(see illustrations)**.

a) On 1.6 litre engine codes BGU, BSE and BSF, a canister-type filter is located from below on the front of the cylinder block.

b) On 2.0 litres engines a filter element is located from below in a housing screwed into the bottom of the oil pump housing.

7 To remove the canister-type oil filter, use an oil filter removal tool **(see illustrations)** and slacken the filter initially, then unscrew it by hand the rest of the way. Empty the oil in the filter into the container.

8 To remove the renewable element type oil filter, unscrew and remove the cap and remove the element and sealing ring. On this type of filter a drain plug is fitted to the filter housing **(see illustration)**. This requires a special drain hose fitting (VAG special tool T40057). However this tool is not essential, but without it anticipate some oil spillage – place a suitable container beneath the filter housing. Use a socket on the housing and unscrew the filter housing.

9 Clean the filter housing and cap as necessary. On canister-type filters, check the old filter to make sure that the rubber sealing ring has not stuck to the engine. If it has, carefully remove it.

10 To fit the canister-type oil filter, apply a light coating of clean engine oil to the sealing ring on the new filter, then screw it into position on the engine **(see illustrations)**.

6.6a Engine oil filter location on 1.6 litre engines...

6.6b ...and on 2.0 litre engines

6.7a Use a strap type tool...

6.7b ...or a claw type tool

6.8 The filter drain plug

6.10a Clean the filter mounting...

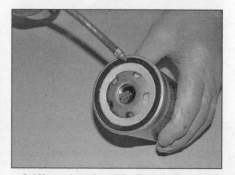

6.10b..and apply a thin film of oil to the filter sealing ring before fitting

6.12 Add oil to bring the level above the minimum mark on the dipstick. Do not overfill

Tighten the filter firmly by hand only – do not use any tools.

11 On 2.0 litre engines recover the sealing ring, using pliers on the tag provided. Clean the threads on the housing, fit the the new seal and lubricate it with clean engine oil. Fit the filter to the housing and screw the assembly back onto the engine. Tighten the cap to the specified torque.

12 Remove the dipstick, and then unscrew the oil filler cap from the cylinder head cover. Fill the engine, using the correct grade and type of oil (see Lubricants and fluids 0 Section 6). A funnel may help to reduce spillage. Pour in half the specified quantity of oil first **(see illustration)**, then wait a few minutes for the oil to run to the sump (see *Weekly checks* Section 5). Continue adding oil a small quantity at a time until the level is above the minimum mark on the dipstick – ideally the level will be between the maximum and minimum mark. Refit the filler cap.

Caution: Because the front of the vehicle is still raised at this point, do not overfill the engine with oil. The level on the dipstick will increase when the vehicle is lowered to the ground.

13 Start the engine and run it for a few minutes; check for leaks around the oil filter seal and the sump drain plug. Note that there may be a few seconds delay before the oil pressure warning light goes out when the engine is started, as the oil circulates through the engine oil galleries and the new oil filter (where fitted) before the pressure builds-up.

Warning: Do not increase the engine speed above idling while the oil pressure light is illuminated, as considerable damage can be caused to the engine and turbocharger.

14 If all is well, refit the engine undershield. Lower the vehicle to the ground and where fitted replace the engine cover.

15 Finally, recheck the level on the dipstick, and add more oil to bring the level up to the maximum mark on the dipstick.

16 Dispose of the used engine oil safely, with reference to *General repair procedures* in the Reference section of this manual.

7 Brake pad check

1 The outer brake pads can be checked (on most models) without removing the wheels, by observing the brake pads through the holes in the wheels **(see illustration)**. If necessary, remove the wheel trim. The thickness of the pad lining must not be less than the dimension given in the Specifications.

2 If the outer pads are worn near their limits, it is worthwhile checking the inner pads as well. Apply the handbrake then jack up the vehicle and support it on axle stands (see *Jacking and vehicle support*). Remove the roadwheels.

3 Use a steel rule to check the thickness of the brake pads, and compare with the minimum thickness given in the Specifications **(see illustrations)**.

4 For a comprehensive check, the brake pads should be removed and cleaned. The operation of the caliper can then also be checked, and the condition of the brake disc itself can be fully examined on both sides. Refer to Chapter.

5 If any pad's friction material is worn to the specified minimum thickness or less, all four pads at the front or rear, as applicable, must be renewed as a set. If there is any doubt as to the condition of the brake pads (and the brake discs) always err on the side of caution and replace them.

6 On completion of the check, refit the road wheels and lower the vehicle to the ground.

8 Resetting the service interval display

1 After all necessary maintenance work has been completed the service interval display must be reset. The Service Interval Display (SID) can only be reset on models that are on fixed service intervals. Models that are factory set to variable service intervals must have the SID reset using diagnostic equipment. A Seat dealer will have the factory scan tool, but many independent garages will also have an aftermarket tool capable of resetting the SID.

2 To continue with the 'variable' or 'longlife' service intervals which take into consideration the number of starts, length of journeys, vehicle speeds, brake pad wear, bonnet opening frequency, fuel consumption, oil level and oil temperature, the display must be reset by a Seat dealership (or suitably equipped garage) using a diagnostic tool.

3 For the home mechanic it is recommended that models have the service schedule changed from 'variable' to 'fixed' either by a Seat dealer or suitably equipped garage.

4 To reset the SID:

● Turn the ignition off.
● Press and hold down the odometer button in the centre of the instrument panel.
● Turn the ignition on (whilst still holding down the button).
● Rotate the button to the right (clockwise) for approximately 1 second.
● Release the button and turn the ignition off.
● The service message will disappear. Note that on some models the message will clear when the engine is restarted.

9 Exhaust system check

1 With the engine cold (at least an hour after the vehicle has been driven), check the complete exhaust system from the engine to the end of the tailpipe. The exhaust system is most easily checked with the vehicle raised on

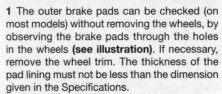

7.1 The outer brake pads can be observed through the holes in the wheels

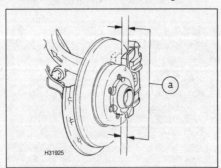

7.3a The thickness (a) of the brake pad linings must not be less than the specified amount

7.3b Specialist tools are available that can easily measure the remaining friction material

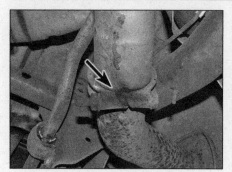

9.2a A typical exhaust gas leak at a clamp...

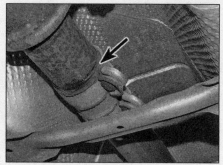

9.2b ...and at a sleeve type clamp

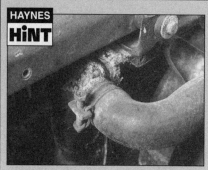

A leak in the cooling system will usually show up as white- or antifreeze-coloured deposits on the area adjoining the leak.

a hoist, or supported on axle stands, so that the exhaust components are readily visible and accessible (see *Jacking and vehicle support*).

2 Check the exhaust pipes and connections for evidence of leaks, severe corrosion and damage. Make sure that all brackets and mountings are in good condition, and that all relevant nuts and bolts are tight. Leakage at any of the joints or in other parts of the system will usually show up as a black sooty stain in the vicinity of the leak **(see illustrations)**.

3 Rattles and other noises can often be traced to the exhaust system, especially the brackets and mountings. Try to move the pipes and silencers. If the components are able to come into contact with the body or suspension parts, secure the system with new mountings. Otherwise separate the joints (if possible) and twist the pipes as necessary to provide additional clearance.

10 Hose and fluid leak check

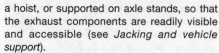

1 Visually inspect the engine joint faces, gaskets and seals for any signs of water or oil leaks. Pay particular attention to the areas around the camshaft cover, cylinder head, oil filter and sump joint faces. Bear in mind that, over a period of time, some very slight seepage from these areas is to be expected – what you are really looking for is any indication of a serious leak. Should a leak be found, renew the offending gasket or oil seal by referring to the appropriate Chapters in this manual.

2 Also check the security and condition of all the engine-related pipes and hoses. Ensure that all cable-ties or securing clips are in place and in good condition. Clips which are broken or missing can lead to chafing of the hoses, pipes or wiring, which could cause more serious problems in the future.

3 Carefully check the radiator hoses and heater hoses along their entire length. Renew any hose which is cracked, swollen or deteriorated. Cracks will show up better if the hose is squeezed. Pay close attention to the hose clips that secure the hoses to the cooling system components. Hose clips can

pinch and puncture hoses, resulting in cooling system leaks.

4 Inspect all the cooling system components (hoses, joint faces, etc) for leaks (see **Haynes Hint**). Where any problems of this nature are found on system components, renew the component or gasket with reference to Chapter.

5 Where applicable, inspect the transmission fluid cooler hoses for leaks or deterioration.

6 With the vehicle raised, inspect the petrol tank and filler neck for punctures, cracks and other damage. The connection between the filler neck and tank is especially critical. Sometimes a rubber filler neck or connecting hose will leak due to loose retaining clamps or deteriorated rubber.

7 Carefully check all rubber hoses and metal fuel lines leading away from the petrol tank. Check for loose connections, deteriorated hoses, crimped lines, and other damage. Pay particular attention to the vent pipes and hoses, which often loop up around the filler neck and can become blocked or crimped. Follow the lines to the front of the vehicle, carefully inspecting them all the way. Renew damaged sections as necessary.

8 From within the engine compartment, check the security of all fuel hose attachments and pipe unions, and inspect the fuel hoses and vacuum hoses for kinks, chafing and deterioration.

9 Where applicable, check the condition of the power steering fluid hoses and pipes.

11 Auxiliary drivebelt check – petrol models

1 Apply the handbrake, then jack up the front of the vehicle and support it on axle stands (see *Jacking and vehicle support*).

2 Using a socket on the crankshaft pulley bolt, turn the engine slowly clockwise so that the full length of the auxiliary drivebelt can be examined. Look for cracks, splitting and fraying on the surface of the belt; check also for signs of glazing (shiny patches) and separation of the belt plies. Use a mirror to check the underside of the drivebelt **(see illustration)**. If damage or wear is visible, or if there are traces of oil or grease on it, the belt should be renewed (see Section 30).

12 Antifreeze check

1 The cooling system should be filled with the recommended G12 antifreeze and corrosion protection fluid – do not mix this antifreeze with any other type. Over a period of time, the concentration of fluid may be reduced due to topping-up (this can be avoided by topping-up with the correct antifreeze mixture – see Specifications) or fluid loss. If loss of coolant has been evident, it is important to make the necessary repair before adding fresh fluid.

2 With the engine cold, carefully remove the cap from the expansion tank. If the engine is not completely cold, place a cloth rag over the cap before removing it, and remove it slowly to allow any pressure to escape.

3 Antifreeze checkers are available from car accessory shops. Draw some coolant from the expansion tank and observe how many plastic balls are floating in the checker. Usually, 2 or 3 balls must be floating for the correct concentration of antifreeze, but follow the manufacturer's instructions.

4 If the concentration is incorrect, it will be necessary to either withdraw some coolant and add antifreeze, or alternatively drain the old coolant and add fresh coolant of the correct concentration (see Section 33).

11.2 Checking the underside of the auxiliary drivebelt with a mirror

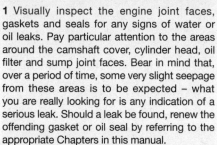

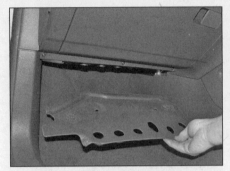

15.2 Remove the facia lower trim from beneath the glovebox

15.3 Remove the access cover...

15.4 ... then slide out the pollen filter element

13 Brake hydraulic circuit check

1 Check the entire brake hydraulic circuit for leaks and damage. Start by checking the master cylinder in the engine compartment. At the same time, check the vacuum servo unit and ABS units for signs of fluid leakage.
2 Raise the front and rear of the vehicle and support it on axle stands (see *Jacking and vehicle support*). Check the rigid hydraulic brake lines for corrosion and damage. Also check the brake pressure regulator in the same manner.
3 At the front of the vehicle, check that the flexible hydraulic hoses to the calipers are not twisted or chafing on any of the surrounding suspension components. Turn the steering on full lock to make this check. Also check that the hoses are not brittle or cracked.
4 Lower the vehicle to the ground after making the checks.

14 Headlight beam adjustment

1 Accurate adjustment of the headlight beam is only possible using optical beam-setting equipment, and this work should therefore be carried out by a dealer or garage with the necessary facilities. All MOT test centres will have the required equipment.
2 Basic adjustments can be carried out in an

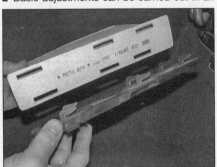

15.5 Observe the air flow direction marked on the filter

emergency, and further details are given in Chapter 12 Section 9.

15 Pollen filter element renewal

1 The pollen filter is located in the heater unit and is accessed from inside the car, on the passenger's side.
2 Where fitted, remove the clips and withdraw the facia lower trim from beneath the glovebox **(see illustration)**.
3 Slide the cover to the left to remove the access cover **(see illustrations)**. On some models there maybe a screw securing the cover in place.
4 Slide out the pollen filter element downwards from the heater unit **(see illustration)**.
5 Fit the new element then **(see illustration)** refit the access cover, making sure that the locating lugs are fitted correctly in the housing. If any of the lugs have been broken off, fit a screw into the cover to make sure it is secure.
6 Refit the trim beneath the facia/glovebox.

16 Manual transmission oil level check

1 There is no requirement to check the manual transmission fluid level – the transmission is effectively sealed for life. However the

19.1 Check the condition of the driveshaft gaiters

transmission should be checked for leaks. Full details of refilling and checking the transmission oil level are given in Chapter 7A Section 4.

17 DSG transmission oil renewal

Renewal

1 The DSG transmission oil renewal is described in Chapter 7B Section 6.

18 Underbody protection check

1 Raise and support the vehicle on axle stands (see *Jacking and vehicle support*). Using an electric torch or lead light, inspect the entire underside of the vehicle, paying particular attention to the wheel arches. Look for any damage to the flexible underbody coating, which may crack or flake off with age, leading to corrosion. Also check that the wheel arch liners are securely attached with any clips provided – if they come loose, dirt may get in behind the liners and defeat their purpose. If there is any damage to the underseal, or any corrosion, it should be repaired before the damage gets too serious.

19 Driveshaft gaiter check

1 With the vehicle raised and securely supported on stands, slowly rotate the roadwheel. Inspect the condition of the outer constant velocity (CV) joint rubber gaiters, squeezing the gaiters to open out the folds. Check for signs of cracking, splits or deterioration of the rubber, which may allow the grease to escape, and lead to water and grit entry into the joint. Also check the security and condition of the retaining clips. Repeat these checks on the inner joints **(see illustration)**. If any damage or deterioration is found, the gaiters should be renewed (see Chapter 8 Section 3).
2 At the same time, check the general condition

20.2 Check the rack gaiters for spilts

20.4 Check for wear in the bearings by grasping the wheel top and bottom and trying to rock it

20.5 Check for wear in the bearings and ball joints by grasping the wheel at the sides and trying to rock it

of the CV joints themselves by first holding the driveshaft and attempting to rotate the wheel. Repeat this check by holding the inner joint and attempting to rotate the driveshaft. Any appreciable movement indicates wear in the joints, wear in the driveshaft splines, or a loose driveshaft retaining nut.

20 Steering and suspension check

1 Raise the front and rear of the vehicle, and securely support it on axle stands (see *Jacking and vehicle support*).
2 Visually inspect the track rod end balljoint dust cover, the lower front suspension balljoint dust cover, and the steering rack-and-pinion gaiters for splits **(see illustration)** chafing or deterioration. Any wear of these components will cause loss of lubricant, together with dirt and water entry, resulting in rapid deterioration of the balljoints or steering gear.
3 Check the power steering fluid hoses for chafing or deterioration, and the pipe and hose unions for fluid leaks. Also check for signs of fluid leakage under pressure from the steering gear rubber gaiters, which would indicate failed fluid seals within the steering gear.
4 Grasp the roadwheel at the 12 o'clock and 6 o'clock positions, and try to rock it **(see illustration)**. Very slight free play may be felt, but if the movement is appreciable, further investigation is necessary to determine the source. Continue rocking the wheel while an assistant depresses the footbrake. If the

movement is now eliminated or significantly reduced, it is likely that the hub bearings are at fault. If the free play is still evident with the footbrake depressed, then there is wear in the suspension joints or mountings.
5 Now grasp the wheel at the 9 o'clock and 3 o'clock positions, and try to rock it as before **(see illustration)**. Any movement felt now may again be caused by wear in the hub bearings or the steering track rod balljoints. If the inner or outer balljoint is worn, the visual movement will be obvious.
6 Using a large screwdriver or flat bar, check for wear in the suspension mounting bushes by levering between the relevant suspension component and its attachment point. Some movement is to be expected as the mountings are made of rubber, but excessive wear should be obvious. Also check the condition of any visible rubber bushes, looking for splits, cracks or contamination of the rubber.
7 With the car standing on its wheels, have an assistant turn the steering wheel back-and-forth about an eighth of a turn each way. There should be very little, if any, lost movement between the steering wheel and roadwheels. If this is not the case, closely observe the joints and mountings previously described, but in addition, check the steering column universal joints for wear, and the rack-and-pinion steering gear itself.
8 Check for any signs of fluid leakage around the front suspension struts and rear shock absorber. Should any fluid be noticed, the suspension strut or shock absorber is defective internally, and should be renewed.

Note: *Suspension struts/shock absorbers should always be renewed in pairs on the same axle to ensure correct vehicle handling.*
9 The efficiency of the suspension strut/shock absorber may be checked by bouncing the vehicle at each corner. Generally speaking, the body will return to its normal position and stop after being depressed. If it rises and returns on a rebound, the suspension strut/shock absorber is probably suspect. Examine also the suspension strut/shock absorber upper and lower mountings for any signs of wear.

21 Battery check

1 The battery is located on the left-hand side of the engine compartment. Remove the cover **(see illustration)** to gain access to the battery. Note that some models may have a soft insulated cover fitted.
2 Check that both battery terminals and all the fuse holder connections are securely attached **(see illustration)** and are free from corrosion. **Note:** *Before disconnecting the terminals from the battery, refer to Chapter 5A Section 3.*
3 Check the battery casing for signs of damage or cracking and check the battery retaining clamp bolt is securely tightened **(see illustration)**. If the battery casing is damaged in any way the battery must be renewed (see Chapter 5A Section 3).
4 On completion refit the battery cover.

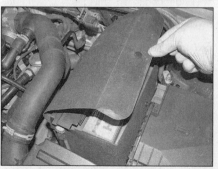

21.1 Unclip the cover

21.2 Check that the terminals are secure

21.3 Check that the battery clamp is secure

25.1 The diagnostic socket

22 Hinge and lock lubrication

1 Lubricate the hinges of the bonnet, doors and tailgate with a light general-purpose oil. Similarly, lubricate all latches, locks and lock strikers. At the same time, check the security and operation of all the locks, adjusting them if necessary (see Chapter 11 Section 13).
2 Lightly lubricate the bonnet release mechanism and cable with a suitable grease.

23 Airbag unit check

1 The airbag system will perform a self-check when the ignition is turned on. The airbag warning light will illuminate and then go out after several seconds. If a system fault is found the airbag warning light will remain on and a fault code will be logged. The fault code can be read by a suitable diagnostic tool and appropriate action taken.
2 Inspect the exterior condition of the airbag(s) for signs of damage or deterioration. If an airbag shows signs of damage, it must be renewed (Chapter 12 Section 27). Note that it is not permissible to attach any stickers to the surface of the airbag, as this may affect the deployment of the unit.

24 Windscreen/tailgate/headlight washer system check

1 Check that each of the washer jet nozzles are clear and that each nozzle provides a strong jet of washer fluid.
2 The tailgate jet should be aimed to spray at the centre of the screen, using a pin.
3 The windscreen washer nozzles should be aimed slightly above the centre of the screen using a small screwdriver to turn the jet eccentric.
4 The headlight inner jet should be aimed slightly above the horizontal centreline of the headlight, and the outer jet should be aimed

slightly below the centreline. Seat technicians use a special tool to adjust the headlight jet after pulling the jet out onto its stop.
5 Especially during the winter months, make sure that the washer fluid frost concentration is sufficient.

25 Engine management self-diagnosis memory fault check

1 This work should be carried out by a Seat dealer or a diagnostic specialist using special equipment. Note however that simple diagnostic tools are now widely available. Most of the lower priced tools will only access the mandatory OBD (On Board Diagnostics) emissions related fault codes. The diagnostic socket is located below the facia, at the right-hand side behind a panel **(see illustration)**. The 16 pin socket is often referred to as the Data Link Connector (or DLC).

26 Sunroof check and lubrication

1 Check the operation of the sunroof, and leave it in the fully open position.
2 Wipe clean the guide rails on each side of the sunroof opening, then apply lubricant to them. Seat recommend VAG lubricant spray G 052 778.

27 Road test and exhaust emissions check

Instruments and electrical equipment

1 Check the operation of all instruments and electrical equipment including the air conditioning system.
2 Make sure that all instruments read correctly, and switch on all electrical equipment in turn, to check that it functions properly.

Steering and suspension

3 Check for any abnormalities in the steering, suspension, handling or road 'feel'.
4 Drive the vehicle, and check that there are no unusual vibrations or noises which may indicate wear in the driveshafts, wheel bearings, etc.
5 Check that the steering feels positive, with no excessive 'sloppiness', or roughness, and check for any suspension noises when cornering and driving over bumps.

Drivetrain

6 Check the performance of the engine, clutch (where applicable), gearbox/transmission and driveshafts.

7 Listen for any unusual noises from the engine, clutch and gearbox/transmission.
8 Make sure the engine runs smoothly at idle, and there is no hesitation on accelerating.
9 Check that, where applicable, the clutch action is smooth and progressive, that the drive is taken up smoothly, and that the pedal travel is not excessive. Also listen for any noises when the clutch pedal is depressed.
10 On manual gearbox models, check that all gears can be engaged smoothly without noise, and that the gear lever action is smooth and not abnormally vague or 'notchy'.
11 On automatic transmission models, make sure that all gearchanges occur smoothly, without snatching, and without an increase in engine speed between changes. Check that all the gear positions can be selected with the vehicle at rest. If any problems are found, they should be referred to a Seat dealer.
12 Listen for a metallic clicking sound from the front of the vehicle, as the vehicle is driven slowly in a circle with the steering on full-lock. Carry out this check in both directions. If a clicking noise is heard, this indicates wear in a driveshaft joint, in which case renew the joint if necessary.

Braking system

13 Make sure that the vehicle does not pull to one side when braking, and that the wheels do not lock when braking hard.
14 Check that there is no vibration through the steering when braking.
15 Check that the handbrake operates correctly without excessive movement of the lever, and that it holds the vehicle stationary on a slope.
16 Test the operation of the brake servo unit as follows. With the engine off, depress the footbrake four or five times to exhaust the vacuum. Hold the brake pedal depressed, then start the engine. As the engine starts, there should be a noticeable 'give' in the brake pedal as vacuum builds-up. Allow the engine to run for at least two minutes, and then switch it off. If the brake pedal is depressed now, it should be possible to detect a hiss from the servo as the pedal is depressed. After about four or five applications, no further hissing should be heard, and the pedal should feel considerably harder.
17 Under controlled emergency braking, the pulsing of the ABS unit must be felt at the footbrake pedal.

Exhaust emissions check

18 Although not part of the manufacturer's maintenance schedule, this check will normally be carried out on a regular basis according to the country the vehicle is operated in. Currently in the UK, exhaust emissions testing is included as part of the annual MOT test after the vehicle is 3 years old. In Germany the test is made when the vehicle is 3 years old, then repeated every 2 years.

28.2 Release the inlet duct

28.3a Disconnect the wiring from the sensor...

28.3b ...and release the spring clips

28 Air filter element renewal – petrol models

Engine codes BWA, BWJ, CDLD and CDLA

1 The air filter element is located in the engine top cover.
2 Compress and slide back the hose clip from the air inlet duct **(see illustration)**.
3 Disconnect the wiring from the mass air flow sensor and then release the spring clips **(see illustrations)**.
4 Where fitted unclip the hose from the air filter housing **(see illustration)**.
5 Carefully work the filter housing free from the top of the engine. Work around the housing, lifting each corner a fraction at a time until it is free. If this gently approach is not taken the locating pegs will break off the housing. Remove the housing and place it upside down on a soft surface.
6 With the assembly inverted on the bench, remove the screws and unhook the filter cover and then remove the filter **(see illustrations)**.
7 Fit the new filter element using a reversal of the removal procedure. Lubricate the the locating pegs on the cover/housing with silicone grease or petroleum jelly (vaseline).

Engine codes BGU, BLR, BLY, BSE, BSF, BVY and BVZ

8 The air cleaner is located in front of the battery on the left-hand side of the engine compartment.
9 Undo the screws and remove the lid from the air cleaner **(see illustration)**.

28.4 Unclip crankcase ventilation hose

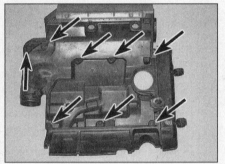

28.6a Remove the screws

28.6b Noting how the cover interlocks, remove it and...

28.6c ...lift out the filter

28.9 Undo the screws and remove the cover

28.10a Undo the screw…

28.10b …and lift out the clamp…

28.10c …followed by the air filter

10 Undo the screw and slide out the clamp, then remove the filter element from the housing **(see illustrations)**.

11 Fit the new filter element using a reversal of the removal procedure.

29 Spark plug renewal

1 The correct functioning of the spark plugs is vital for the correct running and efficiency of the engine. It is essential that the plugs fitted are appropriate for the engine (a suitable type is specified at the beginning of this Chapter). If this type is used and the engine is in good condition, the spark plugs should not need attention between scheduled renewal intervals. Spark plug cleaning is rarely necessary, and should not be attempted unless specialised equipment is available, as damage can easily be caused to the firing ends.

1.6 litre engines

Note: *The spark plugs are located beneath the inlet manifold upper section, and are very difficult to access. Seat technicians use a special tool (T10112) to disconnect the HT leads, together with a universally-jointed spark plug socket to unscrew the spark plugs.*

2 Where fitted, remove the engine top cover.
3 Disconnect the HT lead connectors from the spark plugs. To do this, VW technicians use special tool T10112 to disconnect Nos 2 and 3 leads, and ordinary spark plug lead pliers to disconnect Nos 2 and 3. The special tool is approximately 30 cm in length with a bayonet-type claw at the bottom, which engages the HT lead connectors. If this tool is not available, the plug leads can be lassoed with cable ties. The cable tie must fit

around the plug shroud and not the lead **(see illustration)**.
4 Use a spark plug socket with a universal joint to remove the spark plugs **(see illustration)**.

2.0 litre engine

5 Remove the engine cover/air filter housing.
6 Unscrew the two bolts securing the coil wiring loom to the camshaft cover and then disconnect the wiring loom plugs from the ignition coils **(see illustration)**.
7 Remove the ignition coils from the tops of the spark plugs. Seat technicians use special tool T40039 to pull out the coils. However it is possible to pull them out with a twisting motion. Where this is not possible, a length of welding rod bent to hook under the connectors may be used instead **(see illustration)**.
8 Use a spark plug spanner and remove the spark plugs **(see illustrations)**.

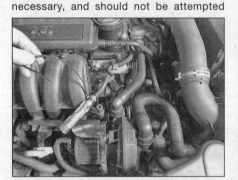

29.3 Remove the HT leads

29.4 Remove the spark plugs

29.6 Remove the wiring plugs from the coils

29.7 Remove the ignition coils

29.8a Remove the plugs

29.8b A magnet may be needed to rescue the plugs

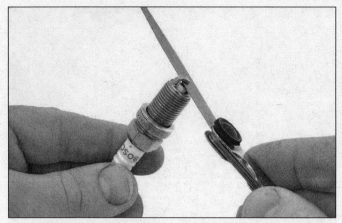

29.14 If single electrode plugs are being fitted, check the electrode gap using a feeler gauge

29.15 Use a rubber hose to avoid cross-threading the spark plugs

All engines

9 It is advisable to remove the dirt from the spark plug recesses using a clean brush, vacuum cleaner or compressed air before removing the plugs, to prevent dirt dropping into the cylinders.

10 Unscrew the plugs using a spark plug spanner, suitable box spanner or a deep socket and extension bar. Keep the socket aligned with the spark plug – if it is forcibly moved to one side, the ceramic insulator may be broken off. The use of a universal joint socket will be helpful. As each plug is removed, examine it as follows.

11 Examination of the spark plugs will give a good indication of the condition of the engine. If the insulator nose of the spark plug is clean and white, with no deposits, this is indicative of a weak mixture or too hot a plug (a hot plug transfers heat away from the electrode slowly, a cold plug transfers heat away quickly).

12 If the tip and insulator nose are covered with hard black-looking deposits, then this is indicative that the mixture is too rich. Should the plug be black and oily, then it is likely that the engine is fairly worn, as well as the mixture being too rich.

13 If the insulator nose is covered with light tan to greyish-brown deposits, then the mixture is correct and it is likely that the engine is in good condition.

14 The spark plug electrode gap is of considerable importance as, if it is too large or too small, the size of the spark and its efficiency will be seriously impaired. On engines covered in this manual the plug gaps should not be adjusted. Where single electrode plugs are fitted the gap can be checked, but if the gap is outside the specifications the plugs should be replaced **(see illustration)**.

15 Before fitting the spark plugs, check that the threaded connector sleeves are tight, and that the plug exterior surfaces and threads are clean. It's often difficult to screw in new spark plugs without cross-threading them – this can be avoided using a piece of rubber hose **(see illustration and Haynes Hint)**.

16 Remove the rubber hose (if used), and tighten the plug to the specified torque using the spark plug socket and a torque wrench **(see illustration)**. Refit the remaining spark plugs in the same manner.

17 Reconnect and refit the HT leads/ignition coils using a reversal of the removal procedure.

30 Auxiliary drivebelt check and renewal

1 The poly-vee drivebelt drives the alternator and where fitted, the air conditioning compressor.

2 On all engines, the drivebelt tension is adjusted automatically by a spring-tensioned idler.

Checking

3 See Section 11.

Renewal

4 Where fitted, remove the engine top cover.

5 For improved access, apply the handbrake, then jack up the front of the vehicle and support it on axle stands (see *Jacking and vehicle support*). Remove the right-hand front roadwheel, then remove the access panel from the inner wheel arch. Note however that it is also possible to remove the belt with the

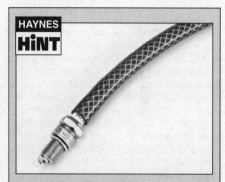

HAYNES HINT

It is very often difficult to insert spark plugs into their holes without cross-threading them. To avoid this possibility, fit a short length of rubber hose over the end of the spark plug. The flexible hose acts as a universal joint to help align the plug with the plug thread, the hose will slip on the spark plug, preventing thread damage to the aluminium cylinder head.

vehicle on the ground, despite the limited access.

6 To improve access, unclip the EVAP canister and move it to the side.

7 If the drivebelt is to be re-used, mark it for clockwise direction to ensure it is refitted the same way round **(see illustration)**.

29.16 Tighten the plugs to the specified torque

30.7 Mark the direction of rotation

30.8a The holes must align to lock the tensioner

30.8b Rotate the tensioner until the belt...

8 Use a spanner on the lug provided and turn the tensioner clockwise. Lock the tensioner in its released position by inserting a drill through the lug into the tensioner body **(see illustrations)**.
9 Note how the drivebelt is routed, then remove it from the crankshaft pulley, alternator pulley and air conditioning compressor pulley **(see illustration)**.
10 Locate the new drivebelt on the pulleys, then release the tensioner. Check that the belt is located correctly in the multi-grooves in the pulleys.
11 Refit the access panel and roadwheel, and lower the vehicle to the ground. Refit the engine top cover.

30.8c ...can be locked in position

30.9 Remove the belt

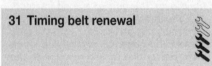

31 Timing belt renewal

Renewal

1 Refer to Chapter 2A Section 7 or Chapter 2B Section 4 for details.

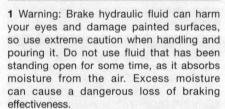

32 Brake (and clutch) fluid renewal – petrol models

1 Warning: Brake hydraulic fluid can harm your eyes and damage painted surfaces, so use extreme caution when handling and pouring it. Do not use fluid that has been standing open for some time, as it absorbs moisture from the air. Excess moisture can cause a dangerous loss of braking effectiveness.
2 The procedure is similar to that for the bleeding of the hydraulic system as described in Chapter 9 Section 2, except that the brake fluid reservoir should be emptied by syphoning, using a clean poultry baster or similar before starting, and allowance should be made for the old fluid to be expelled when

bleeding a section of the circuit. Since the clutch hydraulic system also uses fluid from the brake system reservoir, it should also be bled at the same time by referring to Chapter 6 Section 2.
3 Working as described in Chapter 9 Section 2, open the first bleed screw in the sequence, and pump the brake pedal gently until nearly all the old fluid has been emptied from the master cylinder reservoir.

 HAYNES HiNT *Old hydraulic fluid is often much darker in colour than the new, making it easy to distinguish the two.*

4 Top-up to the MAX level with new fluid, and continue pumping until only the new fluid remains in the reservoir, and new fluid can be seen emerging from the bleed screw. Tighten the screw, and top the reservoir level up to the MAX level line.
5 Work through all the remaining bleed screws in the sequence until new fluid can be seen at all of them. Be careful to keep the master cylinder reservoir topped-up to above the MIN level at all times, or air may enter the system and greatly increase the length of the task.
6 When the operation is complete, check that

all bleed screws are securely tightened, and that their dust caps are refitted. Wash off all traces of spilt fluid, and recheck the master cylinder reservoir fluid level.
7 On models with manual transmission, once the brake fluid has been changed the clutch fluid should also be renewed. Referring to Chapter 6 Section 2, bleed the clutch until new fluid is seen to be emerging from the slave cylinder bleed screw, keeping the master cylinder fluid level above the MIN level line at all times to prevent air entering the system. Once the new fluid emerges, securely tighten the bleed screw then disconnect and remove the bleeding equipment. Securely refit the dust cap then wash off all traces of spilt fluid.
8 On all models, ensure the master cylinder fluid level is correct (see *Weekly checks*) and thoroughly check the operation of the brakes and (where necessary) clutch before taking the car on the road.

33 Coolant renewal – petrol models

 Warning: Wait until the engine is cold before starting this procedure. Do not allow antifreeze to come in contact with your skin,

or with the painted surfaces of the vehicle. Rinse off spills immediately with plenty of water. Never leave antifreeze lying around in an open container, or in a puddle in the driveway or on the garage floor. Children and pets are attracted by its sweet smell, but antifreeze can be fatal if ingested.

Note: *This work is not included in the Seat schedule and should not be required if the recommended VAG G12 LongLife coolant antifreeze/inhibitor is used. However, if standard antifreeze/inhibitor is used, the work should be carried out at the recommended interval.*

Cooling system draining

1 With the engine completely cold, unscrew the expansion tank cap.

2 Firmly apply the handbrake then jack up the front of the vehicle and support it on axle stands (see *Jacking and vehicle support*). Undo the retaining screws and remove the engine undertray to gain access to the base of the radiator.

3 Position a suitable container beneath the coolant drain outlet which is fitted to the coolant bottom hose end fitting. Loosen the drain plug (there is no need to remove it completely) and allow the coolant to drain into the container. If desired, a length of tubing can be fitted to the drain outlet to direct the flow of coolant during draining. Where no drain outlet is fitted to the hose end fitting, prise out the retaining clip a little, and disconnect the bottom hose from the radiator to drain the coolant (see Chapter 3 Section 2) **(see illustrations)**.

4 On engines with an oil cooler, to fully drain the system also disconnect one of the coolant hoses from the oil cooler which is located at the front of the cylinder block.

5 If the coolant has been drained for a reason other than renewal, then provided it is clean, it can be re-used.

6 Once all the coolant has drained, securely tighten the radiator drain plug or reconnect the bottom hose to the radiator (as applicable).

33.3a Releasing the bottom radiator hose...

Where necessary, also reconnect the coolant hose to the oil cooler and secure it in position with the retaining clip. Refit the undertray, and tighten the retaining screws securely.

Cooling system flushing

7 If the recommended Seat coolant has not been used and coolant renewal has been neglected, or if the antifreeze mixture has become diluted, the cooling system may gradually lose efficiency, as the coolant passages become restricted due to rust, scale deposits, and other sediment. The cooling system efficiency can be restored by flushing the system clean.

8 The radiator should be flushed separately from the engine, to avoid excess contamination.

Radiator flushing

9 To flush the radiator, first tighten the radiator drain plug – where fitted.

10 Disconnect the top and bottom hoses and any other relevant hoses from the radiator (see Chapter 3 Section 3).

11 Insert a garden hose into the radiator top inlet. Direct a flow of clean water through the radiator, and continue flushing until clean water emerges from the radiator bottom outlet.

12 If after a reasonable period, the water still does not run clear, the radiator can be flushed with a good proprietary cleaning agent. It is important that their manufacturer's instructions are followed carefully. If the contamination is particularly bad, insert the hose in the radiator bottom outlet, and reverse-flush the radiator.

Engine flushing

13 To flush the engine, remove the thermostat (see Chapter 3 Section 4).

14 With the bottom hose disconnected from the radiator, insert a garden hose into the coolant housing. Direct a clean flow of water through the engine, and continue flushing until clean water emerges from the radiator bottom hose.

33.3b ...to drain the coolant

15 When flushing is complete, refit the thermostat and reconnect the hoses.

Cooling system filling

16 Before attempting to fill the cooling system, ensure the drain plug is securely closed and make sure that all hoses are securely connected and their retaining clips are in good condition. If the recommended Seat coolant is not being used, ensure that a suitable antifreeze mixture is used all year round, to prevent corrosion of the engine components (see following sub-Section).

Note: *Seat recommend that only distilled water should be used.*

17 Remove the expansion tank filler cap and slowly fill the system with the coolant. Continue to fill the cooling system until bubbles stop appearing in the expansion tank. Help to bleed the air from the system by repeatedly squeezing the radiator bottom hose.

18 When no more bubbles appear, top the coolant level up to the MAX level mark then securely refit the cap to the expansion tank.

19 Run the engine at a fast idle speed until the cooling fan cuts in. Wait for the fan to stop then switch the engine off and allow the engine to cool.

20 When the engine has cooled, check the coolant level with reference to *Weekly checks*. Top-up the level if necessary, and refit the expansion tank cap.

Antifreeze mixture

21 If the recommended Seat coolant is not being used, the antifreeze should always be renewed at the specified intervals. This is necessary not only to maintain the antifreeze properties, but also to prevent corrosion which would otherwise occur as the corrosion inhibitors become progressively less effective.

22 Always use an ethylene-glycol based antifreeze which is suitable for use in mixed-metal cooling systems. The quantity of antifreeze and levels of protection are indicated in the Specifications.

23 Before adding antifreeze, the cooling system should be completely drained, preferably flushed, and all hoses checked for condition and security.

24 After filling with antifreeze, a label should be attached to the expansion tank, stating the type and concentration of antifreeze used, and the date installed. Any subsequent topping-up should be made with the same type and concentration of antifreeze.

Caution: Do not use engine antifreeze in the windscreen/tailgate washer system, as it will damage the vehicle paintwork. A screenwash additive should be added to the washer system in the quantities stated on the bottle.

Chapter 1 Part B
Routine maintenance and servicing – diesel engines

Contents

Degrees of difficulty

| **Easy,** suitable for novice with little experience | | **Fairly easy,** suitable for beginner with some experience | | **Fairly difficult,** suitable for competent DIY mechanic | | **Difficult,** suitable for experienced DIY mechanic | | **Very difficult,** suitable for expert DIY or professional | 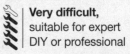 |

1 Servicing specifications – diesel engines

Lubricants and fluids
Refer to Lubricants, fluids and tyre pressures

Engine codes*
Manufacturer's engine codes*:

1599 cc (1.6 litre), CR (Common Rail)	CAYC
1896 cc (1.9 litre) PD (Pumpe Duse)...........................	BJB, BKC, BLS, BXE and BXF
1968 cc (2.0 litre) PD (Pumpe Duse)..........................	BKD, BMN, and BMM
1968 cc (2.0 litre) CR (Common Rail)	CFJA, CLCB, CEGA and CFHC

See 'Vehicle identification' at the end of this manual for the location of engine code markings.

Capacities

Engine oil (including filter)

PD engines (BJB, BKC, BLS, BXE and BXF)	3.8 litres
PD engines (BKD, BMM and BMN)	4.3 litres
CR engines (all versions)	4.3 litres

Cooling system

All engines	8.0 litres

Transmission*
Manual transmission:

Type 02S...	1.9 litres
Type 02Q	2.3 litres

* See Chapter for application details
DSG transmission:

Type 02E...	5.2 litres (refill capacity)

* See Chapter for application details

Fuel tank (approximate

All models..	60 litres

Washer reservoirs

Models with headlight washers	5.5 litres
Models without headlight washers...........................	3.0 litres

Cooling system
Antifreeze mixture:

40% antifreeze	Protection down to -25°C
50% antifreeze	Protection down to -35°C

Note: *Refer to antifreeze manufacturer for latest recommendations.*

Brakes
Brake pad lining minimum thickness:

Front	2.0 mm
Rear	2.0 mm

Torque wrench settings

	Nm	lbf ft
Manual gearbox filler/drain plug:		
Multi-point socket head	30	22
Hexagon socket head...................................	45	31
02E transmission:		
Drain/filler plug	45	33
Oil filter cap	25	18
Roadwheel bolts..	120	89
Sump drain plug...	30	22

2 Maintenance schedule – diesel models

1 The maintenance intervals in this manual are provided with the assumption that you, not the dealer, will be carrying out the work. These are the minimum intervals recommended by us for vehicles driven daily. If you wish to keep your vehicle in peak condition at all times, you may wish to perform some of these procedures more often. We encourage frequent maintenance, since it enhances the efficiency, performance and resale value of your vehicle.

2 When the vehicle is new, it should be serviced by a dealer service department (or other workshop recognised by the vehicle manufacturer as providing the same standard of service), in order to preserve the warranty. The vehicle manufacturer may reject warranty claims if you are unable to prove that servicing has been carried out as and when specified, using only original equipment parts or parts certified to be of equivalent quality.

3 Depending on the model specification service intervals will either be at standard times (or distance) or set to 'LongLife' service intervals. On models set to the LongLife display, the service interval is variable according to the number of starts, length of journeys, vehicle speeds, brake pad wear, bonnet opening frequency, fuel consumption, oil level and oil temperature, however the vehicle must be serviced at least every two years. Note that if the variable (LongLife) service interval is being used, the engine must only be filled with the recommended long-life engine oil (see Lubricants and fluids).

4 All Seat Leon models are equipped with a Service Interval Display (SID) indicator in the instrument panel. When a service date is approaching a message will appear in the message centre or a spanner symbol will appear in the instrument panel. The message or spanner symbol will also show the distance (or time) to the next service and will count down as the due date approaches. A minus symbol or a 'Service – - days ago' (on models with a message centre) will be shown when the due date (or mileage) has passed. Note that on models on Longlife service intervals that have the battery disconnected for a long time the service interval will not be correctly calculated. Where this is the case the vehicle should be serviced at the standard intervals.

5 After completing a service, Seat technicians use a special instrument to reset the service display to the next service interval, and a print-out is put in the vehicle service record. The display can be reset by the owner as described in Section 8.

Every 250 miles

☐ Refer to 'Weekly checks '.

Every 10 000 miles or 12 months, whichever comes first

☐ Renew the engine oil and filter (Section 6)
Note: *Frequent oil and filter changes are good for the engine. We recommend changing the oil at least once a year.*
☐ Check the front and rear brake pad thickness (Section 7)
☐ Reset the service interval display (Section 8)

Every 20 000 miles or 2 years, whichever comes first

In addition to the items listed above, carry out the following:
☐ Check the condition of the exhaust system and its mountings (Section 9)
☐ Check all underbonnet components and hoses for fluid and oil leaks (Section 10)
☐ Renew the fuel filter* (Section 29)
☐ Check the condition of the auxiliary drivebelt (Section 11)
☐ Check the coolant antifreeze concentration (Section 12)
☐ Check the brake hydraulic circuit for leaks and damage (Section 13)
☐ Check the headlight beam adjustment (Section 14)
☐ Renew the pollen filter element (Section 15)
☐ Check the underbody protection for damage (Section 17)
☐ Check the condition of the driveshaft gaiters (Section 18)
☐ Check the steering and suspension components for condition and security (Section 19)
☐ Check the battery condition, security and electrolyte level (Section 20)
☐ Lubricate all hinges and locks (Section 21)
☐ Check the condition of the airbag unit(s) (Section 22)
☐ Check the operation of the windscreen/tailgate/headlight washer system(s) (as applicable) (Section 23)
☐ Check the engine management self-diagnosis memory for faults (Section 24)
☐ Check the operation of the sunroof and lubricate the guide rails (Section 25)
☐ Carry out a road test and check exhaust emissions (Section 26)
* *Only when using diesel fuel not conforming to DIN EN 590 or when using RME fuel (diester)*

Every 40 000 miles or 4 years, whichever comes first

Note: *Many dealers perform these tasks at every second service.*
☐ Renew DSG transmission oil (Section 27)
☐ Renew the air filter element (Section 28)
☐ Renew the fuel filter* (Section 29)
☐ Check the condition of the auxiliary drivebelt (Section 30)
* *Only when using diesel fuel conforming to DIN EN 590*

Every 60 000 miles or 4 years

☐ Renew the timing belt and tensioner roller (Section 31)
Note: *Seat specify a timing belt renewal interval of 130 000 miles for all common rail engines and 75 000 miles for PD engines up to 2006. After 2006 the interval for PD engines is increased to 93 000 miles. They specify a tensioner roller renewal with the belt change. However, if the vehicle is used mainly for short journeys, we recommend that a shorter renewal interval be adhered to. The belt and tensioner renewal interval is very much up to the individual owner but, bearing in mind that severe engine damage will result if the belt breaks in use, we recommend a shorter interval.*
Note: *Always check with the vehicle manufacturer for the latest belt change intervals – they are often subject to change without notice.*
Note: *It is increasingly considered best practise to always replace the idlers, tensioner and coolant pump whenever the timing belt is replaced. Note also that most manufacturers will not guarantee a belt unless the tensioners and rollers have been replaced at the same time.*

Every 95 000 miles, then every 19 000 miles

☐ Check the particulate filter ash deposit mass (Section 34)

Every 2 years

☐ Renew the brake (and clutch) fluid (Section 32)
☐ Renew the coolant* (Section 33)
Note: *This work is not included in the schedule and should not be required if the recommended G12 Plus LongLife coolant antifreeze/inhibitor is used.*

3 Component locations – diesel models

Underbonnet view of a 2.0 litre CR model (engine cover removed)

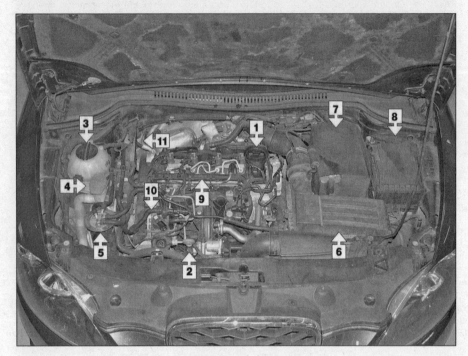

1 Engine oil filler cap
2 Engine oil dipstick
3 Coolant expansion reservoir
4 Windscreen/headlight/rear window washer fluid reservoir
5 Fuel filter
6 Air filter housing
7 Battery (under cover)
8 Fusebox
9 Common rail
10 High pressure pump
11 Brake fluid reservoir

Front underbody view (undershield removed)

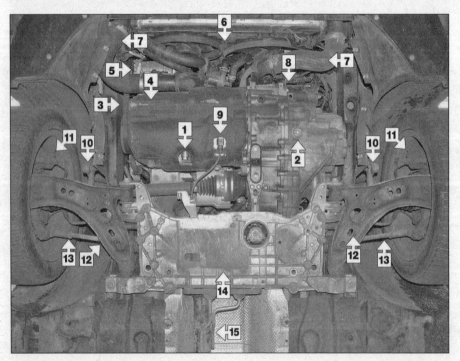

1 Sump drain plug
2 Manual transmission drain plug
3 Auxiliary belt
4 Sump insulator cover
5 Air conditioning compressor
6 Radiator and electric cooling fans
7 Air duct/hoses to intercooler
8 Starter motor
9 Engine oil level/temperature sensor
10 Front anti-roll bar drop-links
11 Front brake calipers
12 Front suspension lower arms
13 Steering track rod arms
14 Front suspension subframe
15 Exhaust pipe

Rear underbody view

1 Exhaust rear silencer
2 Rear track control lower arm
3 Rear anti-roll bar
4 Rear suspension transverse links
5 Spare wheel well
6 Rear suspension subframe beam
7 Rear trailing arm and bracket
8 Fuel tank
9 Handbrake cables

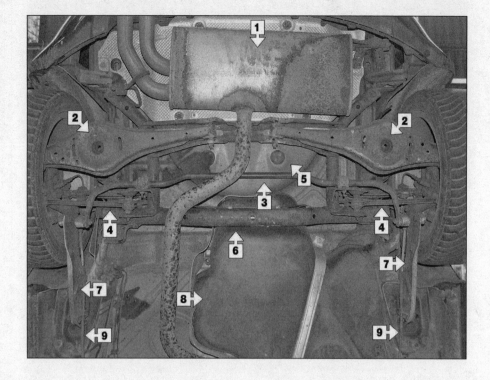

4 Introduction

1 This Chapter is designed to help the home mechanic maintain his/her vehicle for safety, economy, long life and peak performance.
2 The Chapter contains a master maintenance schedule, followed by Sections dealing specifically with each task in the schedule. Visual checks, adjustments, component renewal and other helpful items are included. Refer to the accompanying illustrations of the engine compartment and the underside of the vehicle for the locations of the various components.
3 Servicing your vehicle will provide a planned maintenance programme, which should result in a long and reliable service life. This is a comprehensive plan, so maintaining some items but not others, will not produce the same results.
4 As you service your vehicle, you will discover that many of the procedures can – and should – be grouped together, because of the particular procedure being performed, or because of the proximity of two otherwise unrelated components to one another. For example, if the vehicle is raised for any reason, the exhaust can be inspected at the same time as the suspension and steering components.
5 The first step in this maintenance programme is to prepare yourself before the

actual work begins. Read through all the Sections relevant to the work to be carried out, then make a list and gather all the parts and tools required. If a problem is encountered, seek advice from a parts specialist, or a dealer service department.

5 Regular maintenance

1 If, from the time the vehicle is new, the routine maintenance schedule is followed closely, and frequent checks are made of fluid levels and high-wear items, as suggested throughout this manual, the engine will be kept in relatively good running condition, and the need for additional work will be minimised.
2 It is possible that there will be times when the engine is running poorly due to the lack of regular maintenance. This is even more likely if a used vehicle, which has not received regular and frequent maintenance checks, is purchased. In such cases, additional work may need to be carried out, outside of the regular maintenance intervals.
3 If engine wear is suspected, a compression test (refer to Chapter 2C Section 2, Chapter 2D Section 2 or Chapter 2E Section 2) will provide valuable information regarding the overall performance of the main internal components. Such a test can be used as a basis to decide on the extent of the work to be carried out. If, for example, a compression test indicates serious internal engine wear,

conventional maintenance as described in this Chapter will not greatly improve the performance of the engine, and may prove a waste of time and money, unless extensive overhaul work is carried out first.
4 The following series of operations are those most often required to improve the performance of a generally poor-running engine:

Primary operations

a) Clean and inspect the battery (See 'Weekly checks').
b) Check all the engine-related fluids (See 'Weekly checks').
c) Drain the water from the fuel filter (Section 29).
d) Check the condition and tension of the auxiliary drivebelt (Section 11).
e) Check the condition of the air filter, and renew if necessary (Section 28).
f) Check the condition of all hoses, and check for fluid leaks (Section 10).
5 If the above operations do not prove fully effective, carry out the following secondary operations:

Secondary operations

6 All items listed under Primary operations, plus the following:
a) Check the charging system (see Chapter 5A Section 4).
b) Check the preheating system (see Chapter 5C Section 2).
c) Renew the fuel filter (Section 29) and check the fuel system (see Chapter 4B).

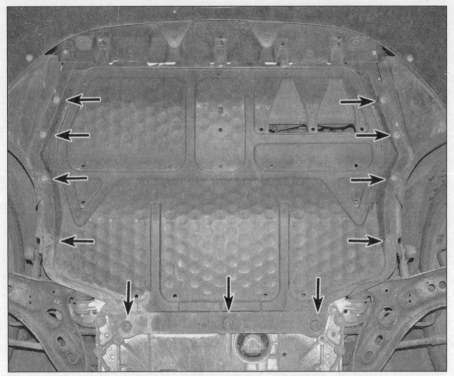

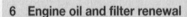

6.2a The engine undertray is secured by a number of screws

Keep the drain plug pressed into the sump while unscrewing it by hand the last couple of turns. As the plug releases, move it away sharply so the stream of oil issuing from the sump runs into the container, not up your sleeve.

6 Engine oil and filter renewal

1 Frequent oil and filter changes are the most important preventative maintenance procedures, which can be undertaken by the DIY owner. As engine oil ages, it becomes diluted and contaminated, which leads to premature engine wear.

2 Before starting this procedure, gather all the necessary tools and materials. Also make sure that you have plenty of clean rags and newspapers handy, to mop-up any spills. Ideally, the engine oil should be warm, as it will drain better, and more built-up sludge will be removed with it. Take care, however,

not to touch the exhaust or any other hot parts of the engine when working under the vehicle. To avoid any possibility of scalding, and to protect yourself from possible skin irritants and other harmful contaminants in used engine oils, it is advisable to wear gloves when carrying out this work. Access to the underside of the vehicle will be greatly improved if it can be raised on a lift, driven onto ramps, or jacked up and supported on axle stands (see *Jacking and vehicle support*). Whichever method is chosen, make sure that the vehicle remains level, or if it is at an angle, that the drain plug is at the lowest point. Undo the retaining screws and remove the engine undertray, then also remove the engine top cover **(see illustrations)**.

3 Slacken the sump drain plug about half a

turn. Position the draining container under the drain plug, and then remove the plug completely **(see illustration and Haynes Hint)**. To drain all oil from the engine, loosen the cap from the top of the oil filter housing using a socket or spanner – this will allow the oil to drain from the filter housing into the sump.

4 Allow some time for the old oil to drain, noting that it may be necessary to reposition the container as the oil flow slows to a trickle.

5 After all the oil has drained, wipe off the drain plug with a clean rag, and fit a new sealing washer. Clean the area around the drain plug opening, and refit the plug. Tighten the drain plug to the specified torque. **Note:** *On some engines, the sealing washer is integral with the drain plug. On these engines, the drain plug must be renewed.*

6 Place absorbent cloths around the oil filter housing to catch any spilt oil. Where necessary, unbolt the bracket and unclip the solenoid valve from above the oil filter **(see illustration)**.

7 Fully unscrew the cap from the top of the oil filter and remove it together with the filter

6.2b Pull up the engine cover upwards to release it

6.3 Undo the sump drain plug

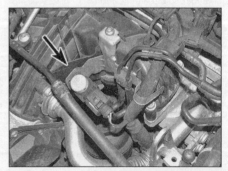

6.6 Remove the solenoid valve

6.7a Use a 32 mm socket to unscrew the filter cap …

6.7b … and withdraw the filter and cap

6.8 Remove the O-ring

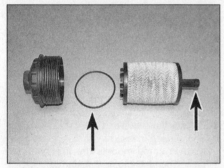

6.9a The new oil filter comes with two O-ring seals. Check that the smaller O-ring is already fitted or supplied with the filter

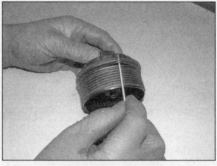

6.9b Fit the main seal to the oil filter cap in the position shown …

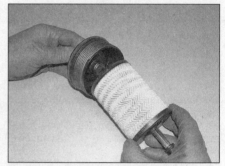

6.9c … and then snap the new filter into the cap

6.9d Lubricate the O-ring seals with clean engine …

6.9e …then refit to the oil filter housing

6.9f Tighten to the specified torque

element. Unclip the filter element from the cap and dispose of it **(see illustrations)**.

8 Remove the O-ring seal from the cap **(see illustration)** and then using a clean rag, wipe all oil and sludge from the inside of the filter housing and cap.

9 The new filter will be supplied with a large sealing O-ring and a small O-ring (already fitted in most cases). Some makes of filter may require the fitting of the small O-ring. Note that common rail engines use a different oil filter to that shown in the illustrations, however replacement is essentially the same. Fit the new sealing ring(s) and then refit the assembly and tighten the cap to the specified torque. (The correct torque is shown on the cap) Make sure the filter element is engaged with the cap and the correct way up **(see illustrations)**. Wipe up any spilt oil before refitting the engine top cover.

10 Remove the old oil and all tools from under the car.

11 Remove the dipstick, and then unscrew the oil filler cap from the cylinder head cover. Fill the engine, using the correct grade and type of oil (see *Lubricants and fluids*). A funnel may help to reduce spillage. Pour in half the specified quantity of oil first **(see illustration)**, then wait a few minutes for the oil to run to the sump (see *Weekly checks*). Continue adding oil a small quantity at a time until the level is above the minimum mark on the dipstick – ideally the level will be between the maximum and minimum mark. Refit the filler cap.

Caution: Because the front of the vehicle is still raised at this point, do not overfill the engine with oil. The level on the dipstick will increase when the vehicle is lowered to the ground.

12 Start the engine and run it for a few

minutes; check for leaks around the oil filter cap. Note that there may be a few seconds delay before the oil pressure warning light

6.11 Pour in half the specified quantity of oil first, wait, then add enough oil to bring the level to above the minimum mark on the dipstick

7.1 On some models the outer brake pads can be observed through the holes in the wheels

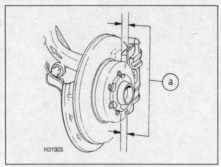

7.3a The thickness (a) of the brake pad linings must not be less than the specified amount

7.3b Specialist tools are also available to check the the friction material

goes out when the engine is started, as the oil circulates through the engine oil galleries and the new oil filter before the pressure builds-up.

⚠ *Warning: Do not increase the engine speed above idling while the oil pressure light is illuminated, as considerable damage can be caused to the turbocharger.*

13 Switch off the engine, and wait a few minutes for the oil to settle in the sump once more.

14 Check around the sump drain plug for oil leaks. If all is well, refit the engine undershield. Lower the vehicle to the ground and fit the engine top cover.

15 Finally, recheck the level on the dipstick, and add more oil to bring the level up to the maximum mark on the dipstick.

16 Dispose of the used engine oil safely, with reference to *General repair procedures* in the Reference section of this manual.

7 Brake pad check 🔧

1 On some models the outer brake pads can be checked without removing the wheels, by observing the brake pads through the holes in the wheels **(see illustration)**. If necessary, remove the wheel trim. The thickness of the pad lining must not be less than the dimension given in the Specifications.

2 If the outer pads are worn near their limits, it is worthwhile checking the inner pads as well. Apply the handbrake then jack up vehicle and support it on axle stands (see *Jacking and vehicle support*). Remove the roadwheels.

3 Use a steel rule to check the thickness of the brake pads, and compare with the minimum thickness given in the Specifications **(see illustrations)**.

4 For a comprehensive check, the brake pads should be removed and cleaned. The operation of the caliper can then also be checked, and the condition of the brake disc itself can be fully examined on both sides. Refer to Chapter 9 Section 4 (front) or Chapter 9 Section 8 (rear).

5 If any pad's friction material is worn to the

specified minimum thickness or less, all four pads at the front or rear, as applicable, must be renewed as a set. If there is any doubt as to the condition of the brake pads (and the brake discs) always err on the side of caution and replace them.

6 On completion of the check, refit the roadwheels and lower the vehicle to the ground.

8 Resetting the service interval display 🔧

1 After all necessary maintenance work has been completed the service interval display must be reset. The Service Interval Display (SID) can only be reset on models that are on fixed service intervals. Models that are factory set to variable service intervals must have the SID reset using diagnostic equipment. A Seat dealer will have the factory scan tool, but many independent garages will also have an aftermarket tool capable of resetting the SID.

2 To continue with the 'variable' or 'longlife' service intervals which take into consideration the number of starts, length of journeys, vehicle speeds, brake pad wear, bonnet opening frequency, fuel consumption, oil level and oil temperature, the display must be reset by a Seat dealership (or suitably equipped garage) using a diagnostic tool.

3 For the home mechanic it is recommended that models have the service schedule changed from 'variable' to 'fixed' either by a Seat dealer or suitably equipped garage.

4 To reset the SID:
● Turn the ignition off.
● Press and hold down the odometer button in the centre of the instrument panel.
● Turn the ignition on (whilst still holding down the button).
● Rotate the button to the right (clockwise) for approximately 1 second.
● Release the button and turn the ignition off.
● The service message will disappear. Note that on some models the message will clear when the engine is restarted.

9 Exhaust system check 🔧

1 With the engine cold (at least an hour after the vehicle has been driven), check the complete exhaust system from the engine to the end of the tailpipe. The exhaust system is most easily checked with the vehicle raised on a hoist, or suitably supported on axle stands, so that the exhaust components are readily visible and accessible (see *Jacking and vehicle support*).

2 Check the exhaust pipes and connections for evidence of leaks, severe corrosion and damage. Make sure that all brackets and mountings are in good condition, and that all relevant nuts and bolts are tight. Leakage at any of the joints or in other parts of the system will usually show up as a black sooty stain in the vicinity of the leak **(see illustrations)**.

3 Rattles and other noises can often be traced to the exhaust system, especially the

9.2a Typical minor exhaust leak from a clamp type fitting...

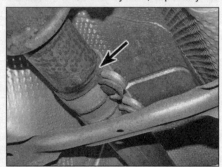

9.2b ...and a sleeve type fitting

brackets and mountings. Try to move the pipes and silencers. If the components are able to come into contact with the body or suspension parts, secure the system with new mountings. Otherwise separate the joints (if possible) and twist the pipes as necessary to provide additional clearance.

10 Hose and fluid leak check

1 Visually inspect the engine joint faces, gaskets and seals for any signs of water or oil leaks. Pay particular attention to the areas around the camshaft cover, cylinder head, oil filter and sump joint faces. Bear in mind that, over a period of time, some very slight seepage from these areas is to be expected – what you are really looking for is any indication of a serious leak. Should a leak be found, renew the offending gasket or oil seal by referring to the appropriate Chapters in this manual.

2 Also check the security and condition of all the engine-related pipes and hoses. Ensure that all cable ties or securing clips are in place and in good condition. Clips, which are broken or missing, can lead to chafing of the hoses, pipes or wiring, which could cause more serious problems in the future.

3 Carefully check the radiator hoses and heater hoses along their entire length. Renew any hose that is cracked, swollen or deteriorated. Cracks will show up better if the hose is squeezed. Pay close attention to the hose clips that secure the hoses to the cooling system components. Hose clips can pinch and puncture hoses, resulting in cooling system leaks.

4 Inspect all the cooling system components (hoses, joint faces, etc) for leaks (see **Haynes Hint**). Where any problems of this nature are found on system components, renew the component or gasket with reference to Chapter 3.

5 Where applicable, inspect the automatic transmission fluid cooler hoses for leaks or deterioration.

6 With the vehicle raised, inspect the fuel tank and filler neck for punctures, cracks and other damage. The connection between the filler neck and tank is especially critical. Sometimes a rubber filler neck or connecting hose will leak due to loose retaining clamps or deteriorated rubber.

7 Carefully check all rubber hoses and metal fuel lines leading away from the tank. Check for loose connections, deteriorated hoses, crimped lines, and other damage. Pay particular attention to the vent pipes and hoses, which often loop up around the filler neck and can become blocked or crimped. Follow the lines to the front of the vehicle, carefully inspecting them all the way. Renew damaged sections as necessary.

8 From within the engine compartment, check the security of all fuel hose attachments and pipe unions, and inspect the fuel hoses

HAYNES HiNT

A leak in the cooling system will usually show up as white- or antifreeze coloured deposits on the area adjoining the leak.

and vacuum hoses for kinks, chafing and deterioration.

11 Auxiliary drivebelt check

1 Apply the handbrake, then jack up the front of the vehicle and support it on axle stands (see *Jacking and vehicle support*).

2 Using a socket on the crankshaft pulley bolt, turn the engine slowly clockwise so that the full length of the auxiliary drivebelt can be examined. Look for cracks, splitting and fraying on the surface of the belt; check also for signs of glazing (shiny patches) and separation of the belt plies. If damage or wear is visible, or if there are traces of oil or grease on it, the belt should be renewed (see Section 30).

12 Antifreeze check

1 The cooling system should be filled with the recommended G12 antifreeze and corrosion protection fluid – do not mix this antifreeze with any other type. Over a period of time, the concentration of fluid may be reduced due to topping-up (this can be avoided by topping-up with the correct antifreeze mixture – see Section 1) or fluid loss. If loss of coolant has been evident, it is important to make the necessary repair before adding fresh fluid.

2 With the engine cold, carefully remove the cap from the expansion tank. If the engine is not completely cold, place a cloth rag over the cap before removing it, and remove it slowly to allow any pressure to escape.

3 Antifreeze checkers are available from car accessory shops. Draw some coolant from the expansion tank and observe how many plastic balls are floating in the checker. Usually, 2 or 3 balls must be floating for the correct concentration of antifreeze, but follow the manufacturer's instructions.

4 If the concentration is incorrect, it will be necessary to either withdraw some coolant

and add antifreeze, or alternatively drain the old coolant and add fresh coolant of the correct concentration (see Section 33).

13 Brake hydraulic circuit check

1 Check the entire brake hydraulic circuit for leaks and damage. Start by checking the master cylinder in the engine compartment. At the same time, check the vacuum servo unit and ABS units for signs of fluid leakage.

2 Raise the front and rear of the vehicle and support it on axle stands (see *Jacking and vehicle support*). Check the rigid hydraulic brake lines for corrosion and damage.

3 At the front of the vehicle, check that the flexible hydraulic hoses to the calipers are not twisted or chafing on any of the surrounding suspension components. Turn the steering on full lock to make this check. Also check that the hoses are not brittle or cracked.

4 Lower the vehicle to the ground after making the checks.

14 Headlight beam adjustment

1 Accurate adjustment of the headlight beam is only possible using optical beam-setting equipment, and this work should therefore be carried out by a Seat dealer or service station with the necessary facilities. MOT test centres have this equipment.

2 Basic adjustments can be carried out in an emergency, and further details are given in Chapter 12 Section 10.

15 Pollen filter element renewal

1 The pollen filter is located in the heater unit and is accessed from inside the car, on the passenger's side.

2 Where fitted, remove the clips and withdraw the facia lower trim from beneath the glovebox **(see illustration)**.

15.2 Remove the facia lower trim from beneath the glovebox

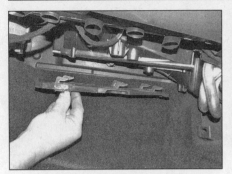

15.3 Remove the access cover...

15.4 ... then slide out the pollen filter element

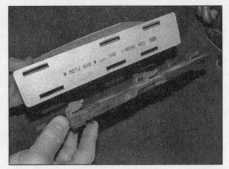

15.5 Observe the air flow direction marked on the filter

3 Slide the cover to the left to remove the access cover **(see illustration)**. On some models there maybe a screw securing the cover in place.

4 Slide out the pollen filter element downwards from the heater unit **(see illustration)**.

5 Fit the new element then **(see illustration)** refit the access cover, making sure that the locating lugs are fitted correctly in the housing. If any of the lugs have been broken off, fit a screw into the cover to make sure it is secure.

6 Refit the trim beneath the facia/glovebox.

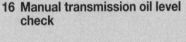

16 Manual transmission oil level check

1 There is no requirement to check the manual transmission fluid level – the transmission is effectively sealed for life. However the transmission should be checked for leaks. Full details of refilling and checking the transmission oil level are given in Chapter 7A Section 4.

17 Underbody protection check

1 Raise and support the vehicle on axle stands (see *Jacking and vehicle support*). Using an electric torch or lead light, inspect the entire underside of the vehicle, paying particular attention to the wheel arches. Look for any damage to the flexible underbody coating, which may crack or flake off with age, leading to corrosion. Also check that the wheel arch liners are securely attached with any clips provided – if they come loose, dirt may get in behind the liners and defeat their purpose. If there is any damage to the under seal, or any corrosion, it should be repaired before the damage gets too serious.

18 Driveshaft gaiter check

1 With the vehicle raised and securely supported on stands, slowly rotate the roadwheel. Inspect the condition of the outer constant velocity (CV) joint rubber gaiters, squeezing the gaiters to open out the folds. Check for signs of cracking, splits or deterioration of the rubber, which may allow the grease to escape, and lead to water and grit entry into the joint. Also check the security and condition of the retaining clips. Repeat these checks on the inner joints **(see illustration)**. If any damage or deterioration is found, the gaiters should be renewed (see Chapter 8 Section 3).

2 At the same time, check the general condition of the CV joints themselves by first holding the driveshaft and attempting to rotate the wheel. Repeat this check

by holding the inner joint and attempting to rotate the driveshaft. Any appreciable movement indicates wear in the joints, wear in the driveshaft splines, or a loose driveshaft retaining nut.

19 Steering and suspension check

1 Raise the front and rear of the vehicle, and securely support it on axle stands (see *Jacking and vehicle support*).

2 Visually inspect the track rod end balljoint dust cover, the lower front suspension balljoint dust cover, and the steering rack-and-pinion gaiters for splits **(see illustration)** chafing or deterioration. Any wear of these components will cause loss of lubricant, together with dirt and water entry, resulting in rapid deterioration of the balljoints or steering gear.

3 Check the power steering fluid hoses for chafing or deterioration, and the pipe and hose unions for fluid leaks. Also check for signs of fluid leakage under pressure from the steering gear rubber gaiters, which would indicate failed fluid seals within the steering gear.

4 Grasp the roadwheel at the 12 o'clock and 6 o'clock positions, and try to rock it **(see illustration)**. Very slight free play may be felt, but if the movement is appreciable, further investigation is necessary to determine the source. Continue rocking the wheel while

18.1 Check the condition of the driveshaft gaiters

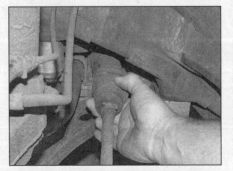

19.2 Check the rack gaiters for spilts

19.4 Check for wear in the bearings by grasping the wheel top and bottom and trying to rock it

19.5 Check for wear in the bearings and ball joints by grasping the wheel at the sides and trying to rock it

an assistant depresses the footbrake. If the movement is now eliminated or significantly reduced, it is likely that the hub bearings are at fault. If the free play is still evident with the footbrake depressed, then there is wear in the suspension joints or mountings.

5 Now grasp the wheel at the 9 o'clock and 3 o'clock positions, and try to rock it as before **(see illustration)**. Any movement felt now may again be caused by wear in the hub bearings or the steering track rod balljoints. If the inner or outer balljoint is worn, the visual movement will be obvious.

6 Using a large screwdriver or flat bar, check for wear in the suspension mounting bushes by levering between the relevant suspension component and its attachment point. Some movement is to be expected as the mountings are made of rubber, but excessive wear should be obvious. Also check the condition of any visible rubber bushes, looking for splits, cracks or contamination of the rubber.

7 With the car standing on its wheels, have an assistant turn the steering wheel back-and-forth about an eighth of a turn each way. There should be very little, if any, lost movement between the steering wheel and roadwheels. If this is not the case, closely observe the joints and mountings previously described, but in addition, check the steering column universal joints for wear, and the rack-and-pinion steering gear itself.

8 Check for any signs of fluid leakage around the front suspension struts and rear shock absorber. Should any fluid be noticed,

the suspension strut or shock absorber is defective internally, and should be renewed. **Note:** *Suspension struts/shock absorbers should always be renewed in pairs on the same axle to ensure correct vehicle handling.*

9 The efficiency of the suspension strut/shock absorber may be checked by bouncing the vehicle at each corner. Generally speaking, the body will return to its normal position and stop after being depressed. If it rises and returns on a rebound, the suspension strut/shock absorber is probably suspect. Examine also the suspension strut/shock absorber upper and lower mountings for any signs of wear.

20 Battery check

1 The battery is located on the left-hand side of the engine compartment. Remove the cover **(see illustration)** to gain access to the battery. Note that some models may have a soft insulated cover fitted.

2 Check that both battery terminals and all the fuse holder connections are securely attached **(see illustration)** and are free from corrosion. **Note:** *Before disconnecting the terminals from the battery, refer to Chapter 5A Section 3.*

3 Check the battery casing for signs of damage or cracking and check the battery retaining clamp bolt is securely tightened **(see illustration)**. If the battery casing is damaged in any way the battery must be renewed (see Chapter 5A Section 3).

4 On completion refit the battery cover.

21 Hinge and lock lubrication

1 Lubricate the hinges of the bonnet, doors and tailgate with light general-purpose oil. Similarly, lubricate all latches, locks and lock strikers. At the same time, check the security and operation of all the locks, adjusting them if necessary (see Chapter 11 Section 13).

2 Lightly lubricate the bonnet release mechanism and cable with suitable grease.

22 Airbag unit check

1 The airbag system will perform a self-check when the ignition is turned on. The airbag warning light will illuminate and then go out after several seconds. If a system fault is found the airbag warning light will remain on and a fault code will be logged. The fault code can be read by a suitable diagnostic tool and appropriate action taken.

2 Inspect the exterior condition of the airbag(s) for signs of damage or deterioration. If an airbag shows signs of damage, it must be renewed (Chapter 12 Section 27). Note that it is not permissible to attach any stickers to the surface of the airbag, as this may affect the deployment of the unit.

23 Windscreen/tailgate/headlight washer system check

1 Check that each of the washer jet nozzles are clear and that each nozzle provides a strong jet of washer fluid.

2 The tailgate jet should be aimed to spray at the centre of the screen, using a pin.

3 The windscreen washer nozzles should be aimed slightly above the centre of the screen using a small screwdriver to turn the jet eccentric (see Chapter 12 Section 21).

4 The headlight inner jet should be aimed slightly above the horizontal centreline of the headlight, and the outer jet should be aimed slightly below the centreline. Seat technicians use a special tool to adjust the headlight jet after pulling the jet out onto its stop.

5 Especially during the winter months, make sure that the washer fluid frost concentration is sufficient.

24 Engine management self-diagnosis memory fault check

1 This work should be carried out by a Seat dealer or a diagnostic specialist using

20.1 Unclip the cover

20.2 Check that the terminals are secure

20.3 Check that the battery clamp is secure

24.1 The diagnostic socket

special equipment. Note however that simple diagnostic tools are now widely available. Most of the lower priced tools will only access the mandatory OBD (On Board Diagnostics) emissions related fault codes. The diagnostic socket is located below the facia, at the right-hand side behind a panel **(see illustration)**. The 16 pin socket is often referred to as the Data Link Connector (or DLC).

25 Sunroof check and lubrication

1 Check the operation of the sunroof, and leave it in the fully open position.
2 Wipe clean the guide rails on each side of the sunroof opening, then apply lubricant to them. Seat recommend VAG lubricant spray G 052 778.

26 Road test and exhaust emissions check

Instruments and electrical equipment

1 Check the operation of all instruments and electrical equipment including the air conditioning system.
2 Make sure that all instruments read correctly, and switch on all electrical equipment in turn, to check that it functions properly.

Steering and suspension

3 Check for any abnormalities in the steering, suspension, handling or road 'feel'.
4 Drive the vehicle, and check that there are no unusual vibrations or noises, which may indicate wear in the driveshafts, wheel bearings, etc.
5 Check that the steering feels positive, with no excessive 'sloppiness', or roughness, and check for any suspension noises when cornering and driving over bumps.

Drivetrain

6 Check the performance of the engine, clutch (where applicable), gearbox/transmission and driveshafts.
7 Listen for any unusual noises from the engine, clutch and gearbox/transmission.
8 Make sure the engine runs smoothly at idle, and there is no hesitation on accelerating.
9 Check that, where applicable, the clutch action is smooth and progressive, that the drive is taken up smoothly, and that the pedal travel is not excessive. Also listen for any noises when the clutch pedal is depressed.
10 On manual gearbox models check that all gears can be engaged smoothly without noise, and that the gear lever action is smooth and not abnormally vague or 'notchy'.
11 On automatic (DSG) transmission models, make sure that all gearchanges occur smoothly, without snatching, and without an increase in engine speed between changes. Check that all the gear positions can be selected with the vehicle at rest. If any problems are found, they should be referred to a Seat dealer.
12 Listen for a metallic clicking sound from the front of the vehicle, as the vehicle is driven slowly in a circle with the steering on full-lock. Carry out this check in both directions. If a clicking noise is heard, this indicates wear in a driveshaft joint, in which case renew the joint if necessary.

Braking system

13 Make sure that the vehicle does not pull to one side when braking, and that the wheels do not lock when braking hard.
14 Check that there is no vibration through the steering when braking.
15 Check that the handbrake operates correctly without excessive movement of the

lever, and that it holds the vehicle stationary on a slope.
16 Test the operation of the brake servo unit as follows. With the engine off, depress the footbrake four or five times to exhaust the vacuum. Hold the brake pedal depressed, and then start the engine. As the engine starts, there should be a noticeable 'give' in the brake pedal as vacuum builds-up. Allow the engine to run for at least two minutes, and then switch it off. If the brake pedal is depressed now, it should be possible to detect a hiss from the servo as the pedal is depressed. After about four or five applications, no further hissing should be heard, and the pedal should feel considerably harder.
17 Under controlled emergency braking, the pulsing of the ABS unit must be felt at the footbrake pedal.

Exhaust emissions check

18 Although not part of the manufacturer's maintenance schedule, this check will normally be carried out on a regular basis according to the country the vehicle is operated in. Currently in the UK, exhaust emissions testing is included as part of the annual MOT test after the vehicle is 3 years old. In Germany the test is made when the vehicle is 3 years old, then repeated every 2 years.

27 DSG transmission oil renewal

1 Renewal of the DSG transmission oil is described in Chapter 7B Section 6.

28 Air filter element renewal

1 The air cleaner is located in front of the battery in the left-hand front corner of the engine compartment.
2 Undo the cover retaining screws and lift the air cleaner cover, taking care not to put any strain on the intake ducting and air mass meter **(see illustration)**.
3 With the cover raised, withdraw the air filter element, noting how it is fitted **(see illustrations)**.

28.2 Undo the retaining screws...

28.3a ...and lift off the upper cover and remove the filter

28.3b Where fitted remove the primary filter

28.4 Clean out the housing. Use a vacuum cleaner if necessary

29.2a Clean around the top of the fuel filter ...

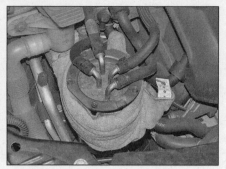

29.2b ... and place some cloth to catch spilt fuel

4 Remove any debris that may have collected inside the air cleaner **(see illustration)**.

5 Fit a new air filter element in position, ensuring that the edges are securely seated.

6 Refit the lid and tighten the screws, then refit the engine top cover.

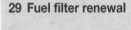

29 Fuel filter renewal

Note: *Carry out this procedure at this interval only when using diesel fuel conforming to DIN EN 590 (standard fuel in the UK).*

Note: *There are several versions of the fuel filter fitted to the Leon diesel range. All are replaced in a similar manner, with the differences confined to the number and type of seals fitted.*

1 The fuel filter is mounted in the right-hand front corner of the engine compartment. Place rags around the filter to absorb any fuel that may be spilt.

2 Before removing the fuel filter, clean around the fuel pipes on the top of the filter housing to prevent any dirt entering the fuel system **(see illustrations)**. Place some cloth around the filter housing to catch any fuel spillage.

3 Undo the screws and lift the cover complete with fuel hoses from the top of the filter housing and move it to one side **(see illustrations)**. Remove the seal and discard, as a new one must be used on refitting.

4 Withdraw the old filter element out from the housing and discard **(see illustration)**.

5 Use a pipette/syringe to draw out the dirty fuel and water residue from inside the fuel filter lower housing. Then insert the new filter element, pressing it down fully into position **(see illustration)**.

6 To help the engine start easier (to prevent having to bleed the system), top up the lower filter housing with clean diesel **(see illustration)**, making sure that no dirt or water enters the system.

7 Fit a new seal to the cover, then refit the cover back on the lower filter housing, making

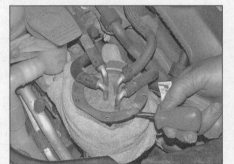

29.3a Carefully prise up the top ...

29.3b ... and remove it from the filter housing

29.4 Remove the old filter element

29.5 Insert the new filter element

29.6 Top up the housing with clean diesel until the level is approximately 20 mm below the top

29.7a Fit new seal ...

29.7b ... then refit the upper cover

29.8 Bleeding the fuel system with a vacuum pump

sure the top is located in the filter correctly (see illustrations). Press down firmly and evenly, then insert the screws and tighten them in a diagonal sequence.

8 Start and run the engine at idle, then check around the fuel filter for fuel leaks. **Note:** *It may take a few seconds of cranking before the engine starts. If the engine does not start the system must be bleed using a vacuum tool or hand primer* (see illustration). *Identify the* fuel return line from the injectors and remove it from the filter. Connect the vacuum pump or hand primer and draw fuel up to the filter. Reconnect the filter fuel line and start the vehicle.

30 Auxiliary drivebelt check and renewal

Check

1 See Section 11, for information on checking the condition of the drivebelt.

2 There are two different types of tensioner fitted to diesel engines; the routing of the belt is the same (except for models with no AC) although the procedure for belt renewal is different.

3 When the belt is removed, check all pulleys are free from any damage and are secure. Also check that the alternator and air-conditioning compressor (where fitted) are mounted securely.

Renewal

4 For improved access, apply the handbrake, and then jack up the front of the vehicle and support it on axle stands (see *Jacking and vehicle support*).

5 Remove the right-hand front roadwheel, then remove the wing liner from the inner wheel arch.

With tensioner spring element

6 Use a spanner on the centre bolt and turn the tensioner clockwise. Lock the tensioner in its released position by inserting a locking pin (Allen key or similar) through the lug into the tensioner body. An alternative tensioner maybe fitted, but the principle is identical **(see illustrations)**.

7 Note how the drivebelt is routed, then remove it from the crankshaft pulley **(see illustration)**, alternator pulley, and air conditioning compressor pulley (as applicable).

8 Locate the new drivebelt on the pulleys, then by holding the pressure of the tensioner with a spanner, remove the locking pin. Slowly release the pressure on the spanner **(see illustration)**, so that the tensioner takes up

30.6a Turn the tensioner clockwise...

30.6b ...or use an adjustable spanner (alternative type of tensioner)

30.6c ... and lock the tensioner in position

30.7a If the belt is to be refitted, mark the direction of rotation

30.7b Note the fitted position of the belt before removal

30.8 With the locking pin removed, slowly take up the slack in the belt

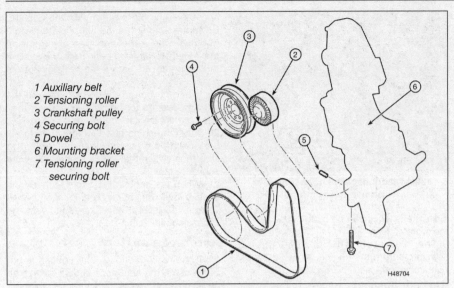

1 Auxiliary belt
2 Tensioning roller
3 Crankshaft pulley
4 Securing bolt
5 Dowel
6 Mounting bracket
7 Tensioning roller
   securing bolt

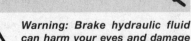

30.11 Auxiliary drivebelt with tensioning roller

the slack in the belt. Check that the belt is located correctly in the multi-grooves in the pulleys.

9 Start the engine and check that the drivebelt runs, as it should over the pulleys. Make sure that all tools and hands are kept clear of the drivebelt with the engine running.

10 With the engine stopped, refit the access panel and roadwheel, and then lower the vehicle to the ground. Refit the engine top cover if removed.

With tensioner roller

11 Undo the tensioner roller retaining bolt, and remove the tensioner roller from the engine. Note the retaining bolt is below the tensioner roller, and goes up through the mounting bracket in a vertical direction **(see illustration)**. Discard the retaining bolt, as a new one will be required for refitting.

12 Note how the drivebelt is routed, then remove it from the crankshaft pulley, alternator pulley, and air-conditioning compressor pulley (as applicable).

13 Locate the new drivebelt on the pulleys, then fit the tensioner roller, making sure its guide pin is located correctly in the mounting bracket.

14 Fit the new retaining bolt, and then tighten to the following five stages:
a) *Tighten bolt by hand at this point.*
b) *Tighten bolt until it reaches stop.*
c) *Turn the bolt back through 90°*
d) *Tighten bolt to 30Nm.*
e) *Tighten a further 90°*

15 When tensioner roller is fitted and the belt is tensioned, check the tensioner retaining bolt. The part of the bolt that has protruded out through the upper end of the mounting bracket (behind the tensioner roller) must not protrude more than 2.5mm higher than the outer surface of the tensioner roller. This ensures that the bolt has been tightened to its end stop.

16 Start the engine and check that the drivebelt runs, as it should over the pulleys. Make sure that all tools and hands are kept clear of the drivebelt with the engine running.

17 With the engine stopped, refit the access panel and roadwheel, and then lower the vehicle to the ground. Refit the engine top cover if removed.

31 Timing belt and tensioner roller renewal

1 Refer to Chapter 2C Section 7 (1.6 litre common rail engines), Chapter 2D Section 7 (2.0 litre common rail engines) or Chapter 2E Section 7 (PD engines) for details of renewing the timing belt and tensioner roller.

32 Brake (and clutch) fluid renewal

⚠️ *Warning: Brake hydraulic fluid can harm your eyes and damage painted surfaces, so use extreme caution when handling and pouring it. Do not use fluid that has been standing open for some time, as it absorbs moisture from the air. Excess moisture can cause a dangerous loss of braking effectiveness.*

1 The procedure is similar to that for the bleeding of the hydraulic system as described in Chapter 9 Section 2, except that the brake fluid reservoir should be emptied by siphoning, using an old, clean antifreeze tester or similar before starting, and allowance should be made for the old fluid to be expelled when bleeding a section of the circuit. Since the clutch hydraulic system also uses fluid from the brake system reservoir, it should also be

bled at the same time by referring to Chapter 6 Section 2.

2 Working as described in Chapter 9 Section 2, open the first bleed screw in the sequence, and pump the brake pedal gently until nearly all the old fluid has been emptied from the master cylinder reservoir.

3 Top-up to the MAX level with new fluid, and continue pumping until only the new fluid remains in the reservoir, and new fluid can be seen emerging from the bleed screw. Tighten the screw, and top the reservoir level up to the MAX level line.

4 Work through all the remaining bleed screws in the sequence until new fluid can be seen at all of them. Be careful to keep the master cylinder reservoir topped-up to above the MIN level at all times, or air may enter the system and greatly increase the length of the task.

5 When the operation is complete, check that all bleed screws are securely tightened, and that their dust caps are refitted. Wash off all traces of spilt fluid, and recheck the master cylinder reservoir fluid level.

6 On models with manual transmission, once the brake fluid has been changed the clutch fluid should also be renewed. Referring to Chapter 6 Section 2, bleed the clutch until new fluid is seen to be emerging from the slave cylinder bleed screw, keeping the master cylinder fluid level above the MIN level line at all times to prevent air entering the system. Once the new fluid emerges, securely tighten the bleed screw then disconnect and remove the bleeding equipment. Securely refit the dust cap then wash off all traces of spilt fluid.

7 On all models, ensure the master cylinder fluid level is correct (see *Weekly checks*) and thoroughly check the operation of the brakes and (where necessary) clutch before taking the car on the road.

33 Coolant renewal

Note: *This work is not included in the Seat schedule and should not be required if the recommended VAG G12 LongLife coolant antifreeze/inhibitor is used. However, if standard antifreeze/inhibitor is used, the work should be carried out at the recommended interval.*

⚠️ *Warning: Wait until the engine is cold before starting this procedure. Do not allow antifreeze to come in contact with your skin, or with the painted surfaces of the vehicle. Rinse off spills immediately with plenty of water. Never leave antifreeze lying around in an open container, or in a puddle in the driveway or on the garage floor. Children and pets are attracted by its sweet smell, but antifreeze can be fatal if ingested.*

33.3 Draining the coolant

33.4 Disconnect the hose from the circulation pump

Cooling system draining

1 With the engine completely cold, unscrew the expansion tank cap.

2 Firmly apply the handbrake then jack up the front of the vehicle and support it on axle stands (see *Jacking and vehicle support*). Undo the retaining screws and remove the engine undertray(s) to gain access to the base of the radiator.

3 Position a suitable container beneath the coolant drain outlet, which is fitted to the coolant bottom hose end fitting. Loosen the drain plug (there is no need to remove it completely) and allow the coolant to drain into the container. If desired, a length of tubing can be fitted to the drain outlet to direct the flow of coolant during draining. Where no drain outlet is fitted to the hose end fitting, remove the retaining clip and disconnect the bottom hose from the radiator to drain the coolant **(see illustration)**. Also see Chapter 3 Section 1.

4 Where fitted, release the clamps and disconnect the hoses from the electric coolant circulation pump (see Chapter 3 Section 7), located at the front side of the cylinder block, below the oil cooler **(see illustration)**.

5 To fully drain the system also disconnect one of the coolant hoses from the oil cooler, which is located at the front of the cylinder block.

6 If the coolant has been drained for a reason other than renewal, then provided it is clean, it can be re-used, though this is not recommended.

7 Once all the coolant has drained, securely tighten the radiator drain plug or reconnect the bottom hose to the radiator (as applicable). Also reconnect the coolant hose to the oil cooler and secure it in position with the retaining clip. Refit the undertray(s), tighten the retaining screws securely.

Cooling system flushing

8 If the recommended Seat coolant has not been used and coolant renewal has been neglected, or if the antifreeze mixture has become diluted, the cooling system may gradually lose efficiency, as the coolant passages become restricted due to rust, scale deposits, and other sediment. The cooling system efficiency can be restored by flushing the system clean.

9 The radiator should be flushed separately from the engine, to avoid excess contamination.

Radiator flushing

10 To flush the radiator, first tighten the radiator drain plug.

11 Disconnect top and bottom hoses and any other relevant hoses from the radiator (see Chapter 3 Section 2).

12 Insert a garden hose into the radiator top inlet. Direct a flow of clean water through the radiator, and continue flushing until clean water emerges from the radiator bottom outlet.

13 If after a reasonable period, the water still does not run clear, the radiator can be flushed with a good proprietary cleaning agent. It is important that their manufacturer's instructions are followed carefully. If the contamination is particularly bad, insert the hose in the radiator bottom outlet, and reverse-flush the radiator.

Engine flushing

14 To flush the engine, remove the thermostat (see Chapter 3 Section 4).

15 With the bottom hose disconnected from the radiator, insert a garden hose into the coolant housing. Direct a clean flow of water through the engine, and continue flushing until clean water emerges from the radiator bottom hose.

16 When flushing is complete, refit the thermostat and reconnect the hoses.

Cooling system filling

17 Before attempting to fill the cooling system, ensure the drain plug (where fitted) is securely closed and make sure that all hoses are connected and are securely retained by their clips. If the recommended Seat coolant is not being used, ensure that a suitable antifreeze mixture is used all year round, to prevent corrosion of the engine components (see following sub-Section). **Note:** *Seat recommend that only distilled water should be used.*

18 Remove the expansion tank filler cap and slowly fill the system with the coolant. Continue to fill the cooling system until bubbles stop appearing in the expansion tank. Help to bleed the air from the system by repeatedly squeezing the radiator bottom hose.

19 When no more bubbles appear, top the coolant level up to the MAX level mark then securely refit the cap to the expansion tank.

20 Run the engine at a fast idle speed until the cooling fan cuts in. Wait for the fan to stop then switch the engine off and allow the engine to cool.

21 When the engine has cooled, check the coolant level with reference to *Weekly checks*. Top-up the level if necessary, and refit the expansion tank cap.

Antifreeze mixture

22 If the recommended Seat coolant is not being used, the antifreeze should always be renewed at the specified intervals. This is necessary not only to maintain the antifreeze properties, but also to prevent corrosion, which would otherwise occur as the corrosion inhibitors become progressively less effective.

23 Always use an ethylene-glycol based antifreeze which is suitable for use in mixed-metal cooling systems. The quantity of antifreeze and levels of protection are indicated in the Specifications.

24 Before adding antifreeze, the cooling system should be completely drained, preferably flushed, and all hoses checked for condition and security.

25 After filling with antifreeze, a label should be attached to the expansion tank, stating the type and concentration of antifreeze used, and the date installed. Any subsequent topping-up should be made with the same type and concentration of antifreeze.

Caution: Do not use engine antifreeze in the windscreen/tailgate washer system, as it will damage the vehicle paintwork. A screenwash additive should be added to the washer system in the quantities stated on the bottle.

34 Particulate filter ash deposit mass check

1 Eventually, the amount of ash deposited in the particle filter by the filtration process will cause a blockage, and engine running problems. Seat state that the maximum amount of ash is 60g. At this point, the particle filter must be renewed. Unfortunately, the mass of the ash can only be established using dedicated diagnostic equipment, connected to the vehicle through the diagnostic plug under the drivers side of the facia. Consequently, we recommend this task is entrusted to a Seat dealer or suitably equipped specialist.

Chapter 2 Part A
1.6 litre petrol engine in-car repair procedures

Contents

Degrees of difficulty

Easy, suitable for novice with little experience	**Fairly easy,** suitable for beginner with some experience	**Fairly difficult,** suitable for competent DIY mechanic	**Difficult,** suitable for experienced DIY mechanic	**Very difficult,** suitable for expert DIY or professional

Specifications

General

Manufacturer's engine codes*	BGU, BSE and BSF
Maximum power output.......................................	75 kW at 5600 rpm
Maximum torque output.......................................	148 Nm at 3800 rpm
Bore	81.0 mm
Stroke	77.4 mm
Compression ratio	10.5 : 1
Compression pressures:	
Minimum compression pressure	Approximately 7.0 bar
Maximum difference between cylinders......................	Approximately 3.0 bar
Firing order..	1 – 3 – 4 – 2
No 1 cylinder location..	Timing belt end

* Note: *See 'Vehicle identification' at the end of this manual for the location of engine code markings.*

Lubrication system

Oil pump type..	Gear type, chain-driven from crankshaft
Oil pressure (oil temperature 80°C):	
At 2000 rpm	2.7 to 4.5 bar

Camshaft

Camshaft endfloat (maximum)	0.17 mm
Camshaft bearing running clearance (maximum)	0.1 mm
Camshaft run-out (maximum)	0.04 mm

Torque wrench settings

	Nm	lbf ft
Ancillary (alternator, etc) bracket mounting bolts	45	33
Auxiliary drivebelt tensioner securing bolt	23	17
Big-end bearing cap bolts*:		
Stage 1	30	22
Stage 2	Angle-tighten a further 90°	
Camshaft bearing retaining frame nuts	23	17
Camshaft cover bolts	9	7
Camshaft sprocket bolt	100	74
Coolant outlet elbow bolts	10	7
Coolant pump bolts	15	11
Crankshaft oil seal housing bolts	15	11
Crankshaft position sensor wheel-to-crankshaft bolts*:		
Stage 1	10	7
Stage 2	Angle-tighten a further 90°	
Crankshaft pulley bolts	25	18
Crankshaft sprocket bolt*:		
Stage 1	90	66
Stage 2	Angle-tighten a further 90°	
Cylinder block oil gallery plug	100	74
Cylinder head bolts*:		
Stage 1	40	30
Stage 2	Angle-tighten a further 90°	
Stage 3	Angle-tighten a further 90°	
Engine mountings:		
RH engine mounting*:		
Limiter:		
Stage 1	20	15
Stage 2	Angle-tighten a further 90°	
Mounting to engine:		
Stage 1	60	44
Stage 2	Angle-tighten a further 90°	
Mounting to body:		
Stage 1	40	30
Stage 2	Angle-tighten a further 90°	
LH engine mounting*:		
Mounting to body:		
Stage 1	60	44
Stage 2	Angle-tighten a further 90°	
Mounting to transmission:		
Stage 1	40	30
Stage 2	Angle-tighten a further 90°	
Rear pendulum mounting*:		
To transmission (8.8 class bolt):		
Stage 1	40	30
Stage 2	Angle-tighten a further 90°	
To transmission (10.9 class bolt)		
Stage 1	50	37
Stage 2	Angle-tighten a further 90°	
To subframe:		
Stage 1	100	74
Stage 2	Angle-tighten a further 90°	
Exhaust manifold to head nuts	25	18
Exhaust pipe to manifold nuts	25	18
Flywheel*:		
Stage 1	60	44
Stage 2	Angle-tighten a further 90°	
Inlet manifold to head	25	18
Inlet manifold upper section to lower section	3	2
Main bearing cap bolts*:		
Stage 1	40	30
Stage 2	Angle-tighten a further 90°	
Oil baffle plate securing bolts	15	11

Torque wrench settings (continued)

	Nm	lbf ft
Oil cooler securing nut	25	18
Oil drain plug	30	22
Oil filter housing-to-cylinder block bolts*:		
Stage 1	15	11
Stage 2	Angle-tighten a further 90°	
Oil level/temperature sender bolts	10	7
Oil pick-up pipe-to-oil pump bolts	15	11
Oil pressure relief valve plug	40	30
Oil pressure warning light switch	25	18
Oil pump	15	11
Oil pump chain tensioner bolt	15	11
Oil pump sprocket bolt:		
Stage 1	20	15
Stage 2	Angle-tighten a further 90°	
Oil spray jet/pressure relief valve bolts	27	20
Roadwheel bolts	120	89
Sump:		
Sump-to-cylinder block bolts	15	11
Sump-to-transmission bolts	25	18
Thermostat cover bolts	15	11
Timing belt outer cover bolts	10	7
Timing belt rear cover bolts:		
Small bolts	10	7
Large bolt	23	17
Timing belt tensioner nut	23	17

*Do not re-use

---

1 General Information

How to use this Chapter

1 This Part of Chapter 2 describes those repair procedures that can reasonably be carried out on the engine while it remains in the vehicle. If the engine has been removed from the vehicle and is being dismantled as described in Part F, any preliminary dismantling procedures can be ignored.

2 Note that while it may be possible physically to overhaul certain items while the engine is in the vehicle, such tasks are not usually carried out as separate operations, and usually require the execution of several additional procedures (not to mention the cleaning of components and of oilways); for this reason, all such tasks are classed as major overhaul procedures, and are described in Part F of this Chapter.

Engine description

3 Throughout this Chapter, engines are identified by the manufacturer's code letters. A listing of all engines covered, together with their code letters, is given in the Specifications.

4 The engine covered in this Part of the Chapter is of water-cooled, single-overhead camshaft (SOHC), in-line four-cylinder design. The 1595 cc engine has an aluminium alloy cylinder block fitted with cast-iron cylinder liners, and an aluminium alloy cylinder head.

The engine is transversely mounted at the front of the vehicle, with the transmission unit on its left-hand end.

5 The crankshaft is of five-bearing type, and thrustwashers are fitted to the centre main bearing to control crankshaft endfloat.

6 The camshaft is mounted at the top of the cylinder head and is driven by a toothed timing belt from the crankshaft sprocket. It is secured to the cylinder head by a retaining frame.

7 The valves are closed by coil springs, and the valves run in guides pressed into the cylinder head. The camshafts actuate the valves by roller rocker fingers supported by hydraulic tappets.

8 The oil pump is driven via a chain from a sprocket on the crankshaft. Oil is drawn from the sump through a strainer, and then forced through an externally-mounted, renewable filter. From there, it is distributed to the cylinder head, where it lubricates the camshaft journals and hydraulic tappets, and also to the crankcase, where it lubricates the main bearings, connecting rod big-ends, gudgeon pins and cylinder bores. A coolant-fed oil cooler is fitted to all engines.

9 Engine coolant is circulated by a pump, driven by the timing belt. For details of the cooling system, refer to Chapter 3.

Operations with engine in car

10 The following operations can be performed without removing the engine:

a) Compression pressure – testing.
b) Camshaft cover – removal and refitting.
c) Crankshaft pulley – removal and refitting.
d) Timing belt covers – removal and refitting.

e) Timing belt – removal, refitting and adjustment.
f) Timing belt tensioner and sprockets – removal and refitting.
g) Camshaft oil seal – renewal.
h) Camshaft and hydraulic tappets – removal, inspection and refitting.
i) Cylinder head – removal and refitting.
j) Cylinder head and pistons – decarbonising.
k) Sump – removal and refitting.
l) Oil pump – removal, overhaul and refitting.
m) Crankshaft oil seals – renewal.
n) Engine/transmission mountings – inspection and renewal.
o) Flywheel/driveplate – removal, inspection and refitting.

Note: It is possible to remove the pistons and connecting rods (after removing the cylinder head and sump) without removing the engine. However, this is not recommended. Work of this nature is more easily and thoroughly completed with the engine on the bench, as described in Chapter 2F.

2 Compression test – description and interpretation

Caution: The following work may insert fault codes in the engine management ECU. These fault codes must be cleared by a Seat dealer.

Note: A suitable compression tester will be required for this test.

1 When engine performance is down, or if misfiring occurs which cannot be attributed

to the ignition or fuel systems, a compression test can provide diagnostic clues as to the engine's condition. If the test is performed regularly it can give warning of trouble before any other symptoms become apparent.

2 The engine must be fully warmed-up to normal operating temperature, the battery must be fully-charged and the spark plugs must be removed. The aid of an assistant will be required.

3 Disable the ignition system by disconnecting the wiring plug from the coil pack unit (see Chapter 5B Section 3).

4 Referring to Chapter 4A Section 5, disconnect the wiring from the fuel injectors.

5 Fit a compression tester to the No 1 cylinder spark plug hole. The type of tester which screws into the plug thread is preferred.

6 Have the assistant hold the throttle wide open and crank the engine for several seconds on the starter motor. **Note:** *The throttle will not operate until the ignition is switched on. After one or two revolutions, the compression pressure should build-up to a maximum figure and then stabilise. Record the highest reading obtained.*

7 Repeat the test on the remaining cylinders, recording the pressure in each.

8 All cylinders should produce very similar pressures. Any difference greater than that specified indicates the existence of a fault. Note that the compression should build-up quickly in a healthy engine. Low compression on the first stroke, followed by gradually increasing pressure on successive strokes, indicates worn piston rings. A low compression reading on the first stroke, which does not build-up during successive strokes, indicates leaking valves or a blown head gasket (a cracked head could also be the cause). Deposits on the undersides of the valve heads can also cause low compression.

9 If the pressure in any cylinder is reduced to the specified minimum or less, carry out the following test to isolate the cause. Introduce a teaspoonful of clean oil into that cylinder through its spark plug hole and repeat the test.

10 If the addition of oil temporarily improves the compression pressure, this indicates that bore or piston wear is responsible for the pressure loss. No improvement suggests that leaking or burnt valves, or a blown head gasket, may be to blame.

11 A low reading from two adjacent cylinders is almost certainly due to the head gasket having blown between them and the presence of coolant in the engine oil will confirm this.

12 If one cylinder is about 20 percent lower than the others and the engine has a slightly rough idle, a worn camshaft lobe could be the cause.

13 If the compression reading is unusually high, the combustion chambers are probably coated with carbon deposits. If this is the case, the cylinder head should be removed and decarbonised.

14 On completion of the test, refit the spark plugs, and reconnect the ignition coil.

15 Have any fault codes cleared by a Seat dealer or suitably equipped garage.

3 Engine assembly and valve timing marks – general information and usage

General information

1 TDC is the highest point in the cylinder that each piston reaches as it travels up and down when the crankshaft turns. Each piston reaches TDC at the end of the compression stroke and again at the end of the exhaust stroke, but TDC generally refers to piston position on the compression stroke. No 1 piston is at the timing belt end of the engine.

2 Positioning No 1 piston at TDC is an essential part of many procedures, such as timing belt removal and camshaft removal.

3 The design of the engines covered in this Chapter is such that piston-to-valve contact may occur if the camshaft or crankshaft is turned with the timing belt removed. For this reason, it is important to ensure that the camshaft and crankshaft do not move in relation to each other once the timing belt has been removed from the engine.

4 On most models, the crankshaft pulley has a mark which, when aligned with a corresponding reference mark on the timing belt cover, indicates that No 1 piston (and hence also No 4 piston) is at TDC **(see illustration)**.

5 The camshaft sprocket is also equipped with a timing mark. When this mark is aligned with the OT mark on the rear timing belt cover, No 1 piston is at TDC on the compression stroke **(see illustrations)**. Note that not all engines are marked 'OT', some have a simple indentation in the belt rear cover.

Setting No 1 cylinder to TDC

6 Before starting work, make sure that the ignition is switched off.

7 Remove the engine top cover.

8 If desired, the make the engine easier to turn, remove all of the spark plugs as described in Chapter 1A Section 29.

9 Remove the upper timing belt cover as described in Section 6.

10 Turn the engine clockwise, using a spanner on the crankshaft sprocket bolt, until the TDC mark on the crankshaft pulley is aligned with the corresponding mark on the timing belt cover and the mark on the camshaft sprocket is aligned with the corresponding mark on the rear timing belt cover.

4 Camshaft cover – removal and refitting

Removal

1 Remove the upper part of the inlet manifold as described in Chapter 4A Section 10.

2 Loosen the clip and disconnect the breather hose from the rear of the camshaft cover.

3 To improve access the upper timing belt cover can be removed with reference to Section 6.

4 Unscrew the bolts securing the camshaft cover to the cylinder head, starting from the outside and working inwards. Note the location of any support brackets. Examine the special bolts and renew them if there is any indication of leaking.

5 Lift the camshaft cover from the

3.4 Crankshaft pulley TDC mark aligned with mark on timing belt lower cover

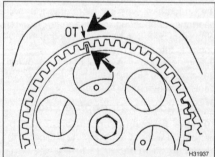

3.5a Camshaft sprocket TDC mark aligned with timing mark on timing belt rear cover

3.5b Engine set to TDC as seen in the car

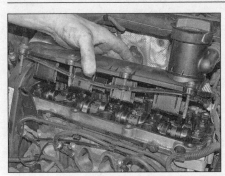

4.5 Remove the cover

5.3 Removing the wheel arch liner access panel

5.5 View of the crankshaft pulley, showing the four securing bolts

cylinder head, and recover the gasket **(see illustration)**.

6 Where fitted, lift the oil deflector from the camshaft cover or the top of the cylinder head.

Refitting

7 Inspect the camshaft cover gasket, and renew if necessary.

8 Thoroughly clean the mating surfaces of the camshaft cover and the camshaft retaining frame, then locate it in position.

9 Tighten the bolts progressively to the specified torque, starting from the inside and working (in a diagonal sequence) outwards.

10 Where removed, refit the upper timing cover, with reference to Section 6.

11 Reconnect the breather hose.

12 Refit the upper part of the inlet manifold as described in Chapter 4A Section 10.

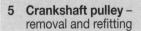

5 Crankshaft pulley –
removal and refitting

Removal

1 Check the ignition and all electrical consumers are switched off.

2 For improved access, raise the front right-hand side of the vehicle, and support

securely on axle stands (see *Jacking and vehicle support*). Remove the road wheel.

3 Remove the securing screws and withdraw the engine undertray(s) and wheel arch liner access panel **(see illustration)**.

4 If necessary (for any later work to be carried out), turn the crankshaft using a socket or spanner on the crankshaft sprocket bolt until the relevant timing marks align (see Section 3).

5 Loosen only the bolts securing the crankshaft pulley to the sprocket **(see illustration)**. If necessary, the pulley can be prevented from turning by counterholding with a spanner or socket on the crankshaft sprocket bolt.

6 Remove the auxiliary drivebelt, as described in Chapter 1A Section 30.

7 Make a note of the fitted position of the pulley, then unscrew the bolts securing the pulley to the sprocket, and remove the pulley.

Refitting

8 Refit the pulley to the sprocket, locating the small offset hole over the sprocket peg as noted on removal, then refit the pulley securing bolts.

9 Refit and tension the auxiliary drivebelt as described in Chapter 1A Section 30.

10 Prevent the crankshaft from turning then tighten the pulley securing bolts to the specified torque.

11 Refit the engine undertray and wheel arch liner as applicable.

12 Refit the road wheel and lower the vehicle to the ground.

6 Timing belt covers –
removal and refitting

Note: *The most likely reason to remove the timing belt covers is to replace the timing belt (and associated parts). The upper cover is easily removed, but removal of the lower covers is more involved procedure. The lower covers can be removed individually (and are described below as such) but removing the lower covers as a complete assembly will be the normal method for timing belt replacement.*

Upper outer cover

Removal

1 Where fitted remove the engine cover, unclip the Evap canister and move it to the side – there is no need to disconnect the fuel/vapour lines.

2 Release the securing clip at the front and rear of the cover, and lift the cover out of the section below it, noting how it fits **(see illustrations)**.

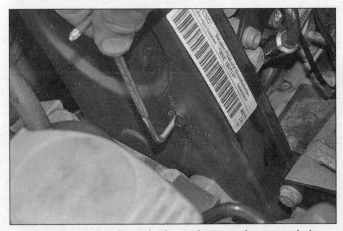

6.2a Rotate the clip at the front of cover using an angled screwdriver (or small coin)...

6.2b ...and then release the locking tabs at the rear (valve cover removed for clarity)

Refitting

3 Refitting is a reversal of removal. Engage the base of the cover correctly before trying to secure the upper clips, or they will not engage.

Centre outer cover

Removal

4 Remove the auxiliary drivebelt as described in Chapter 1A Section 30 and then remove the upper outer cover as described previously in this Section.

5 Where fitted, remove the the right-angled bracket from the auxiliary drivebelt tensioner and then remove the tensioner to improve access (see illustrations).

6 Unless the cover is only to be removed for inspection purposes, the engine mounting should now be removed (as described in Section). With the mounting removed and the engine supported, remove the main bracket from the engine block (see illustration).

7 Unscrew the securing bolts, and withdraw the cover from the engine (see illustration).

Refitting

8 Refitting is a reversal of removal.

Lower outer cover

Removal

9 Remove the upper and centre covers as described previously in this Section.

10 Remove the crankshaft pulley as described in Section 5.

11 Unscrew the securing bolts, and withdraw the cover from the front of the engine (see illustration).

Refitting

12 Refitting is a reversal of removal.

Upper inner cover

Removal

13 Remove the upper outer cover as described previously in this Section, then remove the camshaft sprocket (see Section 5).

14 Unbolt and remove the timing belt inner cover and remove it from the engine.

Refitting

15 Refitting is a reversal of removal.

6.5a Remove the bracket and...

Lower inner cover

Removal

16 Remove the timing belt as described in Section 7.

17 Unscrew the securing bolts and remove the timing belt lower inner cover.

Refitting

18 Refitting is a reversal of removal, but refit and tension the timing belt as described in Section 7.

7 Timing belt – removal and refitting

Removal

1 Where fitted, remove the engine cover.

2 Unbolt the coolant expansion tank, and move it clear of the working area, leaving the hoses connected. Disconnect the vacuum hose from the carbon canister, unclip and the canister and move it to the side.

3 Remove the auxiliary drivebelt as described in Chapter 1A Section 30.

4 Unscrew the securing nut and bolts, and remove the right-angled bracket over the auxiliary drivebelt tensioner; the tensioner is now held by one further bolt at the top – remove the bolt and withdraw the tensioner from the engine.

5 Remove the timing belt upper outer cover, with reference to Section 6.

6 Turn the crankshaft to position No 1 piston at TDC, as described in Section 3.

6.5b ...automatic drivebelt tensioner

7 Use a trolley jack (with a block of wood to spread the load) over the sump and then raise the jack to just take the weight of the engine.

8 Unscrew the securing bolts and remove the right-hand engine mounting assembly, with reference to Section 6 (see illustration 6.6).

9 Remove the crankshaft pulley, with reference to Section 5. Before finally removing the pulley, check that No 1 piston is still positioned at TDC (Section 3).

10 Unbolt the right-hand engine mounting bracket from the engine. Note that it may be necessary to raise the engine slightly, using the hoist, to allow access to unscrew the engine mounting securing bolts (once the bolts have been unscrewed, it will probably be necessary to leave the bolts in position in the bracket until the bracket has been removed).

11 Remove the timing belt centre and lower outer covers, with reference to Section 6.

12 If the timing belt is to be refitted, mark its running direction.

13 Loosen the timing belt tensioner securing nut to release the tensioner, then withdraw the timing belt from the sprockets.

14 Turn the crankshaft a quarter-turn (90°) anti-clockwise to position Nos 1 and 4 pistons slightly down their bores from the TDC position. This will eliminate any risk of piston-to-valve contact if the crankshaft or camshaft is turned whilst the timing belt is removed.

Refitting

15 Check that the camshaft sprocket timing mark is aligned with the corresponding mark on the rear timing belt cover (Section

6.6 Remove the engine mounting support bracket

6.7 Remove the bolts (lower bolts shown)

6.11 Removing the lower cover, complete with the centre cover

3), then turn the crankshaft a quarter-turn (90°) clockwise to reposition Nos 1 and 4 pistons at TDC. Ensure that the appropriate crankshaft timing marks are aligned. If it is not possible to view the flywheel/driveplate timing marks, temporarily refit the crankshaft pulley and timing belt cover, and turn the crankshaft to align the mark on the pulley with the corresponding mark on the belt cover.

16 Fit the timing belt around the crankshaft sprocket, coolant pump sprocket, tensioner, and camshaft sprocket. Where applicable, observe the running direction markings.

17 The timing belt must now be tensioned as follows.

18 Engage a pair of angled circlip pliers (or VAG tool T20197) with the two holes in the centre of the tensioner pulley, then turn the pulley back-and-forth from the clockwise stop to the anti-clockwise stop, five times.

19 Turn the tensioner pulley anti-clockwise to its stop, then slowly release the tension on the pulley until the tension indicator pointer is aligned with the centre of the indicator notch **(see illustration)**. It may be necessary to use a mirror to view the tension indicator alignment.

20 Hold the tensioner pulley in position, with the pointer and notch aligned, and tighten the tensioner nut to the specified torque.

21 Turn the crankshaft through two complete revolutions clockwise until the No 1 piston is positioned at TDC again, with the timing marks aligned (Section 3). It is important to ensure that the last one-eighth of a turn of rotation is completed without stopping.

22 Check that the tension indicator pointer is still aligned with the centre of the indicator notch. If the pointer is not aligned with the centre of the notch, repeat the tensioning procedure given in paragraphs 18 to 22. If the pointer and notch are correctly aligned, proceed as follows.

23 Refit the timing belt lower and centre covers, with reference to Section 6.

24 Refit the crankshaft pulley, with reference to Section 5, and tighten the securing bolts to the specified torque.

25 Refit the right-hand engine mounting bracket, and tighten the securing bolts to the specified torque (slide the securing bolts

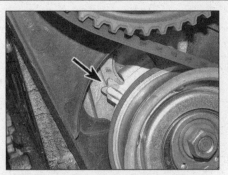

7.19 Tension the timing belt so that the tension indicator pointer (2) is aligned with the centre of the indicator notch (1)

into position in the bracket before offering the bracket up to the engine).

26 Refit the right-hand engine mounting assembly, and check the mounting alignment as described in Section. Once the mounting alignment is correct, tighten the securing bolts to the specified torque.

27 Disconnect the hoist and lifting tackle from the engine.

28 Refit the timing belt upper outer cover.

29 Refit the auxiliary drivebelt tensioner, and tighten the securing bolts to the specified torque, then refit the auxiliary drivebelt as described in Chapter 1A Section 30.

30 Refit the engine top cover.

8 Timing belt tensioner and sprockets – removal and refitting

Camshaft sprocket

Removal

1 Remove the timing belt as described in Section 7.

2 The camshaft must be held stationary as the sprocket bolt is slackened, and this can be achieved by making up a tool, and using it to hold the sprocket stationary by means of the holes in the sprocket face **(see illustration)**.

3 Unscrew the sprocket bolt and withdraw it, then withdraw the sprocket from the end of the camshaft. Recover the Woodruff key if it is loose.

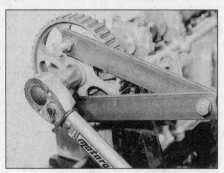

8.2 Using a home-made tool to hold the camshaft sprocket (tool shown being used when tightening bolt)

4 Where necessary, remove the timing belt upper inner cover.

Refitting

5 Prior to refitting, check the camshaft oil seal for signs of leakage, and if necessary renew the seal as described in Section 11.

6 Where removed, refit the timing belt upper inner cover.

7 Refit the Woodruff key to the end of the camshaft, then refit the sprocket.

8 Tighten the sprocket bolt to the specified torque, preventing the sprocket from turning using the method used on removal.

9 Refit the timing belt as described in Section 7.

Crankshaft sprocket

Removal

10 Remove the timing belt as described in Section 7.

11 The crankshaft must be held stationary as the sprocket bolt is slackened. On manual transmission models, engage top gear and apply the footbrake pedal firmly. On automatic transmission models (DSG), unbolt the starter motor and use a wide-bladed screwdriver engaged with the driveplate ring gear to hold the crankshaft stationary.

12 Unscrew the sprocket bolt (note that the bolt is very tight), and withdraw the sprocket from the crankshaft **(see illustrations)**.

Refitting

13 Locate the sprocket on the crankshaft, with the flange against the oil seal housing, then tighten the new securing bolt to the specified torque, whilst holding the crankshaft stationary using the method employed during removal.

⚠ *Warning: Do not turn the crankshaft, as the pistons may hit the valves.*

14 Refit the timing belt as described in Section 7.

Coolant pump sprocket

15 The coolant pump sprocket is integral with the coolant pump, and cannot be removed separately. Refer to Chapter 3 Section 7 for details of coolant pump removal.

8.12a Unscrew the securing bolt...

8.12b ...and withdraw the sprocket from the crankshaft

Tensioner assembly

Removal

16 Remove the timing belt as described in Section 7.

17 Unscrew the securing nut and recover the washer, then withdraw the tensioner assembly from the stud on the engine.

Refitting

18 Offer the tensioner assembly into position over the mounting stud, ensuring that the lug on the tensioner backplate engages with the corresponding cut-out in the cylinder head.

19 Refit the securing nut, ensuring that the washer is in place, but do not fully tighten the nut at this stage.

20 Refit and tension the timing belt as described in Section 7.

9 Camshaft – removal, inspection and refitting

Note: *New camshaft seal(s) should be used on refitting. VAG sealant (D 188 800 A1 or equivalent) will be required to seal the joints between the camshaft retaining frame and the cylinder head on refitting.*

Removal

1 Remove the camshaft cover as described in Section 4.

2 Remove the timing belt as described in Section 7.

3 Remove the camshaft sprocket as described in Section 9, then unbolt the inner belt cover.

4 Progressively loosen the nuts securing numbers 5, 1 and 3 bearing caps, then 2 and 4, using an alternating diagonal sequence (No 1 bearing cap is at the timing belt end of the engine). Note that as the nuts are slackened, the valve springs will push the camshaft up **(see illustration)**.

5 Lift the camshaft retaining frame from the cylinder head.

6 Lift the camshaft from the cylinder head, then remove the oil seal and the sealing cap from the ends of the camshaft and discard them. New seals will be required for refitting.

7 To remove the hydraulic tappets and roller rocker fingers, see Section 10.

Inspection

8 With the camshaft removed, examine the bearing caps/retaining frame and the bearing locations in the cylinder head for signs of obvious wear or pitting. If evident, a new cylinder head will probably be required. Also check that the oil supply holes in the cylinder head are free from obstructions.

9 Visually inspect the camshaft for evidence of wear on the surfaces of the lobes and journals. Normally their surfaces should be smooth and have a dull shine; look for scoring, erosion or pitting and areas that appear highly polished, indicating excessive wear. Accelerated wear will occur once the hardened exterior of the camshaft has been damaged, so always renew worn items. **Note:** *If these symptoms are visible on the tips of the camshaft lobes, check the corresponding tappet/rocker finger, as it may be worn as well.*

10 If the machined surfaces of the camshaft appear discoloured or blued, it is likely that it has been overheated at some point, probably due to inadequate lubrication.

11 To measure the camshaft endfloat, temporarily refit the camshaft to the cylinder head, then fit the frame and tighten the retaining nuts to the specified torque setting. Anchor a DTI gauge to the timing belt end of the cylinder head. Push the camshaft to one end of the cylinder head as far as it will travel, then rest the DTI gauge probe on the end face of the camshaft, and zero the gauge. Push the camshaft as far as it will go to the other end of the cylinder head, and record the gauge reading. Verify the reading by pushing the camshaft back to its original position and checking that the gauge indicates zero again **(see illustration)**. **Note:** *The hydraulic tappets mustnotbe fitted whilst this measurement is being taken.*

Refitting

12 Ensure that the crankshaft has been turned to position Number 1 and 4 pistons slightly down their bores from the TDC position (Section 3). This will eliminate any risk of piston-to-valve contact.

13 Refit the hydraulic tappets/roller rocker fingers as described in Section 10.

14 Lubricate the camshaft and cylinder head

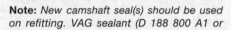

9.4 Layout of the camshaft and roller rocker fingers

1 Bolt	*7 Camshaft*	*12 Valve spring*
2 Camshaft sprocket	*8 Roller rocker finger*	*13 Valve stem seal*
3 Oil seal	*9 Support*	*14 Valve guide*
4 Parallel key	*10 Collets*	*15 Camshaft end sealing cap*
5 Retaining ladder frame	*11 Upper valve spring*	*16 Cylinder head*
6 Retaining nut	*washer*	*17 Valve*

H45326

9.11 Checking the camshaft endfloat using a DTI gauge

bearing journals with clean engine oil (**see illustration**).

15 Carefully lower the camshaft into position in the cylinder head making sure that the cam lobes for No 1 cylinder are pointing upwards.

16 Fit a new oil seal to the camshaft. Make sure that the closed end of the seal faces the camshaft sprocket end of the camshaft, and take care not to damage the seal lip. Locate the seal against the seat in the cylinder head.

17 Oil the upper surfaces of the camshaft bearing journals in the retaining frame, without getting oil onto the sealing surface, where it meets the cylinder head.

18 Ensure that the mating faces/groove of the cylinder head and retaining frame are clean and free from traces of old sealant, then apply an even bead of sealant (VAG sealant – D 188 800 A1 or equivalent) to the groove in the lower surface of the retaining frame. **Note:** *The frame must be ready to tighten down without delay, as soon as it comes into contact with the cylinder head surface the sealant begins to harden immediately.*

19 Before the frame is tightened down, fit a new sealing cap to the transmission end of the camshaft, fitting it flush to the end of the cylinder head.

20 With the frame in position, tighten the retaining nuts for bearing caps 2 and 4, using an alternating diagonal sequence. Then fit and tighten the nuts for 1, 3 and 5 progressively in a diagonal sequence. Note that as the nuts are tightened, the camshaft will be forced down against the pressure of the valve springs.

21 Tighten the retaining nuts to the specified torque in sequence (**see illustration**).

22 Refit the camshaft sprocket as described in Section 8.

23 Refit and tension the timing belt as described in Section 7.

24 Refit the camshaft cover as described in Section 4.

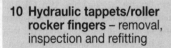

10 Hydraulic tappets/roller rocker fingers – removal, inspection and refitting

Removal

1 Remove the camshaft, as described in Section 4.

2 As the components are removed, keep them in strict order, so that they can be refitted in their original locations. Accelerated wear leading to early failure will result if the tappets and rocker fingers are interchanged.

3 Note the fitted position, then unclip the rocker fingers from the hydraulic tappets and lift them from the cylinder head.

4 Carefully lift the tappets from their bores in the cylinder head. It is advisable to store the tappets (in the correct order) upright in an oil bath whilst they are removed from the engine. Make a note of the position of each tappet, as they must be refitted in their original locations on reassembly.

9.14 Lubricate the camshaft bearings with clean engine oil

Inspection

5 Check the cylinder head bore contact surfaces and the hydraulic tappets for signs of scoring or damage. Also, check that the oil holes in the tappets are free from obstructions. If significant scoring or damage is found, it may be necessary to renew the cylinder head and the complete set of tappets.

6 Check the valve, tappet and camshaft contact faces of the rockers for wear or damage, and also check the rockers for any signs of cracking. Renew any worn or damaged rockers.

7 Inspect the camshaft, as described in Section 9.

Refitting

8 Smear some clean engine oil onto the sides of the hydraulic tappets, and offer them into position in their original bores in the cylinder head. Push them down until they are seated correctly and lubricate the upper surface of the tappet.

9 Oil the rocker contact faces of the tappets, and the tops of the valve stems, then refit the rockers to their original locations, ensuring that the rockers are securely clipped onto the tappets.

10 Lubricate the camshaft lobe contact surfaces and refit the camshaft as described in Section 9.

11 Camshaft oil seal – renewal

Note: *The oil seals are a PTFE (Teflon) type and are fitted dry, without using any grease or oil. These have a wider sealing lip and have been introduced instead of the coil spring type oil seal.*

1 Remove the timing belt as described in Section 7.

2 Remove the camshaft sprocket as described in Section 8.

3 Drill two small holes into the existing oil seal, diagonally opposite each other. Take great care to avoid drilling through into the seal housing or camshaft sealing surface. Thread two self-tapping screws into the holes and, using a pair of pliers, pull on the heads of the screws to extract the oil seal.

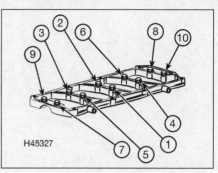

H45327

9.21 Tightening sequence for camshaft retaining frame

4 Clean out the seal housing and the sealing surface of the camshaft by wiping it with a lint-free cloth. Remove any swarf or burrs that may cause the seal to leak.

5 Carefully push the seal over the camshaft until it is positioned above its housing. To prevent damage to the sealing lips, wrap some adhesive tape around the end of the camshaft.

6 Using a hammer and a socket of suitable diameter, drive the seal squarely into its housing. **Note:** *Select a socket that bears only on the hard outer surface of the seal, not the inner lip which can easily be damaged. Remove the adhesive tape from the end of the camshaft after the seal has been located correctly.*

7 Refit the camshaft sprocket with reference to Section 8.

8 Refit and tension the timing belt as described in Section 7.

12 Cylinder head – removal, inspection and refitting

Note: *The cylinder head must be removed with the engine cold. New cylinder head bolts and a new cylinder head gasket will be required on refitting.*

Removal

1 Switch off the ignition and all electrical consumers, and remove the ignition key.

2 Remove the engine top cover.

3 Drain the cooling system as described in Chapter 1A Section 33.

4 Remove the upper inlet manifold as described in Chapter 4A Section 10. Insert clean cloth in the lower manifold ports to prevent entry of dust and dirt.

5 Unbolt the coolant distributor housing from the left-hand end of the cylinder head, and recover the O-ring. There is no need to disconnect the hoses.

6 Loosen the clip and disconnect the breather hose from the outlet on the rear of the cylinder head.

7 Disconnect all wiring from the cylinder head, noting their locations for refitting.

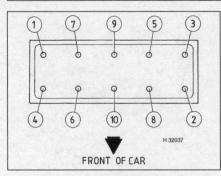

12.19 Cylinder head bolt slackening sequence

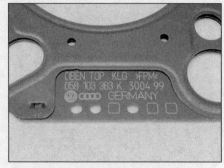

12.32 Typical cylinder head gasket markings

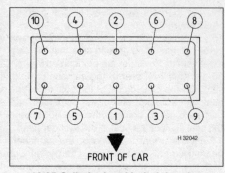

12.35 Cylinder head bolt tightening sequence

8 Remove the exhaust front section, complete with the catalytic converter and the manifold support bracket, as described in Chapter 4C Section 9. Support the exhaust to one side.

9 Remove the auxiliary drivebelt as described in Chapter 1A Section 30.

10 Remove the timing belt as described in Section 7.

11 Remove the camshaft sprocket as described in Section 8, then unbolt and remove the timing belt inner cover.

12 On models with secondary air injection, disconnect and remove the secondary air injection system pipes, and remove the pressure pipe bracket. Remove the secondary air injection pump and mounting bracket, with reference to Chapter 4A Section 10.

13 Disconnect the hose from the charcoal canister solenoid valve at the right-hand side of the engine compartment.

14 Remove the following components:
a) *Spark plug HT leads.*
b) *Camshaft position sensor wiring connector.*
c) *Unscrew the securing nut and bolts, and remove the right-angled bracket and auxiliary drivebelt tensioner from the engine.*

15 As the engine is currently supported using a hoist and lifting tackle attached to the right-hand engine lifting bracket on the cylinder head, it is now necessary to attach a suitable bracket to the cylinder block, so that the engine can still be supported as the cylinder head is removed. Alternatively, the engine can be supported using a trolley jack and a block of wood positioned under the engine sump.

16 If the engine is to be supported using a hoist, bolt a suitable bracket to the cylinder block. Attach a second set of lifting tackle to the hoist, and adjust the lifting tackle to support the engine using the bracket attached to the cylinder block. Once the engine is supported using the bracket attached to the cylinder block, disconnect the lifting tackle from the lifting bracket on the cylinder head.

17 Remove the camshaft cover as described in Section 4.

18 Make a final check to ensure that all relevant wiring, pipes and hoses have been disconnected to facilitate cylinder head removal.

19 Progressively slacken the cylinder head bolts, by one turn at a time, in order **(see illustration)**. Remove the cylinder head bolts.

20 With all the bolts removed, lift the cylinder head from the block, together with the exhaust manifold, and the lower section of the inlet manifold. If the cylinder head is stuck, tap it with a soft-faced mallet to break the joint. Do not insert a lever into the gasket joint.

21 Lift the cylinder head gasket from the block.

22 If desired, the exhaust manifold and the lower section of the inlet manifold can be removed from the cylinder head with reference to Chapter 4C Section 8 and Chapter 4A Section 10 respectively.

Inspection

23 Dismantling and inspection of the cylinder head is covered in Part F of this Chapter.

Refitting

24 The mating faces of the cylinder head and block must be perfectly clean before refitting the head.

25 Use a scraper to remove all traces of gasket and carbon, also clean the tops of the pistons. Take particular care with the aluminium surfaces, as the soft metal is easily damaged.

26 Make sure that debris is not allowed to enter the oil and water passages – this is particularly important for the oil circuit, as carbon could block the oil supply to the camshaft and crankshaft bearings. Using adhesive tape and paper, seal the water, oil and bolt holes in the cylinder block. To prevent carbon entering the gap between the pistons and bores, smear a little grease in the gap. After cleaning a piston, rotate the crankshaft so that the piston moves down the bore, then wipe out the grease and carbon with a cloth rag. Clean the other piston crowns in the same way.

27 Check the head and block for nicks, deep scratches and other damage. If slight, they may be removed carefully with a file. More serious damage may be repaired by machining, but this is a specialist job.

28 If warpage of the cylinder head is suspected, use a straight-edge to check it for distortion, as described in Part G of this Chapter.

29 Ensure that the cylinder head bolt holes in the crankcase are clean and free of oil. Syringe or soak up any oil left in the bolt holes. This is most important in order that the correct bolt tightening torque can be applied, and to prevent the possibility of the block being cracked by hydraulic pressure when the bolts are tightened.

30 Ensure that the crankshaft has been turned to position Nos 1 and 4 pistons slightly down their bores from the TDC position (refer to timing belt refitting in Section 7). This will eliminate any risk of piston-to-valve contact as the cylinder head is refitted.

31 Where applicable, refit the exhaust manifold and the lower section of the inlet manifold to the cylinder head with reference to Chapter 4C Section 8 and Chapter 4A Section 10.

32 Ensure that the cylinder head locating dowels are in place in the cylinder block, then fit a new cylinder head gasket over the dowels, ensuring that the part number is uppermost. Where applicable, the OBEN/TOP marking should also be uppermost **(see illustration)**. Note that Seat recommend that the gasket is only removed from its packaging immediately prior to fitting.

33 Lower the cylinder head into position on the gasket, ensuring that it engages correctly over the dowels.

34 Fit the new cylinder head bolts, and screw them in as far as possible by hand.

35 Working progressively, in sequence, tighten all the cylinder head bolts to the specified Stage 1 torque **(see illustration)**.

36 Again working progressively, in sequence, tighten all the cylinder head bolts through the specified Stage 2 angle.

37 Finally, tighten all the cylinder head bolts, in sequence, through the specified Stage 3 angle.

38 Reconnect the lifting tackle to the engine right-hand lifting bracket on the cylinder head, then adjust the lifting tackle to support the engine. Once the engine is adequately supported using the cylinder head mounting bracket, disconnect the lifting tackle from the bracket bolted to the cylinder block, and unbolt the improvised engine lifting bracket from the cylinder block. Alternatively, remove the trolley jack and block of wood from under the sump.

39 Refit the camshaft cover as described in Section 4.

40 Refit the timing belt inner cover, then refer to Section 8 and refit the camshaft sprocket.

41 Refit and tension the timing belt as described in Section 7.

42 Refit the exhaust front section as described in Chapter 4C Section 8.

43 Where applicable, refit the secondary air injection pump and pipes, with reference to Chapter 4A Section 10.

44 Refit the coolant distribution housing, and reconnect the coolant hoses.

45 Reconnect the charcoal canister solenoid valve hose.

46 Refit the auxiliary drivebelt as described in Chapter 1A Section 30.

47 Refit the following components:
a) Spark plug HT leads.
b) Camshaft position sensor wiring connector.
c) Refit the right-angled bracket and auxiliary drivebelt tensioner to the engine.

48 Reconnect all wiring.

49 Refit the upper part of the inlet manifold, as described in Chapter 4A Section 10.

50 Refill the cooling system as described in Chapter 1A Section 33.

51 Refit the engine top cover.

13 Sump – removal and refitting

Note: VAG sealant (D 176 404 A2 or equivalent) will be required to seal the sump on refitting.

Removal

1 Apply the handbrake, then jack up the front of the vehicle and support securely on axle stands (see Jacking and vehicle support).

2 Remove the securing screws and withdraw the engine undertray(s).

3 Drain the engine oil as described in Chapter 1A Section 6.

4 Where fitted, disconnect the wiring connector from the oil level/temperature sender in the sump **(see illustration)**.

5 Unscrew and remove the bolts securing the sump to the cylinder block, and the bolts securing the sump to the transmission casing, then withdraw the sump. If necessary, release the sump by tapping with a soft-faced hammer.

6 If desired, unbolt the oil baffle plate from the cylinder block.

Refitting

7 Begin refitting by thoroughly cleaning the mating faces of the sump and cylinder block. Ensure that all traces of old sealant are removed.

8 Where applicable, refit the oil baffle plate, and tighten the securing bolts.

9 Ensure that the cylinder block mating face of the sump is free from all traces of old sealant, oil and grease, and then apply a 2.0 to 3.0 mm thick bead of silicone sealant (VAG D 176 404 A2 or equivalent) to the sump **(see illustration)**. Note that the sealant should be run around the inside of the bolt holes in the sump. The sump must be fitted within 5 minutes of applying the sealant.

10 Offer the sump up to the cylinder block, then refit the sump-to-cylinder block bolts, and lightly tighten them by hand, working progressively in a diagonal sequence. **Note:** If the sump is being refitted with the engine and transmission separated, make sure that the sump is flush with the flywheel/driveplate end of the cylinder block.

11 Refit the sump-to-transmission casing bolts, and tighten them lightly, using a socket.

12 Again working in a diagonal sequence, lightly tighten the sump-to-cylinder block bolts, using a socket.

13 Tighten the sump-to-transmission casing bolts to the specified torque.

14 Working in a diagonal sequence, progressively tighten the sump-to-cylinder block bolts to the specified torque.

15 Refit the wiring connector to the oil level/temperature sender (where fitted), then refit the engine undertray(s), and lower the vehicle to the ground.

16 Allow at least 30 minutes from the time of refitting the sump for the sealant to dry, then refill the engine with oil, with reference to Chapter 1A Section 6.

14 Oil pump, drive chain and sprockets – removal, inspection and refitting

Oil pump removal

1 Remove the sump as described in Section 13.

2 Unscrew the securing bolts, and remove the oil baffle from the cylinder block.

3 Unscrew and remove the three mounting bolts, and release the oil pump from the dowels in the crankcase **(see illustration overleaf)**. Unhook the oil pump drive sprocket from the chain and withdraw the oil pump and oil pick-up pipe from the engine. Note that the tensioner will attempt to tighten the chain, and it may be necessary to use a screwdriver to hold it in its released position before releasing the oil pump sprocket from the chain.

4 If desired, unscrew the flange bolts and remove the suction pipe from the oil pump. Recover the O-ring seal. Unscrew the bolts and remove the cover from the oil pump. **Note:** If the oil pick-up pipe is removed from

13.4 Disconnect the wiring connector from the oil level/ temperature sender

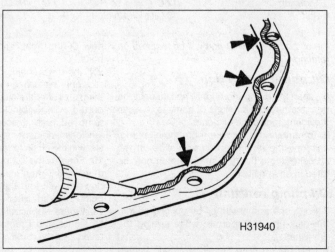

H31940

13.9 Apply the sealant around the inside of the bolt holes

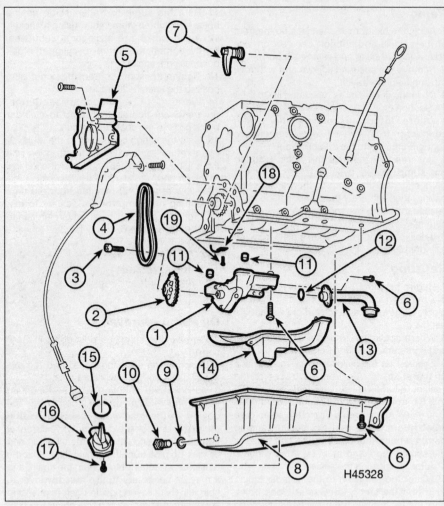

H45328

14.3 Sump and oil pump components

1 Oil pump	7 Drive chain tensioner	14 Oil baffle
2 Oil pump sprocket	8 Sump	15 Seal
3 Bolt	9 Seal	16 Oil level/temperature
4 Oil pump drive chain	10 Sump drain plug	sender
5 Crankshaft oil seal	11 Dowels	17 Bolt
housing	12 O-ring	18 Oil spray jet
6 Bolt	13 Oil pick-up pipe	19 Bolt

the oil pump, a new O-ring will be required on refitting.

Oil pump inspection

5 Clean the pump thoroughly, and inspect the gear teeth/rotors for signs of damage or wear. If evident, renew the oil pump.
6 To remove the sprocket from the oil pump, unscrew the retaining bolt and slide off the sprocket (note that the sprocket can only be fitted in one position).

Oil pump refitting

7 Prime the pump with oil by pouring oil into the pick-up pipe aperture while turning the driveshaft.
8 Refit the cover to the oil pump and tighten the bolts securely. Where applicable, refit the pick-up pipe to the oil pump, using a

new O-ring seal, and tighten the securing bolts.
9 If the drive chain, crankshaft sprocket and tensioner have been removed, delay refitting them until after the oil pump has been mounted on the cylinder block. If they have not been removed, use a screwdriver to press the tensioner against its spring to provide sufficient slack in the chain to refit the oil pump.
10 Engage the oil pump sprocket with the drive chain, then locate the oil pump on the dowels. Refit and tighten the three mounting bolts to the specified torque.
11 Where applicable, refit the drive chain, tensioner and crankshaft sprocket using a reversal of the removal procedure.
12 Refit the oil baffle, and tighten the securing bolts.
13 Refit the sump as described in Section 13.

Oil pump drive chain and sprockets

Note: *VAG sealant (D 176 404 A2 or equivalent) will be required to seal the crankshaft oil seal housing on refitting, and it is advisable to fit a new crankshaft oil seal.*

Removal

14 Proceed as described in paragraphs 1 and 2.
15 To remove the oil pump sprocket, unscrew the securing bolt, then pull the sprocket from the pump shaft, and unhook it from the drive chain.
16 To remove the chain, remove the timing belt as described in Section 7, then unbolt the crankshaft oil seal housing from the cylinder block. Unbolt the chain tensioner from the cylinder block, then unhook the chain from the sprocket on the end of the crankshaft.
17 The oil pump drive sprocket is a press-fit on the crankshaft, and cannot easily be removed. Consult a Seat dealer for advice if the sprocket is worn or damaged.

Inspection

18 Examine the chain for wear and damage. Wear is usually indicated by excessive lateral play between the links, and excessive noise in operation. It is wise to renew the chain in any case if the engine is to be overhauled. Note that the rollers on a very badly worn chain may be slightly grooved. If there is any doubt as to the condition of the chain, renew it.
19 Examine the teeth on the sprockets for wear. Each tooth forms an inverted V. If worn, the side of each tooth under tension will be slightly concave in shape when compared with the other side of the tooth (ie, the teeth will have a hooked appearance). If the teeth appear worn, the sprocket should be renewed (consult a Seat dealer for advice if the crankshaft sprocket is worn or damaged).

Refitting

20 If the oil pump has been removed, refit the oil pump as described previously in this Section before refitting the chain and sprocket.
21 Refit the chain tensioner to the cylinder block, and tighten the securing bolt to the specified torque. Make sure that the tensioner spring is correctly positioned to pretension the tensioner arm.
22 Engage the oil pump sprocket with the chain, then engage the chain with the crankshaft sprocket. Use a screwdriver to press the tensioner against its spring to provide sufficient slack in the chain to engage the sprocket with the oil pump. Note that the sprocket will only fit in one position.
23 Refit the oil pump sprocket bolt, and tighten to the specified torque.
24 Fit a new crankshaft oil seal to the housing, and refit the housing as described in Section 16.
25 Where applicable, refit the oil baffle, and tighten the securing bolts.
26 Refit the sump as described in Section 13.

15.2a Tool used to hold the flywheel stationary

15.2b Unscrew the securing bolts...

15.3 ...and remove the flywheel

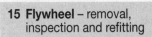

15 Flywheel – removal, inspection and refitting

Note: *New flywheel securing bolts will be required on refitting.*

Removal

1 Remove the gearbox (see Chapter 7A Section 3) and clutch (Chapter 6 Section 6).

2 The flywheel bolts are offset to ensure correct fitment. Unscrew the bolts while holding the flywheel stationary. Temporarily insert a bolt in the cylinder block, and use a screwdriver to hold the flywheel or make up a holding tool **(see illustrations)**.

3 Lift the flywheel from the crankshaft **(see illustration)**.

Inspection

4 Check the flywheel for wear and damage. Examine the starter ring gear for excessive wear to the teeth. If the driveplate or its ring gear are damaged, the complete driveplate must be renewed. The flywheel ring gear, however, may be renewed separately from the flywheel, but the work should be entrusted to a Seat dealer. If the clutch friction face is discoloured or scored excessively, it may be possible to regrind it, but this work should also be entrusted to a Seat dealer.

Refitting

5 Refitting is a reversal of removal, bearing in mind the following points.

a) Ensure that the engine-to-transmission plate is in place before fitting the flywheel/ driveplate.

b) Use new bolts when refitting the flywheel and coat the threads of the bolts with locking fluid before inserting them. Tighten the securing bolts to the specified torque.

16 Crankshaft oil seals – renewal

Note: *The oil seals are a PTFE (Teflon) type and are fitted dry, without using any grease or oil. These have a wider sealing lip and have*

been introduced instead of the coil spring type oil seal.

Timing belt end oil seal

Note: *If the oil seal housing is removed, VAG sealant (D 176 404 A2, or equivalent) will be required to seal the housing on refitting.*

1 Remove the timing belt as described in Section 7, and the crankshaft sprocket with reference to Section 8.

2 To remove the seal without removing the housing, drill two small holes diagonally opposite each other, insert self-tapping screws, and pull on the heads of the screws with pliers.

3 Alternatively, to remove the oil seal complete with its housing, proceed as follows.

a) Remove the sump as described in Section 13. This is necessary to ensure a satisfactory seal between the sump and oil seal housing on refitting.

b) Unbolt and remove the oil seal housing.

c) Working on the bench, lever the oil seal from the housing using a suitable screwdriver. Take care not to damage the seal seating in the housing.

4 Thoroughly clean the oil seal seating in the housing.

5 Wind a length of tape around the end of the crankshaft to protect the oil seal lips as the seal (and housing, where applicable) is fitted.

6 Fit a new oil seal to the housing, pressing or driving it into position using a socket or tube of suitable diameter. Ensure that the socket or tube bears only on the hard outer ring of the seal, and take care not to damage the seal lips. Press or drive the seal into position until it is seated on the shoulder in the housing. Make sure that the closed end of the seal is facing outwards.

7 If the oil seal housing has been removed, proceed as follows, otherwise proceed to paragraph 9.

8 Clean all traces of old sealant from the crankshaft oil seal housing and the cylinder block. Apply a bead of VAG sealant (D 176 404 A2, or equivalent), 2.0 to 3.0 mm thick along the cylinder block mating face of the oil seal housing **(see illustration)**. Note that the seal housing must be refitted within 5 minutes of applying the sealant.

Caution: DO NOT put excessive amounts of sealant onto the housing as it may get into the sump and block the oil pick-up pipe.

9 Refit the oil seal housing, taking care not to damage the seal (see paragraph 5), then tighten the bolts progressively to the specified torque.

10 Refit the sump as described in Section 13.

11 Refit the crankshaft sprocket with reference to Section 8, and the timing belt as described in Section 7.

Flywheel end oil seal

Note: *If the original seal housing was fitted using sealant, VAG sealant (D 176 404 A2, or equivalent) will be required to seal the housing on refitting.*

12 Remove the flywheel as described in Section 15.

13 Remove the sump as described in Section 13. This is necessary to ensure a satisfactory seal between the sump and oil seal housing on refitting.

14 Unbolt and remove the oil seal housing, complete with the oil seal.

15 The new oil seal will be supplied ready-fitted to a new oil seal housing.

16 Thoroughly clean the oil seal housing mating face on the cylinder block.

17 New oil seal/housing assemblies are supplied with a fitting tool to prevent damage to the oil seal as it is being fitted. Locate the tool over the end of the crankshaft.

18 If the original oil seal housing was fitted using sealant, apply a thin bead of VAG

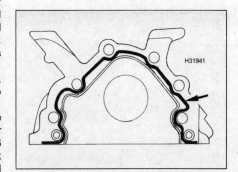

16.8 Apply sealant to the cylinder block mating face of the crankshaft oil seal housing

sealant (D 176 404 A2, or equivalent) to the cylinder block mating face of the oil seal housing. Note that the seal housing must be refitted within 5 minutes of applying the sealant.

Caution: DO NOT put excessive amounts of sealant onto the housing as it may get into the sump and block the oil pick-up pipe.

19 Carefully fit the oil seal/housing assembly over the rear of the crankshaft, and tighten the bolts progressively, in a diagonal sequence, to the specified torque.

20 Remove the oil seal protector tool from the end of the crankshaft.

21 Refit the sump as described in Section 13.

22 Refit the flywheel as described in Section 15.

17 Engine mountings – inspection and renewal

Inspection

1 If improved access is required, jack up the front of the vehicle, and support it securely on axle stands (see *Jacking and vehicle support*). Remove the engine top cover which also incorporates the air filter, then remove the engine undertray(s).

2 Check the mounting rubbers to see if they are cracked, hardened or separated from the metal at any point; renew the mounting if any such damage or deterioration is evident.

3 Check that all the mountings are securely tightened; use a torque wrench to check if possible.

4 Using a large screwdriver or a crowbar, check for wear in the mounting by carefully levering against it to check for free play. Where this is not possible, enlist the aid of an assistant to move the engine/transmission back-and-forth, or from side-to-side, whilst you observe the mounting. While some free play is to be expected, even from new components, excessive wear should be obvious. If excessive free play is found, check first that the fasteners are correctly secured, then renew any worn components as described in the following paragraphs.

Renewal

Right-hand mounting

5 Remove the engine cover/air filter housing.

6 Attach a hoist and lifting tackle to the engine lifting brackets on the cylinder head, and raise the hoist to just take the weight of the engine. Alternatively the engine can be supported on a trolley jack under the engine. Use a block of wood between the sump and the head of the jack, to prevent any damage to the sump.

7 For improved access, unbolt the coolant reservoir and move it to one side. Disconnect the small hose from the top, but leave the rest of the hoses connected **(see illustration)**.

8 Where applicable, move any wiring harnesses, pipes or hoses to one side to enable removal of the engine mounting.

9 Unclip the Evap canister and remove the mounting bracket **(see illustration)**.

10 Unscrew the bolts securing the mounting to the engine, then unscrew the bolts securing it to the body. Withdraw the mounting from the engine compartment **(see illustrations)**.

11 Refitting is a reversal of removal, bearing in mind the following points.

a) Use new securing bolts.
b) Align the mounting as shown **(see illustration)**.
c) Tighten all fixings to the specified torque.

Left-hand mounting

Note: *New mounting bolts will be required on refitting (there is no need to renew the smaller mounting-to-body bolts).*

12 Remove the engine top cover which also incorporates the air filter.

13 Attach a hoist and lifting tackle to the engine lifting brackets on the cylinder head, and raise the hoist to just take the weight of the engine and transmission. Alternatively the engine can be supported on a trolley jack under the transmission. Use a block of wood between the transmission and the head of the jack, to prevent any damage to the transmission.

14 Remove the battery, as described in Chapter 5A Section 3, then disconnect the main starter motor feed cable from the positive battery terminal box.

15 Release any relevant wiring or hoses from the clips on the battery tray, then unscrew

17.7 Move the reservoir to the side

17.9 Remove the bracket

17.10a Remove the small bracket...

17.10b ...and then remove the bolts

17.10c Lift out the mounting

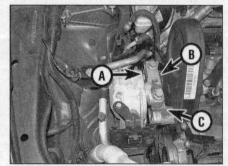
17.11 Align the mounting so that the gap (A) = 16 mm with the bracket (B) parallel to the lower bracket (C)

17.15 Remove the battery tray

17.16 Remove the trunking

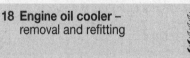

17.17a Remove the bolts

17.17b Lift out the mounting

17.20 The rear torque arm

22 Refitting is a reversal of removal, but use new mounting securing bolts, and tighten all fixings to the specified torque.

18 Engine oil cooler – removal and refitting

Note: *The oil cooler is only fitted to engine code BGU. A new oil filter and a new oil cooler O-ring will be required on refitting.*

Removal

1 The oil cooler is mounted above the oil filter, at the front of the cylinder block **(see illustrations)**.
2 Position a container beneath the oil filter to catch escaping oil and coolant, then remove the oil filter, with reference to Chapter 1A Section 6 if necessary.
3 Clamp the oil cooler coolant hoses to

the bolts and remove the battery tray **(see illustration)**.
16 Remove the wiring loom from the trunking and then remove the trunking **(see illustration)**.
17 Unscrew the bolts securing the mounting to the transmission, and then the bolts securing the mounting to the body. Lift the mounting from the engine compartment **(see illustrations)**.
18 Refitting is a reversal of removal, bearing in mind the following points:
a) *Use new mounting bolts.*
b) *Tighten all fixings to the specified torque.*

Rear mounting (torque arm)

19 Apply the handbrake, then jack up the front of the vehicle and support securely on axle stands (see *Jacking and vehicle support*). Remove the engine undertray(s) for access to the rear mounting (torque arm).

20 Working under the vehicle, unscrew and remove the bolt securing the mounting to the subframe **(see illustration)**.
21 Unscrew the two bolts securing the mounting to the transmission, then withdraw the mounting from under the vehicle.

18.1a Oil cooler

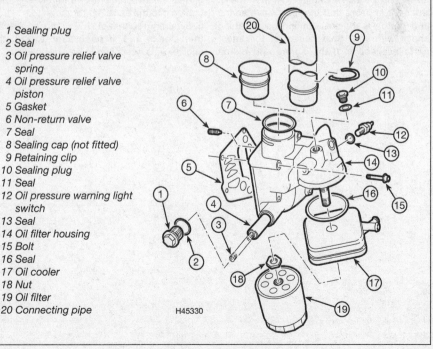

1 Sealing plug
2 Seal
3 Oil pressure relief valve spring
4 Oil pressure relief valve piston
5 Gasket
6 Non-return valve
7 Seal
8 Sealing cap (not fitted)
9 Retaining clip
10 Sealing plug
11 Seal
12 Oil pressure warning light switch
13 Seal
14 Oil filter housing
15 Bolt
16 Seal
17 Oil cooler
18 Nut
19 Oil filter
20 Connecting pipe

H45330

18.1b Oil cooler details

minimise coolant spillage, then remove the clips, and disconnect the hoses from the oil cooler. Be prepared for coolant spillage.

4 Where applicable, release the oil cooler pipes from any retaining brackets or clips.

5 Unscrew the oil cooler securing nut from the oil filter mounting threads, then slide off the oil cooler. Recover the O-ring from the top of the oil cooler.

Refitting

6 Refitting is a reversal of removal, bearing in mind the following points.

a) *Use a new oil cooler O-ring.*

b) *Fit a new oil filter.*

c) *On completion, check and if necessary top-up the oil and coolant levels.*

19 Oil pressure relief valve – removal, inspection and refitting

Removal

1 The oil pressure relief valve is fitted to the right-hand side of the oil filter housing.

2 Wipe clean the area around the relief valve plug then slacken and remove the plug and sealing ring from the filter housing. Withdraw the valve spring and piston, noting their correct fitted positions. If the valve is to be left removed from the engine for any length of time, plug the hole in the oil filter housing.

Inspection

3 Examine the relief valve piston and spring for signs of wear or damage. At the time of writing it appears that the relief valve spring and piston were not available separately; check with your Seat dealer for the latest parts availability. If the spring and piston

are worn it will be necessary to renew the complete oil filter housing assembly. The valve plug and sealing ring are listed as separate components.

Refitting

4 Fit the piston to the inner end of the spring then insert the assembly into the oil filter housing. Ensure the sealing ring is correctly fitted to the valve plug then fit the plug to the housing, tightening it to the specified torque.

5 On completion, check and, if necessary, top-up the engine oil as described in *Weekly checks,*.

20 Oil pressure warning light switch – removal and refitting

Removal

1 The oil pressure warning light switch is fitted to the left-hand side of the oil filter housing.

2 Disconnect the wiring connector and wipe clean the area around the switch.

3 Unscrew the switch from the filter housing and remove it, along with its sealing washer. If the switch is to be left removed from the engine for any length of time, plug the oil filter housing aperture.

Refitting

4 Examine the sealing washer for signs of damage or deterioration and if necessary renew.

5 Refit the switch, complete with washer, and tighten it to the specified torque.

6 Securely reconnect the wiring connector then check and, if necessary, top-up the engine oil as described in *Weekly checks*.

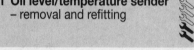

21.1 Oil level/temperature sender – located in the base of the sump

21 Oil level/temperature sender – removal and refitting

Removal

1 The oil level/temperature sender is fitted to bottom of the sump **(see illustration)**.

2 Drain the engine oil as described in Chapter 1A Section 6.

3 Disconnect the wiring connector and wipe clean the area around the sender.

4 Undo the three retaining bolts and remove the sender.

Refitting

5 Examine the sealing washer for signs of damage or deterioration and if necessary renew.

6 Refit the sender and tighten the retaining bolts to the specified torque.

7 Securely reconnect the wiring connector then refill the engine with oil, with reference to Chapter 1A Section 6.

8 On completion, check and, if necessary, top-up the engine oil as described in *Weekly checks*.

Chapter 2 Part B
2.0 litre direct injection petrol engine in-car repair procedures

Contents

Degrees of difficulty

Easy, suitable for novice with little experience	Fairly easy, suitable for beginner with some experience	Fairly difficult, suitable for competent DIY mechanic	Difficult, suitable for experienced DIY mechanic	Very difficult, suitable for expert DIY or professional

Specifications

General

Manufacturer's engine codes*:

1984 cc non-turbo	BLR, BVY, BLY, BVZ,
1984 cc turbo	BWA, BWJ, CDLA and CDLD

Maximum power output:

Engine codes BLR, BVY, BLY, and BVZ	110 kW at 6000 rpm
Engine code BWA..	147 kW at 5700 rpm
Engine codes BWJ and CDLD	177 kW at 6000 rpm
Engine code CDLA.......................................	195 kW at 6000 rpm
Bore	82.5 mm
Stroke	92.8 mm

Compression ratio:

Engine codes BLR and BVY...............................	11.5 : 1
Engine code BWA..	10.3 : 1
Engine codes BLY and BVZ	10.5 : 1
Engine codes BWJ, CDLD and CDLA	9.8 : 1

Compression pressures:

Minimum compression pressure:	Approximately 7.0 bar
Minimum compression pressure (BLR and BVY engine codes):	Approximately 8.0 bar
Maximum difference between cylinders......................	Approximately 3.0 bar
Firing order...	1 – 3 – 4 – 2
No 1 cylinder location...................................	Timing belt end

Note: *See 'Vehicle identification' at the end of this manual for the location of engine code markings.*

Camshafts

Camshaft endfloat	0.17 mm
Camshaft bearing running clearance	0.10 mm
Camshaft run-out	0.035 mm

Lubrication system

Oil pump type	Gear type, chain-driven from crankshaft
Oil pressure (oil temperature 80°C):	
At idling	1.2 to 1.6 bar
At 2000 rpm	2.7 to 4.5 bar

Torque wrench settings

	Nm	lbf ft
Ancillaries bracket to engine block	45	33
Auxiliary drivebelt idler pulley.	40	30
Auxiliary drivebelt tensioner bolts	23	17
Balancer shaft housing bolts*:		
Stage 1	15	11
Stage 2	Angle-tighten a further 90°	
Balancer shaft/oil pump drive chain tensioner bolts	15	11
Big-end bearing caps bolts (BWA, BLR, BVY, BLY and BVZ)*:		
Stage 1	30	22
Stage 2	Angle-tighten a further 90°	
Big-end bearing caps bolts (BWJ, CDLA and CDLA)*:		
Stage 1	45	33
Stage 2	Angle-tighten a further 90°	
Camshaft adjuster chain tensioner.	10	7
Camshaft adjuster rear cover	10	7
Camshaft adjuster-to-exhaust camshaft bolt:		
Stage 1	20	15
Stage 2	Angle-tighten a further 45°	
Camshaft bearing ladder*:		
Stage 1	8	6
Stage 2	Angle-tighten a further 90°	
Camshaft cover	10	7
Camshaft position sensor	10	7
Camshaft sprocket:		
Stage 1	50	37
Stage 2	Angle-tighten a further 180°	
Coolant elbow	15	11
Crankshaft timing belt end oil seal housing	35	26
Crankshaft pulley bolts*:		
Stage 1	10	7
Stage 2	Angle-tighten a further 90°	
Crankshaft sprocket centre bolt*:		
Stage 1	90	66
Stage 2	Angle-tighten a further 90°	
Crankshaft flywheel end oil seal housing	15	11
Cylinder head bolts*:		
Stage 1	40	30
Stage 2	Angle-tighten a further 90°	
Stage 3	Angle-tighten a further 90°	
Driveplate mounting bolts*:		
Stage 1	60	44
Stage 2	Angle-tighten a further 90°	
Engine mountings:		
RH engine mounting:		
Limiter:		
Stage 1	20	15
Stage 2	Angle-tighten a further 90°	
Mounting to engine:		
Stage 1	60	44
Stage 2	Angle-tighten a further 90°	
Mounting to body:		
Stage 1	40	30
Stage 2	Angle-tighten a further 90°	

Torque wrench settings (continued)

	Nm	lbf ft
Engine mountings: (continued)		
LH engine mounting:		
Mounting to body:		
Stage 1	60	44
Stage 2	Angle-tighten a further 90°	
Mounting to transmission:		
Stage 1	40	30
Stage 2	Angle-tighten a further 90°	
Rear mounting link:		
To transmission:		
Stage 1	40	30
Stage 2	Angle-tighten a further 90°	
To subframe:		
Stage 1	100	74
Stage 2	Angle-tighten a further 90°	
Engine speed sender	10	7
Flywheel mounting bolts*:		
Stage 1	60	44
Stage 2	Angle-tighten a further 90°	
Intake manifold support	40	30
Knock sensor	20	15
Main bearing cap bolts*:		
Stage 1	65	48
Stage 2	Angle-tighten a further 90°	
Oil filter housing	15	11
Oil jets 27	20	
Oil pressure relief valve plug	15	11
Oil pump cover-to-balancer shaft housing.	8	6
Oil pump sprocket bolt*:		
Stage 1	20	15
Stage 2	Angle-tighten a further 90°	
Roadwheel bolts.	120	89
Sump:		
Sump-to-block bolts	15	11
Sump-to-transmission bolts.	40	30
Timing belt covers	10	7
Timing belt inner cover.	10	7
Timing belt idler roller:		
Large	40	30
Small	25	18
Timing belt tensioner roller:		
Engines without a split cover	23	17
Engines with a split cover.	25	18

Do not re-use

1 General Information

How to use this Chapter

1 This Part of Chapter 2 describes those repair procedures that can reasonably be carried out on the engine while it remains in the vehicle. If the engine has been removed from the vehicle and is being dismantled as described in Part F, any preliminary dismantling procedures can be ignored.

2 Note that while it may be possible physically to overhaul certain items while the engine is in the vehicle, such tasks are not usually carried out as separate operations, and usually require the execution of several additional procedures (not to mention the cleaning of components and of oilways); for this reason, all such tasks are classed as major overhaul procedures, and are described in Part F of this Chapter.

Engine description

3 Throughout this Chapter, engines are identified by the manufacturer's code letters. A listing of all engines covered, together with their code letters, is given in the Specifications.

4 The engines are water-cooled, double overhead camshaft (DOHC), in-line four-cylinder direct injection units, with aluminium cylinder blocks and cylinder heads. All are mounted transversely at the front of the vehicle, with the transmission bolted to the left-hand end of the engine.

5 The crankshaft is of five-bearing type, and thrust washers are fitted to the centre main bearing to control crankshaft endfloat.

6 The cylinder head carries the double camshafts. It also houses the intake and exhaust valves, which are closed by single coil springs, and which run in guides pressed into the cylinder head. The camshaft actuates the valves by cam followers and hydraulic tappets mounted in the cylinder head. The cylinder head contains integral oilways which supply and lubricate the tappets.

7 The timing belt drives the exhaust camshaft, and the intake camshaft is driven from the exhaust camshaft by chain at the left-hand end of the camshafts. A hydraulic tensioner is fitted to the chain, to automatically vary the intake camshaft valve timing.

8 The valves are operated from the camshafts through rocker arms and hydraulic adjusters – the valve clearances are adjusted automatically.

9 The engine coolant pump is driven by the toothed timing belt.

10 Lubricant is circulated under pressure by a pump, driven by a chain from the crankshaft. Oil is drawn from the sump through a strainer, and then forced through an externally-mounted, renewable screw-on filter. From there, it is distributed to the cylinder head, where it lubricates the camshaft journals and hydraulic tappets, and also to the crankcase, where it lubricates the main bearings, connecting rod big-ends, gudgeon pins and cylinder bores. An oil pressure switch is located on the oil filter housing, operating at 1.4 bars. An oil cooler mounted above the oil filter is supplied with coolant from the cooling system to reduce the temperature of the oil before it re-enters the engine.

11 A balancer shaft housing is fitted below the cylinder block. The two contra-rotating shafts are driven by the same chain that drives the oil pump. The shafts have integral weights fitted along their length, and as they spin, the forces created cancel out almost all of the vibration generated by the engine.

Operations with engine in car

12 The following operations can be performed without removing the engine:

a) Compression pressure – testing.
b) Auxiliary drivebelt – removal and refitting.
c) Camshafts – removal and refitting.
d) Camshaft oil seals – renewal.
e) Camshaft sprocket – removal and refitting.
f) Coolant pump – removal and refitting (refer to Chapter 3 Section 7).
g) Crankshaft oil seals – renewal.
h) Crankshaft sprocket – removal and refitting.
i) Cylinder head – removal and refitting*.
j) Engine mountings – inspection and renewal.
k) Balancer shaft housing – renewal.
l) Oil pump and pick-up assembly – removal and refitting.
m) Sump – removal and refitting.
n) Timing belt, sprockets and cover – removal, inspection and refitting.
o) Flywheel/driveplate – removal and refitting

13 * Cylinder head dismantling procedures are detailed in Chapter 2F Section 6.

Note: It is possible to remove the pistons and connecting rods (after removing the cylinder head and sump) without removing the engine.

However, this is not recommended. Work of this nature is more easily and thoroughly completed with the engine on the bench, as described in Chapter 2F Section 9.

2 Engine valve timing marks – general information and usage

General information

1 The crankshaft and camshaft sprockets are driven by the timing belt, and rotate in phase with each other. When the timing belt is removed during servicing or repair, it is possible for the shafts to rotate independently of each other, and the correct phasing is then lost.

2 The design of the engines covered in this Chapter is such that piston-to-valve contact will occur if the crankshaft is turned with the timing belt removed. For this reason, it is important that the correct phasing between the camshaft and crankshaft is preserved whilst the timing belt is off the engine. This is achieved by setting the engine in a reference condition (known as Top Dead Centre or TDC) before the timing belt is removed, and then preventing the shafts from rotating until the belt is refitted. Similarly, if the engine has been dismantled for overhaul, the engine can be set to TDC during reassembly to ensure that the correct shaft phasing is restored. **Note:** The coolant pump is also driven by the timing belt, but the pump alignment is not critical.

3 TDC is the highest position a piston reaches within its respective cylinder – in a four-stroke engine, each piston reaches TDC twice per cycle; once on the compression stroke, and once on the exhaust stroke. In general, TDC normally refers to No 1 cylinder on the compression stroke. Note that the cylinders are numbered one to four, starting from the timing belt end of the engine.

4 The crankshaft pulley has a mark which, when aligned with a reference mark on the timing belt cover, indicates that No 1 cylinder (and hence also No 4 cylinder) is at TDC **(see illustration)**.

5 The exhaust camshaft sprocket is also equipped with a timing mark **(see illustration)** – when this is aligned with a mark on the small

upper timing belt cover or camshaft cover, No 1 cylinder is at TDC compression.

Setting TDC on No 1 cylinder

6 Before starting work, make sure that the ignition is switched off.

7 Pull the engine top cover upwards from its fasteners.

8 Remove all of the spark plugs as described in Chapter 1A Section 29.

9 Remove the timing belt upper, outer cover as described in Section 4.

10 Turn the engine clockwise with a spanner on the crankshaft pulley until the timing mark on the outer circumference of the camshaft sprocket aligns with the mark on the timing belt cover. With this aligned, the timing mark on the crankshaft pulley should align with the mark on the lower timing belt cover.

3 Cylinder compression test

1 When engine performance is down, or if misfiring occurs which cannot be attributed to the ignition or fuel systems, a compression test can provide diagnostic clues as to the engine's condition. If the test is performed regularly, it can give warning of trouble before any other symptoms become apparent.

2 The engine must be fully warmed-up to normal operating temperature, the battery must be fully charged, and all the spark plugs must be removed (refer to Chapter 1A Section 29). The aid of an assistant will also be required. Pull the plastic cover on the top of the engine upwards from its fasteners.

3 Remove the spark plugs as described in Chapter 1A Section 29.

4 Disable the injectors by disconnecting the wiring plug under the intake manifold.

5 Fit a compression tester to the No 1 cylinder spark plug hole – the type of tester which screws into the plug thread is preferable.

6 Have an assistant hold the throttle wide open. Crank the engine on the starter motor several seconds. After one or two revolutions, the compression pressure should build-up to a maximum figure, and then stabilise. Record the highest reading obtained.

7 Repeat the test on the remaining cylinders, recording the pressure in each. Keep the throttle wide open.

8 All cylinders should produce very similar pressures; a difference of more than 3 bars between any two cylinders indicates a fault. Note that the compression should build-up quickly in a healthy engine. Low compression on the first stroke, followed by gradually-increasing pressure on successive strokes, indicates worn piston rings. A low compression reading on the first stroke, which does not build-up during successive strokes, indicates leaking valves or a blown head gasket (a cracked head could also be the cause).

2.4 Crankshaft pulley TDC marks

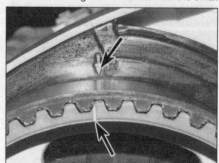

2.5 Exhaust camshaft TDC marks

4.5 Remove the auxiliary drivebelt tensioner

9 Refer to the Specifications of this Chapter, and compare the recorded compression figures with those stated by the manufacturer.
10 On completion of the test, refit the spark plugs, injector wiring plug and top cover. Note that in some cases, disconnecting the wiring plugs from the coils and injectors then cranking the engine may cause fault codes to be stored by the engine management ECM – have these codes erased by means of a suitable diagnostic tool/fault code reader. See your Seat dealer or specialist.

4 Timing belt – removal, inspection and refitting

General information

1 The primary function of the toothed timing belt is to drive the camshafts. Should the belt slip or break in service, the valve timing will be disturbed and piston-to-valve contact will occur, resulting in serious engine damage. For this reason, it is important that the timing belt is tensioned correctly, and inspected regularly for signs of wear or deterioration.

Removal

2 Before starting work, remove the fuel pump fuse with reference to Chapter 14. **Note:** *The fuel pump will be activated by the driver's side door contact switch, if the battery remains connected. Alternatively, disconnect the battery negative lead as described in Chapter 5A Section 3.*

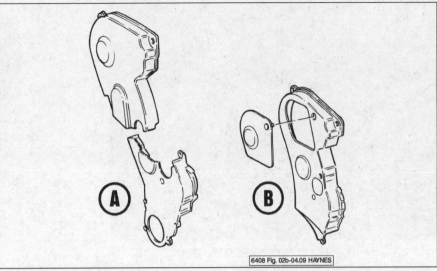

4.8 The timing belt covers. Two piece cover (A) and single piece cover with access panel (B)

3 Apply the handbrake, then jack up the front of the vehicle and support it on axle stands (see *Jacking and vehicle support* Section 5). Undo the fasteners and remove the engine undertray, then remove the right-hand front roadwheel and the wheel arch liner. On turbo models, also remove the charge air hose.
4 Pull the plastic cover/air filter (as applicable) on the top of the engine upwards to release it from the fasteners.
5 Remove the auxiliary drivebelt with reference to Chapter 1A Section 30. Also unbolt the tensioner from the front of the engine **(see illustration)**.
6 Disconnect the fuel supply and breather line located next to the coolant expansion tank.
7 Remove the activated charcoal filter and hoses (Chapter 4C Section 2).
8 Loosen the screenwash filler neck, then undo the screws and place the coolant expansion tank to one side. **Note:** *On some models, it will be necessary to disconnect one of the coolant hoses.*
Note: *There are two versions of the timing belt cover fitted – a single one piece cover with an access panel and a two piece cover with no access panel* **(see illustration)**.
9 On engine with a split cover, remove the

4.9 Timing belt upper cover screws

timing belt upper cover, and unbolt the auxiliary drivebelt tensioner **(see illustration)**.
10 Set the engine to TDC as described in Section 2.
11 Hold the crankshaft stationary with a spanner on the centre bolt, then unscrew the pulley bolts and remove the pulley from the crankshaft sprocket **(see illustrations)**.
12 Remove the lower timing belt cover, then check that the engine is still at TDC, and mark the crankshaft sprocket in relation to the cylinder block to indicate its TDC position **(see illustration)**.

4.11a Crankshaft pulley bolts...

4.11b ...and pulley location hole

4.12 Timing belt cover retaining bolts

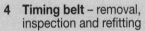

4.19 Remove the timing belt tensioner

4.24 Timing belt routing

13 On engine codes with a split cover, carry out the following:

a) *Refer to Chapter 4C Section 8 and disconnect the exhaust front pipe from the exhaust manifold/turbocharger. Unbolt the exhaust mounting bracket from the underbody, and support the exhaust pipe on an axle stand.*

b) *Disconnect the driveshafts from the transmission with reference to Chapter 8 Section 2 and tie them to the underbody.*

c) *Release the air conditioning pipes from the supports on the right-hand side of the engine compartment. Do not open the air conditioning refrigerant lines.*

14 Support the right-hand end of the engine with a hoist.

15 Beneath the vehicle, unbolt the rear engine torque arm from the bottom of the transmission.

16 Unbolt the right-hand engine mounting and bracket from the engine and body (see Section 17), and raise the engine as far as possible to provide access to the timing belt.

17 On engine codes BLR, BVY and BVZ, unbolt and remove the single timing belt cover. On engines with a split cover, unbolt and remove the lower timing belt cover, then remove the mounting bracket from the engine.

18 Check again that the engine is at TDC. If the timing belt is to be refitted, mark its normal direction of travel with chalk or a marker pen.

Engines without a split cover

19 Insert VAG special tool T10020 into the holes in the tensioner hub arm, then slacken the tensioner nut and rotate the hub clockwise to relieve the tension **(see illustration)**. Temporarily tighten the nut to retain the tensioner in the released position. In the absence of the special tool, a sturdy pair of right-angle circlip pliers will suffice.

Engines with a split cover

20 Slacken the nut in the centre of the tensioner pulley to relieve the tension in the belt.

All engines

21 Slip the timing belt off of the crankshaft, camshaft, and coolant pump sprockets, and remove it from the engine. Do not bend the timing belt sharply if it is to be re-used.

Inspection

22 Examine the belt for evidence of contamination by coolant or lubricant. If this is the case, find the source of the contamination before progressing any further. Check the belt for signs of wear or damage, particularly around the leading edges of the belt teeth. Renew the belt if its condition is in doubt; the cost of belt renewal is negligible compared with potential cost of the engine repairs,

should the belt fail in service. The belt must be renewed if it has covered the mileage stated by the manufacturer (see Chapter 1A Section 31), however, even if it has covered less, it is recommended to renew it regardless of condition as a precautionary measure. **Note:** *If the timing belt is not going to be refitted for some time, it is a wise precaution to hang a warning label on the steering wheel, to remind yourself (and others) not to turn the engine.*

Refitting

23 Ensure that the timing mark on the camshaft and crankshaft sprockets are correctly aligned with the corresponding TDC reference marks on the timing belt cover; refer to Section 2 for details.

Engines without a split cover

24 Loop the timing belt under the crankshaft sprocket loosely, observing the direction of rotation markings if the old timing belt is being refitted. Engage the timing belt teeth with the crankshaft sprocket, then manoeuvre it into position over the coolant pump and camshaft sprockets, tensioner pulley and finally the idler pulleys **(see illustration)**.

25 Refit the lower timing cover and tighten the bolts, then locate the pulley for the auxiliary drivebelt on the crankshaft sprocket, and tighten the new bolts to the specified torque and angle. Note that the pulley will only fit in one fitting position – with the hole in the pulley over the projection on the crankshaft sprocket. Make sure that the TDC marks are correctly aligned.

26 Check the tensioner roller locating arm is correctly located in the backplate, then insert the VAG special tool T10020 (or circlip pliers) into the holes in the tensioner hub arm.

27 Slacken the retaining nut, and rotate the tensioner hub anti-clockwise until the notch in the hub is past the indicator (over-tensioned), then slowly release the tension until the notch aligns with the indicator **(see illustration)**. Tighten the tensioner nut to the specified torque.

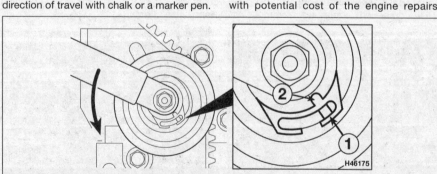

4.27 The notch (2) in the tensioner hub must align with the indicator (1) – engine without split cover

4.29 Align the notch (2) with the indicator (1) – engines with a split cover

Engines with a split cover

28 Loop the timing belt under the crankshaft sprocket loosely, observing the direction of rotation markings if the old timing belt is being refitted. Engage the timing belt teeth with the crankshaft sprocket, then manoeuvre it into position over the coolant pump and camshaft sprockets, tensioner pulley and finally the idler pulleys.

29 With the tensioner centre nut loose, insert an 8 mm Allen key into the tensioner arm in the centre of the pulley, and rotate the arm clockwise until the notch in the arm is past the indicator (over-tensioned), then slowly release the tensioner until the notch is aligned with the indicator (see illustration). Tighten the tensioner nut to the specified torque.

All engines

30 Using a spanner or wrench and socket on the crankshaft pulley centre bolt, rotate the crankshaft through two complete revolutions. Reset the engine to TDC on No 1 cylinder with reference to Section 2.Check that the crankshaft pulley and camshaft sprocket timing marks are correctly aligned. Recheck the timing belt tension and adjust it, if necessary.

31 The remaining procedure is a reversal of removal.

5 Timing belt tensioner and sprockets – removal, inspection and refitting

Removal

1 Remove the timing belt as described in Section 4.

Tensioner/roller

2 Undo the tensioner hub nut and withdraw the assembly from position (see illustrations).

Camshaft sprocket

3 To prevent any accidental piston-to-valve contact, rotate the crankshaft 90° anti-clockwise.

4 Unscrew the camshaft sprocket bolt, while holding the sprocket stationary using a tool as shown. Remove the bolt, washer (where fitted), sprocket, and the key (see illustrations). If necessary, use a 2-legged puller to remove the sprocket. If necessary, a VAG special puller (T40001) is available.

Crankshaft sprocket

5 To prevent any accidental piston-to-valve contact, rotate the crankshaft 90° anti-clockwise.

6 Unscrew and discard the crankshaft sprocket bolt, and remove the sprocket (see illustrations). The bolt is very tight, and the crankshaft must be held stationary. Remove the starter motor (Chapter 5A Section 8) and use a wide-bladed screwdriver in the ring gear to hold the crankshaft stationary.

Inspection

7 Clean all the sprockets and examine them for wear and damage. Spin the tensioner roller, and check that it runs smoothly.

8 Check the tensioner for signs of wear and/or damage and renew if necessary.

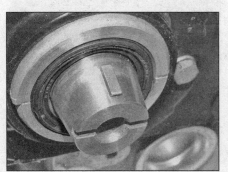

5.2a Tensioner hub and nut

5.2b Tensioner hub location recess in the cylinder head

5.4a Unscrew the camshaft sprocket bolt

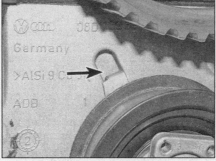

5.4b Camshaft sprocket location key

5.6a Unscrew the bolt...

5.6b ...and remove the crankshaft sprocket

7.4 Disconnect the engine breather hoses from the camshaft cover

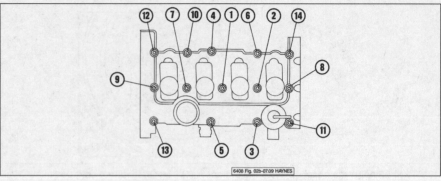

7.9 Tighten the bolts in the order shown

Refitting

Tensioner/roller

9 Refit the tensioner roller and spring assembly using a reversal of the removal procedure.

10 Refit the timing belt as described in Section 5.

Camshaft sprocket

11 Locate the key on the camshaft and refit the sprocket, and bolt. Tighten the bolt to the specified torque while holding the sprocket using the method employed on removal. Note the sprocket must be refitted with the narrow web facing forwards.

12 Carefully turn the crankshaft 90° clockwise, back to the TDC position.

13 Refit the timing belt as described in Section 4.

Crankshaft sprocket

14 Locate the sprocket on the crankshaft, then install the new bolt (do not oil the threads), and tighten the bolt to the specified torque while holding the crankshaft stationary using the method employed on removal. **Note:** *Do not allow the crankshaft to rotate.*

15 Carefully turn the crankshaft 90° clockwise, back to the TDC position.

16 Refit the timing belt as described in Section 4.

6 Auxiliary drivebelt – removal and refitting

1 Refer to Chapter 1A Section 30.

7 Camshaft cover – removal and refitting

Removal

1 Remove the engine top cover.

2 Remove the ignition coils as described in Chapter 5B Section 3.

3 Undo the 2 bolts securing the timing belt upper cover to the camshaft cover.

4 Disconnect the engine breather hose(s) from the camshaft cover **(see illustration)**.

5 Progressively and evenly unscrew the bolts securing the camshaft cover to the cylinder head.

6 Lift the camshaft cover from the cylinder head and recover the gasket. On turbo models, unbolt the valve body from the camshaft cover and recover the gasket.

Refitting

7 Clean the surfaces of the camshaft cover and cylinder head. On turbo models, refit the valve body to the camshaft cover together with a new gasket, and tighten the bolts securely.

8 Check the condition of the camshaft cover gasket and renew if necessary.

9 Carefully fit the gasket to the camshaft cover, ensuring the bolts locate correctly in the gasket, and position the assembly on the cylinder head. Progressively and evenly tighten the bolts to the specified torque **(see illustration)**.

10 Reconnect the breather hoses to the camshaft cover.

11 Apply a little thread-locking compound, then refit the timing belt cover bolts and tighten them to the specified torque.

12 Refit the ignition coils (Chapter 5B Section 3) then refit the engine top cover.

8 Camshaft oil seals – renewal

Exhaust camshaft

1 Remove the camshaft sprocket as described in Section 5. Recover the key from the camshaft.

2 Drill two small holes into the existing oil seal, diagonally opposite each other. Thread two self-tapping screws into the holes, and using two pairs of pliers, pull on the heads of the screws to extract the oil seal. Take great care to avoid drilling through into the seal housing or camshaft sealing surface.

3 Clean out the seal housing and sealing surface of the camshaft by wiping it with a lint-free cloth. Remove any swarf or burrs that may cause the seal to leak.

4 Carefully wrap adhesive tape around the end of the camshaft, to protect the seal lip from the edges of the shaft and keyway **(see illustration)**.

5 Do not lubricate the lip of the new oil seal – it must be fitted dry. Push the new seal over the end of the camshaft, ensuring the lip of the seal is not damaged by the sharp edges of the camshaft and keyway. Use a suitable tubular spacer (socket or similar) that bears only on the hard outer edge of the seal, and a hammer to gently and gradually/evenly push the seal into place. Try to install the seal as squarely as possible **(see illustration)**.

8.4 Wrap adhesive tape around the camshaft to protect the oil seal when fitting

8.5 Fit the exhaust camshaft oil seal

8.8 Remove the rubber cap from the intake camshaft

8.9 The new rubber cap in position

6 Refit the camshaft sprocket with reference to Section 5.

Intake camshaft

7 Remove the upper section of the timing belt cover as described in Section 5.

8 Using a screwdriver or pry bar, pierce the centre of the rubber cap, and lever it from place **(see illustration)**.

9 Ensure the bore of the cylinder head/ bearing ladder is clean, then drive the new cap into place, until it's flush with the casing **(see illustration)**. Note that the rubber cap should be fitted dry, no oil, grease or sealant is to be used.

10 Refit the timing belt cover as described in Section 4.

9 Crankshaft oil seals – renewal

Timing belt end oil seal

1 Remove the timing belt and crankshaft sprocket, with reference to Section 5.

2 The seal may be renewed without removing the housing by drilling two small holes diagonally opposite each other, inserting self-tapping screws, and pulling on the heads

of the screws with pliers. Alternatively, remove the auxiliary belt idler pulley (where applicable) unbolt and remove the housing (including the relevant sump bolts) and remove the gasket then lever out the oil seal on the bench **(see illustrations)**.

3 If the seal housing is still in place, position the new seal over the end of the crankshaft, ensuring the closed side of the seal faces outwards. Ease the seal lip over the shoulder of the crankshaft to prevent any damage to the seal. Drive the new oil seal into place using a suitable tubular spacer that bears only on the hard outer edge of the seal, until it's flush with the housing.

4 If the housing has been removed, remove all traces of the sealant from the mating faces of the housing, cylinder block and sump. Use a socket (or similar) to drive the new seal into place (flush with the housing), then apply a thin bead (2.0 mm) of suitable sealant (available from Seat/VAG dealers/parts specialists) to the housing mating faces. Ensure the bead of sealant is routed 'inside' the bolts holes. Do not apply too much sealant, as any excess may find its way into the oil system. Apply a bead of sealant to the joint between the sump and cylinder block. Refit the housing, easing the seal over the end of the crankshaft, and tighten the bolts to the specified torque. Tighten the bolts securing the housing to the

9.2a Drill two small holes and fit self-tapping screws in the oil seal…

9.2b …then pull out the oil seal

9.2c Remove the auxiliary drivebelt idler pulley

9.2d Unbolt the housing…

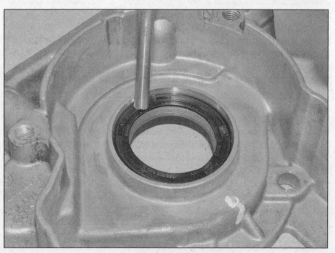

9.2e …and drive out the oil seal

9.4a Use a socket to drive in the new oil seal

9.4b Fitted oil seal

9.4c Apply sealant 'inside' the bolt holes as shown

9.11a Position the housing and seal over the crankshaft…

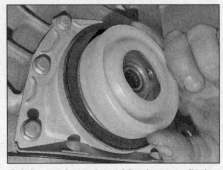

9.11b …and use the guide sleeve to fit the oil seal

9.12 Refit the adapter plate, locating it as shown

cylinder block first, followed by the sump bolts **(see illustrations)**. Note that the housing must be in position with the bolts tightened within 5 minutes of applying the sealant.
5 Refit the timing belt and crankshaft sprocket, with reference to Section 5.

Flywheel end oil seal

6 Remove the flywheel with reference to Section 13.
7 It is recommended that the sump be removed as described in Section 14, however, on engine codes BVY, BLY and BLR, this will necessitate renewing the sump-to-block intermediate plate. On all other engine codes, the sump is sealed to the cylinder block with sealant which can be renewed.
8 Pull the adapter plate from the locating dowels on the rear of the cylinder block, and remove it from the engine.
9 Unbolt and remove the housing (including the relevant sump bolts, where applicable). The seal is only available complete with the housing.
10 Carefully remove any sealant residue from the cylinder block and sump mating surfaces.
11 Where applicable, apply a little sealant to the joint between the sump and cylinder block, then apply a thin layer of sealant to the base of the new housing. The new seal is supplied with a guide sleeve fitted to the centre of the seal. Position the housing and seal over the end of the crankshaft and gently, evenly, push it into position, and tighten the bolts evenly in diagonal sequence to the specified torque, then remove the tool **(see illustrations)**.

12 Refit the adapter plate, locating it over the oil seal housing, and onto the 2 dowels at the back of the cylinder head **(see illustration)**.
13 Where applicable, refit the sump with reference to Section 14.
14 Refit the flywheel with reference to Section 14.

10 Cylinder head – removal and refitting

Note: *Cylinder head dismantling and overhaul is covered in Chapter 2F Section 6.*

Removal

1 Before starting work, switch off the ignition and all electrical consumers and remove the ignition key.
2 Apply the handbrake, then jack up the front of the vehicle and support it on axle stands (see *Jacking and vehicle support*).
3 Remove the engine top cover by pulling it upwards from its fasteners.
4 To improve access, remove the plenum chamber bulkhead at the rear of the engine compartment.
5 Drain the cooling system as described in Chapter 1A Section 33.
6 Working beneath the vehicle, unscrew the nuts securing the front exhaust pipe to the exhaust manifold, then unbolt the exhaust mounting from the crossmember and support the exhaust on an axle stand. **Note:** *Do not*

bend the exhaust pipe flexible joint more than 10°.
7 Unbolt and remove the exhaust manifold supports.
8 Remove the air filter housing and intake hoses.
9 Disconnect the coolant elbow from the left-hand end of the cylinder head.
10 Remove the inlet manifold as described in Chapter 4A Section 10.
11 Note their fitted positions, then disconnect all electrical connectors from the cylinder head, including the injector loom connectors, oil pressure switch plug and the electrically-controlled thermostat plug. Release the wiring harness from the small coolant pipe on the left-hand side of the engine.
12 On engine codes BLR, BVY, BLY and BVZ, disconnect the coolant hoses from the throttle valve control module. Also, where applicable, remove the EGR pipe between the inlet manifold and the EGR valve, and disconnect the coolant hoses.
13 Remove the auxiliary drivebelt as described in Chapter 1A Section 30.
14 Remove the timing belt as described in Section 4.
15 Remove the camshaft cover as described in Section 7.
16 Make a final check to ensure all relevant electrical connectors and coolant hoses have been disconnected. Plug or cover any openings to prevent fluid spillage and contamination.
17 Unscrew the M10 Ribe cylinder head

10.17 Remove the cylinder head bolts

10.23a Fit the new cylinder head gasket with the part number facing upwards

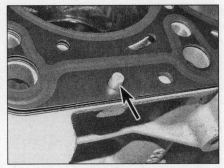

10.23b Head gasket location dowel

bolts a turn at a time, in reverse order to the tightening sequence and remove them **(see illustration)**. Discard the bolts, new ones must be fitted.

18 With the help of an assistant, lift the cylinder head from the block together with the exhaust manifold. If it is stuck, tap it free with a wooden mallet. Do not insert a lever into the gasket joint.

19 Remove the cylinder head gasket from the block.

20 If required, remove the exhaust manifold from the cylinder head with reference to Chapter 4C Section 8.

Refitting

21 Thoroughly clean the contact faces of the cylinder head and block. Also clean any oil or coolant from the bolt holes in the block – if this precaution is not taken, not only will the tightening torque be incorrect but there is the possibility of damaging the block.

22 If removed, refit the exhaust manifold to the cylinder head together with a new gasket with reference to Chapter 4C Section 8.

23 Locate a new gasket on the block, with the part number facing upwards, and readable from the intake side of the engine. Make sure that the location dowels are in position **(see illustrations)**. Seat recommend that the gasket is removed from its packaging just prior to fitting it. Handle the gasket with great

care – damage to the silicone or indented areas will lead to leaks.

24 At this point the crankshaft should still be set at TDC on cylinders 1 and 4. In order to prevent any accidental piston-to-valve contact, rotate the crankshaft a quarter of a turn anti-clockwise.

25 Carefully lower the head onto the block, making sure that it engages the location dowels correctly. Do not use any jointing compound on the cylinder head joint. Insert the new cylinder head bolts (the washers should still be in place on the cylinder head), and initially hand-tighten them.

26 Using the sequence shown **(see illustration)** tighten all the bolts to the Stage 1 torque given in the Specifications.

27 Angle-tighten the bolts in the same sequence to the Stage 2 and Stage 3 angles given in the Section **(see illustration)**.

28 Rotate the crankshaft a quarter of a turn clockwise, back to TDC on cylinders 1 and 4.

29 Refit the camshaft cover with reference to Section 7.

30 The remainder of refitting is a reversal of removal, noting the following points:

a) *Ensure all electrical connectors are securing reconnected, and the harnesses are correctly routed – refit any cable clips removed during dismantling.*

b) *Ensure all coolant hoses are reconnected, and all retaining clips are refitted in their original positions.*

c) *Apply a little thread-locking compound to the tensioner roller mounting plate bolts.*

d) *If any of the coolant hose clips appear weak or corroded – renew them.*

e) *On completion, refill the cooling system with new antifreeze mixture (see Chapter 1A Section 33).*

f) *If you suspect the engine oil has been contaminated with coolant, change the oil and filter as described in Chapter 1A Section 6.*

11 Camshafts – removal, inspection and refitting

Removal

1 Remove the inlet manifold and high-pressure fuel pump as described in Chapter 4A Section 10.

2 Remove the camshaft cover (Section 7) then undo the retaining bolts and remove the cover from the chain adjuster at the left-hand end of the cylinder head.

3 Using VAG tool T10092 and an M5 nut, lock the tensioner in place, then undo the tensioner retaining Torx bolts. In the absence of the VAG tool, use a 5 mm bolt and locknut **(see illustration)**.

4 Paint alignment marks between the chain links, the adjuster, and the intake sprocket

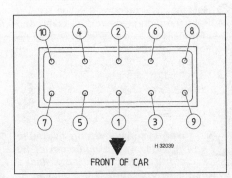

10.26 Cylinder head bolt tightening sequence

FRONT OF CAR

H 32039

10.27 Angle-tightening the cylinder head bolts

11.3 Lock the tensioner in place using a 5 mm bolt and locknut

11.4 Paint alignment marks as shown to ensure correct refitting

11.13 Refit the hydraulic tappets and rockers

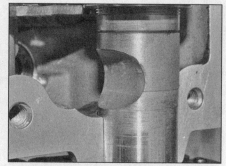

11.14 The cut-outs in the camshafts must face inwards

(see illustration). Count the number of links between the alignment marks and make a note in case the chain needs to be renewed, in which case the alignment marks can be transferred to the new chain.

5 Remove the camshaft sprocket as described in Section 5.

6 Working gradually and evenly, from the outside-in, slacken and remove the camshaft bearing ladder Torx bolts. Remove the bearing ladder.

7 Lift the camshafts together with the tensioner and chain from the cylinder head. Disengage the camshafts from the chain and discard the oil seal.

8 Lift the rocker arms and hydraulic tappets from their bores and store them with the valve contact surface facing downwards, to prevent the oil from draining out. It is recommended that the tappets are kept immersed in oil for the period they are removed from the cylinder head. Make a note of the position of each tappet and rocker arm, as they must be fitted to the same valves on reassembly – accelerated wear leading to early failure will result if they are interchanged.

Inspection

9 Visually inspect each camshaft for evidence of wear on the surfaces of the lobes and journals. Normally their surfaces should be smooth and have a dull shine; look for scoring, erosion or pitting and areas that appear highly polished, indicating excessive wear. Accelerated wear will occur once the hardened exterior of the camshaft has been damaged, so always renew worn items. **Note:** *If these symptoms are visible on the tips of the camshaft lobes, check the corresponding tappet, as it will probably be worn as well.*

10 If the machined surfaces of the camshaft appear discoloured or blued, it is likely that it has been overheated at some point, probably due to inadequate lubrication. This may have distorted the shaft, so have it checked by a Seat dealer or engine reconditioning specialist.

11 To measure the camshaft endfloat, temporarily refit the relevant camshaft to the cylinder head, then fit the bearing ladder, and tighten the retaining nuts to the specified first stage torque setting. Anchor a DTI gauge to

the timing belt end of the cylinder head and align the gauge probe with the camshaft axis. Push the camshaft to one end of the cylinder head as far as it will travel, then rest the DTI gauge probe on the end of the camshaft, and zero the gauge display. Push the camshaft as far as it will go to the other end of the cylinder head, and record the gauge reading. Verify the reading by pushing the camshaft back to its original position and checking that the gauge indicates zero again. Repeat the check on the remaining camshaft. **Note:** *The hydraulic tappets must not be fitted whilst this measurement is being taken.*

12 Check that the camshaft endfloat measurement is within the limit listed in the Specifications. Wear outside of this limit is unlikely to be confined to any one component, so renewal of the camshaft, cylinder head and bearing ladder must be considered.

Refitting

13 Smear some clean engine oil onto the sides of the hydraulic tappets/rocker arms, and fit them into position in their bores in the cylinder head. Push them down until they contact the valves, then lubricate the camshaft lobe contact surfaces **(see illustration)**.

14 Refit the chain onto the camshaft sprockets, aligning the previously-made marks, and position the tensioner in the chain. Ensure the cylinder head bolt cut-outs in the camshafts face each other **(see illustration)**. Note that the side surfaces of the cut-outs should be exactly vertical.

15 Lubricate the running surfaces of the cylinder head and camshafts with clean

11.22 Refit the camshaft adjuster rear cover together with a new gasket

engine oil, then lay the camshafts in position in the cylinder head.

16 Apply a 2.0 mm wide bead of sealant (available from VAG dealers/parts specialists) to the grooves on the underside of the bearing ladder. As the sealant starts to harden immediately, no time should be allowed to elapse before fitting the bearing ladder.

17 Refit the bearing ladder to the cylinder head, then fit the new bolts and working from the inside-out tighten the bolts in several stages until the bearing ladder make contact the with cylinder head over the complete area of the mating surface. Now tighten the bolts to the specified torque.

18 Fit VAG special tool T10252 over the camshafts to check the alignment. It may be necessary to rotate the camshafts slightly to fit the tool. In the absence of the tool, have an assistant hold the exhaust camshaft stationary, and rotate the intake camshaft so the chain free play is equal on the top and lower run of the chain. The sides of the cut-outs in the camshafts must be vertical.

19 Fit new camshaft seal(s) as described in Section 8.

20 Refit the camshaft sprocket with reference to Section 5.

21 Remove the VAG tool (where applicable), then apply a little thread-locking compound and tighten the tensioner retaining bolts. Remove the bolt used to lock the tensioner.

22 Refit the camshaft adjuster rear cover with a new gasket, then apply a little thread-locking compound and tighten the retaining bolts to the specified torque **(see illustration)**.

23 Refit the camshaft cover (Section 7).

24 Refit the high-pressure fuel pump and the inlet manifold as described in Chapter 4A Section 10.

12 Hydraulic tappets – operational check

⚠ *Warning: After fitting hydraulic tappets, wait a minimum of 30 minutes (or preferably, leave overnight) before starting the engine, to allow the tappets time to settle, otherwise the valve heads will strike the pistons.*

1 The hydraulic tappets are self-adjusting, and require no attention whilst in service.

2 If the hydraulic tappets become excessively noisy, their operation can be checked as described below.

3 Run the engine until it reaches its normal operating temperature. Switch off the engine, then refer to Section 7 and remove the camshaft cover.

4 Rotate the camshaft by turning the crankshaft with a socket and wrench, until the first cam lobe over No 1 cylinder is pointing upwards.

5 Using a non-metallic tool, press the tappet downwards then use a feeler blade to check the free travel. If this is more than 0.2 mm before the valve starts to open, the tappet should be renewed.

6 Hydraulic tappet removal and refitting is described as part of the cylinder head overhaul sequence – see Chapter 2F Section 6 for details.

7 If hydraulic tappet noise occurs repeatedly when travelling short distances, renew the oil retention valve located in the rear of the oil filter mounting housing. It will be necessary to remove the oil filter, then unbolt the housing from the cylinder block and recover the gasket. Use a suitable key to unscrew the valve, and tighten the new valve securely. Refit the housing together with a new gasket.

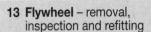

13 Flywheel – removal, inspection and refitting

Removal

1 On manual gearbox models, remove the transmission (see Chapter 7A Section 3) and clutch (Chapter 6 Section 6).

2 On semi-automatic (DSG) transmission models, remove the transmission as described in Chapter 7B Section 2.

3 All models are fitted with a dual-mass flywheel. Begin by making alignment marks between the flywheel and the crankshaft.

4 Rotate the outside of the dual-mass flywheel so that the bolts align with the holes **(see illustration)**.

5 Unscrew the bolts and remove the flywheel.

13.4 Align the bolts with the holes

Use a locking tool to counterhold the flywheel **(see illustration)**. Discard the bolts, new ones must be fitted. **Note:** *In order not to damage the flywheel, do not use a pneumatic or impact driver to unscrew the bolts ñ only use hand tools.*

Inspection

6 Check the flywheel for wear and damage. Examine the starter ring gear for excessive wear to the teeth; if evident, the flywheel must be renewed complete as the ring gear is not supplied separately.

7 The following are guidelines only, but should indicate whether professional inspection is necessary. There should be no cracks in the drive surface of the flywheel. If cracks are evident, the flywheel may need renewing. The dual-mass flywheel should be checked as follows.

8 Warpage: Place a straight-edge across the face of the drive surface, and check by trying to insert a feeler gauge between the straight-edge and the drive surface **(see illustration)**. The flywheel will normally warp like a bowl – ie, higher on the outer edge. If the warpage is more than 0.40 mm, the flywheel may need renewing.

9 Free rotational movement: This is the distance the drive surface of the flywheel can be turned independently of the flywheel primary element, using finger effort alone. Move the drive surface in one direction and make a mark where the locating pin aligns with the flywheel edge. Move the drive surface in the other direction (finger pressure only) and make another mark **(see illustration)**. The total

13.5 Using a locking tool to hold the flywheel

of free movement should not exceed 10.0 mm. If it's more, the flywheel may need renewing.

10 Total rotational movement: This is the total distance the drive surface can be turned independently of the flywheel primary element. Insert two bolts into the clutch pressure plate/damper unit mounting holes, and with the crankshaft/flywheel held stationary, use a lever/pry bar between the bolts and use some effort to move the drive surface fully in one direction – make a mark where the locating pin aligns with the flywheel edge. Now force the drive surface fully in the opposite direction, and make another mark. The total rotational movement should not exceed 44.00 mm. If it does, have the flywheel professionally inspected.

11 Lateral movement: The lateral movement (up and down) of the drive surface in relation to the primary element of the flywheel, should not exceed 2.0 mm. If it does, the flywheel may need renewing. This can be checked by pressing the drive surface down on one side into the flywheel (flywheel horizontal) and making an alignment mark between the drive surface and the inner edge of the primary element. Now press down on the opposite side of the drive surface, and make another mark above the original one. The difference between the two marks is the lateral movement **(see illustration)**.

Refitting

12 Refitting is a reversal of removal, but use new bolts and tighten them to the specified torque.

13.8 Check the dual mass flywheel for warpage

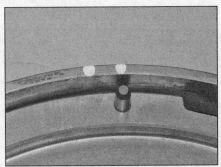

13.9 Check the free rotational movement of the dual mass flywheel

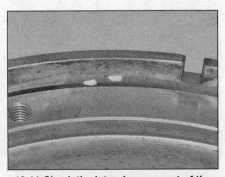

13.11 Check the lateral movement of the dual mass flywheel

14.16 Rear sump bolts

14.19 Position the sump in relation to the cylinder block

14 Sump – removal and refitting

Note: *Seat insist that on engine codes BLR, BVY, BVZ and BLY, the intermediate plate between the oil pump/balancer shaft housing and the engine block must be renewed when the sump is removed. Oil pump/balancer shaft renewal is described in Section 15.*

Removal

1 Apply the handbrake, then jack up the front of the vehicle and support it on axle stands (see *Jacking and vehicle support*).
2 Release the fasteners, remove the engine undertray, then undo the bolts and detach the undertray bracket.
3 Position a container beneath the sump, then unscrew the drain plug and drain the engine oil. Clean the plug and if necessary renew the washer, then refit and tighten the plug after all the oil has drained. Remove the dipstick from the engine.
4 Remove the auxiliary drivebelt as described in Chapter 1A Section 30.
5 Disconnect the oil temperature sensor wiring plug from the base of the sump.
6 Undo the bolt and detach the refrigerant support pipe from the sump.
7 Disconnect the wiring plug, then undo

the bolts and position the air conditioning compressor to one side (where applicable). Support the compressor by suspending it from the vehicle body with wire/string – there is no need to disconnect the refrigerant pipes, but ensure they are not damaged or kinked during the process.
8 Unbolt the stop bracket/torque arm bracket from the front of the engine.
9 On vehicles with gas discharge headlights, unclip the vehicle height sensor actuator rod from the front-left lower transverse link arm.
10 Undo the bolts and detach the inboard clamps from the front-anti roll bar.
11 Detach the starter motor cables from under the engine mounting by cutting the plastic cable ties and easing them from the plastic ducting.
12 Unscrew and remove the nuts from the bottom of each engine mounting.
13 Remove the engine top cover, then release the clips and remove the intake ducting form the air cleaner to the throttle body.
14 Connect a suitable hoist to the engine, then raise it as far as possible without damaging or stretching the coolant hoses and wiring.
15 Mark the fitted position of the subframe, then position a workshop trolley jack to take the weight, then undo the bolts and lower the front subframe.
16 Gradually unscrew and remove the sump bolts working in a diagonal pattern. Note that

on manual transmission models, the two rear sump bolts are accessed through a cut-out in the flywheel – turn the flywheel as necessary to align the cut-out **(see illustration)**.
17 Remove the sump. If it is stuck, tap it gently with a mallet to free it.

Refitting

18 Thoroughly clean the contact faces of the sump. It is recommended that a rotary wire brush is used to clean away the sealant.
19 Refit the sump and tighten the retaining bolts hand-tight initially. If the engine is out of the car, make sure that the rear edge of the sump overhangs the rear edge of the cylinder block by 0.8 mm, so that it is flush when the adapter plate is fitted **(see illustration)**. Progressively tighten the sump bolts in a diagonal pattern to the specified torque. It is advisable to wait at least 30 minutes for the intermediate plate sealant to set before filling the engine with oil.
20 The remaining refitting procedure is a reversal removal, but tighten the nuts and bolts to the specified torque where given in the Specifications. On completion, fill the engine with the correct quantity of oil as described in Chapter 1A Section 6.

15 Oil pump and balancer shaft housing – removal, inspection and refitting

Removal

1 Set the engine to TDC on No 1 cylinder as described in Section 2.
2 Remove the sump as described in Section 14.
3 Undo the 2 bolts securing the oil pipe at the rear of the engine block **(see illustration)**.
4 Using a small screwdriver, depress the tabs to release the retaining clips, and remove the chain/sprocket cover horizontally **(see illustrations)**.
5 Use a Torx bit to slacken the oil pump drive

15.3 Oil pipe on the rear of the block

15.4a Depress the tabs...

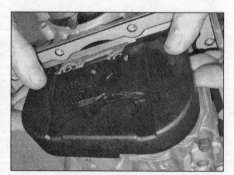

15.4b ...and remove the chain/sprocket cover

15.5 Loosen the oil pump drive sprocket retaining bolt

15.6 Lock the drive chain tensioner with a drill bit

15.9 Unbolt the drive chain tensioner

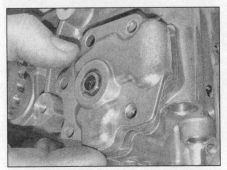

15.10 Remove the cover from the oil pump

15.14a Outer rotor alignment mark

15.14b Remove the inner rotor

sprocket retaining bolt approximately 1 turn **(see illustration)**.

6 Using a screwdriver, push the drive chain tensioner blade to relieve the tension on the chain. Insert a 3 mm drill bit into the hole in the tensioner assembly to lock the blade in this position **(see illustration)**

7 Remove the Torx bolt and pull the sprocket from the oil pump shaft. Disengage the sprockets from the chain.

8 Working from the outside-in, gradually and evenly slacken and remove the bolts securing the balance shaft housing and intermediate

plate. Note the fitted positions of the bolts – some are longer than others. Remove the balancer shaft housing and plate from the engine block, noting that the housing locates on dowels. Discard the retaining bolts, new ones must be fitted. Seat insist that the intermediate plate is renewed.

9 Undo the retaining bolts and remove the drive chain tensioner from the balance shaft housing **(see illustration)**.

10 If required, undo the 5 bolts and remove the oil pump cover. Pull the inner and outer rotors from the pump **(see illustration)**

Inspection

11 Examine the drive chain for wear and damage. To remove the chain, the timing belt must first be removed (see Section 4), then the crankshaft oil seal housing unbolted from the cylinder block.

12 It is not advisable to dismantle the balancer shafts and housing. No separate parts are available. If faulty, the complete assembly must be renewed.

13 If the oil pump has been dismantled, clean the components and check them for wear and damage. Examine the inner and outer rotors for scoring or any signs of wear/damage. If evident, renew the oil pump.

14 If the oil pump components are re-usable, fit the outer, and inner rotors to the pump body, with the marks on the ends of the inner rotor facing inwards and the one on the outer rotor facing outwards **(see illustrations)**. Refit the pump cover and tighten the retaining bolts to the specified torque.

Refitting

15 Prime the pump with oil by pouring oil into the suction pipe aperture while turning the driveshaft.

16 Refit the chain tensioner assembly to the balancer shaft housing, and tighten the retaining bolts to the specified torque. Ensure the tensioner blade is in the locked position, as described in paragraph 6.

17 Apply a bead of suitable sealant, approximately 2.0 mm thick, to the cylinder block side of the intermediate plate **(see illustration)**. Take great care not to apply the sealant too thickly, as any excess may find its way in the oil galleries.

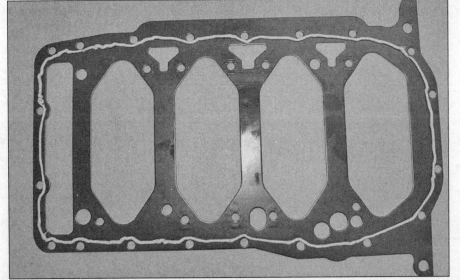

15.17 Apply sealant to the intermediate plate as shown

15.20 Lock the balancer shaft sprocket in its TDC position

15.21 Oil pump sprocket locating flat

18 Position the intermediate plate over the locating dowels on the cylinder block sealing surface. Feed the drive chain through the intermediate plate.

19 Position the balancer shaft housing on the base of the cylinder block/intermediate plate, then insert the new bolts (ensure the new O-ring is fitted to the appropriate bolt). Working from the inside tighten the new bolts to the Stage one torque setting, then in the same order, tighten them to the Stage two setting. Ensure the correct length bolt is fitted to the correct positions.

20 Ensure the crankshaft pulley is still in the TDC, then rotate the balancer shaft sprocket until the mark on the sprocket face is aligned with the locating hole, Insert an 5.0 mm drill bit into the hole to lock the sprocket in this position (**see illustration**).

21 Engage the drive chain with the balancer shaft sprocket, then fit the oil pump sprocket into the chain. Fit the sprocket on to the oil pump shaft, noting that the sprocket will only fit in one position – if necessary, rotate the oil pump shaft to enable the fitment of the sprocket (**see illustration**). Fit the new sprocket retaining bolt and tighten it to the specified torque.

22 Remove the balancer shaft sprocket locking tool (drill bit), and the Allen key locking the tensioner blade.

23 Refit the cover over the drive sprockets, and secure it with the retaining clips.

24 On non-turbocharged models, refit the oil pipe to the housing, and tighten the bolts securely.

25 Refit the sump with reference to Section 14.

16 Oil pressure switch –
removal and refitting

Removal

1 Jack up and support the front of the vehicle – *Jacking and vehicle support*. Remove the engine undershield.

2 Remove the charge air pipe on the left-hand side. The switch is fitted to the rear of the oil filter housing, close to the oil cooler.

3 Disconnect the wiring plug and using a deep reach 24mm socket (with a universal joint fitted) remove the switch.

Refitting

4 Refitting is a reversal of removal.

17 Engine mountings –
inspection and renewal

Inspection

1 If improved access is required, jack up the front of the vehicle, and support it securely on axle stands (see *Jacking and vehicle support*). Remove the engine top cover which also incorporates the air filter, then remove the engine undertray(s).

2 Check the mounting rubbers to see if they are cracked, hardened or separated from the metal at any point; renew the mounting if any such damage or deterioration is evident.

3 Check that all the mountings are securely tightened; use a torque wrench to check if possible.

4 Using a large screwdriver or a crowbar, check for wear in the mounting by carefully levering against it to check for free play.

Where this is not possible, enlist the aid of an assistant to move the engine/transmission back-and-forth, or from side-to-side, whilst you observe the mounting. While some free play is to be expected, even from new components, excessive wear should be obvious. If excessive free play is found, check first that the fasteners are correctly secured, then renew any worn components as described in the following paragraphs.

Renewal

Right-hand mounting

5 Remove the engine cover/air filter housing.

6 Attach a hoist and lifting tackle to the engine lifting brackets on the cylinder head, and raise the hoist to just take the weight of the engine. Alternatively the engine can be supported on a trolley jack under the engine. Use a block of wood between the sump and the head of the jack, to prevent any damage to the sump.

7 For improved access, unbolt the coolant reservoir and move it to one side. Disconnect the small hose from the top, but leave the rest of the hoses connected (**see illustration**).

8 Where applicable, move any wiring harnesses, pipes or hoses to one side to enable removal of the engine mounting.

9 Unclip the Evap canister and remove the mounting bracket (**see illustration**).

10 Unscrew the bolts securing the mounting

17.7 Move the reservoir to the side

17.9 Remove the bracket

17.10a Remove the small bracket...

17.10b ...and then remove the bolts

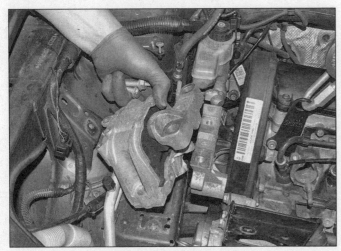

17.10c Lift out the mounting

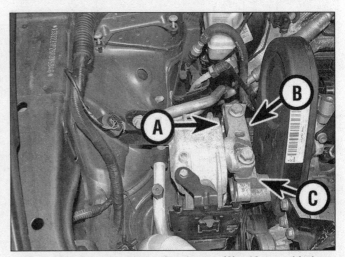

17.11 Align the mounting so that the gap (A) = 16 mm with the bracket (B) parallel to the lower bracket (C)

to the engine, then unscrew the bolts securing it to the body. Withdraw the mounting from the engine compartment **(see illustrations)**.

11 Refitting is a reversal of removal, bearing in mind the following points.

a) Use new securing bolts.

b) Align the mounting as shown **(see illustration)**.

c) Tighten all fixings to the specified torque.

Left-hand mounting

Note: *New mounting bolts will be required on refitting (there is no need to renew the smaller mounting-to-body bolts).*

12 Remove the engine top cover which also incorporates the air filter.

13 Attach a hoist and lifting tackle to the engine lifting brackets on the cylinder head, and raise the hoist to just take the weight of the engine and transmission. Alternatively the engine can be supported on a trolley jack under the transmission. Use a block of wood between the transmission and the head

of the jack, to prevent any damage to the transmission.

14 Remove the battery, as described in Chapter 5A Section 3, then disconnect the main starter motor feed cable from the positive battery terminal box.

15 Release any relevant wiring or hoses from

the clips on the battery tray, then unscrew the bolts and remove the battery tray **(see illustration)**.

16 Remove the wiring loom from the trunking and then remove the trunking **(see illustration)**.

17 Unscrew the bolts securing the mounting

17.15 Remove the battery tray

17.16 Remove the trunking

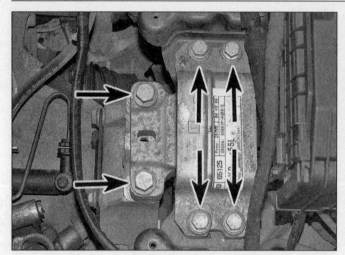

17.17a Remove the bolts

17.17b Lift out the mounting

to the transmission, and then the bolts securing the mounting to the body. Lift the mounting from the engine compartment **(see illustrations)**.

18 Refitting is a reversal of removal, bearing in mind the following points:

a) *Use new mounting bolts.*

b) *Tighten all fixings to the specified torque.*

Rear mounting (torque arm)

19 Apply the handbrake, then jack up the front of the vehicle and support securely on axle stands (see *Jacking and vehicle support*). Remove the engine undertray(s) for access to the rear mounting (torque arm).

20 Working under the vehicle, unscrew and remove the bolt securing the mounting to the subframe **(see illustration)**.

21 Unscrew the two bolts securing the mounting to the transmission, then withdraw the mounting from under the vehicle.

22 Refitting is a reversal of removal, but use new mounting securing bolts, and tighten all fixings to the specified torque.

17.20 The rear torque arm

Chapter 2 Part C
1.6 litre common rail (CR) diesel engine in-car repair procedures

Contents

Degrees of difficulty

Easy, suitable for novice with little experience	Fairly easy, suitable for beginner with some experience	Fairly difficult, suitable for competent DIY mechanic	Difficult, suitable for experienced DIY mechanic	Very difficult, suitable for expert DIY or professional

Specifications

General

Manufacturer's engine codes*:
1599 cc (1.6 litre), 16-valve, DOHC . . . . CAYC

Maximum outputs:	Power	Torque
Engine code:		
CAYC	77 kW @ 4400 rpm	250 Nm @ 1500 to 2500 rpm

Bore . . . . 79.5 mm
Stroke . . . . 80.5 mm
Compression ratio . . . . 16.5 : 1
Firing order. . . . . 1 – 3 – 4 – 2
No.1 cylinder location. . . . . Timing belt end

* **Note:** *See 'Vehicle identification' at the end of this manual for the location of engine code markings.*

Lubrication system

Oil pump type. . . . . Gear type, belt-driven from crankshaft
Oil pressure switch (green) . . . . 0.5 bar
Oil pressure (oil temperature 80°C):
Minimum @ idling . . . . 0.6 bar
Minimum @ 2000 rpm. . . . . 1.0 bar
Maximum @ high rpm. . . . . 5.0 bar

Torque wrench settings

	Nm	lbf ft
Ancillary (alternator, etc) bracket mounting bolts*:		
Stage 1 (all six bolts)	40	30
Stage 2 (for two lower bolts)	Angle-tighten a further 45°	
Stage 2 (for four upper bolts)	Angle-tighten a further 90°	
Auxiliary drivebelt tensioner securing bolt:		
Stage 1	20	15
Stage 2	Angle-tighten a further 180°	
Big-end bearing caps bolts*:		
Stage 1	30	22
Stage 2	Angle-tighten a further 90°	
Camshaft retaining frame bolts	10	7
Camshaft cover bolts	10	7
Camshaft sprocket hub centre bolt	100	74
Camshaft sprocket-to-hub bolts*:		
Stage 1	20	15
Stage 2	Angle-tighten a further 45°	
Common rail bolts	22	16
Coolant pump bolts	15	11
Crankshaft oil seal housing bolts	15	11
Crankshaft pulley-to-sprocket bolts*:		
Stage 1	10	7
Stage 2	Angle-tighten a further 90°	
Crankshaft sprocket bolt*:		
Stage 1	120	89
Stage 2	Angle-tighten a further 90°	
Cylinder head bolts*:		
Stage 1	35	26
Stage 2	60	44
Stage 3	Angle-tighten a further 90°	
Stage 4	Angle-tighten a further 90°	
Engine mountings:		
RH engine mounting*:		
Mounting bracket to engine:		
Stage 1	40	30
Stage 2	Angle-tighten a further 180°	
Mounting to body:		
Stage 1	40	30
Stage 2	Angle-tighten a further 90°	
Mounting to bracket:		
Stage 1	60	44
Stage 2	Angle-tighten a further 90°	
LH engine/transmission mounting*:		
Mounting to body:		
Stage 1	40	30
Stage 2	Angle-tighten a further 90°	
Mounting to bracket on transmission:		
Stage 1	60	44
Stage 2	Angle-tighten a further 90°	
Rear mounting link*:		
Link-to-transmission:		
Short (front) bolt:		
Stage 1	40	30
Stage 2	Angle-tighten a further 90°	
Long (rear) bolt:		
Stage 1	60	44
Stage 2	Angle-tighten a further 90°	
Link-to-subframe:		
Stage 1	100	74
Stage 2	Angle-tighten a further 90°	
Flywheel*:		
Stage 1	60	44
Stage 2	Angle-tighten a further 90°	
Fuel pump hub nut	95	70
Fuel pump sprocket bolts*	20	15
Main bearing cap bolts*:		
Stage 1	65	48
Stage 2	Angle-tighten a further 90°	

Torque wrench settings (continued)

	Nm	lbf ft
Oil cooler screws	11	7
Oil drain plug*	30	22
Oil filter housing-to-cylinder block bolts*:		
Stage 1	14	10
Stage 2	Angle-tighten a further 180°	
Oil filter cover	25	18
Oil level/temperature sensor-to-sump bolts	9	7
Oil pick-up pipe securing bolts	9	7
Oil pressure warning light switch	22	16
Oil pump securing bolts	16	11
Piston oil spray jet bolt	27 19	
Sump:		
Sump-to-cylinder block bolts	13	9
Sump-to-transmission bolts	40	30
Thermostat housing	15	11
Timing belt outer cover bolts	10	7
Timing belt tensioner roller securing nut:		
Stage 1	20	15
Stage 2	Angle-tighten a further 45°	
Timing belt idler pulleys:		
Lower idler roller nut	20	15
Upper idler roller (small) bolt	15	11
Upper idler roller (large) bolt*:		
Stage 1	50	37
Stage 2	Angle-tighten a further 90°	

*Do not re-use fasteners

1 General Information

How to use this Chapter

1 This Part of Chapter 2 describes those repair procedures that can reasonably be carried out on the engine while it remains in the vehicle. If the engine has been removed from the vehicle and is being dismantled as described in Part F, any preliminary dismantling procedures can be ignored.

2 Note that while it may be possible physically to overhaul certain items while the engine is in the vehicle, such tasks are not usually carried out as separate operations, and usually require the execution of several additional procedures (not to mention the cleaning of components and of oilways); for this reason, all such tasks are classed as major overhaul procedures, and are described in Part F of this Chapter.

Engine description

3 Throughout this Chapter, engines are referred to by type, and are identified and referred to by the manufacturer's code letters. A listing of all engines covered, together with their code letters, is given in the Specifications at the start of this Chapter.

4 The engines are water-cooled, double overhead camshafts (DOHC), in-line four-cylinder units, with cast-iron cylinder blocks and aluminium-silicone alloy cylinder heads. All are mounted transversely at the front of the vehicle, with the transmission bolted to the left-hand end of the engine.

5 The crankshaft is of five-bearing type, and thrustwashers are fitted to the centre main bearing (No.3) to control crankshaft endfloat.

6 Drive for the exhaust camshaft is by a toothed timing belt from the crankshaft, with the intake camshaft driven by interlocking gears at the left-hand end of both camshafts. The gears incorporate a toothed backlash compensator element. Each camshaft is mounted at the top of the cylinder head, and is secured by a bearing frame/ladder.

7 The valves are closed by coil springs, and run in guides pressed into the cylinder head. The valves are operated by roller rocker arms incorporating hydraulic tappets.

8 The gear-type oil pump is driven by a belt from the right-hand (timing belt) end of the crankshaft. Oil is drawn from the sump through a strainer, and then forced through an externally mounted, renewable filter. From there, it is distributed to the cylinder head, where it lubricates the camshaft journals and hydraulic tappets, and also to the crankcase, where it lubricates the main bearings, connecting rod big-ends, gudgeon pins and cylinder bores. A coolant-fed oil cooler is fitted to the oil filter housing on all engines. Oil jets are fitted to the base of each cylinder – these spray oil onto the underside of the pistons, to improve cooling.

9 All engines are fitted with a combined brake servo vacuum pump, driven by the camshaft on the transmission end of the cylinder head.

10 On all engines, engine coolant is circulated by a pump, driven by the timing belt. For details of the cooling system, refer to Chapter 3.

Operations with engine in car

11 The following operations can be performed without removing the engine:

a) Compression pressure – testing.
b) Camshaft cover – removal and refitting.
c) Crankshaft pulley – removal and refitting.
d) Timing belt covers – removal and refitting.
e) Timing belt – removal, refitting and adjustment.
f) Timing belt tensioner and sprockets – removal and refitting.
g) Camshaft oil seals – renewal.
h) Camshafts and hydraulic tappets – removal, inspection and refitting.
i) Cylinder head – removal and refitting.
j) Cylinder head and pistons – decarbonising.
k) Sump – removal and refitting.
l) Oil pump – removal, overhaul and refitting.
m) Crankshaft oil seals – renewal.
n) Engine/transmission mountings – inspection and renewal.
o) Flywheel/driveplate – removal, inspection and refitting.

Note: *It is possible to remove the pistons and connecting rods (after removing the cylinder head and sump) without removing the engine. However, this is not recommended. Work of this nature is more easily and thoroughly completed with the engine on the bench, as described in Chapter 2F.*

2 Compression and leakdown tests – description and interpretation

Compression test

Note: *A compression tester suitable for use with diesel engines will be required for this test.*

1 When engine performance is down, or if misfiring occurs which cannot be attributed to the ignition or fuel systems, a compression test can provide diagnostic clues as to the engine's condition. If the test is performed regularly, it can give warning of trouble before any other symptoms become apparent.

2 The engine must be fully warmed-up to normal operating temperature, the battery must be fully charged, and you will require the aid of an assistant.

3 Remove the glow plugs as described in Chapter 5C Section 2, and then fit a compression tester to the No.1 cylinder glow plug hole. The type of tester that screws into the plug thread is preferred. **Note:** *Part of the glow plug removal procedure is to disconnect the fuel injector wiring plugs. As a result of the plugs being disconnected and the engine cranked, faults may be stored in the ECU memory. These must be erased after the compression test.*

4 Have your assistant crank the engine for several seconds on the starter motor. After one or two revolutions, the compression pressure should build-up to a maximum figure and then stabilise. Record the highest reading obtained.

5 Repeat the test on the remaining cylinders, recording the pressure in each.

6 The cause of poor compression is less easy to establish on a diesel engine than on a petrol engine. The effect of introducing oil into the cylinders (wet testing) is not conclusive, because there is a risk that the oil will sit in the recess on the piston crown, instead of passing to the rings. However, the following can be used as a rough guide to diagnosis.

7 All cylinders should produce very similar pressures. Any difference greater than that specified indicates the existence of a fault. Note that the compression should build-up quickly in a healthy engine. Low compression on the first stroke, followed by gradually increasing pressure on successive strokes, indicates worn piston rings. A low compression reading on the first stroke, which does not build-up during successive strokes, indicates leaking valves or a blown head gasket (a cracked head could also be the cause).

8 A low reading from two adjacent cylinders is almost certainly due to the head gasket having blown between them and the presence of coolant in the engine oil will confirm this.

9 On completion, remove the compression tester, and refit the glow plugs, with reference to Chapter 5C Section 2.

10 Reconnect the wiring to the injector solenoids. Finally, have a Seat dealer or suitably equipped specialist erase any fault codes from the ECU memory.

Leakdown test

11 A leakdown test measures the rate at which compressed air fed into the cylinder is lost. It is an alternative to a compression test, and in many ways it is better, since the escaping air provides easy identification of where pressure loss is occurring (piston rings, valves or head gasket).

12 The equipment required for leakdown testing is unlikely to be available to the home mechanic. If poor compression is suspected, have the test performed by a suitably equipped garage.

3 Engine assembly and valve timing marks – general information and usage

General information

1 TDC is the highest point in the cylinder that each piston reaches as it travels up-and-down when the crankshaft turns. Each piston reaches TDC at the end of the compression stroke and again at the end of the exhaust stroke, but TDC generally refers to piston position on the compression stroke. No 1 piston is at the timing belt end of the engine.

2 Positioning No 1 piston at TDC is an essential part of many procedures, such as timing belt removal and camshaft removal.

3 The design of the engines covered in this Chapter is such that piston-to-valve contact may occur if the camshaft or crankshaft is turned with the timing belt removed. For this reason, it is important to ensure that the camshaft and crankshaft do not move in relation to each other once the timing belt has been removed from the engine.

Setting TDC on No 1 cylinder

Note: *VAG special tool T10050 is required to lock the crankshaft sprocket in the TDC position. An aftermarket version is widely available.*

4 Raise the front of the vehicle and support it securely on axle stands (see *Jacking and vehicle support*). Remove the front right-hand road wheel, then release the fasteners and remove the lower section of the wheelarch liner.

5 Remove the auxiliary drivebelt as described in Chapter 1B Section 30

6 Remove the crankshaft pulley/vibration damper as described in Section 5.

7 Remove the timing belt outer covers as described in Section 6.

8 Using a spanner or socket on the crankshaft sprocket bolt, turn the crankshaft in the normal direction of rotation (clockwise) until the alignment mark on the face of the sprocket is almost vertical, and the hole in the camshaft sprocket hub aligns with the hole in the cylinder head **(see illustration)**.

9 While in this position it should be possible to insert the VAG tool T10050 to lock the crankshaft, and a 6 mm diameter rod/drill bit to lock the camshafts **(see illustrations)**.

3.8 The alignment mark on the crankshaft sprocket should be almost vertical

3.9a Fit the tool to the hole in the oil seal housing…

3.9b …so the marks on the tool and sprocket align

3.9c Insert a 6 mm locking tool into the camshaft hub …

3.9d …with the arrow almost at the 12 o'clock position

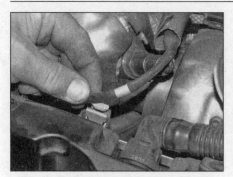

4.3 Unclip the wiring from the retaining clips

4.4 Squeeze together the sides of the collar to disconnect the breather hose

4.5 Undo the bolts and lift away the camshaft cover

Note: *The mark on the crankshaft sprocket and the mark on the VAG tool must align, whilst at the same time the shaft of tool must engage in the drilling in the crankshaft oil seal housing.*

10 The engine is now set to TDC on No.1 cylinder.

4 Camshaft cover – removal and refitting

Removal

1 Remove the fuel injectors (Chapter 4B Section 5) and fuel rail (Chapter 4B Section 4).
2 Remove the timing belt upper cover as described in Section 6.
3 Note their fitted positions, then disconnect the vacuum hoses from the camshaft cover, and release them, and the wiring loom from the retaining clips at the rear of the cover **(see illustration)**.
4 Squeeze together the sides of the collar, and disconnect the breather hose from the camshaft cover **(see illustration)**.
5 Release the wiring from the clips at the rear of the cover, then unscrew the camshaft cover retaining bolts and lift the cover away. If the cover sticks, do not attempt to lever it off – instead free it by working around the cover and tapping it lightly with a soft-faced mallet **(see illustration)**.

6 Recover the camshaft cover gasket. Inspect the gasket carefully, and renew it if damage or deterioration is evident – note that the retaining bolts and seals must be pushed fully through the cover **(see illustrations)**.
7 Clean the mating surfaces of the cylinder head and camshaft cover thoroughly, removing all traces of oil – take care to avoid damaging the surfaces as you do this.

Refitting

8 Refit the camshaft cover by following the removal procedure in reverse, tightening the cover retaining bolts to the specified torque, starting with the centre bolts and working outwards.

5 Crankshaft pulley – removal and refitting

Removal

1 Switch off the ignition and all electrical consumers and remove the ignition key.
2 Raise the front right-hand side of the vehicle, and support securely on axle stands (see *Jacking and vehicle support*). Remove the roadwheel.
3 Remove the securing fasteners and withdraw the lower section of the front wheel arch liner.
4 Slacken the bolts securing the crankshaft

pulley to the sprocket **(see illustration)**. If necessary, the pulley can be prevented from turning by counterholding with a spanner or socket on the crankshaft sprocket bolt.
5 Remove the auxiliary drivebelt, as described in Chapter 1B Section 30.
6 Unscrew the bolts securing the pulley to the sprocket, and remove the pulley. Discard the bolts – new ones must be fitted.

Refitting

7 Refit the pulley over the locating peg on the crankshaft sprocket, then fit the new pulley securing bolts.
8 Refit and tension the auxiliary drivebelt as described in Chapter 1B Section 30.
9 Prevent the crankshaft from turning as during removal, then fit the pulley securing bolts, and tighten to the specified torque.
10 Refit the wheel arch liner.
11 Refit the roadwheel and lower the vehicle to the ground.

6 Timing belt covers – removal and refitting

Upper outer cover

1 Pull the engine top cover upwards to release it from the mountings.
2 Release the hoses from the retaining clips

4.6a Renew the cover seal if necessary

4.6b Bolts and seals must be pushed fully through the cover before fitting the gasket

5.4 Undo the pulley bolts, counterholding it with a socket on the centre sprocket bolt

6.2 Unclip the hoses from the retaining clips

6.3a Disconnect the wiring connector …

6.3b …undo the two retaining bolts …

6.3c … and withdraw the mounting bracket

6.4 Remove the washer reservoir filler neck

6.5a Undo the filter retaining bolts/nut…

6.5b … and release the fuel pipe retaining clip

on the right-hand side of the cylinder head and move them to one side **(see illustration)**.
3 Disconnect the wiring plug, undo the retaining bolts, and remove the pressure differential sender (for the particulate filter) including bracket from the top of the engine mounting and move it to one side **(see illustrations)**.
4 Undo the retaining bolt, then twist the washer reservoir filler neck to remove it from engine compartment **(see illustration)**.

5 Undo the retaining bolts and nut, then remove the fuel filter from the engine mounting and move it to one side **(see illustrations)**. The fuel hoses do not need to be disconnected from the filter.
6 To make access easier, undo the retaining bolt, disconnect the wiring connector and move the coolant reservoir to one side **(see illustration)**.
7 Disconnect the radiator outlet temperature sensor wiring plug **(see illustration)**.

6.6 Move the coolant reservoir to one side

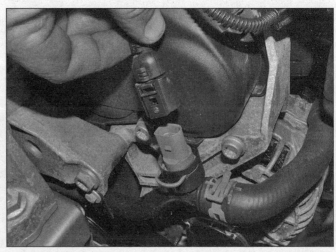

6.7 Disconnect the temperature sensor wiring connector

6.8a Release the rear lower clip...

6.8b ... the two upper front clips...

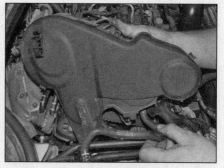

6.8c ...and manoeuvre the upper cover from place

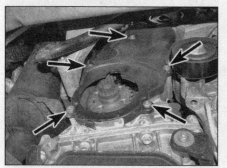

6.12a Undo the lower cover retaining bolts

6.12b ...and manoeuvre the lower cover from place

8 Release the 3 clips and remove the timing belt upper cover **(see illustrations)**.
9 Refitting is a reversal of removal, noting that the lower edge of the upper cover engages with the lower cover.

Lower outer cover

10 Remove the upper cover as described previously.
11 If not already done, remove the crankshaft pulley as described in Section 5.
12 Unscrew the five bolts securing the lower cover, and remove it **(see illustrations)**.
13 Refitting is a reversal of removal; noting that the upper edge of the lower cover engages with the upper cover.

Rear cover

14 Remove the timing belt, tensioner and sprockets as described in Section 7 and Section 8.

6.16a Remove the retaining bolts...

15 Undo the retaining bolt and remove the rear hub cover from the end of the camshaft.
16 Slacken and withdraw the retaining bolts and lift the timing belt inner cover from the studs on the end of the engine, and remove it from the engine compartment **(see illustrations)**. It may be required to remove the coolant pump (Chapter 3 Section 7), before the rear cover can be removed.
17 Refitting is a reversal of removal.

| 7 | Timing belt – removal, inspection and refitting | 🔧 |

Note: It is consider best practise to always replace the tensioner, idler and coolant pump whenever the timing belt is replaced. Any mounting studs should also be replaced at the same time. Most of the components

6.16b ...and remove the rear cover

(apart from the coolant pump) will be supplied with a timing belt kit. Note also that most manufactures will not guarantee a belt against premature failure if the idler(s) and tensioner are not replaced with the belt.
Note: *There are two types of tensioner fitted to this engine(see illustration). Check to see which type is fitted before removing the timing belt. Type A tensioner requires a locking pin for installation and it tensions the belt by rotating clockwise. Type B tensioner does not require a locking pin for installation and it tensions the belt by rotating anti-clockwise (see illustration). See text.*

Removal

1 The primary function of the toothed timing belt is to drive the camshaft, but it also drives the coolant pump and high-pressure fuel pump. Should the belt slip or break in service, the valve timing will be disturbed and piston-to-valve contact may occur, resulting

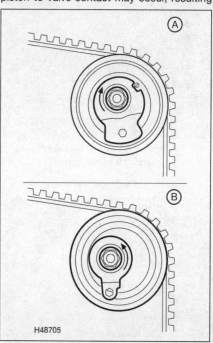

H48705

7.0 Two types of tensioner type (A) and (B)

A Roller tensions clockwise
B Roller tensions anti-clockwise

7.4 Slacken the sprocket-to-hub bolts

7.5 Slacken the high-pressure fuel pump sprocket bolts

7.6 Insert an Allen key, slacken the nut, and rotate the hub anti-clockwise until a 2 mm rod/drill bit can be inserted to lock the hub to the pulley

in serious engine damage. For this reason, it is important that the timing belt is tensioned correctly, and inspected regularly for signs of wear or deterioration.

2 Switch off the ignition and all electrical consumers and remove the ignition key.

3 Set the engine to TDC on No. 1 cylinder as described in Section 3.

4 Slacken the three bolts securing the sprocket to the camshaft hub by 90° **(see illustration)**.

5 Slacken the three bolts securing the sprocket to the high-pressure fuel pump by 90° **(see illustration)**.

Type A tensioner

6 Insert a suitable Allen key into the tensioner hub, then slacken the retaining nut and rotate the tensioner hub anti-clockwise until it can

be locked in place using a 2.0 mm pin/drill bit **(see illustration)**.

7 Leaving the pin in place, now rotate the tensioner hub clockwise to the stop, and hand-tighten the retaining nut **(see illustration)**.

Type B tensioner

8 Insert a suitable Allen key into the tensioner hub, then slacken the retaining nut and rotate the tensioner hub clockwise, until the tensioner hub is loosened. When in position hand-tighten the retaining nut.

All engines

9 If the original timing belt is to be refitted, mark the running direction of the belt, to ensure correct refitting.
Caution: If the belt appears to be in good condition and can be re-used, it is essential

that it is refitted the same way around, otherwise accelerated wear will result, leading to premature failure.

10 Slide the belt from the sprockets, taking care not to twist or kink the belt excessively if it is to be re-used.

Inspection

11 Examine the belt for evidence of contamination by coolant or lubricant. If this is the case, find the source of the contamination before progressing any further. Check the belt for signs of wear or damage, particularly around the leading edges of the belt teeth. Renew the belt if its condition is in doubt; the cost of belt renewal is negligible compared with potential cost of the engine repairs, should the belt fail in service. The belt must be renewed if it has covered the mileage given in Chapter 1B Section 31, however, if it has covered less, it is prudent to renew it regardless of condition, as a precautionary measure.

12 If the timing belt is not going to be refitted for some time, it is a wise precaution to hang a warning label on the steering wheel, to remind yourself (and others) not to attempt to start the engine. Have the battery disconnected to prevent any engine damage.

13 If the tensioner roller is to be renewed, the engine mounting will also need to be removed, as described in Section 16. Then the tensioner removed as described in Section 8.

Refitting

14 Ensure that the crankshaft and camshaft are still set to TDC on No 1 cylinder, as described in Section 3. The camshaft sprocket bolts should be renewed, and slackened at this point.

15 Renew the high-pressure fuel pump sprocket bolts one at a time. They should also be slackened.

16 Using a screwdriver on the bolts heads, rotate the high-pressure fuel pump clockwise until a 6.0 mm locking pin/drill bit can be inserted into the housing adjacent to the sprocket, locking the pump in place **(see illustration)**.

17 Rotate the camshaft sprocket and high-pressure fuel pump sprocket fully clockwise so that the securing bolts are at the end of the elongated holes **(see illustrations)**.

7.7 Rotate the tensioner hub clockwise until it hits the stop

7.16 Rotate the high-pressure fuel pump clockwise until a 6 mm drill bit/rod can be inserted into the housing and hub

7.17a Rotate the sprockets fully clockwise until the fuel pump sprocket...

7.17b ...and camshaft sprocket bolts are at the end of the elongated holes

7.20 Timing belt routing

7.21a Rotate the tensioner clockwise ...

18 Loop the timing belt loosely under the crankshaft sprocket. **Note:** *Observe any direction of rotation markings on the belt.*

19 Fit the belt around the tensioner pulley, engage the timing belt teeth with the camshaft sprockets, then manoeuvre it into position around the coolant pump sprocket and the fuel pump sprocket. Make sure that the belt teeth seat correctly on the sprockets. **Note:** *Slight adjustment to the position of the camshaft sprocket may be necessary to achieve this. Avoid bending the belt back on itself or twisting it excessively as you do this.*

20 Finally, fit the belt around the idler roller **(see illustration)**. Ensure that any slack in the belt is in the section of belt that passes over the tensioner roller.

Type A tensioner

21 Loosen the timing belt tensioner securing nut, and pull out the tensioner locking pin. Turn the tensioner clockwise with an Allen key until the pointer is just past the middle of the gap in the tensioner base plate **(see illustrations)**. With the tensioner held in this position, tighten the securing nut to the specified torque and angle.

Type B tensioner

22 Loosen the timing belt tensioner securing nut, and turn the tensioner anti-clockwise with an Allen key until the pointer is just past the middle of the gap in the tensioner base plate **(see illustration)**. With the tensioner held in this position, tighten the securing nut to the specified torque and angle.

All engines

23 Counterhold the camshaft sprocket and fuel pump sprocket with a home made tool to prevent any rotation, and then tighten the camshaft sprocket and fuel pump sprocket bolts to 20 Nm **(see illustrations)**.

24 Remove the sprockets' locking tools and the crankshaft locking tool **(see illustrations)**.

25 Using a spanner or wrench and socket on the crankshaft pulley centre bolt, rotate the crankshaft clockwise through two complete

7.21b ...until the pointer is just past the gap in the base plate

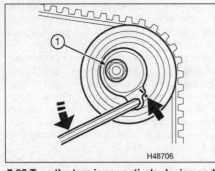

7.22 Turn the tensioner anti-clockwise and secure the retaining nut (1)

7.23a Counterhold the camshaft sprocket ...

7.23b ... and the pump sprocket, while the bolts are tightened

7.24a Remove the upper locking pins ...

7.24b ... and crankshaft locking tool

7.26 Check the pointer is centred or within 5mm to the right of the gap in the base plate

7.27 Slight misalignment of the pump sprocket timing hole is acceptable

8.4 Coolant pipe upper mounting bolt and lower mounting nut

revolutions. Reset the engine to TDC on No.1 cylinder, with reference to Section 3 and refit the crankshaft locking tool.

26 Check that the tensioner roller indicator arm is centred, or within a maximum of 5 mm to the right of the notch in the base plate **(see illustration)**. If not, hold the tensioner hub stationary with an Allen key, slacken the retaining nut and position the arm in the centre of the notch. Tighten the retaining nut to the specified torque. Remove the Allen key.

27 Check that the camshaft sprocket locking pin can still be inserted. **Note:** *It's very difficult to align the locking point of the fuel pump hub again* **(see illustration)**. *However, a slight misalignment of holes will not affect engine performance.*

28 If the camshaft sprocket locking pin cannot be inserted, pull the crankshaft locking tool slight away from the engine, and rotate the crankshaft anti-clockwise slightly past TDC. Now slowly rotate the crankshaft clockwise until the camshaft sprocket locking tool can be inserted.

29 If the locating pin of the crankshaft locking tool is to the left of the corresponding hole, slacken the camshaft sprocket bolts, slowly rotate the crankshaft clockwise until the locking tool can be fully inserted. Tighten the camshaft sprocket bolts to 20 Nm.

30 If the locating pin of the crankshaft locking tool is to the right of the corresponding hole, slacken the camshaft sprocket bolts, rotate the crankshaft anti-clockwise slightly until the pin is to the left of the hole, then slowly rotate it clockwise until the lock tool can be fully

inserted. Tighten the camshaft sprocket bolts to 20 Nm.

31 Remove the crankshaft and camshaft locking tools, then rotate the crankshaft 2 complete revolutions clockwise and check the locking tools can be reinserted. If necessary, repeat the adjustment procedure described previously.

32 Tighten the camshaft and fuel pump sprocket bolts to the specified torque.

33 The remainder of refitting is a reversal of removal.

<table>
<tr><td>8</td><td>**Timing belt tensioner and sprockets** – removal and refitting</td></tr>
</table>

Timing belt tensioner

Removal

1 In order to remove the timing belt tensioner, the engine mounting bracket must first be removed. Either support the engine from above using a crossbeam or an engine hoist or support if from underneath with a trolley jack and block of wood.

2 Remove the timing belt as described in Section 7.

3 Undo the bolts and remove the right-hand engine mounting.

4 Undo the bolt securing the coolant pipe to the mounting bracket **(see illustration)**.

5 Working in the wheelarch area, undo the nut securing the lower end of the coolant pipe.

6 Undo the 3 retaining bolts and remove the engine mounting bracket **(see illustration)**.

7 Unscrew the timing belt tensioner nut, and remove the tensioner from the engine.

Refitting

8 When refitting the tensioner to the engine, ensure that the lug on the tensioner backplate engages with the corresponding cut-out in the rear timing belt cover, then refit the tensioner nut **(see illustration)**.

9 The remainder of refitting is a reversal of removal.

Idler pulleys

Removal

10 Remove the timing belt as described in Section 7.

11 Unscrew the relevant idler pulley/roller securing bolt/nut, and then withdraw the pulley **(see illustration)**.

Refitting

12 Refit the pulley and tighten the securing bolt or nut to the specified torque. **Note:** *Renew the large roller/pulley retaining bolt (where applicable).*

13 Refit and tension the timing belt as described in Section 7.

Crankshaft sprocket

Note: *A new crankshaft sprocket securing bolt must be used on refitting.*

Removal

14 Remove the timing belt as described in Section 7.

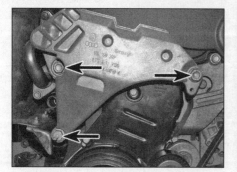

8.6 Engine mounting bracket bolts

8.8 Ensure the lug on the backplate engages with the cut-out in the timing belt cover

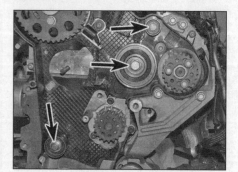

8.11 Timing belt idler pulleys

8.17 Using a puller to remove the crankshaft sprocket

8.27 Fabricate a home made tool to counterhold the hub. Undo the bolt...

8.28 ...and slide the hub from the camshaft

15 The sprocket securing bolt must now be slackened, and the crankshaft must be prevented from turning as the sprocket bolt is unscrewed. To hold the sprocket, make up a suitable tool, and screw it to the sprocket using a two bolts screwed into two of the crankshaft pulley bolt holes.

16 Hold the sprocket using the tool, then slacken the sprocket securing bolt. Take care, as the bolt is very tight. Do not allow the crankshaft to turn as the bolt is slackened.

17 Unscrew the bolt, and slide the sprocket from the end of the crankshaft, noting which way round the sprocket's raised boss is fitted. If required, use a puller to withdraw the sprocket from the end of the crankshaft **(see illustration)**.

Refitting

18 Commence refitting by positioning the sprocket on the end of the crankshaft.

19 Fit a new sprocket securing bolt, then counterhold the sprocket using the method employed on removal, and tighten the bolt to the specified torque in the two stages given in the Specifications.

20 Refit the timing belt as described in Section 7.

Camshaft sprocket

Removal

21 Remove the timing belt as described in Section 7, then rotate the crankshaft 90° anti-clockwise to prevent any accidental piston-to-valve contact.

22 Unscrew and remove the three retaining

bolts and remove the camshaft sprocket from the camshaft hub.

Refitting

23 Refit the sprocket ensuring that it is fitted the correct way round, as noted before removal, then insert the new sprocket bolts, and tighten by hand only at this stage.

24 If the crankshaft has been turned, turn the crankshaft clockwise 90° back to TDC.

25 Refit and tension the timing belt as described in Section 7.

Camshaft hub

Note: *Seat technicians use special tool T10051 to counterhold the hub, however it is possible to fabricate a suitable alternative.*

Removal

26 Remove the camshaft sprocket as described previously in this Section.

27 Engage special tool T10051 with the three locating holes in the face of the hub to prevent the hub from turning. If this tool is not available, fabricate a suitable alternative. Whilst holding the tool, undo the central hub retaining bolt about two turns **(see illustration)**.

28 Slide the hub from the camshaft. If necessary, attach VAG tool T10052 (or a similar three-legged puller) to the hub, and evenly tighten the puller until the hub is free of the camshaft taper **(see illustration)**.

Refitting

29 Ensure that the camshaft taper and the hub centre is clean and dry, locate the hub

on the taper, noting that the built-in key in the hub taper must align with the keyway in the camshaft taper **(see illustration)**.

30 Hold the hub in this position with tool T10051 (or similar home-made tool), and tighten the central bolt to the specified torque.

31 Refit the camshaft sprocket as described previously in this Section.

Coolant pump sprocket

32 The coolant pump sprocket is integral with the coolant pump. Refer to Chapter 3 Section 7 for details of coolant pump removal.

9 Camshaft and hydraulic tappets – removal, inspection and refitting

Note: *A new camshaft oil seal(s) will be required on refitting. VAG removal tool T40094 (or similar tool) will be required to refit the camshafts – this is necessary to prevent damage to the retaining frame and cylinder head as the camshafts are refitted.*

Removal

1 With the timing set at TDC, remove the camshaft hub (see Section 8).

2 Remove the camshaft cover (see Section 4).

3 Remove the brake vacuum pump as described in Chapter 9 Section 21.

4 If the camshafts are not going to be renewed, use two flat pieces of metal flat bar and cable ties and secure the two camshafts to the retaining frame **(see illustrations)**.

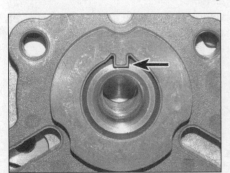

8.29 Ensure the integral key aligns with the keyway in the camshaft

9.4a Using flat metal bar and cable ties to secure the camshafts ...

9.4b ... to the upper ladder frame for removal

9.7 Make alignment marks for the position of the sprockets

9.10 Check the tappets for wear

9.12 VAG special tool no. T40094 shown

Progressively unscrew the camshaft retaining frame bolts in the reverse of the sequence shown in illustration 9.27, and carefully remove the retaining frame, complete with camshafts. Remove the oil seal from the end of the camshaft and discard it – a new one will be required for refitting.

5 If the camshafts are going to be renewed, progressively unscrew the camshaft retaining frame bolts in the reverse of the sequence shown in illustration 9.27, and carefully remove the retaining frame. Then carefully lift the camshafts from the cylinder head, keeping them identified for location. Remove the oil seal from the end of the camshaft and discard it – a new one will be required for refitting.

6 Lift the rocker arms and hydraulic tappets

from place. Store the rockers and tappets in a container with numbered compartments to ensure they are refitted to their correct locations. It is recommended that the tappets are kept immersed in oil for the period they are removed from the cylinder head.

Inspection

7 If the camshafts are still secured to the retaining frame, mark the two sprockets in relation with each other, then cut the cable ties to remove them from the frame **(see illustration)**. With the camshafts removed, examine the retaining frame and the bearing locations in the cylinder head for signs of obvious wear or pitting. If evident, a new cylinder head will probably be required. Also

check that the oil supply holes in the cylinder head are free from obstructions.

8 Visually inspect the camshafts for evidence of wear on the surfaces of the lobes and journals. Normally their surfaces should be smooth and have a dull shine; look for scoring, erosion or pitting and areas that appear highly polished, indicating excessive wear. Accelerated wear will occur once the hardened exterior of the camshaft has been damaged, so always renew worn items. **Note:** *If these symptoms are visible on the tips of the camshaft lobes, check the corresponding rocker arm, as it will probably be worn as well.*

9 If the machined surfaces of the camshaft appear discoloured or blued, it is likely that it has been overheated at some point, probably due to inadequate lubrication. This may have distorted the shaft, so have the camshaft runout and endfloat checked by an automotive engine reconditioning specialist.

10 Inspect the hydraulic tappets for obvious signs of wear or damage **(see illustration)**, and renew if necessary. Check that the oil holes in the tappets are free from obstructions.

Refitting

11 Oil the rocker arms and hydraulic tappets, and then refit them to their original positions.

⚠ *Warning: After fitting hydraulic tappets, wait a minimum of 30 minutes (or preferably, leave overnight) before starting the engine; to allow the tappets time to settle, otherwise the valve heads will strike the pistons.*

Using VAG special tool No.T40094

12 If the camshafts were removed from the retaining frame, use the VW special tool No.T40094, as shown in the following procedures **(see illustration)**.

13 Position the inlet camshaft as shown with the cylinder head bolt indents facing outwards, then slide the end support into the slot in the end of the camshaft **(see illustrations)**.

14 Position the exhaust camshaft on the supports, again with the cylinder head bolt indents facing outwards, and then fit the locating clamp into the slot in the end of the camshaft **(see illustrations)**.

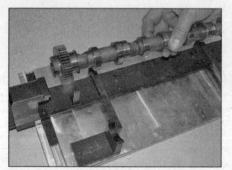

9.13a Position the inlet camshaft on the tool with the head bolt indent facing outwards …

9.13b …then slide the end of the special tool, in to the slot in the end of the camshaft

9.14a Position the exhaust camshaft on the tool with the head bolt indent facing outwards …

9.14b … then fit special tool clamp into the slot in the end of the exhaust camshaft

9.15 Tighten the thumbwheel to align the gear teeth. Ensure the clamping jaw with the arrow is seated on the wider gear

9.16 Slide the exhaust camshaft to the inlet camshaft until the teeth are in mesh

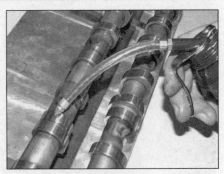

9.17a Lubricate the bearing mountings …

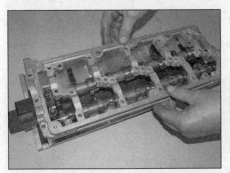

9.17b …and fit the upper ladder frame to the camshafts

9.18 Secure the camshafts in place in the frame using tool No. T40095

9.19a Slide the locking tool out from the end of the inlet camshaft …

15 Fit the VAG clamping tool No.T40096 to the double gear on the exhaust camshaft, tightening the knurled thumb wheel until the faces of the gear teeth are in alignment **(see illustration)**. Note some camshafts only have a single sprocket, so will not require this procedure.

16 Slide the exhaust camshaft towards the inlet camshaft until the gear teeth engage **(see illustration)**.

17 Ensure the gasket faces of the retaining frame are clean, then apply a smear of clean engine oil to the bearing surfaces and lower the frame into position over the camshafts **(see illustrations)**. Ensure the bearing surfaces locate correctly on the camshafts.

18 Fit the clamping tool No.T40095 over the camshafts and frame, and tighten the thumbwheels to hold the camshafts in position in the frame **(see illustration)**.

19 Slide out locking clamps from each end of the camshafts **(see illustrations)**, and then lift the camshafts, retaining frame and clamping tool from the VW special tool no. 40094.

Without VAG special tool

20 Ensure the gasket faces of the retaining frame are clean, then apply a smear of clean engine oil to the bearing surfaces and turn it upside down on a clean surface.

21 Position the camshafts on the retaining frame with the cylinder head bolt indents facing outwards. Align the marks made on

the camshaft sprockets on removal **(see illustration 9.7)**.

22 If a double gear is fitted to the exhaust camshaft, fit the VAG clamping tool No.T40096 (or similar) to align the faces of the teeth. If the VW clamping tool is not available use two flat bladed screwdrivers and a pair of pliers, this will need an assistant to keep firm pressure on the camshafts to make sure they locate in the retaining frame correctly **(see illustrations)**. Note some camshafts only have a single sprocket, so will not require this procedure.

23 With the marks on the sprockets aligned and the cylinder head bolt indents facing outwards, the slot in the end of the inlet camshaft (that drives the vacuum pump)

9.19b … and remove the locking clamp from the end of the exhaust camshaft

9.22a VAG special tool for aligning teeth on sprocket

9.22b Using two flat bladed screwdrivers and a pair of pliers to align teeth

9.23a Inlet camshaft slot needs to be horizontal

9.23b Slot in end of exhaust camshaft should be at the 12 o'clock position

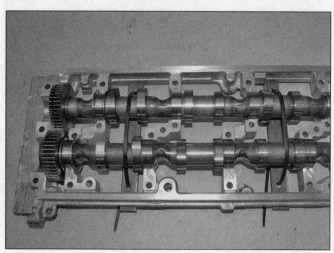

9.24 Camshafts are secured to the ladder frame with two pieces of flat bar and cable ties

9.25a Apply a 2.0 mm thick bead of sealant ...

should be horizontal and the slot in the end of the exhaust camshaft (that locates the timing belt sprocket) should be at the top when fitted **(see illustrations)**.

24 With the camshafts correctly in position, ensuring the camshafts locate correctly in the bearing surfaces on the retaining frame, use two flat pieces of metal flat bar and cable ties to secure them in position **(see illustration)**.

Caution: Make sure the camshafts are located correctly in the bearing surfaces on the retaining frame, otherwise damage can occur, and this could damage the retaining frame when tightening down onto the cylinder head. The retaining frame is matched to the cylinder head and can only be purchased with a new cylinder head. If there is any doubt, then the VAG special tool should be used – see previous refitting procedure.

All

25 Ensure the sealing surfaces of the cylinder head are clean, and then apply a 2.0 mm wide bead of sealant (D 176 501 A1 or equivalent) as shown. Take care not to apply too much sealant, ensuring the oil holes supply holes are not blocked **(see illustrations)**.

26 Fit a new sealing cap to the timing belt end of the cylinder head and make sure the

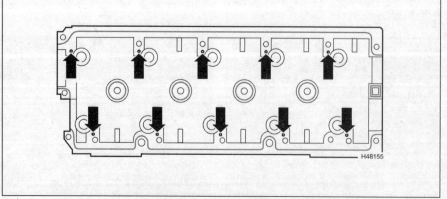

9.25b ... to the area shown by the thick, black line, taking care not to block the oil holes

9.26a Fit a new seal/end cap to the cylinder head

9.26b Check that the locating dowels are in place

locating dowels are fitted to the cylinder head **(see illustrations)**.

27 Apply a smear of clean engine oil to the bearing surfaces and place the camshafts, frame and clamping tool in place on the cylinder head. Progressively, carefully, hand tighten the frame retaining bolts in the sequence shown **(see illustration)**, until the retaining frame makes contact with the cylinder head over the complete surface, then tighten the bolts to the specified torque, again in the correct sequence.

28 Remove the gear aligning tool (T40096) and the clamping tool (T40095), or the metal flat bars and cable ties from the top of the camshaft retaining frame **(see illustration)**.

29 Renew the camshaft oil seal as described in Section 10.

30 The remainder of refitting is a reversal of removal.

10 Camshaft oil seals – renewal

Right-hand oil seal

1 Remove the camshaft sprocket and hub, as described in Section 8.

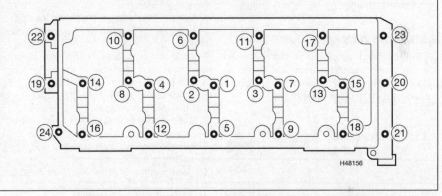

9.27 Camshaft retaining frame bolt tightening sequence

2 Drill two small holes into the existing oil seal, diagonally opposite each other. Take great care to avoid drilling through into the seal housing or camshaft sealing surface. Thread two self-tapping screws into the holes, and using a pair of pliers, pull on the heads of the screws to extract the oil seal **(see illustration)**.

3 Clean out the seal housing and the sealing surface of the camshaft by wiping it with a lint-free cloth. Remove any

swarf or burrs that may cause the seal to leak.

4 Do not lubricate the lip and outer edge of the new oil seal, push it over the camshaft until it is positioned in place above its housing. To prevent damage to the sealing lips, wrap some adhesive tape around the end of the camshaft **(see illustration)**.

5 Using a hammer and a socket of suitable diameter, drive the seal squarely into its housing. Make sure the seal is fitted the

9.28 Remove the cable ties and flat bar from the top of the ladder frame

10.2 Screw in a self-tapping screw, then pull the screw and seal from place

10.4 Using a plastic sleeve to slide seal over end of camshaft

10.5 Note some seals have 'OUTSIDE' to show fitted position

11.5 Battery tray securing bolts

11.11 Remove the charge air pipe from the turbocharger

correct way around, some have 'OUTSIDE' stamped on the seal **(see illustration)**. **Note:** *Select a socket that bears only on the hard outer surface of the seal, not the inner lip that can easily be damaged.*

6 Refit the camshaft sprocket and its hub, as described in Section 8.

Left-hand oil seal

7 The left-hand camshaft oil seal is formed by the brake vacuum pump seal. Refer to Chapter 9 Section 21, for details of brake vacuum pump removal and refitting.

---

11 Cylinder head – removal, inspection and refitting

Note: *The cylinder head must be removed with the engine cold. New cylinder head bolts and a new cylinder head gasket will be required on refitting, and suitable studs will be required to guide the cylinder head into position – see text.*

Removal

1 Remove the battery as described in Chapter 5A Section 3.

2 Drain the cooling system (Chapter 1B Section 33) and engine oil (Chapter 1B Section 6).

3 Pull the plastic cover on the top of the engine upwards to release it from its mountings.

4 Remove the air filter housing as described in.

5 Undo the bolts and remove the battery tray **(see illustration)**.

6 Remove the radiator cooling fan(s) and shroud as described in Chapter 3 Section 3.

7 Undo the bolts and remove the air hose/ duct from the intercooler to the turbocharger. Release the wiring looms from the clips as necessary to enable the duct to be manoeuvred from place.

8 Remove the camshaft cover as described in Section 4.

9 Remove the camshaft sprocket and hub as described in Section 8.

10 Remove the inlet manifold as described in.

11 Undo the retaining bolts, and remove the charge air pipe from the turbocharger **(see illustration)**.

12 Remove the EGR pipes from the cylinder head as described in Chapter 4D Section 3.

13 Remove the exhaust manifold as described in Chapter 4D Section 5. There is no need to completely remove the particulate filter from the vehicle. Slacken the Allen bolt and release the clamp securing the diesel particulate filter/catalytic converter to the turbocharger, then undo the bolts/ nuts securing the brackets to the cylinder

block/head and lay the filter/converter to one side.

14 Apply a little lubrication spray to the rubber sleeve, pull up the pipe from the vacuum pump, then undo the 4 retaining bolts and remove the vacuum pump from the left-hand end of the cylinder head **(see illustration)**. Renew the pump-to-cylinder head seal/gasket (Chapter 9 Section 21).

15 Disconnect the coolant temperature sensor wiring plug at the left-hand end of the cylinder head, and release the wiring loom from any retaining clips.

16 Disconnect the gearchange cables from the levers on the transmission as described in (manual transmission) or (DSG transmission).

17 Undo the bolts/nut, securing the gearchange bracket to the top of the transmission. Move the bracket and cables to one side.

18 Note their fitted locations, then release the clamps and disconnect the various coolant hoses from the cylinder head.

19 Disconnect the wiring connector from the oil pressure switch, undo the bolt securing the pipe bracket at the left-hand end of the cylinder head **(see illustrations)**, and the bolt securing the lifting bracket on the rear of the head.

20 Undo the bolt securing the timing belt guard adjacent to the timing belt tensioner, and the bolt securing the camshaft position

11.14 Disconnect the vacuum pipe

11.19a Disconnect the oil pressure switch wiring connector

11.19b Undo the retaining bolt/screw and remove the lifting bracket

11.20 Undo the bolt securing the timing belt guard

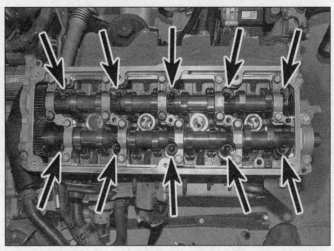

11.21 Undo the cylinder head bolts using a M12 multi-splined (12 pointed star) tool

sensor, then remove the tensioner retaining nut **(see illustration)**.

21 Using an M12 multi-splined tool (12 pointed star), undo the cylinder head bolts, working from the outside-in, evenly and gradually **(see illustration)**. Remove the bolts and recover the washers. Check that nothing remains connected, and starting at the gearbox side, lift the cylinder head from the engine block, sliding the belt tensioner from the mounting stud as the cylinder head is removed. Seek assistance if possible, as it is a heavy assembly, especially as it is being removed complete with the manifolds.

22 Remove the gasket from the top of the block, noting the locating dowels. If the dowels are a loose fit, remove them and store them with the head for safekeeping. Do not discard the gasket yet – it will be needed for identification purposes.

Inspection

23 Dismantling and inspection of the cylinder head is covered in Chapter 2F Section 6.

Cylinder head gasket selection

Note: *A dial test indicator (DTI) will be required for this operation.*

24 Examine the old cylinder head gasket for manufacturer's identification markings **(see illustration)**. These will be in the form of holes, and a part number on the edge of the gasket. Unless new pistons have been fitted, the new cylinder head gasket must be of the same type as the old one.

25 If new piston assemblies have been fitted as part of an engine overhaul, or if a new short engine is to be fitted, the projection of the piston crowns above the cylinder head mating face of the cylinder block at TDC must be measured. This measurement is used to determine the thickness of the new cylinder head gasket required.

26 Anchor a dial test indicator (DTI) to the top face (cylinder head gasket mating face) of the cylinder block, and zero the gauge on the gasket mating face.

27 Rest the gauge probe on No 1 piston crown, and turn the crankshaft slowly by hand until the piston reaches TDC. Measure and record the maximum piston projection at TDC **(see illustration)**.

28 Repeat the measurement for the remaining pistons, and record the results.

29 If the measurements differ from piston-to-piston, take the highest figure, and use this to determine the thickness of the head gasket required as follows.

Piston projection	Gasket identification (number of holes)
0.91 to 1.00 mm	1
1.01 to 1.10 mm	2
1.11 to 1.20 mm	3

30 Purchase a new gasket according to the results of the measurements.

Refitting

31 The mating faces of the cylinder head and block must be perfectly clean before refitting the head. Use a scraper to remove all traces of gasket and carbon, also clean the tops of the pistons. Take particular care with the aluminium surfaces, as the soft metal is easily damaged.

32 Make sure that debris is not allowed to enter the oil and water passages – this is particularly important for the oil circuit, as carbon could block the oil supply to the camshaft and crankshaft bearings. Using adhesive tape and paper, seal the water, oil and bolt holes in the cylinder block.

33 To prevent carbon entering the gap between the pistons and bores, smear a little grease in the gap. After cleaning a piston, rotate the crankshaft to that the piston moves down the bore, and then wipe out the grease and carbon with a cloth rag. Clean the other piston crowns in the same way.

34 Check the head and block for nicks, deep scratches and other damage. If slight, they may be removed carefully with a file. More serious damage may be repaired by machining, but this is a specialist job.

35 If warpage of the cylinder head is suspected, use a straight-edge to check it for distortion, as described in Chapter 2F Section 7.

36 Ensure that the cylinder head bolt holes in the crankcase are clean and free of oil. Syringe or soak up any oil left in the bolt holes. This is most important in order that the correct bolt tightening torque can be applied, and to prevent the possibility of the block being cracked by hydraulic pressure when the bolts are tightened.

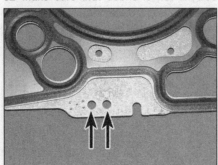

11.24 The holes identify the thickness of the cylinder head gasket

11.27 Measure the piston protrusion using a DTI gauge

11.39 Ensure the dowels are in place, then fit the new gasket with the part number uppermost

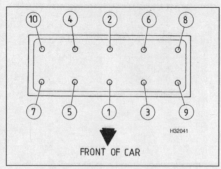

11.42a Cylinder head bolt tightening sequence

11.42b Tighten the cylinder head bolts to the Stage 1 torque

11.44 Use an angle-tightening gauge

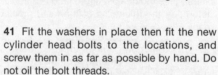

12.4 Slacken the two retaining clips

a) *Tighten all fasteners to their specified torque where given.*
b) *Renew all seals and gaskets.*
c) *Refill the cooling system as described, in Chapter 1B Section 33.*
d) *Refill the engine oil, as described in Chapter 1B Section 6.*
e) *Ensure all wiring is correctly routed.*
f) *Run the vehicle and make sure the cooling fans operate when the engine gets up to temperature.*

12 Sump – removal and refitting

Removal

1 Apply the handbrake, then jack up the front of the vehicle and support securely on axle stands (see *Jacking and vehicle support*).
2 Remove the securing screws and remove the engine undershield(s).
3 Drain the engine oil as described in Chapter 1B Section 6.
4 Release the retaining clips and remove the air hose from the intercooler outlet to the charge air pipe **(see illustration)**.
5 Undo the retaining bolts, release the clamp and move the charger air pipe from the front of the cylinder block **(see illustrations)**. Disconnect the charger air pressure sensor wiring plug.

37 Turn the crankshaft anti-clockwise all the pistons at an equal height, approximately half-way down their bores from the TDC position (see Section 3). This will eliminate any risk of piston-to-valve contact as the cylinder head is refitted.
38 Where applicable, refit the manifolds with reference to and/or Chapter 4D Section 5.
39 Ensure that the cylinder head locating dowels are in place in the cylinder block, and then fit the new cylinder head gasket over the dowels, ensuring that the part number is uppermost **(see illustration)**. Note that Seat recommend that the gasket is only removed from its packaging immediately prior to fitting.
40 Lower the cylinder head into position on the gasket, ensuring that it engages correctly over the dowels. Refit the timing belt tensioner as the cylinder head is refitted.

41 Fit the washers in place then fit the new cylinder head bolts to the locations, and screw them in as far as possible by hand. Do not oil the bolt threads.
42 Working progressively, in sequence, tighten all the cylinder head bolts to the specified Stage 1 torque **(see illustrations)**.
43 Again working progressively, in sequence, tighten all the cylinder head bolts to the specified Stage 2 torque.
44 Tighten all the cylinder head bolts, in sequence, through the specified Stage 3 angle **(see illustration)**.
45 Finally, tighten all the cylinder head bolts, in sequence, through the specified Stage 4 angle.
46 The remainder of the refitting procedure is a reversal of the removal procedure, noting the following points:

12.5a Undo the retaining bolt on the end of the sump…

12.5b …and the bolt on the front of the cylinder block …

12.5c …then disconnect the pressure switch wiring connector

12.7a Release the two clips at the rear of the sump

12.7b At the front, prise down the centre pin and pull the clip downwards

12.7c Remove the insulation cover from the around the sump

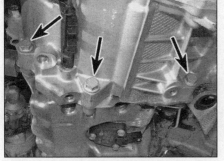

12.8a Undo the three transmission to sump bolts…

12.8b … and also the four bolts at the transmission end of the sump

12.8c Lower the sump from the cylinder block

6 Disconnect the wiring connector from the oil level/temperature sender in the sump.

7 Release the retaining clips and remove the sump insulation cover from the sump **(see illustrations)**.

8 Unscrew and remove the bolts securing the sump to the cylinder block, and the bolts securing the sump to the transmission casing, then withdraw the sump **(see illustrations)**. If necessary, release the sump by tapping with a soft-faced hammer.

Refitting

9 Begin refitting by thoroughly cleaning the mating faces of the sump and cylinder block. Ensure that all traces of old sealant are removed.

10 Ensure that the cylinder block mating face of the sump is free from all traces of old sealant, oil and grease, and then apply a 2.0 to 3.0 mm thick bead of silicone sealant (VW D 176 404 A2 or equivalent) to the sump **(see illustration)**. Note that the sealant should be run around the inside of the bolt holes in the sump. The sump must be fitted within 5 minutes of applying the sealant.

11 Offer the sump up to the cylinder block, then refit the sump-to-cylinder block bolts, and lightly tighten them by hand, working progressively in a diagonal sequence. **Note:** *If the sump is being refitted with the engine and transmission separated, make sure that the sump is flush with the flywheel end of the cylinder block.*

12 Refit the sump-to-transmission casing bolts, and tighten them lightly, using a socket.

13 Again working in a diagonal sequence, lightly tighten the sump-to-cylinder block bolts, using a socket.

14 Tighten the sump-to-transmission casing bolts to the specified torque.

15 Working in a diagonal sequence, progressively tighten the sump-to-cylinder block bolts to the specified torque.

16 The remainder of refitting is a reversal of removal, noting to allow at least 30 minutes from the time of refitting the sump for the sealant to dry, then refill the engine with oil, with reference to Chapter 1B Section 6.

13 Oil pump and drive belt – removal, inspection and refitting

Oil pump removal

1 Remove the sump as described in Section 12.

2 Unscrew the flange bolts and remove the oil pick-up pipe/filter from the oil pump **(see illustration)**. Recover the O-ring seal and discard, as a new one will be required for refitting.

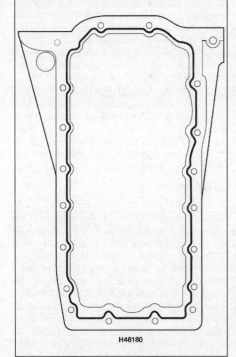

12.10 Apply a bead of sealant around the inside of the bolt holes

H46180

13.2 Remove the oil pick-up pipe/filter

13.3 Remove the oil baffle plate bolt (which is also one of the oil pump mounting bolts)

13.4 Oil pump mounting bolts

13.11 Fit a new seal to the end of the oil pick-up pipe

3 Unscrew the securing bolt, and remove the oil baffle from the cylinder block (see illustration).

4 Unscrew and remove the mounting bolts, and release the oil pump from the dowels in the crankcase (see illustration). Unhook the oil pump drive sprocket from the belt and withdraw the oil pump and oil pick-up pipe from the engine. Note, the bolt holding the baffle plate is also one of the pump mounting bolts.

Oil pump inspection

5 Clean the pump thoroughly, and inspect for signs of damage or wear. If evident, renew the oil pump.

Oil pump refitting

6 Prime the pump with oil by pouring oil into the pick-up pipe aperture while turning the driveshaft.

7 If the drive belt and crankshaft sprocket have been removed, delay refitting them until after the oil pump has been mounted on the cylinder block.

8 Engage the oil pump sprocket with the drive belt, and then locate the oil pump on the dowels. Refit and tighten the mounting bolts to the specified torque.

9 Where applicable, refit the drive belt and crankshaft sprocket using a reversal of the removal procedure.

10 Refit the oil baffle plate, and tighten the securing bolt.

11 Refit the pick-up pipe to the oil pump,

using a new O-ring seal, and tighten the securing bolts (see illustration).

12 Refit the sump as described in Section 12.

Oil pump drive belt and sprockets

Note: VAG sealant (D 176 404 A2 or equivalent) will be required to seal the crankshaft oil seal housing on refitting, and it is advisable to fit a new crankshaft oil seal.

Removal

13 Proceed as described in paragraphs 1 and 2.

14 To remove the belt, remove the timing belt as described in Section 7, then unbolt the crankshaft oil seal housing from the cylinder block, as described in Section 15. Unhook the belt from the sprocket on the end of the crankshaft.

15 The oil pump drive sprocket is a press-fit on the crankshaft, and cannot easily be removed. Consult a Seat dealer for advice if the sprocket is worn or damaged.

Inspection

16 It is wise to renew the belt in any case if the engine is to be overhauled, If there is any doubt as to the condition of the belt, renew it.

Refitting

17 If the oil pump has been removed, refit the oil pump as described previously in this Section before refitting the belt and sprocket.

18 Engage the oil pump sprocket with the belt, then engage the belt with the crankshaft sprocket.

19 Fit a new crankshaft oil seal to the housing, and refit the housing as described in Section 15.

20 Where applicable, refit the oil baffle and pick-up pipe, and tighten the securing bolts.

21 Refit the sump as described in Section 12.

14 Flywheel – removal, inspection and refitting

Removal

1 On manual gearbox models, remove the gearbox (see Chapter 7A Section 3) and clutch (see Chapter 6 Section 6).

2 On semi-automatic (DSG) transmission models, remove the transmission as described in Chapter 7B Section 2.

3 The flywheel can only be fitted in one position due to the offset of the flywheel mounting holes in the end of the crankshaft (see illustration). Note, manual models are fitted with a dual-mass flywheel.

4 Rotate the outside of the dual-mass flywheel so that the bolts align with the holes (if necessary).

5 Unscrew the bolts and remove the flywheel. Using a locking tool, counter-hold the flywheel to prevent it from turning (see illustration). Discard the bolts, as new ones must be fitted. Note: In order not to damage the flywheel, do not allow the bolt heads to make contact with the flywheel during the unscrewing procedure.

Inspection

6 Check the dual-mass flywheel for wear and damage. Examine the starter ring gear for excessive wear to the teeth. If the driveplate or its ring gear are damaged, the complete driveplate must be renewed. The flywheel ring gear, however, may be renewed separately from the flywheel, but the work should be entrusted to a Seat dealer. If the clutch friction face is discoloured or scored excessively, it may be possible to regrind it, but this work should also be entrusted to a Seat dealer.

7 The following are guidelines only, but should indicate whether professional inspection is necessary. The dual-mass flywheel should be checked as follows:

14.3 Flywheel bolts are offset and can only be fitted in one position – DSG transmission shown

14.5 Use a locking tool to counterhold the flywheel

14.8 Flywheel warpage check – see text

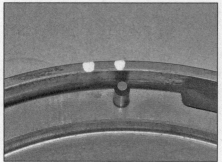

14.9 Flywheel free rotational movement check alignment marks – see text

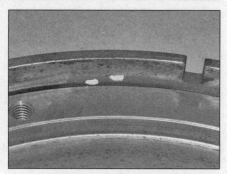

14.11 Flywheel lateral movement check marks – see text

8 Warpage: Place a straight edge across the face of the drive surface, and check by trying to insert a feeler gauge between the straight edge and the drive surface **(see illustration)**. The flywheel will normally warp like a bowl – ie. Higher on the outer edge. If the warpage is more than 0.40 mm, the flywheel may need replacing.

9 Free rotational movement: This is the distance the drive surface of the flywheel can be turned independently of the flywheel primary element, using finger effort alone. Move the drive surface in one direction and make a mark where the locating pin aligns with the flywheel edge. Move the drive surface in the other direction (finger pressure only) and make another mark **(see illustration)**. The total of free movement should not exceed 20.0 mm. If it's more, the flywheel may need replacing.

10 Total rotational movement: This is the total distance the drive surface can be turned independently of the flywheel primary element. Insert two bolts into the clutch pressure plate/damper unit mounting holes, and with the crankshaft/flywheel held stationary, use a lever/pry bar between the bolts and use some effort to move the drive surface fully in one direction – make a mark where the locating pin aligns with the flywheel edge. Now force the drive surface fully in the opposite direction, and make another mark. The total rotational movement should not exceed 44.00 mm. If it does, have the flywheel professionally inspected.

11 Lateral movement: The lateral movement (up and down) of the drive surface in relation to the primary element of the flywheel, should not exceed 2.0 mm. If it does, the flywheel may need replacing. This can be checked by pressing the drive surface down on one side into the flywheel (flywheel horizontal) and making an alignment mark between the drive surface and the inner edge of the primary element. Now press down on the opposite side of the drive surface, and make another mark above the original one. The difference between the two marks is the lateral movement **(see illustration)**.

12 There should be no cracks in the drive surface of the flywheel. If cracks are evident, the flywheel may need replacing.

Refitting

13 Refitting is a reversal of removal. Use new bolts when refitting the flywheel or driveplate **(see illustration)**, and coat the threads of the bolts (if not already coated with locking compound) with locking fluid before inserting them. Tighten them to the specified torque.

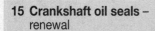

15 Crankshaft oil seals – renewal

Note: *The oil seals are a PTFE (Teflon) type and are fitted dry, without using any grease or oil. These have a wider sealing lip and have*

been introduced instead of the coil spring type oil seal.

Timing belt end oil seal

1 Remove the timing belt as described in Section 7, and the crankshaft sprocket with reference to Section 8.

2 To remove the seal without removing the housing, drill two small holes diagonally opposite each other, insert self-tapping screws, and pull on the heads of the screws with pliers **(see illustration)**.

3 Alternatively, to remove the oil seal complete with its housing, proceed as follows.

a) *Remove the sump as described in Section 12. This is necessary to ensure a satisfactory seal between the sump and oil seal housing on refitting.*

b) *Unbolt and remove the oil seal housing.*

c) *Working on the bench, lever the oil seal from the housing using a suitable screwdriver. Take care not to damage the seal seating in the housing.*

4 Thoroughly clean the oil seal seating in the housing.

5 Wind a length of tape around the end of the crankshaft (or a plastic sleeve) to protect the oil seal lips as the seal (and housing, where applicable) is fitted **(see illustration)**.

6 Fit a new oil seal to the housing, pressing or driving it into position using a socket or tube of suitable diameter. Ensure that the socket or tube bears only on the hard outer ring of the seal, and take care not to damage the seal lips. Press or drive the seal into position until it

14.13 Use new bolts when refitting

15.2 Pull the screw and seal from place using pliers

15.5 Using a plastic sleeve to slide the seal over the end of the crankshaft

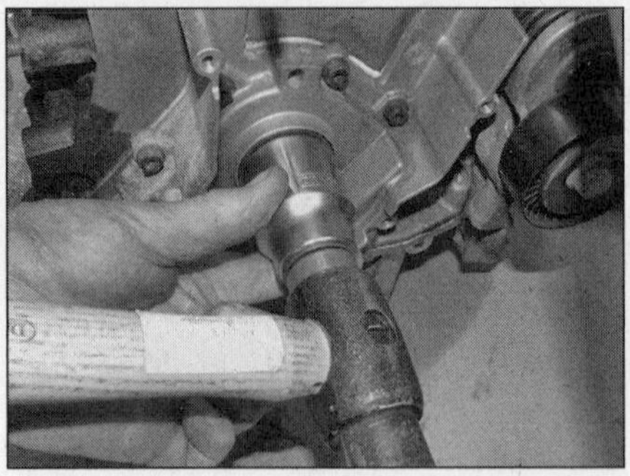
15.6 Carefully tap the seal into position

is seated on the shoulder in the housing **(see illustration)**. Make sure that the closed end of the seal is facing outwards.

7 If the oil seal housing has been removed, proceed as follows, otherwise proceed to paragraph 11.

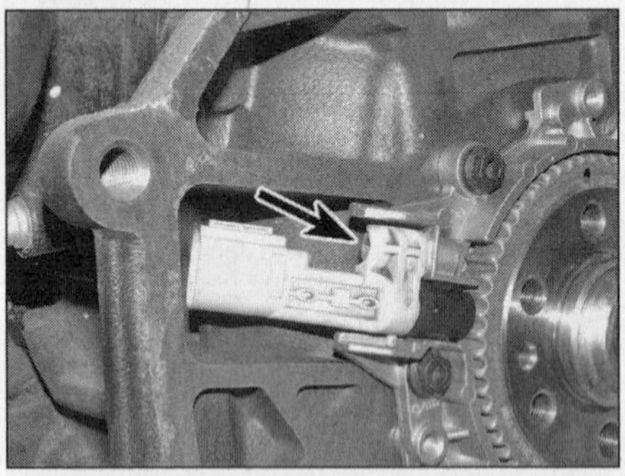
15.13a Undo the crankshaft speed sensor retaining bolt

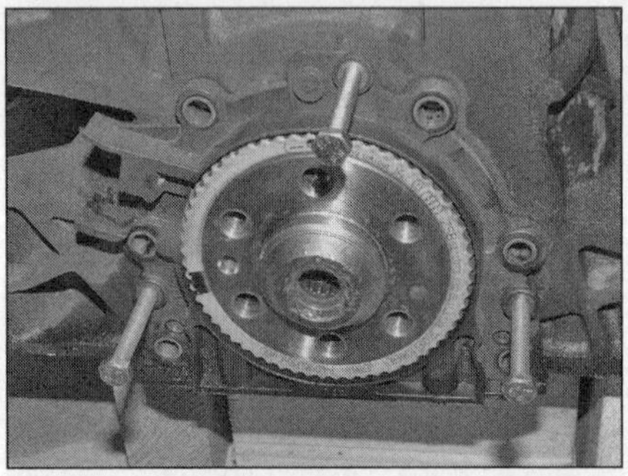
15.14a Screw in three 6 x 35 mm bolts ...

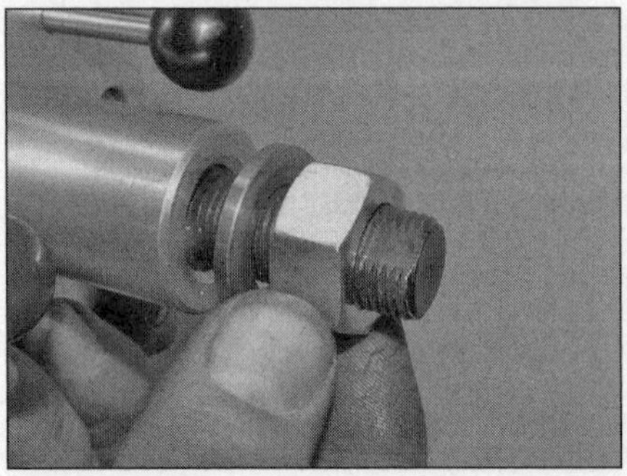
15.16a Rotate the nut until its level with the end of the flat clamping surface...

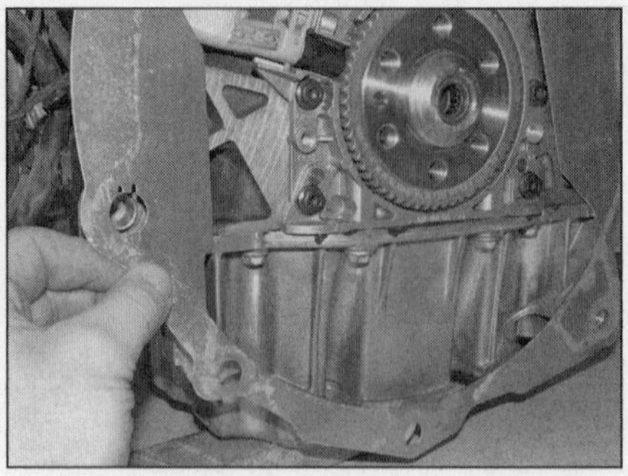
15.12a Remove the intermediate plate from the dowels ...

8 Clean all traces of old sealant from the crankshaft oil seal housing and the cylinder block, then coat the cylinder block mating faces of the oil seal housing with a 2.0 to 3.0 mm thick bead of silicone sealant (VW D 176 404 A2, or equivalent). Note that the seal housing must be refitted within 5 minutes of applying the sealant.

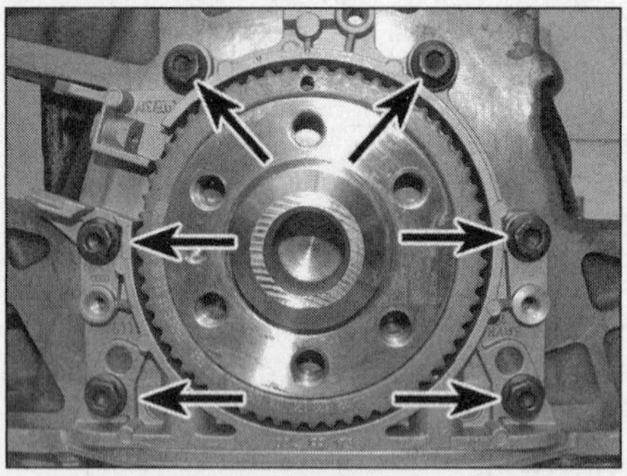
15.13b Sealing flange bolts

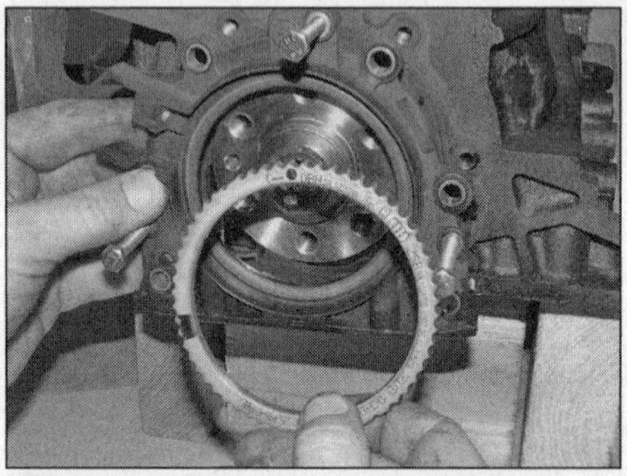
15.14b ...and draw the sealing flange and sender wheel from place

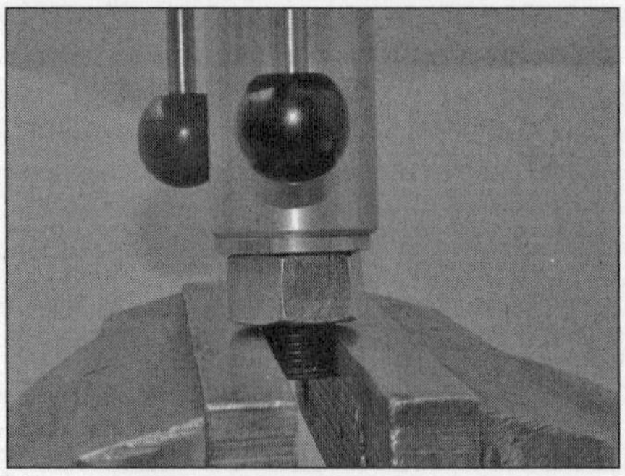
15.16b ...then clamp it in a vice

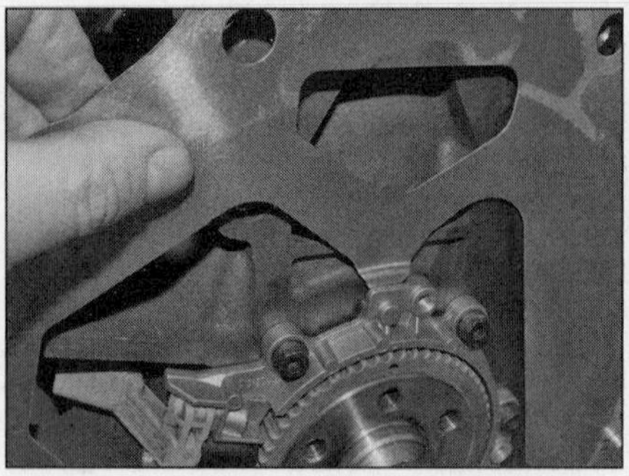
15.12b ... and from behind the top of the crankshaft seal housing

Caution: DO NOT put excessive amounts of sealant onto the housing as it may get into the sump and block the oil pick-up pipe.

9 Refit the oil seal housing, and tighten the bolts progressively to the specified torque.

10 Refit the sump as described in Section 12.

11 Refit the crankshaft sprocket with reference to Section 8, and the timing belt as described in Section 7.

Flywheel end oil seal

Note: *In these engines, the seal, sealing flange and sender wheel are a complete unit. Special tools are required to refit the sealing flange, and press the sender wheel onto the end of the crankshaft. It is not possible to accurately fit these parts without the tools, which may be available from VAG (part no. T10134) and are available from aftermarket automotive tool specialists. E.g. Draper tools).*

12 Remove the flywheel as described in Section 14, then prise the intermediate plate from the locating dowels on the cylinder block and unhook it from behind the top of the seal housing **(see illustrations)**.

13 Undo the bolt securing the crankshaft speed sensor and remove it from the seal housing, then undo the bolts securing the sealing flange to the cylinder block **(see illustrations)**.

14 Insert three 6 x 35 mm bolts into the threaded holes in the sealing flange. Tighten the bolts gradually and evenly, and press the sealing flange, and sender wheel from the crankshaft/cylinder block **(see illustrations)**. The seal, sender wheel and sealing flange are supplied as a complete unit.

15 Ensure the mating face of the cylinder block is clean and free from debris. The new sealing flange/seal/sender wheel assembly is supplied with a sealing lip support ring, which serves as a fitting sleeve, and must not be removed prior to installation. Equally, the sender wheel must not be separated from the assembly.

16 If using the VW tool, proceed as follows. If using an aftermarket tool specialist's product, follow the instructions supplied with the tool. Rotate the large spindle nut until it's level with the end of the clamping surface of the spindle, then clamp the spindle in a vice **(see illustrations)**.

15.17a Rotate the nut until the inner part of the tool ...

15.17b ...is flush with the flat surface of the housing

15.18a Remove the securing clip ...

17 Press the tool housing downwards until it rests on the nut and washer. Rotate the nut until the inner part of the tool is at the same height as the housing **(see illustrations)**.

18 Remove the seal securing clip. The hole on the sender wheel must align with the marking on the sealing flange **(see illustrations)**.

19 Place the flange outer side down on a clean, flat surface, then press the seal guide fitting sleeve (supplied ready fitted), housing, and sender wheel downwards until all the components are flat on the surface. In this position the upper edge of the sender wheel should be level with the edge of the sealing flange **(see illustrations)**.

20 Place the sealing flange on the assembly tool, so the pin locates in the hole in the sender wheel **(see illustration)**.

21 Push the sealing flange and guide fitting sleeve against the tool whilst tightening the 3

knurled screws. Ensure the pin is still located in the sender wheel **(see illustration)**.

22 Ensure the end of the crankshaft is clean, and is locked at TDC on No. 1 cylinder as described in Section 3.

23 Unscrew the large nut to the end of

the spindle threads, then press the spindle inwards as far as possible **(see illustrations)**.

24 Align the flat side of the assembly with the sump flange, then secure the tool to the crankshaft using the integral Allen bolts **(see illustration)**. Only hand tighten the bolts.

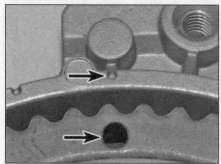

15.18b ...the hole in the sender wheel should align with the marking on the flange

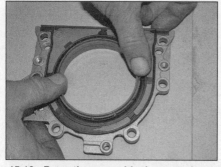

15.19a Press the assembly downwards on a clean, flat surface...

15.19b ...so the upper edge of the sender wheel is level with the edge of the flange

15.20 Fit the flange to the tool, ensuring the pin locates in the hole

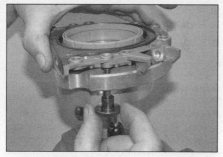

15.21 With the pin engaged in the hole, tighten the 3 knurled screws to secure the flange to the tool

15.23a Unscrew the nut to the end of the thread ...

15.23b ...and push the spindle in as far as possible

15.24 Hand-tighten the hex bolts to secure the tool to the crankshaft

15.25 Use two M7 x 35 mm bolts to guide the sealing flange

15.26 Push the black knob into the hole in the crankshaft

15.27a After tightening the spindle nut to 35 Nm …

15.27b …there should be an air gap between the sealing flange and the cylinder block

15.28 Remove the tool and seal fitting guide sleeve

15.29 Measure the fitted depth of the sender wheel in relation to the end of the crankshaft

25 Insert two M7x 35 mm bolts to guide the sealing flange to the cylinder block **(see illustration)**.
26 Using hand pressure alone, push the tool assembly onto the crankshaft until the seal guide fitting sleeve contacts the crankshaft flange, then push the guide pin (black knob) into the hole in the crankshaft. This is to ensure the sender wheel reaches its correct installation position **(see illustration)**.
27 Rotate the large nut until it makes contact with the tool housing, then tighten it to 35 Nm. After tightening this nut, a small air gap must still be present between the sealing flange and cylinder block **(see illustrations)**.
28 Unscrew the large nut; the two M7 x 35 Nm screws, the three knurled screws and the Allen bolts securing the tool to the crankshaft. Remove the tool, and pull the seal guide fitting sleeve from place (if it didn't come out with the tool) **(see illustration)**.
29 Use a vernier caliper or feeler gauge to measure the fitted depth of the sender wheel in relation to the crankshaft flange **(see illustration)**. The correct depth is 0.5 mm.
30 If the gap is correct, fit the sealing flange bolts and tighten them to the specified torque.
31 If the gap is too small, re-attach the tool to the sealing flange and crankshaft, then refit the two M7 x 35 mm guide bolts to the flange. Tighten the large spindle nut to 40 Nm, remove the tool and re-measure the air gap. If the gap is still too small, re-attach the tool and

tighten the spindle nut to 45 Nm. Re-measure the gap. When the gap is correct, refit the flange retaining bolts, and tighten them to the specified torque.
32 The remainder of refitting is a reversal of removal.

16 Engine/transmission mountings – inspection and renewal

Inspection

1 If improved access is required, jack up the front of the vehicle, and support it securely on axle stands (see *Jacking and vehicle support*). Remove the engine top cover that also incorporates the air filter, and then remove the engine undershield(s).
2 Check the mounting rubbers to see if they are cracked, hardened or separated from the metal at any point; renew the mounting if any such damage or deterioration is evident.
3 Check that all the mountings are securely tightened; use a torque wrench to check if possible.
4 Using a large screwdriver or a crowbar, check for wear in the mounting by carefully levering against it to check for free play. Where this is not possible, enlist the aid of an assistant to move the engine/transmission

back-and-forth, or from side-to-side, whilst you observe the mounting. While some free play is to be expected, even from new components, excessive wear should be obvious. If excessive free play is found, check first that the fasteners are correctly secured, and then renew any worn components as described in the following paragraphs.

Renewal

Right-hand mounting

Note: *New mounting securing bolts will be required on refitting.*
5 Attach a hoist and lifting tackle to the engine lifting brackets on the cylinder head, and raise the hoist to just take the weight of the engine. Alternatively the engine can be supported on a trolley jack under the engine. Use a block of wood between the sump and the head of the jack, to prevent any damage to the sump.
6 For improved access, unscrew the coolant reservoir and move it to one side, leaving the coolant hoses connected.
7 Disconnect the wiring plug, undo the retaining bolts, and remove the pressure differential sender (for particulate filter) including bracket from the top of the engine mounting and move it to one side **(see illustrations 6.3a & 6.3b)**.
8 Undo the retaining bolts and move the fuel filter to one side. Where applicable, move any

16.9 Removing the right-hand engine mounting

16.10 There must be at least 10 mm between the bracket and the chassis member (arrowed)

16.14 Left-hand engine/transmission mounting

16.19 Rear mounting arm-to-transmission bolts

17.4a Undo the engine oil cooler retaining screws

17.4b Renew the seals/gaskets

wiring harnesses, pipes or hoses to one side to enable removal of the engine mounting.

9 Unscrew the bolts securing the mounting to the engine bracket, and then unscrew the bolts securing it to the body **(see illustration)**. Withdraw the mounting from the engine compartment.

10 Refitting is a reversal of removal, bearing in mind the following points.

a) Use new securing bolts.

b) There must be at least 10 mm between the engine mounting bracket and the right-hand side chassis member **(see illustration)**.

c) The side of the mounting support arm must be parallel to the side of the engine mounting bracket.

d) Tighten all fixings to the specified torque.

Left-hand mounting

Note: New mounting bolts will be required on refitting.

11 Remove the engine top cover.

12 Attach a hoist and lifting tackle to the engine lifting brackets on the cylinder head, and raise the hoist to just take the weight of the engine and transmission. Alternatively the engine can be supported on a trolley jack under the transmission. Use a block of wood between the transmission and the head of the jack, to prevent any damage to the transmission.

13 Remove the battery and battery tray, as described in (Chapter 5A Section 3).

14 Unscrew the bolts securing the mounting to the transmission, and the remaining bolts securing the mounting to the body **(see illustration)**, then lift the mounting from the engine compartment.

15 Refitting is a reversal of removal, bearing in mind the following points:

a) Use new mounting bolts.

b) The edges of the mounting support arm must be parallel to the edge of the mounting.

c) Tighten all fixings to the specified torque.

Rear mounting (torque arm)

Note: New mounting bolts will be required on refitting.

16 Apply the handbrake, then jack up the front of the vehicle and support securely on axle stands (see Jacking and vehicle support). Remove the engine undershield(s) for access to the rear mounting (torque arm).

17 Support the rear of the transmission beneath the final drive housing. To do this, use a trolley jack and block of wood, or alternatively wedge a block of wood between the transmission and the subframe.

18 Working under the vehicle, unscrew and remove the bolt securing the mounting to the subframe.

19 Unscrew the two bolts securing the mounting to the transmission, then withdraw the mounting from under the vehicle **(see illustration)**.

20 Refitting is a reversal of removal, but use new mounting securing bolts, and tighten all fixings to the specified torque.

---

17 Engine oil cooler/filter housing – removal and refitting

Removal

1 The oil cooler is mounted on the lower part of the oil filter housing on the front of the cylinder block.

2 Position a container beneath the oil filter housing to catch escaping oil and coolant.

3 Clamp the oil cooler coolant hoses to minimise coolant spillage, or drain the cooling system as described in Chapter 1B Section 33.

4 Unscrew the oil cooler retaining screws and remove the oil cooler from the front of the oil filter housing **(see illustrations)**. Recover the O-rings from between the cooler and the oil filter housing, new ones will be required for refitting.

5 If required, disconnect the coolant hose, unclip the dipstick from the side of the housing, then undo the retaining bolts and remove the oil filter housing from the cylinder block **(see illustrations)**.

17.5a Disconnect the coolant hoses ...

17.5b ...release the oil dipstick guide tube retaining clip...

17.5c ... and undo the filter housing retaining bolts

Refitting

6 Refitting is a reversal of removal, bearing in mind the following points:

a) *Use new oil cooler and housing O-rings (see illustration).*
b) *Tighten the oil cooler and filter housing bolts to the correct torque.*
c) *On completion, check and if necessary top-up the oil and coolant levels.*

18 Oil pressure warning light switch – removal and refitting

Removal

1 The oil pressure warning light switch is fitted to the left-hand rear of the cylinder head

17.6 Renew the seals/gaskets

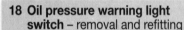

18.2 Remove the air intake hose

(see illustration). Remove the engine top cover to gain access to the switch.
2 Slacken the retraining clips and remove the air intake hose from the turbocharger (see illustration).
3 Disconnect the wiring connector and wipe clean the area around the switch.
4 Unscrew the switch and remove it, along with its sealing washer (where fitted). If the switch is to be left removed from the engine for any length of time, plug the aperture in the cylinder head.

Refitting

5 Examine the sealing washer for signs of damage or deterioration and if necessary renew.
6 Refit the switch, complete with washer (where fitted), and tighten it to the specified torque.

18.1 Oil pressure warning light switch location

19.1 Oil level/temperature sensor location

7 Securely reconnect the wiring connector, refit the air intake hose and then, if necessary, top-up the engine oil as described in. On completion, refit the engine top cover.

19 Oil level/temperature sender – removal and refitting

Removal

1 The oil level/temperature sender is fitted to bottom of the sump (see illustration).
2 Drain the engine oil as described in Chapter 1B Section 6.
3 Disconnect the wiring connector from the sender.
4 Release the retaining clips and remove the insulation cover from the sump (see illustration).
5 Wipe clean the area around the sender, then undo the three retaining bolts and remove the sender.

Refitting

6 Examine the sealing washer for signs of damage or deterioration and if necessary renew.
7 Refit the switch and tighten the retaining bolts to the specified torque.
8 Refit the insulation cover, then reconnect the wiring connector and refill the engine with oil (refer to Chapter 1B Section 6).

19.4 Remove the insulation cover to access the bolts

Chapter 2 Part D
2.0 litre common rail (CR) diesel engine in-car repair procedures

Contents

Degrees of difficulty

Easy, suitable for novice with little experience	**Fairly easy,** suitable for beginner with some experience	**Fairly difficult,** suitable for competent DIY mechanic	**Difficult,** suitable for experienced DIY mechanic	**Very difficult,** suitable for expert DIY or professional

Specifications

General

Manufacturer's engine codes: *
   1968 cc (2.0 litre), 16-valve, DOHC . . . . CFJA, CFHC, CFJB and CFHF

Maximum outputs:

	Power	Torque
Engine code CEGA.	125 kW @ 4200 rpm	350 Nm @ 1750 to 2500 rpm
Engine code CFJA	125 kW @ 4200 rpm	350 Nm @ 1750 to 2500 rpm
Engine code CLCB.	103 kW @ 4200 rpm	350 Nm @ 1750 to 2500 rpm
Engine code CFHC.	103 kW @ 4200 rpm	350Nm @ 1750 to 2750 rpm

Bore   . . . . 81.0 mm
Stroke . . . . 95.5 mm
Compression ratio . . . . 16.5 : 1
Compression pressures:
   Minimum compression pressure . . . . Approximately 19.0 bar
   Maximum difference between cylinders. Approximately 5.0 bar
Firing order . . . . 1 – 3 – 4 – 2
No 1 cylinder location. . . . . Timing belt end
* **Note:** *See 'Vehicle identification' at the end of this manual for the location of engine code markings.*

Lubrication system

Oil pump type.	Gear type, chain-driven from crankshaft
Oil pressure switch (green cap)	0.5 bar
Oil pressure switch (brown cap).	0.7 bar

Oil pressure (oil temperature 80°C):
Minimum @ idling	0.6 bar
Minimum @ 2000 rpm:	
CBAA, CBAB & CBBB	2.0 bar
CFFA, CFFB & CFGB	1.0 bar
Maximum @ high rpm:	
CBAA, CBAB & CBBB	7.0 bar
CFFA, CFFB & CFGB	5.0 bar

Torque wrench settings

	Nm	lbf ft
Ancillary (alternator, etc) bracket mounting bolts: *		
Stage 1.	40	30
Stage 2.	Angle-tighten a further 45°	
Auxiliary drivebelt tensioner securing bolt:		
Stage 1.	20	15
Stage 2.	Angle-tighten a further 180°	
Balancer shaft drivegear bolts: *		
Stage 1.	20	15
Stage 2.	Angle-tighten a further 90°	
Balance shaft housing to cylinder block: *		
M7:		
Stage 1.	13	10
Stage 2.	Angle-tighten a further 90°	
M8:		
Stage 1.	20	15
Stage 2.	Angle-tighten a further 90°	
Balance shaft idler gear: *		
Stage 1.	90	66
Stage 2.	Angle-tighten a further 90°	
Big-end bearing caps bolts: *		
Stage 1.	30	22
Stage 2.	Angle-tighten a further 90°	
Camshaft bearing frame bolts/nut	10	7
Camshaft cover bolts.	10	7
Camshaft sprocket hub centre bolt	100	74
Camshaft sprocket-to-hub bolts: *		
Stage 1.	20	15
Stage 2.	Angle-tighten a further 45°	
Common rail bolts	22	16
Coolant pump bolts	15	11
Crankshaft oil seal housing bolts	15	11
Crankshaft pulley-to-sprocket bolts: *		
Stage 1.	10	7
Stage 2.	Angle-tighten a further 90°	
Crankshaft sprocket bolt: *		
Stage 1.	120	89
Stage 2.	Angle-tighten a further 90°	
Cylinder head bolts: *		
Stage 1.	30	22
Stage 2.	50	37
Stage 3.	Angle-tighten a further 90°	
Stage 4.	Angle-tighten a further 90°	
Engine mountings: *		
RH engine mounting:		
Mounting bracket to engine:		
Stage 1.	40	30
Stage 2.	Angle-tighten a further 180°	
Mounting to body:		
Stage 1.	40	30
Stage 2.	Angle-tighten a further 90°	
Mounting to bracket:		
Stage 1.	60	44
Stage 2.	Angle-tighten a further 90°	

Torque wrench settings (continued)

	Nm	lbf ft
Engine mountings: * (continued)		
LH engine/transmission mounting:		
Mounting to body:		
Stage 1	40	30
Stage 2	Angle-tighten a further 90°	
Mounting to bracket on transmission:		
Stage 1	60	44
Stage 2	Angle-tighten a further 90°	
Rear mounting link:		
M10 strength class 8.8:		
Stage 1	40	30
Stage 2	Angle-tighten a further 90°	
M10 strength class 10.9:		
Stage 1	50	37
Stage 2	Angle-tighten a further 90°	
M12:		
Stage 1	60	44
Stage 2	Angle-tighten a further 90°	
Flywheel: *		
Stage 1	60	44
Stage 2	Angle-tighten a further 90°	
Fuel injector:		
Retaining nuts....................	22	16
Cover bolts	5	4
Fuel pump hub nut	95	70
Fuel pump sprocket bolts: *		
Stage 1	20	15
Stage 2	Angle-tighten a further 90°	
Intermediate gear bolt: *		
Stage 1	90	66
Stage 2	Angle-tighten a further 90°	
Main bearing cap bolts: *		
Stage 1	65	48
Stage 2	Angle-tighten a further 90°	
Oil cooler screws (CFFA, CFFB & CFGB).....................	11	7
Oil cooler centre bolt (CBAA, CBAB & CBBB).....................	25	18
Oil drain plug*...	30	22
Oil filter housing-to-cylinder block bolts: *		
Stage 1	15	11
Stage 2	Angle-tighten a further 90°	
Oil filter cover	25	18
Oil level/temperature sensor-to-sump bolts.....................	10	7
Oil pick-up pipe securing bolts	10	7
Oil pressure warning light switch	20	15
Oil pump securing bolts:		
M6	10	7
M7: *		
Stage 1	13	10
Stage 2	Angle-tighten a further 90°	
Piston oil spray jet bolt.....................................	25	18
Sump:		
Sump-to-cylinder block bolts.....................	15	11
Sump-to-transmission bolts.....................	45	33
Thermostat housing.....................	15	11
Timing belt outer cover bolts.....................	10	7
Timing belt tensioner roller securing nut:		
Stage 1	20	15
Stage 2	Angle-tighten a further 45°	
Timing belt idler pulleys:		
Lower idler roller nut.....................	20	15
Upper idler roller (small) bolt.....................	20	15
Upper idler roller (large) bolt: *		
Stage 1	50	37
Stage 2	Angle-tighten a further 90°	

Do not re-use fasteners

1 General Information

How to use this Chapter

1 This Part of Chapter 2 describes those repair procedures that can reasonably be carried out on the engine while it remains in the vehicle. If the engine has been removed from the vehicle and is being dismantled as described in Part F, any preliminary dismantling procedures can be ignored.

2 Note that while it may be possible physically to overhaul certain items while the engine is in the vehicle, such tasks are not usually carried out as separate operations, and usually require the execution of several additional procedures (not to mention the cleaning of components and of oilways); for this reason, all such tasks are classed as major overhaul procedures, and are described in Part F of this Chapter.

Engine description

3 Throughout this Chapter, engines are referred to by type, and are identified and referred to by the manufacturer's code letters. A listing of all engines covered, together with their code letters, is given in the Section at the start of this Chapter.

4 The engines are water-cooled, double overhead camshafts (DOHC), in-line four-cylinder units, with cast-iron cylinder blocks and aluminium-silicon alloy cylinder heads. All are mounted transversely at the front of the vehicle, with the transmission bolted to the left-hand end of the engine.

5 The crankshaft is of five-bearing type, and thrustwashers are fitted to the centre main bearing to control crankshaft endfloat.

6 Drive for the exhaust camshaft is by a toothed timing belt from the crankshaft, with the intake camshaft driven by interlocking gears at the left-hand end of both camshafts. The gears incorporate a toothed backlash compensator element. Each camshaft is mounted at the top of the cylinder head, and is secured by a bearing frame/ladder.

7 The valves are closed by coil springs, and run in guides pressed into the cylinder head. The valves are operated by roller rocker arms incorporating hydraulic tappets.

8 A twin, counter-rotating balance shaft assembly is fitted to the base of the cylinder block. The rearmost balance shaft is driven by a gear on the crankshaft, via an intermediate gear bolted to the balance shaft housing. The two balance shafts are geared together.

9 The gear-type oil pump is driven by the front balance shaft. Oil is drawn from the sump through a strainer, and then forced through an externally-mounted, renewable filter. From there, it is distributed to the cylinder head, where it lubricates the camshaft journals and hydraulic tappets, and also to the crankcase,

where it lubricates the main bearings, connecting rod big-ends, gudgeon pins and cylinder bores. A coolant-fed oil cooler is fitted to the oil filter housing on all engines. Oil jets are fitted to the base of each cylinder – these spray oil onto the underside of the pistons, to improve cooling.

10 All engines are fitted with a brake servo vacuum pump driven by the camshaft on the transmission end of the cylinder head.

11 On all engines, engine coolant is circulated by a pump, driven by the timing belt. For details of the cooling system, refer to Chapter 3 Section 1.

Operations with engine in car

12 The following operations can be performed without removing the engine:

a) Compression pressure – testing.
b) Camshaft cover – removal and refitting.
c) Crankshaft pulley – removal and refitting.
d) Timing belt covers – removal and refitting.
e) Timing belt – removal, refitting and adjustment.
f) Timing belt tensioner and sprockets – removal and refitting.
g) Camshaft oil seals – renewal.
h) Camshafts and hydraulic tappets – removal, inspection and refitting.
i) Cylinder head – removal and refitting.
j) Cylinder head and pistons – decarbonising.
k) Sump – removal and refitting.
l) Oil pump – removal, overhaul and refitting.
m) Crankshaft oil seals – renewal.
n) Engine/transmission mountings – inspection and renewal.
o) Flywheel – removal, inspection and refitting.

Note: It is possible to remove the pistons and connecting rods (after removing the cylinder head and sump) without removing the engine. However, this is not recommended. Work of this nature is more easily and thoroughly completed with the engine on the bench, as described in Chapter 2F Section 9.

2 Compression and leakdown tests – description and interpretation

Compression test

Note: A compression tester suitable for use with diesel engines will be required for this test.

1 When engine performance is down, or if misfiring occurs which cannot be attributed to the ignition or fuel systems, a compression test can provide diagnostic clues as to the engine's condition. If the test is performed regularly, it can give warning of trouble before any other symptoms become apparent.

2 The engine must be fully warmed-up to normal operating temperature, the battery

must be fully charged, and you will require the aid of an assistant.

3 Remove the glow plugs as described in Chapter 5C Section 2, and then fit a compression tester to the No 1 cylinder glow plug hole. The type of tester that bolts into the plug thread is preferred. **Note:** Part of the glow plug removal procedure is to disconnect the fuel injector wiring plugs. As a result of the plugs being disconnected and the engine cranked, faults will be stored in the ECM memory. These must be erased after the compression test.

4 Have your assistant crank the engine for several seconds on the starter motor. After one or two revolutions, the compression pressure should build-up to a maximum figure and then stabilise. Record the highest reading obtained.

5 Repeat the test on the remaining cylinders, recording the pressure in each.

6 The cause of poor compression is less easy to establish on a diesel engine than on a petrol engine. The effect of introducing oil into the cylinders (wet testing) is not conclusive, because there is a risk that the oil will sit in the recess on the piston crown, instead of passing to the rings. However, the following can be used as a rough guide to diagnosis.

7 All cylinders should produce very similar pressures. Any difference greater than that specified indicates the existence of a fault. Note that the compression should build-up quickly in a healthy engine. Low compression on the first stroke, followed by gradually increasing pressure on successive strokes, indicates worn piston rings. A low compression reading on the first stroke, which does not build-up during successive strokes, indicates leaking valves or a blown head gasket (a cracked head could also be the cause).

8 A low reading from two adjacent cylinders is almost certainly due to the head gasket having blown between them and the presence of coolant in the engine oil will confirm this.

9 On completion, remove the compression tester, and refit the glow plugs, with reference to Chapter 5C Section 2.

10 Reconnect the wiring to the injector solenoids. Finally, have a Seat dealer or suitably equipped specialist erase the fault codes from the ECM memory.

Leakdown test

11 A leakdown test measures the rate at which compressed air fed into the cylinder is lost. It is an alternative to a compression test, and in many ways it is better, since the escaping air provides easy identification of where pressure loss is occurring (piston rings, valves or head gasket).

12 The equipment required for leakdown testing is unlikely to be available to the home mechanic. If poor compression is suspected, have the test performed by a suitably equipped garage.

3 Engine assembly and valve timing marks – general information and usage

General information

1 TDC is the highest point in the cylinder that each piston reaches as it travels up-and-down when the crankshaft turns. Each piston reaches TDC at the end of the compression stroke and again at the end of the exhaust stroke, but TDC generally refers to piston position on the compression stroke. No 1 piston is at the timing belt end of the engine.

2 Positioning No 1 piston at TDC is an essential part of many procedures, such as timing belt removal and camshaft removal.

3 The design of the engines covered in this Chapter is such that piston-to-valve contact may occur if the camshaft or crankshaft is turned with the timing belt removed. For this reason, it is important to ensure that the camshaft and crankshaft do not move in relation to each other once the timing belt has been removed from the engine.

Setting TDC on No 1 cylinder

Note: *VAG special tool T10050 is required to lock the crankshaft sprocket in the TDC position. Alternatively obtain a tool from automotive tool specialists.*

4 Raise the front of the vehicle and support it securely on axle stands (see *Jacking and vehicle support*). Remove the front right-hand roadwheel, then release the fasteners and remove the lower section of the wheel arch liner.

5 Remove the auxiliary drivebelt as described in Chapter 1B Section 30.

6 Remove the crankshaft pulley/vibration damper as described in Section 5.

7 Remove the timing belt outer covers as described in Section 6.

8 Using a spanner or socket on the crankshaft sprocket bolt, turn the crankshaft in the normal direction of rotation (clockwise) until the alignment mark on the face of the sprocket is almost vertical, and the hole in the camshaft sprocket hub aligns with the hole in the cylinder head **(see illustration)**.

9 While in this position it should be possible

3.8 The alignment mark on the crankshaft sprocket should be almost vertical

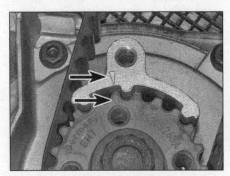

3.9b ... so the marks on the tool and sprocket align

to insert the VAG tool T10050 to lock the crankshaft, and a 6 mm diameter rod/drill bit to lock the camshafts **(see illustrations)**.
Note: *The mark on the crankshaft sprocket and the mark on the VAG tool must align, whilst at the same time the shaft of tool must engage in the drilling in the crankshaft oil seal housing.*

10 The engine is now set to TDC on No 1 cylinder.

4 Camshaft cover – removal and refitting

Removal

1 Remove the fuel injectors (Chapter 4B Section 5) and fuel rail (Chapter 4B Section 4).

3.9a Fit the tool to the hole in the oil seal housing...

3.9c Insert a 6 mm drill bit/rod to lock the camshaft hub

2 Remove the timing belt upper cover as described in Section 6.

3 Note their fitted positions, then disconnect the vacuum hoses from the camshaft cover, and release them and the wiring loom from the retaining clips at the left-hand end of the cover **(see illustration)**.

4 Squeeze together the sides of the collar, and disconnect the breather hose from the camshaft cover **(see illustration)**.

5 Release the wiring from the clips at the rear of the cover, then unscrew the camshaft cover retaining bolts and lift the cover away. If the cover sticks, do not attempt to lever it off – instead free it by working around the cover and tapping it lightly with a soft-faced mallet **(see illustration)**.

6 Recover the camshaft cover gasket. Inspect the gasket carefully, and renew it if

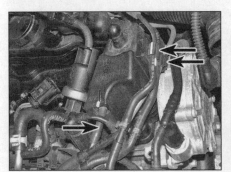

4.3 Disconnect the vacuum hoses from the camshaft cover

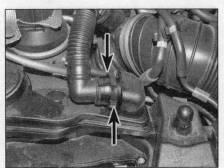

4.4 Squeeze together the sides of the collar and disconnect the breather hose

4.5 Undo the bolts and lift away the camshaft cover

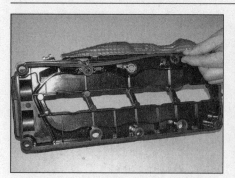

4.6a Renew the cover seal if necessary

4.6b Bolts and seals must be pushed fully through the cover before fitting the gasket

consumers and remove the ignition key.

2 Raise the front right-hand side of the vehicle, and support securely on axle stands (see *Jacking and vehicle support*). Remove the roadwheel.

3 Remove the securing fasteners and withdraw the lower section of the front wheel arch liner.

4 Slacken the bolts securing the crankshaft pulley to the sprocket **(see illustration)**. If necessary, the pulley can be prevented from turning by counterholding with a spanner or socket on the crankshaft sprocket bolt.

5 Remove the auxiliary drivebelt, as described in Chapter 1B Section 30.

6 Unscrew the bolts securing the pulley to the sprocket, and remove the pulley. Discard the bolts – new ones must be fitted.

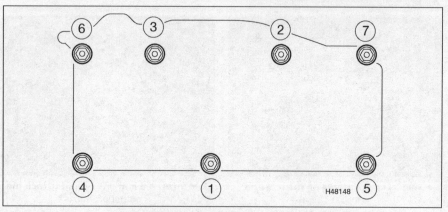

4.8 Cylinder head cover bolt tightening sequence

Refitting

7 Refit the pulley over the locating peg on the crankshaft sprocket, then fit the new pulley securing bolts.

8 Refit and tension the auxiliary drivebelt as described in Chapter 1B Section 30.

9 Prevent the crankshaft from turning as during removal, then fit the pulley securing bolts, and tighten to the specified torque.

10 Refit the wheel arch liner.

11 Refit the roadwheel and lower the vehicle to the ground.

damage or deterioration is evident – note that the retaining bolts and seals must be pushed fully through the cover **(see illustrations)**.

7 Clean the mating surfaces of the cylinder head and camshaft cover thoroughly, removing all traces of oil – take care to avoid damaging the surfaces as you do this.

Refitting

8 Refit the camshaft cover by following the removal procedure in reverse, tightening the

cover retaining bolts to the specified torque in the sequence shown **(see illustration)**.

5 Crankshaft pulley – removal and refitting

Removal

1 Switch off the ignition and all electrical

6 Timing belt covers – removal and refitting

Upper outer cover

1 Pull the engine top cover upwards to release the mountings **(see illustration)**.

2 Disconnect the wiring plug, undo the retaining bolt, and remove the exhaust

5.4 Undo the pulley bolts, counterholding it with a socket on the centre sprocket bolt

6.1 Pull the engine cover upwards from the mountings

6.2 Disconnect the wiring plug and undo the retaining bolt

6.3a Note their positions, then disconnect the hoses from the fuel filter

6.3b Plug the openings to prevent contamination

gas pressure sensor bracket by pushing it forwards **(see illustration)**.

3 Disconnect the wiring plug, unclip the fuel hoses, then release the clamps and disconnect the hoses from the filter **(see illustrations)**. Note the fitted locations of the hoses to aid refitting. Plug the openings to prevent contamination.

4 Undo the bolts/nut and remove the fuel filter assembly **(see illustration)**. Release the hose clamp, as the filter assembly is withdrawn.

5 Disconnect the fuel temperature sensor wiring plug, then release the clamp and disconnect the fuel supply pipe from the high-pressure fuel pump **(see illustration)**. Plug the openings to prevent contamination, and place a rag above the alternator to prevent fuel ingress.

6 Slide up the hose retaining clips, then undo the two retaining bolts and remove the supplementary fuel pump **(see illustration)**.

6.6 Supplementary fuel pump retaining bolts

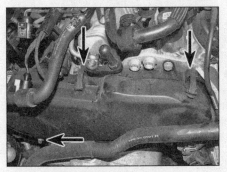

6.7b Release the clips...

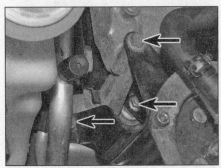

6.4 Undo the nut/bolts and remove the fuel filter

7 Disconnect the radiator outlet temperature sensor wiring plug, then release the three clips and remove the timing belt upper cover **(see illustrations)**.

8 Refitting is a reversal of removal, noting that

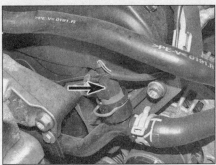

6.7a Disconnect the temperature sensor wiring plug

6.7c ... and manoeuvre the timing belt upper cover from place

6.5 Disconnect the fuel supply hose and temperature sensor

the lower edge of the upper cover engages with the centre cover.

Centre outer cover

9 Remove the crankshaft pulley as described in Section 5. It is assumed that, if the centre cover is being removed, the lower cover will be also – if not, simply remove the components described in Section 5 for access to the crankshaft pulley, and leave the pulley in position.

10 Undo the retaining nuts/bolts and move the coolant pipe that lies across the centre cover, away towards the inner wing.

11 With the upper cover removed (paragraphs 1 to 6), unscrew and remove the three retaining bolts from the centre cover. Withdraw the centre cover from the engine, noting how it fits over the lower cover **(see illustration)**. Note if the auxiliary belt tensioner

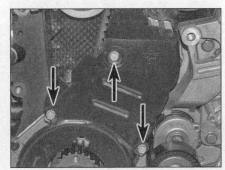

6.11 Centre timing belt cover bolts

6.15 Lower cover retaining bolts

7.4 Slacken the sprocket-to-hub bolts

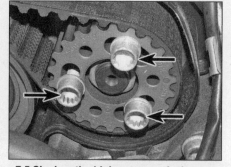

7.5 Slacken the high-pressure fuel pump sprocket bolts

is in the 'locked' position as described in the belt removal procedure, the locking drill bit/rod must be removed for access to the cover retaining bolt.

12 Refitting is a reversal of the removal procedure, using a little thread-locking compound on the retaining bolts.

Lower outer cover

13 Remove the upper and centre covers as described previously.

14 If not already done, remove the crankshaft pulley as described in Section 5.

15 Unscrew the remaining bolts securing the lower cover, and remove it **(see illustration)**.

16 Refitting is a reversal of removal; locate the centre cover in place before fitting the top two bolts.

Rear cover

17 Remove the timing belt, tensioner and sprockets as described in Section 7 and Section 8.

18 Slacken and withdraw the retaining bolts and lift the timing belt inner cover from the studs on the end of the engine, and remove it from the engine compartment.

19 Refitting is a reversal of removal.

---

7 Timing belt – removal, inspection and refitting

Note: *It is considered best practise to always replace the tensioner, idler and coolant pump whenever the timing belt is replaced. Any mounting studs should also be replaced at the same time. Most of the components (apart from the coolant pump) will be supplied with a timing belt kit. Note also that most manufactures will not guarantee a belt against premature failure if the idler(s) and tensioner are not replaced with the belt.*

Removal

1 The primary function of the toothed timing belt is to drive the camshaft, but it also drives the coolant pump and high-pressure fuel pump. Should the belt slip or break in service, the valve timing will be disturbed and piston-to-valve contact may occur, resulting in serious engine damage. For this reason, it

is important that the timing belt is tensioned correctly, and inspected regularly for signs of wear or deterioration.

2 Switch off the ignition and all electrical consumers and remove the ignition key.

3 Set the engine to TDC on No1 cylinder as described in Section 3.

4 Slacken the three bolts securing the sprocket to the camshaft hub **(see illustration)**.

5 Slacken the three bolts securing the sprocket to the high-pressure fuel pump **(see illustration)**.

6 Insert a suitable Allen key into the tensioner hub, then slacken the retaining nut and rotate the tensioner hub anti-clockwise until it can be locked in place using a 2.0 mm pin/drill bit **(see illustration)**.

7 Now rotate the tensioner hub clockwise to the stop, and hand-tighten the retaining nut.

8 If the original timing belt is to be refitted, mark the running direction of the belt, to ensure correct refitting.

Caution: If the belt appears to be in good condition and can be re-used, it is essential that it is refitted the same way around, otherwise accelerated wear will result, leading to premature failure.

9 Slide the belt from the sprockets, taking care not to twist or kink the belt excessively if it is to be re-used.

Inspection

10 Examine the belt for evidence of contamination by coolant or lubricant. If this is

the case, find the source of the contamination before progressing any further. Check the belt for signs of wear or damage, particularly around the leading edges of the belt teeth. Renew the belt if its condition is in doubt; the cost of belt renewal is negligible compared with potential cost of the engine repairs, should the belt fail in service. The belt must be renewed if it has covered the mileage given in Chapter 1B Section 31, however, if it has covered less, it is prudent to renew it regardless of condition, as a precautionary measure.

11 If the timing belt is not going to be refitted for some time, it is a wise precaution to hang a warning label on the steering wheel, to remind your self (and others) not to attempt to start the engine.

Refitting

12 Ensure that the crankshaft and camshaft are still set to TDC on No 1 cylinder, as described in Section 3. The camshaft sprocket bolts should be renewed, and slackened at this point.

13 Renew the high-pressure fuel pump sprocket bolts one at a time. They should be kept loose.

14 Using a screwdriver on the bolts heads, rotate the high-pressure fuel pump clockwise until a 6.0 mm locking pin/drill bit can be inserted into the housing adjacent to the sprocket, locking the pump in place **(see illustration)**.

7.6 Using an Allen key, slacken the nut, and rotate the hub anti-clockwise until a 2 mm rod/drill bit can be inserted to lock the hub to the pulley

7.14 Rotate the high-pressure fuel pump clockwise until a 6 mm drill bit/rod can be inserted into the housing and hub

7.15a Rotate the sprockets fully clockwise until the fuel pump sprocket …

7.15b … and camshaft sprocket bolts are at the end of the elongated holes

7.18 Timing belt routing

7.19 Rotate the tensioner clockwise until the pointer is just past the gap in the backplate

15 Rotate the camshaft sprocket and high-pressure fuel pump sprocket fully clockwise so that the securing bolts are at the end of the elongated holes **(see illustrations)**.

16 Loop the timing belt loosely under the crankshaft sprocket. **Note:** *Observe any direction of rotation markings on the belt.*

17 Fit the belt around the tensioner pulley, engage the timing belt teeth with the camshaft sprockets, then manoeuvre it into position around the coolant pump sprocket and the fuel pump sprocket. Make sure that the belt teeth seat correctly on the sprockets. **Note:** *Slight adjustment to the position of the camshaft sprocket may be necessary to achieve this. Avoid bending the belt back on itself or twisting it excessively as you do this.*

18 Finally, fit the belt around the idler roller **(see illustration)**. Ensure that any slack in the belt is in the section of belt that passes over the tensioner roller.

19 Loosen the timing belt tensioner securing nut, and pull out the tensioner locking pin. Turn the tensioner clockwise with an Allen key until the pointer is just past the middle of the gap in the tensioner backplate **(see illustration)**. With the tensioner held in this position, tighten the securing nut to the specified torque and angle.

20 Counterhold the camshaft sprocket with a homemade tool to prevent any rotation, and then tighten the camshaft sprocket and fuel pump sprocket bolts to 20 Nm (15 lbf ft). Remove the sprockets locking tools and the crankshaft locking tool.

21 Using a spanner or wrench and socket on the crankshaft pulley centre bolt, rotate the crankshaft clockwise through two complete revolutions. Reset the engine to TDC on No1 cylinder, with reference to Section 3 and refit the crankshaft locking tool.

22 Check that the tensioner roller indicator arm is centred, or within a maximum of 5 mm to the right of the notch in the backplate **(see illustration)**. If not, hold the tensioner hub stationary with an Allen key, slacken the retaining nut and position the arm in the centre of the notch. Tighten the retaining nut to the specified torque. Remove the Allen key.

23 Check that the camshaft sprocket locking pin can still be inserted. **Note:** *It's very difficult to align the locking point of the fuel pump hub*

7.22 The pointer should be centred in, or within 5 mm to the right of, the gap in the backplate

again. However, a misalignment of holes will not affect engine performance.

24 If the camshaft sprocket locking pin cannot be inserted, pull the crankshaft locking tool slight away from the engine, and rotate the crankshaft anti-clockwise slightly past TDC. Now slowly rotate the crankshaft clockwise until the camshaft sprocket locking tool can be inserted.

25 If the locating pin of the crankshaft locking tool is to the left of the corresponding hole, slacken the camshaft sprocket bolts, slowly rotate the crankshaft clockwise until the locking tool can be fully inserted. Tighten the camshaft sprocket bolts to 20 Nm (15 lbf ft).

26 If the locating pin of the crankshaft locking tool is to the right of the corresponding hole, slacken the camshaft sprocket bolts, rotate the crankshaft anti-clockwise slightly until the pin is to the left of the hole, then slowly rotate it clockwise until the lock tool can be fully inserted. Tighten the camshaft sprocket bolts to 20 Nm (15 lbf ft).

27 Remove the crankshaft and camshaft locking tools, then rotate the crankshaft two complete revolutions clockwise and check the locking tools can be reinserted. If necessary, repeat the adjustment procedure described previously.

28 Tighten the camshaft and fuel pump sprocket bolts to the specified torque.

29 The remainder of refitting is a reversal of removal.

---

8 Timing belt tensioner and sprockets – removal and refitting

---

Timing belt tensioner

Removal

1 In order to remove the timing belt tensioner, the engine mounting bracket must first be removed. Either support the engine from above using a crossbeam or an engine hoist, or support if from underneath with a trolley jack and block of wood.

2 Remove the timing belt as described in Section 7.

3 Undo the bolts and remove the right-hand engine mounting.

4 Undo the bolt securing the coolant pipe to the mounting bracket **(see illustration)**.

5 Working in the wheel arch area, undo the nut securing the lower end of the coolant pipe.

6 Undo the 3 retaining bolts and remove the engine mounting bracket **(see illustration)**.

7 Unscrew the timing belt tensioner nut, and remove the tensioner from the engine.

Refitting

8 When refitting the tensioner to the engine, ensure that the lug on the tensioner backplate engages with the corresponding cut-out in the rear timing belt cover, then refit the tensioner nut **(see illustration)**.

9 The remainder of refitting is a reversal of removal.

Idler pulleys

Removal

10 Remove the timing belt as described in Section 7.

11 Unscrew the relevant idler pulley/roller securing bolt/nut, and then withdraw the pulley.

Refitting

12 Refit the pulley and tighten the securing bolt or nut to the specified torque. **Note:** *Renew the large roller/pulley retaining bolt (where applicable).*

13 Refit and tension the timing belt as described in Section 7.

Crankshaft sprocket

Note: *A new crankshaft sprocket securing bolt must be used on refitting.*

Removal

14 Remove the timing belt as described in Section 7.

15 The sprocket securing bolt must now be slackened, and the crankshaft must be prevented from turning as the sprocket bolt is unscrewed. To hold the sprocket, make up a suitable tool, and bolt it to the sprocket using two bolts bolted into two of the crankshaft pulley bolt holes.

16 Hold the sprocket using the tool, then slacken the sprocket securing bolt. Take care, as the bolt is very tight. Do not allow the crankshaft to turn as the bolt is slackened.

17 Unscrew the bolt, and slide the sprocket from the end of the crankshaft, noting which way round the sprocket's raised boss is fitted.

Refitting

18 Commence refitting by positioning the sprocket on the end of the crankshaft.

19 Fit a new sprocket securing bolt, then counter hold the sprocket using the method employed on removal, and tighten the bolt to the specified torque in the two stages given in the Specifications.

20 Refit the timing belt as described in Section 7.

Camshaft sprocket

Removal

21 Remove the timing belt as described in Section 7, then rotate the crankshaft 90° anti-clockwise to prevent any accidental piston-to-valve contact.

22 Unscrew and remove the three retaining bolts and remove the camshaft sprocket from the camshaft hub.

Refitting

23 Refit the sprocket ensuring that it is fitted the correct way round, as noted before removal, then insert the new sprocket bolts, and tighten by hand only at this stage.

24 If the crankshaft has been turned, turn the crankshaft clockwise 90° back to TDC.

25 Refit and tension the timing belt as described in Section 7.

Camshaft hub

Note: *Seat technicians use VAG special tool T10051 to counter hold the hub, however it is possible to fabricate a suitable alternative.*

Removal

26 Remove the camshaft sprocket as described previously in this Section.

27 Engage special tool T10051 with the three locating holes in the face of the hub to prevent the hub from turning. If this tool is not available, fabricate a suitable alternative. Whilst holding the tool, undo the

8.4 Coolant pipe upper mounting bolt and lower mounting nut

8.6 Engine mounting bracket bolts

8.8 Ensure the lug on the backplate engages with the cut-out in the timing belt cover

8.27 Fabricate a home-made tool to counterhold the hub. Undo the bolt …

8.28 … and slide the hub from the camshaft

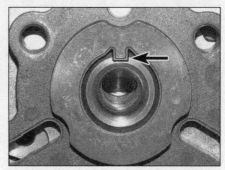

8.29 Ensure the integral key aligns with the keyway in the camshaft

central hub retaining bolt about two turns **(see illustration)**.

28 Remove the central hub retaining bolt, and slide the hub from the camshaft. If necessary, attach VAG tool T10052 (or a similar three-legged puller) to the hub, and evenly tighten the puller until the hub is free of the camshaft taper **(see illustration)**.

Refitting

29 Ensure that the camshaft taper and the hub centre are clean and dry, locate the hub on the taper, noting that the built-in key in the hub taper must align with the keyway in the camshaft taper **(see illustration)**.

30 Hold the hub in this position with tool T10051 (or similar home-made tool), and tighten the central bolt to the specified torque.

31 Refit the camshaft sprocket as described previously in this Section.

Coolant pump sprocket

32 The coolant pump sprocket is integral with the coolant pump. Refer to Chapter 3 Section 7 for details of coolant pump removal.

9 Camshaft and hydraulic tappets – removal, inspection and refitting

Note: *A new camshaft oil seal(s) will be required on refitting. VAG removal tool T40094 (or similar tool) will be required to refit the camshafts – this is necessary to prevent*

damage to the retaining frame and cylinder head as the camshafts are refitted.

Removal

1 Remove the camshaft hub as described in Section 8.

2 Remove the camshaft cover as described in Section 4.

3 Remove the brake vacuum pump as described in Chapter 9 Section 21.

4 Progressively unscrew the camshaft retaining frame bolts in the reverse of the sequence shown in illustration 9.20, and carefully remove the retaining frame.

5 Carefully lift the camshafts from the cylinder head, keeping them identified for location. Remove the oil seal from the end of the camshaft and discard it – a new one will be required for refitting.

6 Lift the rocker arms and hydraulic tappets from place. Store the rockers and tappets in a container with numbered compartments to ensure they are refitted to their correct locations. It is recommended that the tappets are kept immersed in oil for the period they are removed from the cylinder head.

Inspection

7 With the camshafts removed, examine the retaining frame and the bearing locations in the cylinder head for signs of obvious wear or pitting. If evident, a new cylinder head will probably be required. Also check that the oil supply holes in the cylinder head are free from obstructions.

8 Visually inspect the camshafts for evidence

of wear on the surfaces of the lobes and journals. Normally their surfaces should be smooth and have a dull shine; look for scoring, erosion or pitting and areas that appear highly polished, indicating excessive wear. Accelerated wear will occur once the hardened exterior of the camshaft has been damaged, so always renew worn items. **Note:** *If these symptoms are visible on the tips of the camshaft lobes, check the corresponding rocker arm, as it will probably be worn as well.*

9 If the machined surfaces of the camshaft appear discoloured or blued, it is likely that it has been overheated at some point, probably due to inadequate lubrication. This may have distorted the shaft, so have the camshaft runout and endfloat checked by an automotive engine reconditioning specialist.

10 Inspect the hydraulic tappets for obvious signs of wear or damage, and renew if necessary. Check that the oil holes in the tappets are free from obstructions.

Refitting

11 Oil the rocker arms and hydraulic tappets, and then refit them to their original positions.

12 After fitting hydraulic tappets, wait a minimum of 30 minutes (or preferably, leave overnight) before starting the engine, to allow the tappets time to settle, otherwise the valve heads will strike the pistons.

13 To set up the tool, remove the supports number 3, 4 and 5, then install the supports number 1, 2, 9 and 10 as shown **(see illustrations)**.

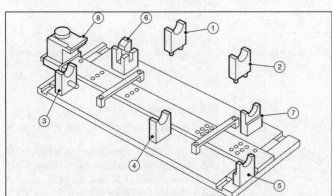

9.13a The different elements of tool No T40094

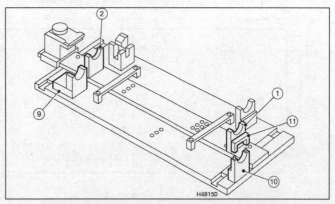

9.13b Position tools No 1, 2, 9 and 10 as shown

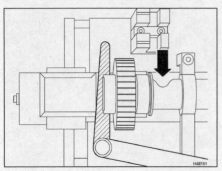

9.14 Position the inlet camshaft on the tool with the bolt indent facing outwards, then slide tool No 8 in to the slot in the end of the camshaft and use a 0.50 mm feeler gauge to remove any free play

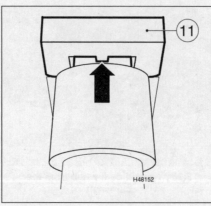

9.15 Fit tool No 11 into the slot in the end of the exhaust camshaft

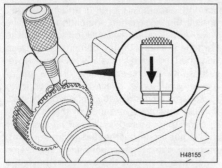

9.16 Tighten the thumbwheel to align the gear teeth. Ensure the clamping jaw with the arrow in seated on the wider gear

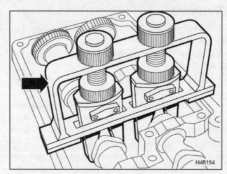

9.19 Secure the camshafts in place in the frame using tool No T40095

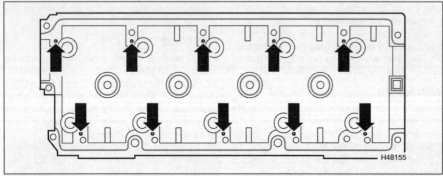

9.20 Apply a 2.0 mm thick bead of sealant to the area shown by the thick, black line. Take care not to block the oil holes

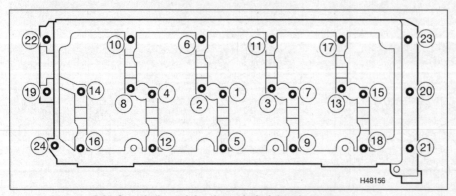

9.21 Camshaft retaining frame bolt tightening sequence

14 Position the inlet camshaft as shown with the cylinder head bolt indent facing outwards, then slide the support number 8 into the slot in the end of the camshaft and remove any free play with a 0.50 mm feeler gauge **(see illustration)**.

15 Position the exhaust camshaft on supports numbers 9 and 10, and fit the tool No 11 into the slot in the end of the camshaft **(see illustration)**.

16 Fit the clamping tool No T40096 to the gear on the exhaust camshaft, tightening the knurled thumb wheel until the faces of the gear teeth are in alignment. If necessary, use a 13 mm spanner **(see illustration)**.

17 Slide the exhaust camshaft towards the inlet camshaft until the gear teeth engage.

18 Ensure the gasket faces of the retaining frame are clean, then apply a smear of clean engine oil to the bearing surfaces and lower the frame into position over the camshafts. Ensure the bearing surfaces locate correctly on the camshafts.

19 Fit the clamping tool No T40095 over the camshafts and frame, and tighten the thumbwheels to hold the camshafts in position in the frame **(see illustration)**.

20 Ensure the sealing surfaces of the cylinder head are clean, and then apply a 2.0 mm wide bead of sealant (D 176 501 A1 or equivalent) as shown. Take care not to apply too much sealant, ensuring the oil holes supply holes are not blocked **(see illustration)**.

21 Slide out tool No's 8 and 11, then lift the camshafts, retaining frame and clamping tool from the tool No T40094. Place the camshafts, frame and tool in place on the cylinder head. Progressively and carefully, hand-tighten the frame retaining bolts in the sequence shown, until the retaining frame makes contact with the cylinder head over the complete surface, then tighten the bolts to the specified torque, again in the correct sequence **(see illustration)**.

22 Remove the gear aligning tool (T40096) and the clamping tool (T40095).

23 Renew the camshaft oil seal (Section 10), then drive in a new sealing cap.

24 The remainder of refitting is a reversal of removal.

10 Camshaft oil seals – renewal

Right-hand oil seal

1 Remove the camshaft sprocket and hub, as described in Section 8.

2 Drill two small holes into the existing oil seal, diagonally opposite each other. Take great care to avoid drilling through into the seal housing or camshaft sealing surface. Thread two self-tapping screws into the holes, and using a pair of pliers, pull on the heads of the screws to extract the oil seal **(see illustration)**.

3 Clean out the seal housing and the sealing surface of the camshaft by wiping it with a lint-free cloth. Remove any swarf or burrs that may cause the seal to leak.

4 Do not lubricate the lip and outer edge of the new oil seal, push it over the camshaft until it is positioned in place above its housing. To prevent damage to the sealing lips, wrap some adhesive tape around the end of the camshaft.

5 Using a hammer and a socket of suitable diameter, drive the seal squarely into its housing. **Note:** *Select a socket that bears only on the hard outer surface of the seal, not the inner lip that can easily be damaged.*

6 Refit the camshaft sprocket and its hub, as described in Section 8.

Left-hand oil seal

7 The left-hand camshaft oil seal is formed, by the brake vacuum pump seal, which is fitted to the cylinder head. Refer to Chapter 9 Section 21 for details of brake vacuum pump removal and refitting.

11 Cylinder head – removal, inspection and refitting

Note: *The cylinder head must be removed with the engine cold. New cylinder head bolts and a new cylinder head gasket will be required on refitting, and suitable studs will be required to guide the cylinder head into position – see text.*

Removal

1 Remove the battery as described in Chapter 5A Section 3.

2 Drain the cooling system (Chapter 1B Section 33) and engine oil (Chapter 1B Section 6).

3 Pull the plastic cover on the top of the engine upwards from its mountings.

4 Remove the air filter housing as described in.

5 Undo the 3 bolts and remove the battery tray **(see illustration)**.

6 Remove the radiator cooling fan(s) and shroud as described in Chapter 3 Section 3.

7 Undo the bolts and remove the air hose/duct from the intercooler to the turbocharger. Release the wiring looms from the clips as necessary to enable the duct to be manoeuvred from place.

8 Remove the camshaft cover as described in Section 4.

9 Remove the camshaft sprocket and hub as described in Section 8.

10 Disconnect the wiring plugs from the EGR valve and throttle body/intake manifold flap.

11 Disconnect the charge air pressure sensor wiring plug, then undo the 2 retaining bolts, release the clamps, and remove the charge air pipe and hose from the front of the engine **(see illustrations)**.

12 Undo the bolt securing the oil level dipstick guide tube to the throttle body/intake manifold flap **(see illustration)**.

10.2 Bolt in a self-tapping screw, then pull the bolt and seal from place

13 Undo the 2 bolts securing the connecting pipe to the EGR valve **(see illustration)**.

14 Undo the front retaining bolt, twist the charge air pipe clockwise and disconnect it from the turbocharger. Note their fitted positions, and unclip the vacuum hoses from the charge air pipe **(see illustration)**.

11.11a Disconnect the pressure sensor wiring plug…

11.12 Oil level dipstick guide tube bolt

11.14 Charge air pipe retaining bolt. Note the position of the vacuum hoses

11.5 Battery tray bolts

15 Apply a little lubrication spray to the rubber sleeve, pull up the pipe from the vacuum pump, then undo the 4 retaining bolts and remove the vacuum pump from the left-hand end of the cylinder head **(see illustration)**. Renew the pump-to-cylinder head seal.

11.11b … then remove the charge air pipe and hose

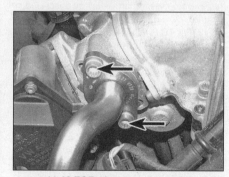

11.13 EGR pipe-to-valve bolts

11.15 With some lubrication, the hose connection pulls up from the vacuum pump

11.19 EGR pipe-to-cooler bolts

11.21 Remove the EGR pipe between the manifold and cooler

11.23a Undo the banjo bolt at the base of the turbocharger support bracket/oil return pipe and the bolt at the top

11.23b Note the 2 O-ring seals at the base of the oil return pipe

16 Disconnect the coolant temperature sensor wiring plug at the left-hand end of the cylinder head, and release the wiring loom from any retaining clips.

17 Disconnect the gear change cables from the levers on the transmission, as described in Chapter 7A Section 2 (manual transmission), or Chapter 7B Section 4 (DSG transmission).

18 Undo the bolts/nut, securing the gear-change bracket to the top of the transmission. Move the bracket and cables to one side.

19 Undo the bolts securing the EGR pipe to the cooler at the left-hand end of the engine, and the nut securing the bracket to the cylinder head, then remove the pipe (see illustration). Recover the gasket at each end.

20 Working underneath the vehicle, slacken the Allen bolt and release the clamp securing the diesel particulate filter/catalytic converter

to the turbocharger, then undo the bolts/nuts securing the brackets to the cylinder block/head and lay the filter/converter to one side.

21 Undo the nuts and multi-spline bolts securing the EGR pipe to the right-hand end of the exhaust manifold and EGR cooler (see illustration). Remove the pipe and recover the gaskets.

22 Trace the exhaust manifold gas temperature sensor wiring back, releasing it from any retaining clips, disconnect its wiring plug at the bulkhead, and slide it from the retaining bracket. Unclip the wiring loom from the top of the turbocharger heat shield.

23 Undo the bolt securing the support bracket/oil return pipe to the underside of the turbocharger, and then undo the bolt securing the top of the support bracket to the underside of the turbocharger. Now pull

and twist the support bracket to disconnect it from the oil return pipe at the top. Remove the pipe/bracket (see illustrations). Note the O-ring and copper sealing washer on the lower banjo bolt, and the two O-rings fitted to the base of the oil return pipe still fitted to the turbocharger.

24 Note their fitted locations, then release the clamps and disconnect the various coolant hoses from the cylinder head.

25 Undo the bolt securing the turbocharger oil supply pipe bracket at the left-hand end of the cylinder head, the nut securing the bracket on the rear of the head, then undo the union bolts and remove the pipe. Renew any seals.

26 Undo the bolt securing the timing belt guard adjacent to the timing belt tensioner, and the bolt securing the camshaft position sensor, then remove the tensioner retaining nut (see illustration).

27 Disconnect the turbocharger wastegate position sensor wiring plug, and the intake manifold changeover valve wiring plug. Note their fitted positions and disconnect the vacuum hoses from the EGR cooler (where the plastic pipe joins the metal pipe at the back of the cylinder head) and the turbocharger wastegate actuator. Undo the nuts securing the boost pressure solenoid valve to the bulkhead. Release the loom wiring plug, lay the wiring loom and vacuum hoses over the front of the engine, clear of the cylinder head (see illustration). Make a final check to ensure all relevant wiring and vacuum hoses have been disconnected. Note the loom/hose routing to aid refitting.

28 Using an M12 multi-splined tool (12-pointed star), undo the cylinder head bolts, working from the outside-in, evenly and gradually (see illustration). Remove the bolts and recover the washers. Check that nothing remains connected, and starting at the gearbox side, lift the cylinder head from the engine block, sliding the belt tensioner from the mounting stud as the cylinder head is removed. Seek assistance if possible, as it is a heavy assembly, especially as it is being removed complete with the manifolds.

29 Remove the gasket from the top of the block, noting the locating dowels. If the dowels are a loose fit, remove them and store

11.26 Undo the bolt securing the timing belt guard

11.27 Undo the boost pressure solenoid valve nuts

11.28 Undo the cylinder head bolts using a M12 multi-splined (12-pointed star) tool

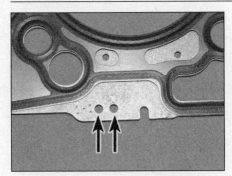

11.31 The holes identify the thickness of the cylinder head gasket

them with the head for safekeeping. Do not discard the gasket yet – it will be needed for identification purposes. If desired, the manifolds can be removed from the cylinder head with reference to Chapter 4B Section 6 (inlet manifold) or Chapter 4D Section 5 (turbocharger/exhaust manifold).

Inspection

30 Dismantling and inspection of the cylinder head is covered in Chapter 2F Section 6.

Cylinder head gasket selection

Note: *A dial test indicator (DTI) will be required for this operation.*

31 Examine the old cylinder head gasket for manufacturer's identification markings **(see illustration)**. These will be in the form of holes, and a part number on the edge of the gasket. Unless new pistons have been fitted, the new cylinder head gasket must be of the same type as the old one.

32 If new piston assemblies have been fitted as part of an engine overhaul, or if a new short engine is to be fitted, the projection of the piston crowns above the cylinder head mating face of the cylinder block at TDC must be measured. This measurement is used to determine the thickness of the new cylinder head gasket required.

33 Anchor a dial test indicator (DTI) to the top face (cylinder head gasket mating face) of the cylinder block, and zero the gauge on the gasket mating face.

34 Rest the gauge probe on No 1 piston

11.34 Measure the piston protrusion using a DTI guage

crown, and turn the crankshaft slowly by hand until the piston reaches TDC. Measure and record the maximum piston projection at TDC **(see illustration)**.

35 Repeat the measurement for the remaining pistons, and record the results.

36 If the measurements differ from piston-to-piston, take the highest figure, and use this to determine the thickness of the head gasket required as follows.

Piston projection	Gasket identification (number of holes)
0.91 to 1.00 mm	1
1.01 to 1.10 mm	2
1.11 to 1.20 mm	3

37 Purchase a new gasket according to the results of the measurements.

Refitting

38 The mating faces of the cylinder head and block must be perfectly clean before refitting the head. Use a scraper to remove all traces of gasket and carbon, also clean the tops of the pistons. Take particular care with the aluminium surfaces, as the soft metal is easily damaged.

39 Make sure that debris is not allowed to enter the oil and water passages – this is particularly important for the oil circuit, as carbon could block the oil supply to the camshaft and crankshaft bearings. Using adhesive tape and paper, seal the water, oil and bolt holes in the cylinder block.

40 To prevent carbon entering the gap between the pistons and bores, smear a little grease in the gap. After cleaning a piston, rotate the crankshaft to that the piston moves down the bore, and then wipe out the grease and carbon with a cloth rag. Clean the other piston crowns in the same way.

41 Check the head and block for nicks, deep scratches and other damage. If slight, they may be removed carefully with a file. More serious damage may be repaired by machining, but this is a specialist job.

42 If warpage of the cylinder head is suspected, use a straight-edge to check it for distortion, as described in Chapter 2F Section 7.

43 Ensure that the cylinder head bolt holes in the crankcase are clean and free of oil. Syringe or soak up any oil left in the bolt holes. This is most important in order that the correct bolt tightening torque can be applied, and to prevent the possibility of the block being cracked by hydraulic pressure when the bolts are tightened.

44 Turn the crankshaft anti-clockwise all the pistons at an equal height, approximately halfway down their bores from the TDC position (see Section 3). This will eliminate any risk of piston-to-valve contact as the cylinder head is refitted.

45 Where applicable, refit the manifolds with reference to Chapter 4B Section 6 (inlet manifold) or Chapter 4D Section 5 (turbocharger/exhaust manifold).

46 Ensure that the cylinder head locating dowels are in place in the cylinder block, and then fit the new cylinder head gasket over the dowels, ensuring that the part number is uppermost **(see illustration)**. Note that Seat recommend that the gasket is only removed from its packaging immediately prior to fitting.

47 Lower the cylinder head into position on the gasket, ensuring that it engages correctly over the dowels. Refit the timing belt tensioner as the cylinder head is refitted.

48 Fit the washers in place then fit the new cylinder head bolts to the locations, and bolt them in as far as possible by hand. Do not oil the bolt threads.

49 Working progressively, in sequence, tighten all the cylinder head bolts to the specified Stage 1 torque **(see illustrations)**.

11.46 Ensure the dowels are in place, then fit the new gasket with the part number uppermost

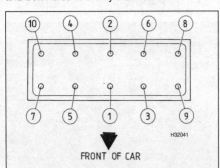

FRONT OF CAR

H32041

11.49a Cylinder head bolt tightening sequence

11.49b Tighten the cylinder head bolts to the Stage 1 torque

11.51 Use an angle-tightening gauge

12.4 Raise the clip and disconnect the air duct from the intercooler

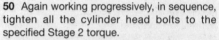

12.6 Electric coolant circulation pump retaining bolt

50 Again working progressively, in sequence, tighten all the cylinder head bolts to the specified Stage 2 torque.

51 Tighten all the cylinder head bolts, in sequence, through the specified Stage 3 angle **(see illustration)**.

52 Finally, tighten all the cylinder head bolts, in sequence, through the specified Stage 4 angle.

53 The remainder of the refitting procedure is a reversal of the removal procedure, noting the following points:
a) *Tighten all fasteners to their specified torque where given.*
b) *Renew all seals and gaskets.*
c) *Refill the cooling system, as described in Chapter 1B Section 33.*
d) *Refill the engine oil, as described in Chapter 1B Section 6.*
e) *Ensure all wiring is correctly routed.*

12 Sump – removal and refitting

Removal

1 Apply the handbrake, then jack up the front of the vehicle and support securely on axle stands (see *Jacking and vehicle support*).

2 Remove the securing bolts and withdraw the engine undershield(s).

3 Drain the engine oil as described in Chapter 1B Section 6.

4 Raise the retaining clip and remove the air duct from the intercooler outlet **(see illustration)**.

5 Undo the retaining bolts, release the clamp and remove the charge air pipe from the front of the cylinder block **(see illustration 11.11a and 11.11b)**. Disconnect the charge air pressure sensor wiring plug as the pipe is withdrawn.

6 Undo the retaining bolt and move the electric coolant circulation pump to one side **(see illustration)**.

7 Undo the bolt securing the turbo-to-intercooler air pipe to the sump.

8 Disconnect the wiring connector from the oil level/temperature sender in the sump.

9 Pull the sump insulation cover downwards at the rear to release the retaining clips, then prise down the centre clip and pull the clip on the front side of the cover downwards (where fitted) **(see illustrations)**.

10 Unscrew and remove the bolts securing the sump to the cylinder block, and the bolts securing the sump to the transmission casing, then withdraw the sump. If necessary, release the sump by tapping with a soft-faced hammer.

Refitting

11 Begin refitting by thoroughly cleaning the mating faces of the sump and cylinder block. Ensure that all traces of old sealant are removed.

12 Ensure that the cylinder block mating face of the sump is free from all traces of old sealant, oil and grease, and then apply a 2.0

to 3.0 mm thick bead of silicone sealant (VAG D 176 404 A2 or equivalent) to the sump **(see illustration)**. Note that the sealant should be run around the inside of the bolt holes in the sump. The sump must be fitted within 5 minutes of applying the sealant.

13 Offer the sump up to the cylinder block, then refit the sump-to-cylinder block bolts, and lightly tighten them by hand, working progressively in a diagonal sequence. **Note:** *If the sump is being refitted with the engine and transmission separated, make sure that the sump is flush with the flywheel end of the cylinder block.*

14 Refit the sump-to-transmission casing bolts, and tighten them lightly, using a socket.

15 Again working in a diagonal sequence, lightly tighten the sump-to-cylinder block bolts, using a socket.

16 Tighten the sump-to-transmission casing bolts to the specified torque.

17 Working in a diagonal sequence, progressively tighten the sump-to-cylinder block bolts to the specified torque.

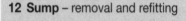

12.9a Pull the insulation down at the rear to release the clips

12.9b At the front, prise down the centre pin and pull the clip downwards

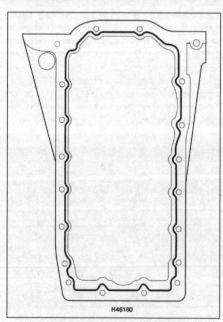

H46180

12.12 Apply a bead of sealant around the inside of the bolt holes

13.2a Remove the circlip …

13.2b … and pull out the oil pump shaft using an M3 bolt

13.3 Oil pump pick-up pipe bolts

18 The remainder of refitting is a reversal of removal, noting to allow at least 30 minutes from the time of refitting the sump for the sealant to dry, then refill the engine with oil, with reference to Chapter 1B Section 6.

13 Oil pump and balance shaft assembly – removal, inspection and refitting

Removal

1 Remove the sump as described in Section 12.

Oil pump

2 Remove the retaining circlip, then pull the oil pump shaft out using an M3 bolt **(see illustrations)**.

3 Undo the retaining bolts, and then detach the pick-up pipe from the pump **(see illustration)**.

4 Undo the bolts and detach the oil pump from the balance shaft assembly **(see illustration)**.

Balance shaft assembly

5 Lock the camshafts and crankshaft at TDC on No 1 cylinder as described in Section 3.

6 Working gradually and evenly, undo the retaining bolts and detach the balance shaft assembly from the base of the cylinder block.

Inspection

7 At the time of writing, it would appear that no parts are available for the oil pump or balance shaft assembly. If defective, the oil pump or

13.4 Oil pump mounting bolts

balance shaft assembly must be renewed. Consult a Seat dealer or parts specialist.

Refitting

Oil pump

8 Refit the pump to the balance shaft assembly and tighten the retaining bolts to the specified torque.

9 Refit the oil pick-up pipe using a new O-ring, then tighten the retaining bolts to the specified torque **(see illustration)**.

10 Push the driveshaft into place, and secure it with the circlip.

11 The remainder of refitting is a reversal of removal.

Balance shaft assembly

Note: *If the original balance shaft assembly is being refitted, it's essential that neither the*

13.9 Renew the pick-up pipe O-ring seal

drivegear on the crankshaft or the crankshaft itself has been renewed, or the idler gear bolt has been slackened. If they have, proceed under the heading for the installation of a new balance shaft assembly.

Refitting the original assembly

12 Rotate the balance shaft until VAG tool No T10255 can be fitted into the groove on the left-hand end of the rear shaft **(see illustrations)**.

13 Ensure the engine is still locked at TDC for No 1 cylinder, and then position the balance shaft assembly over the locating dowels on the base of the cylinder block. The idler gear must engage with the drivegear of the crankshaft, and there must be noticeable backlash.

14 Fit the new balance shaft assembly retaining bolts, and working from the centre

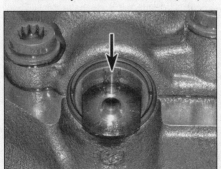

13.12a Rotate the balance shaft until the groove on the end of the rear shaft is vertical …

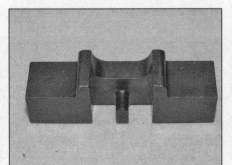

13.12b … and tool No T10255 …

13.12c … can be fitted

outwards, tighten them to the specified torque. Remove the special tool.

15 The remainder of refitting is a reversal of removal.

Fitting a new balance shaft assembly

16 New balance shaft assemblies are supplied with an idler gear with a special coating. Once fitted, the coating wears down to give the correct backlash between the gears.

17 Ensure the engine is still locked at TDC for No 1 cylinder as described in Section 3.

18 Slacken the idler gear retaining bolt 90°.

19 Position the balance shaft assembly over the locating dowels on the base of the cylinder block, ensuring the white mark on the idler gear is centrally aligned with the crankshaft drivegear. Idler gears not marked with a white mark can be installed in any position.

20 Fit the new balance shaft assembly retaining bolts, and working from the centre outwards, tighten them to the specified torque.

21 Rotate the balance shaft until VAG tool No T10255 can be fitted into the groove on the left-hand end of the rear shaft **(see illustration 13.12a, 13.12b and 13.12c).**

22 Fit the balance shaft drivegear onto the shaft so the holes in the gear align with the holes in the shaft. Tighten the retaining bolts to the specified torque.

23 Have an assistant push the idler gear between the two gears to remove any backlash. At the same time, rotate the balance shaft anti-clockwise slightly, and tighten the idler gear retaining bolt to the specified

torque. Remove the balance shaft locking tool.

24 The remainder of refitting is a reversal of removal.

14 Flywheel – removal, inspection and refitting

Removal

1 On manual transmission models, remove the transmission (see Chapter 7A Section 3) and clutch (see Chapter 6 Section 6).

2 On semi-automatic transmission models, remove the DSG transmission as described in Chapter 7B Section 2.

3 Manual models are fitted with a dual-mass flywheel, which can only be fitted in one position due to the offset of the mounting bolt holes in the end of the crankshaft.

4 Rotate the outside of the dual-mass flywheel so that the bolts align with the holes (if necessary).

5 Unscrew the bolts and remove the flywheel. Use a locking tool to counterhold the flywheel **(see illustration)**. Discard the bolts, as new ones must be fitted. **Note:** *In order not to damage the flywheel, do not allow the bolt heads to make contact with the flywheel during the unscrewing procedure.*

Inspection

6 Check the flywheel for wear and damage. Examine the starter ring gear for excessive wear to the teeth. If the driveplate or its ring

gear is damaged, the complete driveplate must be renewed. The flywheel ring gear, however, may be renewed separately from the flywheel, but the work should be entrusted to a Seat dealer. If the clutch friction face is discoloured or scored excessively, it may be possible to regrind it, but this work should also be entrusted to a Seat dealer.

7 The following are guidelines only, but should indicate whether professional inspection is necessary. The dual-mass flywheel should be checked as follows:

8 There should be no cracks in the drive surface of the flywheel. If cracks are evident, the flywheel may need renewing.

Warpage

9 Place a straightedge across the face of the drive surface, and check by trying to insert a feeler gauge between the straightedge and the drive surface **(see illustration)**. The flywheel will normally warp like a bowl – i.e. higher on the outer edge. If the warpage is more than 0.40 mm, the flywheel may need renewing.

Free rotational movement

10 This is the distance the drive surface of the flywheel can be turned independently of the flywheel primary element, using finger effort alone. Move the drive surface in one direction and make a mark where the locating pin aligns with the flywheel edge. Move the drive surface in the other direction (finger pressure only) and make another mark **(see illustration)**. The total of free movement should not exceed 20.0 mm. If it's more, the flywheel may need renewing.

Total rotational movement

11 This is the total distance the drive surface can be turned independently of the flywheel primary element. Insert two bolts into the clutch pressure plate/damper unit mounting holes, and with the crankshaft/flywheel held stationary, use a lever/pry bar between the bolts and use some effort to move the drive surface fully in one direction – make a mark where the locating pin aligns with the flywheel edge. Now force the drive surface fully in the opposite direction, and make another mark. The total rotational movement should not exceed 44.0 mm. If it does, have the flywheel professionally inspected.

Lateral movement

12 The lateral movement (up and down) of the drive surface in relation to the primary element of the flywheel should not exceed 2.0 mm. If it does, the flywheel may need renewing. This can be checked by pressing the drive surface down on one side into the flywheel (flywheel horizontal) and making an alignment mark between the drive surface and the inner edge of the primary element. Now press down on the opposite side of the drive surface, and make another mark above the original one. The difference between the two marks is the lateral movement **(see illustration)**.

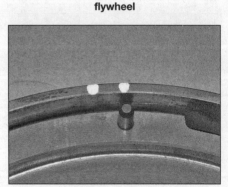

14.5 Use a locking tool to counterhold the flywheel

14.9 Flywheel warpage check – see text

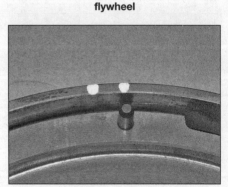

14.10 Flywheel free rotational movement check alignment marks – see text

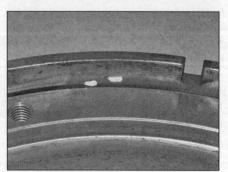

14.12 Flywheel lateral movement check marks – see text

Refitting

13 Refitting is a reversal of removal. Use new bolts when refitting the flywheel or driveplate, and coat the threads of the bolts with locking fluid before inserting them. Tighten them to the specified torque.

15 Crankshaft oil seals – renewal

Note: *The oil seals are a PTFE (Teflon) type and are fitted dry, without using any grease or oil. These have a wider sealing lip and have been introduced instead of the coil spring type oil seal.*

Timing belt end oil seal

1 Remove the timing belt as described in Section 7, and the crankshaft sprocket with reference to Section 8.

2 To remove the seal without removing the housing, drill two small holes diagonally opposite each other, insert self-tapping screws, and pull on the heads of the screws with pliers **(see illustration)**.

3 Alternatively, to remove the oil seal complete with its housing, proceed as follows.

a) *Remove the sump as described in Section 12. This is necessary to ensure a satisfactory seal between the sump and oil seal housing on refitting.*

b) *Unscrew and remove the oil seal housing.*

c) *Working on the bench, lever the oil seal from the housing using a suitable screwdriver. Take care not to damage the seal seating in the housing* **(see illustration)**.

4 Thoroughly clean the oil seal seating in the housing.

5 Wind a length of tape around the end of the crankshaft to protect the oil seal lips as the seal (and housing, where applicable) is fitted.

6 Fit a new oil seal to the housing, pressing or driving it into position using a socket or tube of suitable diameter. Ensure that the socket or tube bears only on the hard outer ring of

15.2 Pull the screw and seal from place using pliers

the seal, and take care not to damage the seal lips. Press or drive the seal into position until it is seated on the shoulder in the housing. Make sure that the closed end of the seal is facing outwards.

7 If the oil seal housing has been removed, proceed as follows, otherwise proceed to paragraph 11.

8 Clean all traces of old sealant from the crankshaft oil seal housing and the cylinder block, then coat the cylinder block mating faces of the oil seal housing with a 2.0 to 3.0 mm thick bead of silicone sealant (D 176 404 A2, or equivalent). Note that the seal housing must be refitted within 5 minutes of applying the sealant.

Caution: DO NOT put excessive amounts of sealant onto the housing as it may get into the sump and block the oil pick-up pipe.

9 Refit the oil seal housing, and tighten the bolts progressively to the specified torque **(see illustration)**.

10 Refit the sump as described in Section 12.

11 Refit the crankshaft sprocket with reference to Section 8, and the timing belt as described in Section 7.

Flywheel end oil seal

Note: *In these engines, the seal, sealing flange and sender wheel are a complete unit. Special tools are required to refit the sealing flange,*

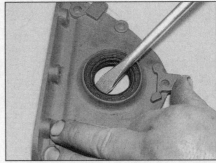
15.3 Prise the oil seal from the crankshaft oil seal housing

and press the sender wheel onto the end of the crankshaft. It is not possible to accurately fit these parts without the tools, which may be available from VAG (part No T10134) and are available from aftermarket automotive tool specialists.

12 Remove the flywheel as described in Section 14, then prise the intermediate plate from the locating dowels on the cylinder block.

13 Undo the bolts securing the sealing flange to the cylinder block **(see illustration)**.

14 Insert three 6 x 35 mm bolts into the threaded holes in the sealing flange. Tighten the bolts gradually and evenly, and press the sealing flange and sender wheel from the crankshaft/cylinder block **(see illustration)**. The seal, sender wheel and sealing flange are supplied as a complete unit.

15 Ensure the mating face of the cylinder block is clean and free from debris. The new sealing flange/seal/sender wheel assembly is supplied with a sealing lip support ring, which serves as a fitting sleeve, and must not be removed prior to installation. Equally, the sender wheel must not be separated from the assembly.

16 If using the VAG tool, proceed as follows. If using an aftermarket tool specialist's product, follow the instructions supplied with the tool. Rotate the large spindle nut until it's level with the end of the clamping surface of

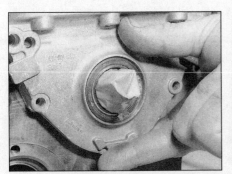
15.9 Slide the oil seal housing over the end of the crankshaft

15.13 Sealing flange bolts

15.14 Bolt in three 6 x 35 mm bolts and draw the sealing flange and sender wheel from place

15.16a Rotate the nut until it is level with the end of the clamping surface...

15.16b ... then clamp it in a vice

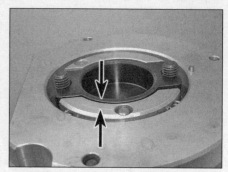

15.17 Rotate the nut until the inner part of the tool is flush with the housing

15.18a Remove the securing clip ...

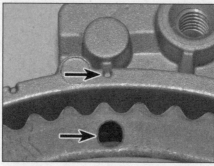

15.18b ... the hole in the sender wheel should align with the marking on the flange

the spindle, then clamp the spindle in a vice **(see illustrations)**.

17 Press the tool housing downwards until it rests on the nut and washer. Rotate the nut until the inner part of the tool is at the same height as the housing **(see illustration)**.

18 Remove the seal securing clip. The hole on the sender wheel must align with the marking on the sealing flange **(see illustrations)**.

19 Place the flange outer side down on a clean, flat surface, then press the seal guide fitting sleeve (supplied ready fitted), housing, and sender wheel downwards until all the components are flat on the surface. In this position the upper edge of the sender wheel should be level with the edge of the sealing flange **(see illustrations)**.

20 Place the sealing flange on the assembly tool, so the pin locates in the hole in the sender wheel **(see illustration)**.

21 Push the sealing flange and guide fitting sleeve against the tool whilst tightening the 3 knurled bolts. Ensure the pin is still located in the sender wheel **(see illustration)**.

22 Ensure the end of the crankshaft is clean, and is locked at TDC on No 1 cylinder as described in Section 3.

23 Unscrew the large nut to the end of the spindle threads, then press the spindle inwards as far as possible **(see illustration)**.

24 Align the flat side of the assembly with the sump flange, then secure the tool to the

15.19a Press the assembly downwards on a clean, flat surface ...

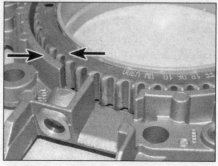

15.19b ... so the upper edge of the sender wheel is level with the edge of the flange

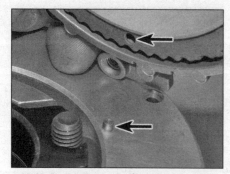

15.20 Fit the flange to the tool, ensuring the pin locates in the hole

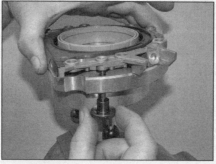

15.21 With the pin engaged in the hole, tighten the 3 knurled bolts to secure the flange to the tool

15.23 Unscrew the nut to the end of the thread, and push the spindle in as far as possible

15.24 Hand-tighten the Allen bolts to secure the tool to the crankshaft

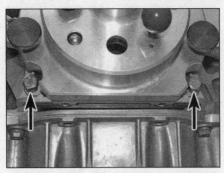

15.25 Use 2 M7 x 35 mm bolts to guide the sealing flange

15.26 Push the black knob into the hole in the crankshaft

crankshaft using the integral Allen bolts **(see illustration)**. Only hand-tighten the bolts.

25 Insert two M7 x 35 mm bolts to guide the sealing flange to the cylinder block **(see illustration)**.

26 Using hand-pressure alone, push the tool assembly onto the crankshaft until the seal guide fitting sleeve contacts the crankshaft flange, then push the guide pin (black knob) into the hole in the crankshaft. This is to ensure the sender wheel reaches its correct installation position **(see illustration)**.

27 Rotate the large nut until it makes contact with the tool housing, then tighten it to 35 Nm (26 lbf ft). After tightening this nut, a small air gap must still be present between the sealing flange and cylinder block **(see illustration)**.

28 Unscrew the large nut; the two M7 x 35 mm bolts, the three knurled bolts and the Allen bolts securing the tool to the crankshaft. Remove the tool, and pull the seal guide fitting sleeve from place (if it didn't come out with the tool).

29 Use a vernier caliper or feeler gauge to measure the fitted depth of the sender wheel in relation to the crankshaft flange **(see illustration)**. The correct depth is 0.5 mm.

30 If the gap is correct, fit the sealing flange bolts and tighten them to the specified torque.

31 If the gap is too small, re-attach the tool to the sealing flange and crankshaft, then refit the two M7 x 35 mm guide bolts to the flange.

Tighten the large spindle nut to 40 Nm (30 lbf ft), remove the tool and measure the air gap. If the gap is still too small, re-attach the tool and tighten the spindle nut to 45 Nm (33 lbf ft). Measure the gap again. When the gap is correct, refit the flange retaining bolts, and tighten them to the specified torque.

32 The remainder of refitting is a reversal of removal.

16 Engine/transmission mountings – inspection and renewal

Inspection

1 If improved access is required, jack up the front of the vehicle, and support it securely on axle stands (see *Jacking and vehicle support*). Remove the engine top cover, which also incorporates the air filter, and then remove the engine undershield(s).

2 Check the mounting rubbers to see if they are cracked, hardened or separated from the metal at any point; renew the mounting if any such damage or deterioration is evident.

3 Check that all the mountings are securely tightened; use a torque wrench to check if possible.

4 Using a large screwdriver or a crowbar, check for wear in the mounting by carefully levering against it to check for free play.

Where this is not possible, enlist the aid of an assistant to move the engine/transmission back-and-forth, or from side-to-side, whilst you observe the mounting. While some free play is to be expected, even from new components, excessive wear should be obvious. If excessive free play is found, check first that the fasteners are correctly secured, and then renew any worn components as described in the following paragraphs.

Renewal

Note: *New mounting securing bolts will be required on refitting.*

Right-hand mounting

5 Attach a hoist and lifting tackle to the engine lifting brackets on the cylinder head, and raise the hoist to just take the weight of the engine. Alternatively the engine can be supported on a trolley jack under the engine. Use a block of wood between the sump and the head of the jack, to prevent any damage to the sump.

6 For improved access, unscrew the coolant reservoir and move it to one side, leaving the coolant hoses connected.

7 Where fitted, undo the retaining bolts and move the supplementary fuel pump to one side **(see illustration)**.

8 Undo the retaining bolts and move the fuel filter to one side **(see illustration)**. Where applicable, move any wiring harnesses, pipes

15.27 After tightening the spindle nut there should be an air gap between the sealing flange and the cylinder block

15.29 Measure the fitted depth of the sender wheel in relation to the end of the crankshaft

16.7 Unbolt the supplementary fuel pump from the top of the mounting

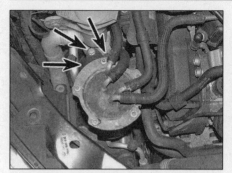

16.8 Undo the filter retaining bolts/nut

16.9 Right-hand engine mounting

16.10 There must be at least 10 mm between the bracket and the chassis member

16.19 Rear lower mounting arm-to-transmission bolts

or hoses to one side to enable removal of the engine mounting.

9 Unscrew the bolts securing the mounting to the engine bracket, and then unscrew the bolts securing it to the body. Also, unscrew the movement limiter. Withdraw the mounting from the engine compartment **(see illustration)**.

10 Refitting is a reversal of removal, bearing in mind the following points.

a) Use new securing bolts.

b) There must be at least 10 mm between the engine mounting bracket and the right-hand side chassis member **(see illustration)**.

c) The side of the mounting support arm must be parallel to the side of the engine mounting bracket.

d) Tighten all fixings to the specified torque.

Left-hand mounting

11 Remove the engine top cover.

12 Attach a hoist and lifting tackle to the engine lifting brackets on the cylinder head, and raise the hoist to just take the weight of the engine and transmission. Alternatively the engine can be supported on a trolley jack under the transmission. Use a block of wood between the transmission and the head of the jack, to prevent any damage to the transmission.

13 Remove the battery and battery tray, as described in Chapter 5A Section 3.

14 Unscrew the bolts securing the mounting to the transmission, and the remaining bolts securing the mounting to the body, then lift the mounting from the engine compartment.

15 Refitting is a reversal of removal, bearing in mind the following points:

a) Use new mounting bolts.

b) The edges of the mounting support arm must be parallel to the edge of the mounting.

c) Tighten all fixings to the specified torque.

Rear mounting (torque arm)

16 Apply the handbrake, then jack up the front of the vehicle and support securely on axle stands (see *Jacking and vehicle support*). Remove the engine undershield(s) for access to the rear mounting (torque arm).

17 Support the rear of the transmission beneath the final drive housing. To do this, use a trolley jack and block of wood, or alternatively wedge a block of wood between the transmission and the subframe.

18 Working under the vehicle, unscrew and remove the bolt securing the mounting to the subframe.

19 Unscrew the two bolts securing the mounting to the transmission, then withdraw the mounting from under the vehicle **(see illustration)**.

20 Refitting is a reversal of removal, but use new mounting securing bolts, and tighten all fixings to the specified torque.

17 Engine oil cooler/filter housing – removal and refitting

Removal

1 The oil cooler is mounted on the lower part of the oil filter housing on the front of the cylinder block, which is the same as the 1.6 litre diesel engine. Refer to Chapter 2C Section 17, for the removal and refitting procedure

18 Oil pressure warning light switch – removal and refitting

1 The oil pressure warning light switch is fitted to the left-hand rear of the cylinder head, which is the same as the 1.6 litre diesel engine. Refer to Chapter 2C Section 18, for the removal refitting procedure.

19 Oil level/temperature sender – removal and refitting

1 The oil level/temperature sender is fitted to bottom of the sump, which is the same as the 1.6 litre diesel engine. Refer to Chapter 2C Section 19, for the removal refitting procedure.

Chapter 2 Part E
1.9 and 2.0 litre unit injector (PD) diesel engine in-car repair procedures

Contents

Degrees of difficulty

Easy, suitable for novice with little experience	**Fairly easy,** suitable for beginner with some experience	**Fairly difficult,** suitable for competent DIY mechanic	**Difficult,** suitable for experienced DIY mechanic	**Very difficult,** suitable for expert DIY or professional

Specifications

General

Manufacturer's engine codes*:
1896 cc (1.9 litre), 8-valve, turbo, SOHC	BKC, BLS, BXE, BXF and BJB
1968 cc (2.0 litre), 8-valve, turbo, SOHC	BMM
1968 cc (2.0 litre), 16-valve, turbo, DOHC	BKD and BMN

Maximum outputs:

	Power	Torque
Engine code BXF	66 kW at 4000 rpm	320 Nm at 1900 rpm
Engine code BKC	77 kW at 4000 rpm	250 Nm at 1900 rpm
Engine code BLS	77 kW at 4000 rpm	250 Nm at 1900 rpm
Engine code BJB	77 kW at 4000 rpm	250 Nm at 1900 rpm
Engine code BXE	77 kW at 4000 rpm	250 Nm at 1900 rpm
Engine code BKD	103 kW at 4000 rpm	320 Nm at 1750 rpm
Engine code BMM	103 kW at 4000 rpm	320 Nm at 1750 to 2500 rpm
Engine code BMN	125 kW at 4200 rpm	350 Nm at 1800 rpm

Bore:
1896 cc engines	79.5 mm
1968 cc engines	81.0 mm
Stroke	95.5 mm

Compression ratio:
Engine code BJB, BKC, BXF and BMM	19.0 : 1
Engine code BLS, BXE, BKD and BMN	18.5 : 1

Compression pressures:
Minimum compression pressure	Approximately 19.0 bar
Maximum difference between cylinders	Approximately 5.0 bar
Firing order	1 – 3 – 4 – 2
No 1 cylinder location	Timing belt end

* Note: *See 'Vehicle identification' at the end of this manual for the location of engine code markings.*

Camshaft

Camshaft endfloat (maximum)	0.15 mm
Camshaft bearing running clearance (maximum)	0.11 mm
Camshaft run-out (maximum)	0.01 mm

Lubrication system

Oil pump type. . . . . Gear type, chain-driven from crankshaft
Oil pressure (oil temperature 80°C, at 2000 rpm). . . . . 2.0 bar

Torque wrench settings

	Nm	lbf ft
Ancillary (alternator, etc) bracket mounting bolts.	45	33
Air conditioning compressor	45	33
Alternator and tensioner.	25	18
Auxiliary drivebelt tensioner securing bolt	25	18
Big-end bearing caps bolts*:		
Stage 1	30	22
Stage 2	Angle-tighten a further 90°	
Camshaft bearing cap bolts*:		
SOHC engines:		
Stage 1	8	6
Stage 2	Angle-tighten a further 90°	
Camshaft bearing frame bolts*:		
DOHC engines	20	15
Camshaft cover nuts/bolts.	10	7
Camshaft sprocket hub centre bolt	100	74
Camshaft sprocket-to-hub bolts	25	18
Coolant pump bolts	15	11
Crankshaft oil seal housing bolts.	15	11
Crankshaft pulley-to-sprocket bolts:		
Stage 1	10	7
Stage 2	Angle-tighten a further 90°	
Crankshaft sprocket bolt*:		
Stage 1	120	89
Stage 2	Angle-tighten a further 90°	
Cylinder head bolts*:		
Stage 1	35	26
Stage 2	60	44
Stage 3	Angle-tighten a further 90°	
Stage 4	Angle-tighten a further 90°	
Driveplate:		
Stage 1	60	44
Stage 2	Angle-tighten a further 90°	
Engine mountings:		
RH engine mounting:		
Mounting to engine:		
Stage 1	60	44
Stage 2	Angle-tighten a further 90°	
Mounting to body:		
Stage 1	40	30
Stage 2	Angle-tighten a further 90°	
LH engine mounting:		
Mounting to body:		
Stage 1	60	44
Stage 2	Angle-tighten a further 90°	
Mounting to transmission:		
Stage 1	40	30
Stage 2	Angle-tighten a further 90°	
Rear mounting link:		
To transmission:		
Stage 1	40	30
Stage 2	Angle-tighten a further 90°	
To subframe:		
Stage 1	100	74
Stage 2	Angle-tighten a further 90°	
Flywheel:		
Stage 1	60	44
Stage 2	Angle-tighten a further 90°	
Glow plugs:		
SOHC engines	15	11
DOHC engines	10	7
Injector rocker arm shafts*:		
Stage 1	20	15
Stage 2	Angle-tighten a further 90°	

Torque wrench settings

	Nm	lbf ft
Main bearing cap bolts*:		
Stage 1	65	48
Stage 2	Angle-tighten a further 90°	
Oil drain plug	30	22
Oil filter housing-to-cylinder block bolts*:		
Stage 1	15	11
Stage 2	Angle-tighten a further 90°	
Oil filter cover	25	18
Oil level/temperature sensor-to-sump bolts	10	7
Oil pick-up pipe securing bolts	15	11
Oil pressure relief valve plug	40	30
Oil pressure warning light switch	20	15
Oil pump chain tensioner bolt	15	11
Oil pump securing bolts	15	11
Oil pump sprocket securing bolt:		
Stage 1	20	15
Stage 2	Angle-tighten a further 90°	
Piston oil spray jet bolt	25	18
Sump:		
Sump-to-cylinder block bolts	15	11
Sump-to-transmission bolts	45	33
Tandem pump	20	15
Thermostat housing	15	11
Timing belt idler pulley bolt	20	15
Timing belt outer cover bolts	10	7
Timing belt rear cover-to-cylinder head bolt	10	7
Timing belt tensioner roller securing nut:		
Stage 1	20	15
Stage 2	Angle-tighten a further 45°	
Timing belt idler pulleys:		
Lower right-hand idler roller (below coolant pump sprocket) bolt*:		
Stage 1	40	30
Stage 2	Angle-tighten a further 90°	
Upper idler roller bolt	20	15
Turbocharger/exhaust manifold	20	15
Turbocharger oil supply line union nut	22	16

* Use new fixings

1 General Information

How to use this Chapter

1 This Part of Chapter 2 describes those repair procedures that can reasonably be carried out on the engine while it remains in the vehicle. If the engine has been removed from the vehicle and is being dismantled as described in Part F, any preliminary dismantling procedures can be ignored.

2 Note that while it may be possible physically to overhaul certain items while the engine is in the vehicle, such tasks are not usually carried out as separate operations, and usually require the execution of several additional procedures (not to mention the cleaning of components and of oilways); for this reason, all such tasks are classed as major overhaul procedures, and are described in Part F of this Chapter.

Engine description

3 Throughout this Chapter, engines are referred to by type, and are identified and referred to by the manufacturer's code letters. A listing of all engines covered, together with their code letters, is given in the Specifications at the start of this Chapter.

4 The engines are water-cooled, single (1896 cc) or double/single (1968 cc) overhead camshaft(s), in-line four-cylinder units, with cast-iron cylinder blocks and aluminium-alloy cylinder heads. All are mounted transversely at the front of the vehicle, with the transmission bolted to the left-hand end of the engine.

5 The crankshaft is of five-bearing type, and thrustwashers are fitted to the centre main bearing to control crankshaft endfloat.

6 Drive for the single or double camshafts is by a toothed timing belt from the crankshaft. Each camshaft is mounted at the top of the cylinder head, and is secured by bearing caps (1896 cc) or a bearing frame/ladder (1968 cc).

7 The valves are closed by coil springs, and run in guides pressed into the cylinder head. On 1896cc engines, the camshaft actuates the valves directly, through hydraulic tappets; on 1968 cc engines, the valves are operated by roller rocker arms incorporating hydraulic tappets.

8 The gear-type oil pump is driven by a chain from a sprocket on the crankshaft. Oil is drawn from the sump through a strainer, and then forced through an externally-mounted, renewable filter. From there, it is distributed to the cylinder head, where it lubricates the camshaft journals and hydraulic tappets, and also to the crankcase, where it lubricates the main bearings, connecting rod big-ends, gudgeon pins and cylinder bores. A coolant-fed oil cooler is fitted to the oil filter housing on all engines. Oil jets are fitted to the base of each cylinder – these spray oil onto the underside of the pistons, to improve cooling.

9 All engines are fitted with a combined brake servo vacuum pump and fuel lift pump (tandem pump), driven by the camshaft on the transmission end of the cylinder head.

10 On all engines, engine coolant is circulated by a pump, driven by the timing belt. For details of the cooling system, refer to Chapter 3.

Operations with engine in car

11 The following operations can be performed without removing the engine:
a) *Compression pressure – testing.*
b) *Camshaft cover – removal and refitting.*
c) *Crankshaft pulley – removal and refitting.*

d) *Timing belt covers – removal and refitting.*
e) *Timing belt – removal, refitting and adjustment.*
f) *Timing belt tensioner and sprockets – removal and refitting.*
g) *Camshaft oil seals – renewal.*
h) *Camshaft(s) and hydraulic tappets – removal, inspection and refitting.*
i) *Cylinder head – removal and refitting.*
j) *Cylinder head and pistons – decarbonising.*
k) *Sump – removal and refitting.*
l) *Oil pump – removal, overhaul and refitting.*
m) *Crankshaft oil seals – renewal.*
n) *Engine/transmission mountings – inspection and renewal.*
o) *Flywheel/driveplate – removal, inspection and refitting.*

Note: *It is possible to remove the pistons and connecting rods (after removing the cylinder head and sump) without removing the engine. However, this is not recommended. Work of this nature is more easily and thoroughly completed with the engine on the bench, as described in Chapter 2F Section 12.*

2 Compression and leakdown tests – description and interpretation

Compression test

Note: *A compression tester suitable for use with diesel engines will be required for this test.*

1 When engine performance is down, or if misfiring occurs which cannot be attributed to the ignition or fuel systems, a compression test can provide diagnostic clues as to the engine's condition. If the test is performed regularly, it can give warning of trouble before any other symptoms become apparent.

2 The engine must be fully warmed-up to normal operating temperature, the battery must be fully-charged, and you will require the aid of an assistant.

3 Disconnect the injector solenoids by disconnecting the connector at the end of the cylinder head **(see illustration)**. **Note:** *As a result of the wiring being disconnected, faults will be stored in the ECU memory. These must be erased after the compression test.*

4 Remove the glow plugs as described in

2.3 Disconnect the injector solenoids wiring plug connector

Chapter 5C Section 2, then fit a compression tester to the No 1 cylinder glow plug hole. The type of tester which screws into the plug thread is preferred.

5 Have your assistant crank the engine for several seconds on the starter motor. After one or two revolutions, the compression pressure should build-up to a maximum figure and then stabilise. Record the highest reading obtained.

6 Repeat the test on the remaining cylinders, recording the pressure in each.

7 The cause of poor compression is less easy to establish on a diesel engine than on a petrol engine. The effect of introducing oil into the cylinders (wet testing) is not conclusive, because there is a risk that the oil will sit in the recess on the piston crown, instead of passing to the rings. However, the following can be used as a rough guide to diagnosis.

8 All cylinders should produce very similar pressures. Any difference greater than that specified indicates the existence of a fault. Note that the compression should build-up quickly in a healthy engine. Low compression on the first stroke, followed by gradually increasing pressure on successive strokes, indicates worn piston rings. A low compression reading on the first stroke, which does not build-up during successive strokes, indicates leaking valves or a blown head gasket (a cracked head could also be the cause).

9 A low reading from two adjacent cylinders is almost certainly due to the head gasket having blown between them and the presence of coolant in the engine oil will confirm this.

10 On completion, remove the compression tester, and refit the glow plugs, with reference to Chapter 5C Section 2.

11 Reconnect the wiring to the injector solenoids. Finally, have a Seat dealer or suitably equipped garage erase the fault codes from the ECU memory.

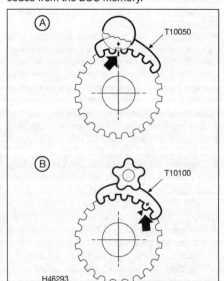

3.8a TDC setting tools

Leakdown test

12 A leakdown test measures the rate at which compressed air fed into the cylinder is lost. It is an alternative to a compression test, and in many ways it is better, since the escaping air provides easy identification of where pressure loss is occurring (piston rings, valves or head gasket).

13 The equipment required for leakdown testing is unlikely to be available to the home mechanic. If poor compression is suspected, have the test performed by a suitably-equipped garage.

3 Engine assembly and valve timing marks – general information and usage

General information

1 TDC is the highest point in the cylinder that each piston reaches as it travels up-and-down when the crankshaft turns. Each piston reaches TDC at the end of the compression stroke and again at the end of the exhaust stroke, but TDC generally refers to piston position on the compression stroke. No 1 piston is at the timing belt end of the engine.

2 Positioning No 1 piston at TDC is an essential part of many procedures, such as timing belt removal and camshaft removal.

3 The design of the engines covered in this Chapter is such that piston-to-valve contact may occur if the camshaft or crankshaft is turned with the timing belt removed. For this reason, it is important to ensure that the camshaft and crankshaft do not move in relation to each other once the timing belt has been removed from the engine.

Setting TDC on No 1 cylinder

Note: *VAG special tool T10050 (early) or T10100 (late) is required to lock the crankshaft sprocket in the TDC position. The early type tool locates directly onto the sprocket teeth, but, the contour of the teeth on the later type means that the tool must be pushed onto the sprocket teeth from the right-hand end of the engine.*

4 Remove the auxiliary drivebelt(s) as described in Chapter 1B Section 30.

5 Remove the crankshaft pulley/vibration damper as described in Section 5.

6 Remove the timing belt covers as described in Section 6.

7 Remove the glow plugs, as described in Chapter 5C Section 2, to allow the engine to turn more easily. **Note:** *On DOHC engines the glow plugs are accessed by removing the camshaft cover, so this may not be considered an option.*

8 Using a spanner or socket on the crankshaft sprocket bolt, turn the crankshaft in the normal direction of rotation (clockwise) until the alignment mark on the face of the sprocket is almost vertical for tool T10050, or at 2 o'clock for tool T10100 **(see illustrations)**.

9 On engine codes BKC, BMM, BLS, BXE

3.8b Position the crankshaft so that the mark on the sprocket is almost vertical…

3.8c … then insert the VAG tool T10050…

3.8d … and align the marks on the tool and sprocket

and BXF, the arrow (marked 4Z) on the rear section of the upper timing belt upper cover aligns between the two lugs on the rear of the camshaft hub sender wheel **(see illustration)**.

10 On engine codes BKD and BMN, the pointer on the rear section of the timing belt inner cover aligns with the camshaft hub sender wheel, and the marks on the camshaft toothed segments and sprockets must be vertical **(see illustration)**.

11 While in this position it should be possible to insert the VAG tool to lock the crankshaft, and a 6 mm diameter rod to lock the camshaft(s) **(see illustrations)**. **Note:** *The mark on the crankshaft sprocket and the mark on the VAG tool must align, whilst at the same time the shaft of tool must engage in the drilling in the crankshaft oil seal housing.*

12 The engine is now set to TDC on No 1 cylinder.

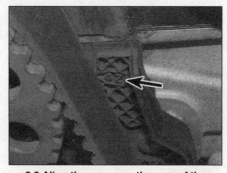

3.9 Align the arrow on the rear of the cover between the lugs on the rear of the camshaft hub sender wheel

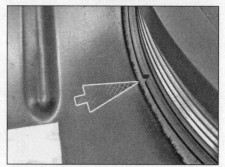

3.10 TDC mark and notch (2.0 litre DOHC engine)

<table><tr><td>4</td><td>**Camshaft cover –** removal and refitting</td></tr></table>

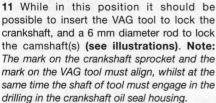

Removal

1 Remove the dipstick and prise off and remove the engine top cover, then disconnect the breather hose from the camshaft cover **(see illustration)**.

2 Unscrew the camshaft cover retaining bolts and lift the cover away. If it sticks, do not attempt to lever it off – instead free it by working around the cover and tapping it lightly with a

3.11a Using the VAG tool to lock the crankshaft at TDC

3.11b Insert a 6 mm drill bit through the camshaft hub into the cylinder head to lock the camshaft

soft-faced mallet. Note on DOHC engines (BKD and BMN engine codes) it will be necessary to unbolt the air ducting from the rear of the

cylinder head, and unbolt the fuel lines from the front of the cylinder head to provide additional clearance **(see illustrations)**.

4.1 One of the cover locating pegs and also breather pipe

4.2a Unbolt the fuel lines to provide additional clearance …

4.2b … when removing the camshaft cover on 2.0 litre DOHC engines

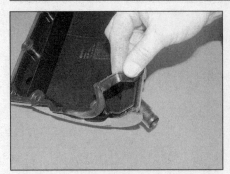

4.3 Ensure the retaining bolts are pushed fully through the gasket before refitting the camshaft cover

4.5a Apply sealant to the points shown on the cylinder head

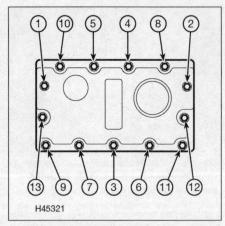

H45321

4.5b Camshaft cover tightening sequence (SOHC engines)

4.5c Camshaft cover tightening sequence (DOHC engines)

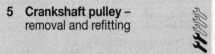

5.3 Removing the intercooler air duct for access to the crankshaft pulley

3 Recover the camshaft cover gasket. Inspect the gasket carefully, and renew it if damage or deterioration is evident – note that the retaining bolts must be pushed fully through the gasket before refitting the cover **(see illustration)**.

4 Clean the mating surfaces of the cylinder head and camshaft cover thoroughly, removing all traces of oil and old gasket – take care to avoid damaging the surfaces as you do this.

Refitting

5 Refit the camshaft cover by following the removal procedure in reverse, noting the following points:

a) *On engine codes BKC, BLS, BXE, BXF and BMM, apply suitable sealant to the points where the camshaft bearing cap contacts the cylinder head* **(see illustration)**.

b) *Tighten the camshaft cover retaining nuts/ bolts progressively to the specified torque in the sequence shown* **(see illustrations)**.

5   Crankshaft pulley – removal and refitting

Removal

1 Switch off the ignition and all electrical consumers and remove the ignition key.

2 For improved access, raise the front right-hand side of the vehicle, and support securely on axle stands (see *Jacking and vehicle support*). Remove the roadwheel.

3 Remove the securing screws and withdraw the engine undertray(s) and/or wheel arch liner panels. Unscrew the nut at the rear, and the washer-type fasteners further forward, then where necessary release the air hose clip and manipulate out the plastic air duct for the intercooler **(see illustration)**.

4 On models fitted with a fuel fired auxiliary heater unbolt the coolant pipes and move them to the side. Where applicable, prise the cover from the centre of the pulley to expose the securing bolts **(see illustration)**.

5 Slacken the bolts securing the crankshaft pulley to the sprocket **(see illustration)**. If necessary, the pulley can be prevented from turning by counterholding with a spanner or socket on the crankshaft sprocket bolt.

6 Remove the auxiliary drivebelt, as described in Chapter 1B Section 30.

7 Unscrew the bolts securing the pulley to the sprocket, and remove the pulley **(see illustration)**.

Refitting

8 Refit the pulley over the locating peg on the crankshaft sprocket, then refit the pulley securing bolts.

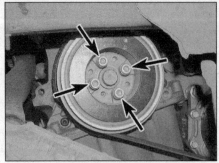

5.4 Prising out the crankshaft pulley centre cap

5.5 Showing the four crankshaft pulley bolts

5.7 Removing the crankshaft pulley

6.2a Release the retaining clips...

6.2b ...and withdraw the upper cover

6.2c Removing the upper timing cover (2.0 litre DOHC engine)

9 Refit and tension the auxiliary drivebelt as described in Chapter 1B Section 30.
10 Prevent the crankshaft from turning as during removal, then fit the pulley securing bolts, and tighten to the specified torque.
11 Refit the engine undertray(s), wheel arch liners and the intercooler air duct, as applicable.
12 Refit the roadwheel and lower the vehicle to the ground.

6 Timing belt covers – removal and refitting

Upper outer cover

1 Where applicable, release the retaining clips and remove the air intake hose from across the top of the timing belt cover.
2 Release the uppermost part of the timing belt outer cover by prising open the metal spring clips, then withdraw the cover away from the engine. On DOHC engines there are three clips, two at the front and a single one at the rear of the cover **(see illustrations)**.
3 Refitting is a reversal of removal, noting that the lower edge of the upper cover engages with the centre cover.

Centre outer cover

4 Remove the auxiliary drivebelt as described in Chapter 1B Section 30.
5 Remove the crankshaft pulley as described in Section 5. It is assumed that, if the centre cover is being removed, the lower cover will be also – if not, simply remove the components described in Section 5 for access to the crankshaft pulley, and leave the pulley in position.
6 With the upper cover removed (paragraphs 1 to 3), unscrew and remove the retaining bolts from the centre cover. Withdraw the centre cover from the engine, noting how it fits over the lower cover **(see illustration)**.
7 Refitting is a reversal of the removal procedure.

Lower outer cover

8 Remove the upper and centre covers as described previously.
9 If not already done, remove the crankshaft pulley as described in Section 5.
10 Unscrew the remaining bolt(s) securing the lower cover, and lift it out **(see illustration)**.
11 Refitting is a reversal of removal; locate the centre cover in place before fitting the top two bolts.

Rear cover

12 Remove the upper, centre and lower covers as described previously.
13 Remove the timing belt, tensioner and sprockets as described in Sections 7 and 8.
14 Slacken and withdraw the retaining bolts and lift the timing belt inner cover from the studs on the end of the engine, and remove it from the engine compartment.
15 Refitting is a reversal of removal.

7 Timing belt – removal, inspection and refitting

Note: *It is consider best practise to always replace the tensioner, idler and coolant pump whenever the timing belt is replaced. Any mounting studs should also be replaced at the same time. Most of the components (apart from the coolant pump) will be supplied with a timing belt kit. Note also that most manufactures will not guarantee a belt against premature failure if the idler(s) and tensioner are not replaced with the belt.*

Removal

1 The primary function of the toothed timing belt is to drive the camshaft, but it also

6.6 Removing the centre outer timing cover (2.0 litre DOHC engine)

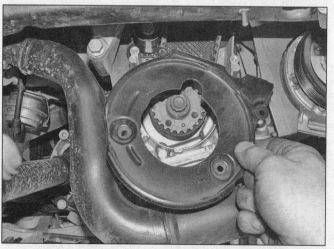

6.10 Removing the lower outer timing cover (2.0 litre DOHC engine)

7.5a Use a spanner to turn the tensioner clockwise...

7.5b ...then lock with a suitable metal rod...

7.5c ...remove the auxiliary drivebelt...

7.5d ...and unbolt the tensioner (2.0 litre DOHC engine)

7.16 Using a home-made tool to hold the camshaft sprocket while loosening the bolts

7.22 Position the camshaft sprocket so that the securing bolts are in the centre part of the elongated holes

drives the coolant pump. Should the belt slip or break in service, the valve timing will be disturbed and piston-to-valve contact may occur, resulting in serious engine damage. For this reason, it is important that the timing belt is tensioned correctly, and inspected regularly for signs of wear or deterioration.

2 Switch off the ignition and all electrical consumers and remove the ignition key.

3 Apply the handbrake, then jack up the front of the vehicle and support securely on axle stands (see *Jacking and vehicle support*).

4 Remove the securing screws and withdraw the engine undertray. Remove the wheel arch liner. Also, remove the engine cover.

5 Remove the auxiliary drivebelt as described in Chapter 1B Section 30, then unbolt and remove the drivebelt tensioner **(see illustrations)**.

6 Remove the crankshaft pulley/vibration damper as described in Section 5.

7 Remove the timing belt covers as described in Section 6.

8 Support the engine and remove the right-hand engine mounting as described in Section 18. Note that on some DOHC engines, the mounting bracket on the engine can only be removed with the timing belt loose.

9 Remove the fuel filter from its bracket and place to one side.

10 On 2.0 litre DOHC engines, disconnect the fuel supply and return lines. Also, where necessary on all engines, remove the intercooler charge air pipe.

11 Unbolt the filler neck from the screen washer reservoir.

12 Where applicable, unbolt the fuel filter bracket from the engine mounting.

13 Unbolt the coolant expansion tank and position it to one side. **Note:** *Do not disconnect the hoses.*

14 Set the engine to TDC on No 1 cylinder as described in Section 3.

15 If the original timing belt is to be refitted, mark the running direction of the belt, to ensure correct refitting.

Caution: If the belt appears to be in good condition and can be re-used, it is essential that it is refitted the same way around, otherwise accelerated wear will result, leading to premature failure.

16 Loosen the bolts securing the sprocket(s) to the camshaft(s) while holding the sprockets with a suitable tool **(see illustration)**.

17 Loosen the timing belt tensioner securing nut, then use circlip pliers or an Allen key (as applicable) to turn the tensioner anti-clockwise until a suitable pin or drill bit can be inserted through the locking holes. Now, turn the tensioner clockwise to the stop and tighten the securing nut.

18 Slide the belt from the sprockets, taking care not to twist or kink the belt excessively if it is to be re-used. If the belt is to be replaced remove the tensioner and idler. Also consider replacing the coolant pump – see the note at the beginning of this Section.

Inspection

19 Examine the belt for evidence of contamination by coolant or lubricant. If this is the case, find the source of the contamination before progressing any further. Check the belt for signs of wear or damage, particularly around the leading edges of the belt teeth. Renew the belt if its condition is in doubt; the cost of belt renewal is negligible compared with potential cost of the engine repairs, should the belt fail in service. The belt must be renewed if it has covered the mileage given in Chapter 1B Section 31, however, if it has covered less, it is prudent to renew it regardless of condition, as a precautionary measure.

20 If the timing belt is not going to be refitted for some time, it is a wise precaution to hang a warning label on the steering wheel, to remind yourself (and others) not to attempt to start the engine.

Refitting

21 Ensure that the crankshaft and camshaft are still set to TDC on No 1 cylinder, as described in Section 3. Fit the new tensioner and idler as described in Section 8.

22 Position the camshaft sprocket(s) so that the securing bolts are in the centre part of the elongated holes **(see illustration)**.

23 Loop the timing belt loosely under the crankshaft sprocket. **Note:** *Observe any direction of rotation markings on the belt.*

24 Engage the timing belt teeth with the

7.24 Fitting the timing belt

7.25 The tensioner correctly tensioned

8.2 Timing belt tensioner nut

8.3 Ensure that the lug on the tensioner backplate engages with the cut-out in the rear timing belt cover

8.4 Lock two nuts together and replace the tensioner and idler mounting studs

8.6 Remove the idler

camshaft sprocket(s) **(see illustration)**, then manoeuvre it into position around the tensioning roller, crankshaft sprocket, and finally around the coolant pump sprocket. Make sure that the belt teeth seat correctly on the sprockets. **Note:** *Slight adjustment to the position of the camshaft sprocket may be necessary to achieve this. Avoid bending the belt back on itself or twisting it excessively as you do this. Ensure that any slack in the belt is in the section of belt that passes over the tensioner roller.*

25 Loosen the timing belt tensioner securing nut, and turn the tensioner anti-clockwise with the circlip pliers or an Allen key (as applicable) until the locking pin can be removed. Now, turn the tensioner clockwise until the pointer is in the middle of the gap in the tensioner base plate **(see illustration)**. With the tensioner held in this position, tighten the securing nut to the specified torque and angle.

26 Tighten the camshaft sprocket bolts to the specified torque, remove the sprocket locking pin(s) and the crankshaft locking tool.

27 Using a spanner or wrench and socket on the crankshaft pulley centre bolt, rotate the crankshaft through two complete revolutions. Reset the engine to TDC on No 1 cylinder, with reference to Section 3 and check that the crankshaft and camshaft sprocket locking pins can still be inserted. If the camshaft sprocket locking pin(s) cannot be inserted, slacken the retaining bolts, turn the hub(s) until the pin(s) fit, and tighten the sprocket retaining bolts to the specified torque.

28 Refit the coolant expansion tank.
29 Where applicable, refit the fuel filter bracket to the engine mounting.
30 Refit the screen washer reservoir filler neck.
31 Refit the fuel filter to its bracket, and where necessary reconnect the fuel lines and intercooler charge air pipe.
32 Where applicable, refit the right-hand engine mounting as described in Section 18.
33 Refit the timing belt covers as described in Section 6.
34 Refit the crankshaft pulley/vibration damper as described in Section 5.
35 Refit the auxiliary drivebelt and tensioner with reference to Chapter 1B Section 30.
36 Refit the engine undertray(s), right-hand wheel arch liner and engine top cover, and lower the vehicle to the ground.

8   Timing belt tensioner and sprockets – removal and refitting

Timing belt tensioner

Note: *Early versions of the tensioner only had holes for a pin type spanner. Later tensioners have provision for the use of a hex key.*

Removal

1 Remove the timing belt as described in Section 7.
2 Unscrew the timing belt tensioner nut, and

remove the tensioner from the engine **(see illustration)**.
3 When refitting the tensioner to the engine, ensure that the lug on the tensioner backplate engages with the corresponding cut-out in the rear timing belt cover, then refit the tensioner nut **(see illustration)**.

Refitting

4 Fitting new studs to the tensioner and idler is recommended **(see illustration)**. Refit and tension the timing belt as described in Section 7, making sure that the tensioner backplate is correctly engaged with the hole in the cylinder head.

Idler pulleys

Removal

5 Remove the timing belt as described in Section 7.
6 Unscrew the relevant idler pulley securing bolt/nut, then withdraw the pulley **(see illustration)**.

Refitting

7 Refit the pulley and tighten the securing bolt or nut to the specified torque. **Note:** *Renew the bolt (where applicable).*
8 Refit and tension the timing belt as described in Section 7.

Crankshaft sprocket

Note: *A new crankshaft sprocket securing bolt must be used on refitting.*

Removal

9 Remove the timing belt as described in Section 7.

10 The sprocket securing bolt must now be slackened, and the crankshaft must be prevented from turning as the sprocket bolt is unscrewed. To hold the sprocket, make up a suitable tool, and screw it to the sprocket using a two bolts screwed into two of the crankshaft pulley bolt holes.

11 Hold the sprocket using the tool, then slacken the sprocket securing bolt. Take care, as the bolt is very tight. Do not allow the crankshaft to turn as the bolt is slackened.

12 Unscrew the bolt, and slide the sprocket from the end of the crankshaft, noting which way round the sprocket's raised boss is fitted.

Refitting

13 Commence refitting by positioning the sprocket on the end of the crankshaft, with the raised boss fitted as noted on removal.

14 Fit a new sprocket securing bolt, then counterhold the sprocket using the method employed on removal, and tighten the bolt to the specified torque in the two stages given **(see illustration)**.

15 Refit the timing belt as described in Section 7.

Camshaft sprocket

Removal

16 Remove the timing belt as described in Section 7.

17 Unscrew and remove the three retaining bolts and remove the camshaft sprocket from the camshaft hub.

Refitting

18 Refit the sprocket ensuring that it is fitted the correct way round, as noted before removal, then insert the sprocket bolts, and tighten by hand only at this stage.

19 If the crankshaft has been turned, turn the crankshaft clockwise 90° back to TDC.

20 Refit and tension the timing belt as described in Section 7.

8.14 Fitting a new crankshaft sprocket securing bolt

Camshaft hub

Note: *Seat technicians use special tool T10051 to counterhold the hub, however it is possible to fabricate a suitable alternative – see below.*

Removal

21 Remove the camshaft sprocket as described previously in this Section.

22 Engage special tool T10051 with the three locating holes in the face of the hub to prevent the hub from turning. If this tool is not available, fabricate a suitable alternative. Whilst holding the tool, undo the central hub retaining bolt about two turns **(see illustration)**.

23 Leaving the central hub retaining bolt in place, attach VAG tool T10052 (or a similar three-legged puller) to the hub, and evenly tighten the puller until the hub is free of the camshaft taper **(see illustration)**. Remove the central hub retaining bolt once the taper is released.

Refitting

24 Ensure that the camshaft taper and the hub centre is clean and dry, locate the hub on the taper, noting that the built-in key in the hub taper must align with the keyway in the camshaft taper **(see illustration)**.

25 Hold the hub in this position with tool

8.22 Using a fabricated tool to counterhold the camshaft hub

T10051 (or similar home-made tool), and tighten the central bolt to the specified torque.

26 Refit the camshaft sprocket as described previously in this Section.

Coolant pump sprocket

27 The coolant pump sprocket is integral with the coolant pump. Refer to Chapter 3 Section 7 for details of coolant pump removal.

9 Pump injector rocker shaft assembly – removal and refitting

Removal

1 Remove the camshaft cover as described in Section 4. In order to ensure that the rocker arms are refitted to their original locations, use a marker pen or paint and number the arms 1 to 4, with No 1 nearest the timing belt end of the engine. If the arms are not fitted to their original locations the injector basic clearance setting procedure must be carried out as described in.

2 Slacken the locknut of the adjustment screw on the end of the rocker arm above the respective injector, and undo the adjustment screw until the rocker arm lies against the plunger pin of the injector. Starting at the

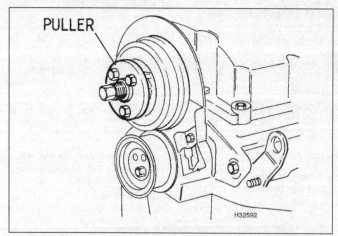
8.23 Attach a three-legged puller to the hub, and evenly tighten the puller until the hub is free of the camshaft taper

8.24 The built in key in the hub taper must align with the keyway in the camshaft taper

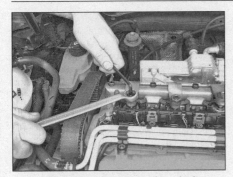

9.2a Slackening the rocker shaft adjustment screws

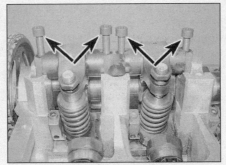

9.2b Starting with the outer bolts first, carefully and evenly slacken the rocker shaft retaining bolts (SOHC engine)

9.2c Rocker shaft retaining bolts – 3 of 4 shown (DOHC engine)

outside and working in, gradually and evenly slacken and remove the rocker shaft retaining bolts. Lift off the rocker shaft. Discard the rocker shaft bolts, new ones must be fitted (see illustrations).

Refitting

3 Thoroughly check the rocker shaft, rocker arms and camshaft bearing cap seating surface for any signs of excessive wear or damage.

4 Smear some grease (VAG No G000 100) onto the contact face of each rocker arm adjustment screw, and refit the rocker shaft assembly, tightening the new retaining bolts as follows. Starting from the inside out, hand-tighten the bolts. Again, from the inside out, tighten the bolts to the Stage one torque setting. Finally, from the inside out, tighten the bolts to the Stage two angle tightening setting.

5 Attach a DTI (Dial Test Indicator) gauge to the cylinder head upper surface, and position the DTI probe against the top of the adjustment screw. Turn the crankshaft until the rocker arm roller is on the highest point of its corresponding camshaft lobe, and the adjustment screw is at its lowest. Once this position has been established, remove the DTI gauge, screw the adjustment screw in until firm resistance is felt, and the injector

spring cannot be compressed further. Turn the adjustment screw anti-clockwise 180°, and tighten the locknut to the specified torque. Repeat this procedure for any other injectors that have been refitted.

6 Refit the camshaft cover and upper timing belt cover, as described in Section 4.

7 Start the engine and check that it runs correctly.

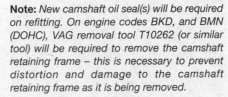

10 Camshaft and hydraulic tappets – removal, inspection and refitting

Note: *New camshaft oil seal(s) will be required on refitting. On engine codes BKD, and BMN (DOHC), VAG removal tool T10262 (or similar tool) will be required to remove the camshaft retaining frame – this is necessary to prevent distortion and damage to the camshaft retaining frame as it is being removed.*

Removal

1 Turn the crankshaft to position No 1 piston at TDC on the firing stroke, and lock the camshaft and the fuel injection sprocket in position, as described in Section 3.

2 Remove the timing belt as described in Section 7.

3 Remove the camshaft sprocket(s) and hubs as described in Section 8.

4 Remove the tandem fuel/brake vacuum pump (where fitted) as described in Chapter 9 Section 21 (see illustration).

5 Remove the injector rocker arms and shaft as described in Section 9.

Engine codes BMM, BJB, BKC, BLS, BXE and BXF

6 Check the camshaft bearing caps for identification markings (see illustration). The bearing caps are normally stamped with their respective cylinder numbers. If no marks are present, make suitable marks using a scriber or punch. The caps should be numbered from 1 to 5, with No 1 at the timing belt end of the engine. Note on which side of the bearing caps the marks are made to ensure that they are refitted the correct way round.

7 The camshaft rotates in shell bearings. As the camshaft bearing caps are removed, recover the shell bearing halves from the camshaft. Number the back of the bearings with a felt pen to ensure that, if re-used, the bearings are fitted to their original locations. **Note:** *Fitted into the cylinder head, under each camshaft bearing cap, is a washer for each cylinder head bolt.*

8 Unscrew the securing nuts, and remove Nos 1, 3 and 5 bearing caps.

10.4 Tandem fuel/brake vacuum pump (DOHC engine)

10.6 Check the camshaft bearing caps for markings

9 Working progressively, in a diagonal sequence, slacken the nuts securing Nos 2 and 4 bearing caps. Note that as the nuts are slackened, the valve springs will push the camshaft up.

10 Once the nuts securing Nos 2 and 4 bearing caps have been fully slackened, lift off the bearing caps.

11 Carefully lift the camshaft from the cylinder head, keeping it level and supported at both ends as it is removed so that the journals and lobes are not damaged. Remove the oil seal from the end of the camshaft and discard it – a new one will be required for refitting **(see illustration).**

12 Lift the hydraulic tappets from their bores in the cylinder head, and store them with the valve contact surfaces facing downwards, to prevent the oil from draining out. It is recommended that the tappets are kept immersed in oil for the period they are removed from the cylinder head. Make a note of the position of each tappet, as they must be refitted in their original locations on reassembly – accelerated wear leading to early failure will result if the tappets are interchanged.

13 Recover the lower shell bearing halves from the cylinder head; number the back of the shells with a felt pen to ensure that, if re-used, the bearings are fitted to their original locations.

Engine codes BKD and BMN

14 Unscrew the upper bolt securing the EGR cooler bracket.

15 Disconnect the wiring from the unit injectors and glow plugs. To disconnect the wiring connector from the left-hand end of the cylinder head, use a screwdriver to pull out the plastic red lock, then unscrew the collar and pull the connector from the pins. To remove the wiring conduit, first undo the screws retaining it to the cylinder head, then release the multi-pin plug by raising the clip and unscrewing the outer threaded collar, ideally using the special VAG tool T10310 which engages the three slots **(see illustrations)**.

16 Progressively unscrew the camshaft retaining frame bolts starting from the outside to inside.

17 The VAG tool is now used to remove the retaining bearing frame. First, fully unscrew the ejector bolts, then fit the tool to the frame and tighten the bolts. Screw in the ejector bolts until they contact the cylinder head bolt(s), then progressively tighten them so that the bearing frame is released from the cylinder head.

18 Carefully lift the camshafts from the cylinder head, keeping them identified for location. Remove the oil seals from the ends of the camshafts and discard them – new ones will be required for refitting.

19 To remove the exhaust roller rocker fingers and hydraulic tappets, carry out the following:
a) *Drain the coolant as described in Chapter 1B Section 33.*
b) *Remove the connecting pipe between the EGR valve and bypass flap.*
c) *Remove the thermostat housing as described in Chapter 3 Section 4.*
d) *Unscrew the plug from the end of the exhaust roller rocker shaft.* **Note:** *If the plug is very tight, the shaft may be deformed while unscrewing it, making the oil supply hole alignment incorrect. If this happens, renew the shaft.*
e) *Undo the screw securing the roller rocker shaft to the cylinder head.*

20 To remove either exhaust or inlet roller rocker fingers and hydraulic tappets, use VAG slide hammer/puller tool T10055 to pull the shaft from the cylinder head while removing the roller rockers. Store the rockers in a container with numbered compartments to ensure they are refitted to their correct locations. It is recommended that the tappets are kept immersed in oil for the period they are removed from the cylinder head.

Inspection – all engines

21 With the camshaft(s) removed, examine the bearing caps/frame and the bearing locations in the cylinder head for signs of obvious wear or pitting. If evident, a new cylinder head will probably be required. Also check that the oil supply holes in the cylinder head are free from obstructions.

22 Visually inspect the camshaft for evidence of wear on the surfaces of the lobes and journals. Normally their surfaces should be smooth and have a dull shine; look for scoring, erosion or pitting and areas that appear highly polished, indicating excessive wear. Accelerated wear will occur once the

10.11 Remove the camshaft oil seal

10.15a Disconnecting the wiring from the unit injectors

10.15b Wiring loom connector, showing the three pins in the outer collar

10.15c Retaining clip on the inside of the wiring loom connector

10.24 Checking camshaft endfloat using a DTI gauge

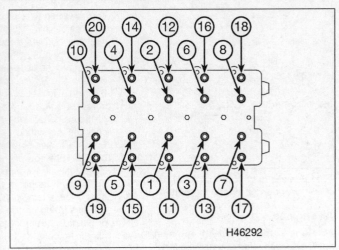

H46292

10.42 Camshaft bearing frame bolt tightening sequence

hardened exterior of the camshaft has been damaged, so always renew worn items. **Note:** *If these symptoms are visible on the tips of the camshaft lobes, check the corresponding tappet, as it will probably be worn as well.*

23 If the machined surfaces of the camshaft appear discoloured or blued, it is likely that it has been overheated at some point, probably due to inadequate lubrication. This may have distorted the shaft, so check the run-out as follows: place the camshaft between two V-blocks and using a DTI gauge, measure the run-out at the centre journal. If it exceeds the figure quoted in the Specifications at the start of this Chapter, renew the camshaft.

24 To measure the camshaft endfloat, temporarily refit the camshaft to the cylinder head, then fit Nos 1 and 5 bearing caps and tighten the retaining nuts to the specified torque setting. Anchor a DTI gauge to the timing belt end of the cylinder head **(see illustration)**. Push the camshaft to one end of the cylinder head as far as it will travel, then rest the DTI gauge probe on the end face of the camshaft, and zero the gauge. Push the camshaft as far as it will go to the other end of the cylinder head, and record the gauge reading. Verify the reading by pushing the camshaft back to its original position and checking that the gauge indicates zero again. **Note:** *The hydraulic tappets must not be fitted whilst this measurement is being taken.*

25 Check that the camshaft endfloat measurement is within the limit listed in the Section. If the measurement is outside the specified limit, wear is unlikely to be confined to any one component, so renewal of the camshaft, cylinder head and bearing caps must be considered.

26 The camshaft bearing running clearance should now be measured. This will be difficult to achieve without a range of micrometers or internal/external expanding calipers, measure the outside diameters of the camshaft bearing surfaces and the internal diameters formed by the bearing caps/frame (and shell bearings

where applicable) and the bearing locations in the cylinder head. The difference between these two measurements is the running clearance.

27 Compare the camshaft running clearance measurements with the figure given in the Specifications; if any are outside the specified tolerance, the camshaft, cylinder head and bearing caps/frame (and shell bearings where applicable) should be renewed.

28 Inspect the hydraulic tappets for obvious signs of wear or damage, and renew if necessary. Check that the oil holes in the tappets are free from obstructions.

Refitting

Engine codes BMM, BJB BKC, BLS, BXE and BXF

29 Smear some clean engine oil onto the sides of the hydraulic tappets, and offer them into position in their original bores in the cylinder head. Push them down until they contact the valves, then lubricate the camshaft lobe contact surfaces.

30 Lubricate the camshaft and cylinder head bearing journals and shell bearings with clean engine oil.

31 Carefully lower the camshaft into position in the cylinder head making sure that the cam lobes for No 1 cylinder are pointing upwards.

32 Refit a new camshaft oil seal on the end of the camshaft. Make sure that the closed end of the seal faces the camshaft sprocket end of the camshaft, and take care not to damage the seal lip. Locate the seal against the seat in the cylinder head.

33 Oil the upper surfaces of the camshaft bearing journals and shell bearings, then fit Nos 2 and 4 bearing caps. Ensure that they are fitted the right way round and in the correct locations, then progressively tighten the retaining nuts in a diagonal sequence to the specified torque. Note that as the nuts are tightened, the camshaft will be forced down against the pressure of the valve springs.

34 Fit bearing caps 1, 3 and 5 over the camshaft and progressively tighten the nuts to the specified torque. Note that it may be necessary to locate No 5 bearing cap by tapping lightly on the end of the camshaft.

35 Refit the injector rocker arms as described in Section 9.

Engine codes BKD and BMN

36 Oil the roller rocker fingers and hydraulic tappets together with the shafts, then insert each shaft into the cylinder head while at the same time fitting the roller rockers in their correct order. The inlet shaft must be inserted until flush with the cylinder head. The exhaust shaft must be correctly aligned with the retaining bolt hole. Insert and tighten the bolt securely.

37 Reverse the procedures listed in paragraph 19.

38 Lubricate the camshaft and cylinder head bearing journals with clean engine oil.

39 Carefully lower the camshafts into position in the cylinder head making sure that the cam lobes for No 1 cylinder are pointing upwards.

40 Fit new camshaft oil seals on the end of the camshafts. Make sure that the closed end of the seal faces the camshaft sprocket end of the camshaft, and take care not to damage the seal lip. Locate the seal against the seat in the cylinder head.

41 Oil the upper surfaces of the camshaft bearing frame journals, then apply sealant to the mating surfaces of the bearing frame and cylinder head.

42 Refit the bearing frame together with the rocker arms and shaft, using the VAG tool, ensuring the frame is the correct way round. Progressively tighten the retaining bolts to the specified torque in the order shown **(see illustration)**, then remove the tool. Note that as the bolts are tightened, the camshafts will be forced down against the pressure of the valve springs.

43 Tighten the rocker arm shaft bolts to the specified angle. Refit the wiring conduit and secure the connectors.

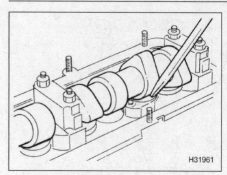

11.8 Press down on the tappet using a wooden or plastic instrument

All engines

44 Renew the camshaft oil seal(s) as applicable with reference to Section 12.

45 Refit the tandem fuel/brake vacuum pump (where fitted) as described in Chapter 9 Section 21.

46 Refit the camshaft sprocket(s) and hubs as described in Section 8.

47 Refit the timing belt as described in Section 7.

11 Hydraulic tappets – testing

⚠️ **Warning: After fitting hydraulic tappets, wait a minimum of 30 minutes (or preferably, leave overnight) before starting the engine, to allow the tappets time to settle, otherwise the valve heads will strike the pistons.**

1 The hydraulic tappets are self-adjusting, and require no attention whilst in service.

2 If the hydraulic tappets become excessively noisy, their operation can be checked as described below.

3 Start the engine, and run it until it reaches normal operating temperature, increase the

engine speed to approximately 2500 rpm for 2 minutes.

4 If any hydraulic tappets are heard to be noisy, carry out the following checks.

5 Remove the camshaft cover as described in Section 4.

6 Using a socket or spanner on the crankshaft sprocket bolt, turn the crankshaft until the tip of the camshaft lobe above the tappet to be checked is pointing vertically upwards.

7 Using feeler blades, check the clearance between the top of the tappet, and the cam lobe. If the play is in excess of 0.1 mm, renew the relevant tappet. If the play is less than 0.1 mm, or there is no play, proceed as follows.

8 Press down on the tappet using a wooden or plastic instrument **(see illustration)**. If free play in excess of 1.0 mm is present before the tappet contacts the valve stem, renew the relevant tappet.

9 On completion, refit the camshaft cover as described in Section 4.

12 Camshaft oil seals – renewal

Right-hand oil seal(s)

1 Remove the timing belt as described in Section 7.

2 Remove the camshaft sprocket and hub, as described in Section 8.

3 Drill two small holes into the existing oil seal, diagonally opposite each other. Take great care to avoid drilling through into the seal housing or camshaft sealing surface. Thread two self-tapping screws into the holes, and using a pair of pliers, pull on the heads of the screws to extract the oil seal.

4 Clean out the seal housing and the sealing surface of the camshaft by wiping it with a lint-free cloth. Remove any swarf or burrs that may cause the seal to leak.

5 Do not lubricate the lip and outer edge of the new oil seal, push it over the camshaft until it is positioned in place above its housing. To prevent damage to the sealing lips, wrap some adhesive tape around the end of the camshaft.

6 Using a hammer and a socket of suitable diameter, drive the seal squarely into its housing. **Note:** *Select a socket that bears only on the hard outer surface of the seal, not the inner lip which can easily be damaged.*

7 Refit the camshaft sprocket and its hub, as described in Section 8.

8 Refit and tension the timing belt as described in Section 7.

Left-hand oil seal

9 The left-hand camshaft oil seal is formed by the brake vacuum pump seal. Refer to Chapter 9 Section 21 for details of brake vacuum pump removal and refitting.

13 Cylinder head – removal, inspection and refitting

Note: *The cylinder head must be removed with the engine cold. New cylinder head bolts and a new cylinder head gasket will be required on refitting, and suitable studs will be required to guide the cylinder head into position – see text.*

Removal

1 Switch off the ignition and all electrical consumers, and remove the ignition key. For improved access on DOHC engines, remove the battery as described in Chapter 5A Section 3 **(see illustration)**.

2 Drain the cooling system and engine oil as described in Chapter 1B Section 33.

3 Remove the front section of the plenum chamber as described in Chapter 12 Section 19 **(see illustration)**.

4 Remove the air filter complete with the

13.1 Removing the battery

13.3 Removing the panel from the rear of the engine compartment

13.4a Disconnect the wiring…

13.4b …and vacuum hose…

13.4c …then release the clip…

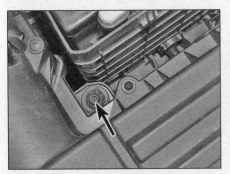

13.4d …undo the screw…

13.4e …remove the upper…

13.4f …and lower covers

air mass meter and air ducts. Also, where applicable, unbolt and remove the duct from the turbocharger **(see illustrations)**.

5 Disconnect the fuel supply and return lines, and also the coolant hoses from the cylinder head. In the interests of safety, it is recommended that the fuel is syphoned from the tandem pump on the left-hand end of the cylinder head. If necessary, the pump may be unbolted and removed **(see illustration)**.

6 Remove the fuel filter from its mounting bracket and position to one side.

7 Remove the front exhaust pipe as described in Chapter 4D Section 9.

8 Remove the turbocharger support and oil return line from the turbocharger. Also, remove the oil supply pipe and place to one side.

9 Remove the camshaft cover as described in Section 4.

10 Remove the timing belt as described in Section 7. Where the right-hand engine mounting has been removed, ensure the engine is supported adequately on a trolley jack and block of wood.

11 Remove the camshaft sprocket and timing belt tensioner as described in Section 8.

12 Where applicable, unscrew the bolt(s) securing the rear timing belt cover to the cylinder head **(see illustrations)**.

13 Remove the camshaft position sensor from the right-hand end of the cylinder head **(see illustration)**. Note that on DOHC engines, the idler roller must be loosened to release the sensor wiring.

13.5 Undo the four tandem pump retaining bolts

13.12a Where applicable, undo the bolt from the inner cover…

13.12b …and the one on the side of the cover

13.13 Unscrew the bolt and remove the camshaft position sensor

13.14 EGR connecting pipe at the flap housing

13.15 Disconnect the central connector for the injectors

13.16a Disconnect the coolant hose from the end of the cylinder head

13.16b Disconnect the vacuum pipes

13.17 The camshaft retaining frame inner row bolts are tightened into the tops of the front cylinder headbolts

14 Remove the exhaust gas recirculation connecting pipe **(see illustration)**.

15 Note the locations of all the electrical wiring, then disconnect them methodically **(see illustration)**.

16 Disconnect all vacuum and coolant hoses **(see illustrations)**.

17 On DOHC engines, unbolt and remove the injector rocker arm shaft as follows. Slacken the locknut of the adjustment screw on the end of the rocker arm above each of the injectors, and undo the adjustment screw until the rocker arm lies against the plunger pin of the injector. Starting at the outside and working in, gradually and evenly slacken and remove the rocker shaft retaining bolts. Lift off the rocker shaft – this will allow access to the rear cylinder head bolts. Now unscrew the row of inner bolts securing the camshaft retaining frame – the

bolts screw into the tops of the front cylinder head bolts, and are located behind the exhaust camshaft **(see illustration)**. As the bolts are removed, recover the large washers.

18 Using a multi-splined tool, undo the cylinder head bolts, working from the outside-in, evenly and gradually **(see illustration)**. Check that nothing remains connected, and lift the cylinder head from the engine block. Seek assistance if possible, as it is a heavy assembly, especially as it is being removed complete with the manifolds.

19 Remove the gasket from the top of the block, noting the locating dowels. If the dowels are a loose fit, remove them and store them with the head for safe-keeping. Do not discard the gasket yet – it will be needed for identification purposes. If desired, the manifolds can be removed from the cylinder

head with reference to (inlet manifold) or Chapter 4D Section 5 (exhaust manifold).

Inspection

20 Dismantling and inspection of the cylinder head is covered in Chapter 2F Section 6.

Cylinder head gasket selection

Note: *A dial test indicator (DTI) will be required for this operation.*

21 Examine the old cylinder head gasket for manufacturer's identification markings **(see illustration)**. These will be in the form of holes or notches, and a part number on the edge of the gasket. Unless new pistons have been fitted, the new cylinder head gasket must be of the same type as the old one.

22 If new piston assemblies have been fitted as part of an engine overhaul, or if a new short engine is to be fitted, the projection of the piston crowns above the cylinder head mating face of the cylinder block at TDC must be measured. This measurement is used to determine the thickness of the new cylinder head gasket required.

23 Anchor a dial test indicator (DTI) to the top face (cylinder head gasket mating face) of the cylinder block, and zero the gauge on the gasket mating face.

24 Rest the gauge probe on No 1 piston crown, and turn the crankshaft slowly by hand until the piston reaches TDC. Measure and record the maximum piston projection at TDC **(see illustration)**.

25 Repeat the measurement for the remaining pistons, and record the results.

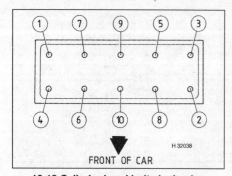

13.18 Cylinder head bolt slackening sequence

13.21 The thickness of the cylinder head gasket can be identified by notches or holes

13.24 Measuring the piston projection at TDC using a dial gauge

26 If the measurements differ from piston-to-piston, take the highest figure, and use this to determine the thickness of the head gasket required as follows.

Piston projection	Gasket identification (number of holes/notches)
0.91 to 1.00 mm	1
0.01 to 1.10 mm	2
1.11 to 1.20 mm	3

27 Purchase a new gasket according to the results of the measurements.

Refitting

Note: *If a Seat exchange cylinder head, complete with camshaft(s), is to be fitted, the manufacturers recommend the following:*

a) *Lubricate the contact surfaces between the tappets and the cam lobes before fitting the camshaft cover.*

b) *Do not remove the plastic protectors from the open valves until immediately before fitting the cylinder head.*

c) *Additionally, if a new cylinder head is fitted, Seat recommend that the coolant is renewed.*

28 The mating faces of the cylinder head and block must be perfectly clean before refitting the head. Use a scraper to remove all traces of gasket and carbon, also clean the tops of the pistons. Take particular care with the aluminium surfaces, as the soft metal is easily damaged.

29 Make sure that debris is not allowed to enter the oil and water passages – this is particularly important for the oil circuit, as carbon could block the oil supply to the camshaft and crankshaft bearings. Using adhesive tape and paper, seal the water, oil and bolt holes in the cylinder block.

30 To prevent carbon entering the gap between the pistons and bores, smear a little grease in the gap. After cleaning a piston, rotate the crankshaft to that the piston moves down the bore, then wipe out the grease and carbon with a cloth rag. Clean the other piston crowns in the same way.

31 Check the head and block for nicks, deep scratches and other damage. If slight, they may be removed carefully with a file. More serious damage may be repaired by machining, but this is a specialist job.

32 If warpage of the cylinder head is suspected, use a straight-edge to check it for distortion, as described in Chapter 2F Section 7.

33 Ensure that the cylinder head bolt holes in the crankcase are clean and free of oil. Syringe or soak up any oil left in the bolt holes. This is most important in order that the correct bolt tightening torque can be applied, and to prevent the possibility of the block being cracked by hydraulic pressure when the bolts are tightened.

34 Turn the crankshaft anti-clockwise all the pistons at an equal height, approximately half-way down their bores from the TDC position (see Section 3). This will eliminate any risk of piston-to-valve contact as the cylinder head is refitted.

35 Where applicable, refit the manifolds with reference to and Chapter 4D Section 5.

36 To guide the cylinder head into position, screw two long studs (or old cylinder head bolts with the heads cut off, and slots cut in the ends to enable the bolts to be unscrewed) into the cylinder block **(see illustration)**.

37 Ensure that the cylinder head locating dowels are in place in the cylinder block, then fit the new cylinder head gasket over the dowels, ensuring that the part number is uppermost. Where applicable, the OBEN/TOP marking should also be uppermost. Note that Seat recommend that the gasket is only removed from its packaging immediately prior to fitting.

38 Lower the cylinder head into position on the gasket, ensuring that it engages correctly over the guide studs and dowels.

39 Fit the new cylinder head bolts to the eight remaining bolt locations, and screw them in as far as possible by hand.

40 Unscrew the two guide studs from the exhaust side of the cylinder block, then screw in the two remaining new cylinder head bolts as far as possible by hand.

41 Working progressively, in sequence, tighten all the cylinder head bolts to the specified Stage 1 torque **(see illustrations)**.

42 Again working progressively, in sequence, tighten all the cylinder head bolts to the specified Stage 2 torque.

43 Tighten all the cylinder head bolts, in sequence, through the specified Stage 3 angle **(see illustration)**.

44 Finally, tighten all the cylinder head bolts,

13.36 Two of the old head bolts can be used as cylinder head alignment guides

in sequence, through the specified Stage 4 angle.

45 After finally tightening the cylinder head bolts, turn the camshaft so that the cam lobes for No 1 cylinder are pointing upwards.

46 Where applicable, reconnect the lifting tackle to the engine lifting brackets on the cylinder head, then adjust the lifting tackle to support the engine. Once the engine is adequately supported using the cylinder head brackets, disconnect the lifting tackle from the bracket bolted to the cylinder block, and unbolt the improvised engine lifting bracket from the cylinder block. Alternatively, remove the trolley jack and block of wood from under the sump.

47 The remainder of the refitting procedure is a reversal of the removal procedure, bearing in mind the following points.

a) *Refit the injector rocker shaft.*

b) *Refit the camshaft cover with reference to Section 4.*

c) *On turbo models, use new sealing rings when reconnecting the turbocharger oil return pipe to the cylinder block.*

d) *Reconnect the exhaust front section to the exhaust manifold or turbocharger, as applicable, with reference to Chapter 4D Section 9.*

e) *Refit the timing belt tensioner with reference to Section 8.*

f) *Refit the camshaft sprocket as described in Section 8, and refit the timing belt as described in Section 7.*

g) *Refill the cooling system (Chapter 1B Section 33) and engine oil as described in Chapter 1B Section 6.*

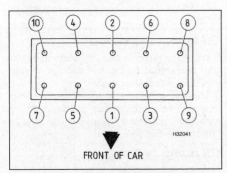

13.41a Cylinder head bolt tightening sequence

13.41b Using a torque wrench to tighten the cylinder head bolts

13.43 Angle-tighten the cylinder head bolts

14 Sump – removal and refitting

Note: *VAG sealant (D 176 404 A2 or equivalent) will be required to seal the sump on refitting.*

Removal

1 Apply the handbrake, then jack up the front of the vehicle and support securely on axle stands (see *Jacking and vehicle support*).
2 Remove the securing screws and withdraw the engine undertray and where fitted remove the sump insulation.
3 Drain the engine oil as described in Chapter 1B Section 6.
4 Where fitted, disconnect the wiring connector from the oil level/temperature sender in the sump **(see illustration)**.
5 Unscrew and remove the bolts securing the sump to the cylinder block, and the bolts securing the sump to the transmission casing, then withdraw the sump. If necessary, release the sump by tapping with a soft-faced hammer.
6 If desired, unbolt the oil baffle plate from the cylinder block.

Refitting

7 Begin refitting by thoroughly cleaning the mating faces of the sump and cylinder block. Ensure that all traces of old sealant are removed.
8 Where applicable, refit the oil baffle plate, and tighten the securing bolts.
9 Ensure that the cylinder block mating face of the sump is free from all traces of old

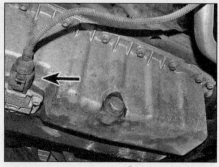

14.4 Disconnect the wiring connector from the oil level/temperature sender

sealant, oil and grease, and then apply a 2.0 to 3.0 mm thick bead of silicone sealant (VAG D 176 404 A2 or equivalent) to the sump **(see illustration)**. Note that the sealant should be run around the inside of the bolt holes in the sump. The sump must be fitted within 5 minutes of applying the sealant.
10 Offer the sump up to the cylinder block, then refit the sump-to-cylinder block bolts, and lightly tighten them by hand, working progressively in a diagonal sequence. **Note:** *If the sump is being refitted with the engine and transmission separated, make sure that the sump is flush with the flywheel/driveplate end of the cylinder block.*
11 Refit the sump-to-transmission casing bolts, and tighten them lightly, using a socket.
12 Again working in a diagonal sequence, lightly tighten the sump-to-cylinder block bolts, using a socket.
13 Tighten the sump-to-transmission casing bolts to the specified torque.

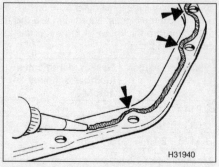

14.9 Apply the sealant around the inside of the bolt holes

14 Working in a diagonal sequence, progressively tighten the sump-to-cylinder block bolts to the specified torque.
15 Refit the wiring connector to the oil level/ temperature sender (where fitted), then refit the engine undertray(s), and lower the vehicle to the ground.
16 Allow at least 30 minutes from the time of refitting the sump for the sealant to dry, then refill the engine with oil, with reference to Chapter 1B Section 6.

15 Oil pump, drive chain and sprockets – removal, inspection and refitting

Oil pump removal

1 Remove the sump as described in Section 14.
2 Unscrew the securing bolts, and remove the oil baffle from the cylinder block.
3 Unscrew and remove the three mounting bolts, and release the oil pump from the dowels in the crankcase **(see illustration)**. Unhook the oil pump drive sprocket from the chain and withdraw the oil pump and oil pick-up pipe from the engine. Note that the tensioner will attempt to tighten the chain, and it may be necessary to use a screwdriver to hold it in its released position before releasing the oil pump sprocket from the chain.
4 If desired, unscrew the flange bolts and remove the suction pipe from the oil pump. Recover the O-ring seal. Unscrew the bolts and remove the cover from the oil pump. **Note:** *If the oil pick-up pipe is removed from the oil pump, a new O-ring will be required on refitting.*

Oil pump inspection

5 Clean the pump thoroughly, and inspect the gear teeth/rotors for signs of damage or wear. If evident, renew the oil pump.
6 To remove the sprocket from the oil pump, unscrew the retaining bolt and slide off the sprocket (note that the sprocket can only be fitted in one position).

Oil pump refitting

7 Prime the pump with oil by pouring oil into the pick-up pipe aperture while turning the driveshaft.

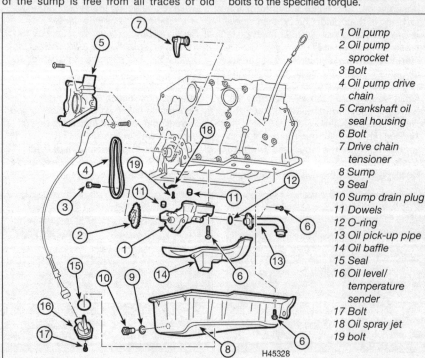

1 Oil pump
2 Oil pump sprocket
3 Bolt
4 Oil pump drive chain
5 Crankshaft oil seal housing
6 Bolt
7 Drive chain tensioner
8 Sump
9 Seal
10 Sump drain plug
11 Dowels
12 O-ring
13 Oil pick-up pipe
14 Oil baffle
15 Seal
16 Oil level/ temperature sender
17 Bolt
18 Oil spray jet
19 bolt

15.3 Sump and oil pump components

8 Refit the cover to the oil pump and tighten the bolts securely. Where applicable, refit the pick-up pipe to the oil pump, using a new O-ring seal, and tighten the securing bolts.

9 If the drive chain, crankshaft sprocket and tensioner have been removed, delay refitting them until after the oil pump has been mounted on the cylinder block. If they have not been removed, use a screwdriver to press the tensioner against its spring to provide sufficient slack in the chain to refit the oil pump.

10 Engage the oil pump sprocket with the drive chain, then locate the oil pump on the dowels. Refit and tighten the three mounting bolts to the specified torque.

11 Where applicable, refit the drive chain, tensioner and crankshaft sprocket using a reversal of the removal procedure.

12 Refit the oil baffle, and tighten the securing bolts.

13 Refit the sump as described in Section 14.

Oil pump drive chain and sprockets

Note: *VAG sealant (D 176 404 A2 or equivalent) will be required to seal the crankshaft oil seal housing on refitting, and it is advisable to fit a new crankshaft oil seal.*

Removal

14 Proceed as described in paragraphs 1 and 2.

15 To remove the oil pump sprocket, unscrew the securing bolt, then pull the sprocket from the pump shaft, and unhook it from the drive chain.

16 To remove the chain, remove the timing belt as described in Section 7, then unbolt the crankshaft oil seal housing from the cylinder block. Unbolt the chain tensioner from the cylinder block, then unhook the chain from the sprocket on the end of the crankshaft.

17 The oil pump drive sprocket is a press-fit on the crankshaft, and cannot easily be removed. Consult a Seat dealer for advice if the sprocket is worn or damaged.

Inspection

18 Examine the chain for wear and damage. Wear is usually indicated by excessive lateral play between the links, and excessive noise in operation. It is wise to renew the chain in any case if the engine is to be overhauled. Note that the rollers on a very badly worn chain

may be slightly grooved. If there is any doubt as to the condition of the chain, renew it.

19 Examine the teeth on the sprockets for wear. Each tooth forms an inverted V. If worn, the side of each tooth under tension will be slightly concave in shape when compared with the other side of the tooth (ie, the teeth will have a hooked appearance). If the teeth appear worn, the sprocket should be renewed (consult a Seat dealer for advice if the crankshaft sprocket is worn or damaged).

Refitting

20 If the oil pump has been removed, refit the oil pump as described previously in this Section before refitting the chain and sprocket.

21 Refit the chain tensioner to the cylinder block, and tighten the securing bolt to the specified torque. Make sure that the tensioner spring is correctly positioned to pretension the tensioner arm.

22 Engage the oil pump sprocket with the chain, then engage the chain with the crankshaft sprocket. Use a screwdriver to press the tensioner against its spring to provide sufficient slack in the chain to engage the sprocket with the oil pump. Note that the sprocket will only fit in one position.

23 Refit the oil pump sprocket bolt, and tighten to the specified torque.

24 Fit a new crankshaft oil seal to the housing, and refit the housing as described in Section 17.

25 Where applicable, refit the oil baffle, and tighten the securing bolts.

26 Refit the sump as described in Section 14.

16 Flywheel – removal, inspection and refitting

Removal

1 On manual transmission models, remove the transmission (see) and clutch (see).

2 On automatic transmission models, remove the semi-automatic (DSG) transmission as described in.

3 All models are fitted with a dual-mass flywheel, which can only be fitted in one position due to the offset of the mounting bolt holes in the end of the crankshaft.

4 Rotate the outside of the dual-mass flywheel so that the bolts align with the holes (if necessary).

5 Unscrew the bolts and remove the flywheel. Use a locking tool to counterhold the flywheel **(see illustration)**. Discard the bolts, as new ones must be fitted. **Note:** *In order not to damage the flywheel, do not allow the bolt heads to make contact with the flywheel during the unscrewing procedure.*

Inspection

6 Check the flywheel for wear and damage. Examine the starter ring gear for excessive wear to the teeth. If the driveplate or its ring gear is damaged, the complete driveplate must be renewed. The flywheel ring gear, however, may be renewed separately from the flywheel, but the work should be entrusted to a Seat dealer. If the clutch friction face is discoloured or scored excessively, it may be possible to regrind it, but this work should also be entrusted to a Seat dealer.

7 The following are guidelines only, but should indicate whether professional inspection is necessary. The dual-mass flywheel should be checked as follows:

8 There should be no cracks in the drive surface of the flywheel. If cracks are evident, the flywheel may need renewing.

Warpage

9 Place a straightedge across the face of the drive surface, and check by trying to insert a feeler gauge between the straightedge and the drive surface **(see illustration)**. The flywheel will normally warp like a bowl – i.e. higher on the outer edge. If the warpage is more than 0.40 mm, the flywheel may need renewing.

Free rotational movement

10 This is the distance the drive surface of the flywheel can be turned independently of the flywheel primary element, using finger effort alone. Move the drive surface in one direction and make a mark where the locating pin aligns with the flywheel edge. Move the drive surface in the other direction (finger pressure only) and make another mark **(see illustration)**. The total of free movement should not exceed 20.0 mm. If it's more, the flywheel may need renewing.

16.5 Use a locking tool to counterhold the flywheel

16.9 Flywheel warpage check – see text

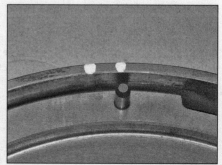

16.10 Flywheel free rotational movement check alignment marks – see text

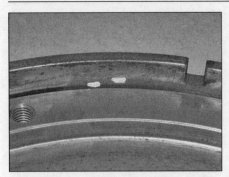

16.12 Flywheel lateral movement check marks – see text

Total rotational movement

11 This is the total distance the drive surface can be turned independently of the flywheel primary element. Insert two bolts into the clutch pressure plate/damper unit mounting holes, and with the crankshaft/flywheel held stationary, use a lever/pry bar between the bolts and use some effort to move the drive surface fully in one direction – make a mark where the locating pin aligns with the flywheel edge. Now force the drive surface fully in the opposite direction, and make another mark. The total rotational movement should not exceed 44.0 mm. If it does, have the flywheel professionally inspected.

Lateral movement

12 The lateral movement (up and down) of the drive surface in relation to the primary element of the flywheel should not exceed 2.0 mm. If it does, the flywheel may need renewing. This can be checked by pressing the drive surface down on one side into the flywheel (flywheel horizontal)

17.2 Removing the crankshaft oil seal using self-tapping screws

17.9 Slide the oil seal housing over the end of the crankshaft

and making an alignment mark between the drive surface and the inner edge of the primary element. Now press down on the opposite side of the drive surface, and make another mark above the original one. The difference between the two marks is the lateral movement **(see illustration)**.

Refitting

13 Refitting is a reversal of removal. Use new bolts when refitting the flywheel or driveplate, and coat the threads of the bolts with locking fluid before inserting them. Tighten them to the specified torque.

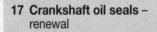

17 Crankshaft oil seals – renewal

Note: *The oil seals are a PTFE (Teflon) type and are fitted dry, without using any grease or oil. These have a wider sealing lip and have been introduced instead of the coil spring type oil seal.*
Note: *If the oil seal housing is removed, suitable sealant (VAG D 176 404 A2, or equivalent) will be required to seal the housing on refitting.*

Timing belt end oil seal

1 Remove the timing belt as described in Section 7, and the crankshaft sprocket with reference to Section 8.
2 To remove the seal without removing the housing, drill two small holes diagonally opposite each other, insert self-tapping screws, and pull on the heads of the screws with pliers **(see illustration)**.

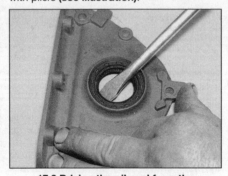

17.3 Prising the oil seal from the crankshaft oil seal housing

17.17 Locate the crankshaft oil seal fitting tool over the end of the crankshaft

3 Alternatively, to remove the oil seal complete with its housing, proceed as follows.
a) Remove the sump as described in Section 14. This is necessary to ensure a satisfactory seal between the sump and oil seal housing on refitting.
b) Unbolt and remove the oil seal housing.
c) Working on the bench, lever the oil seal from the housing using a suitable screwdriver. Take care not to damage the seal seating in the housing **(see illustration)**.
4 Thoroughly clean the oil seal seating in the housing.
5 Wind a length of tape around the end of the crankshaft to protect the oil seal lips as the seal (and housing, where applicable) is fitted.
6 Fit a new oil seal to the housing, pressing or driving it into position using a socket or tube of suitable diameter. Ensure that the socket or tube bears only on the hard outer ring of the seal, and take care not to damage the seal lips. Press or drive the seal into position until it is seated on the shoulder in the housing. Make sure that the closed end of the seal is facing outwards.
7 If the oil seal housing has been removed, proceed as follows, otherwise proceed to paragraph 11.
8 Clean all traces of old sealant from the crankshaft oil seal housing and the cylinder block, then coat the cylinder block mating faces of the oil seal housing with a 2.0 to 3.0 mm thick bead of sealant (VAG D 176 404 A2, or equivalent). Note that the seal housing must be refitted within 5 minutes of applying the sealant.
Caution: DO NOT put excessive amounts of sealant onto the housing as it may get into the sump and block the oil pick-up pipe.
9 Refit the oil seal housing, and tighten the bolts progressively to the specified torque **(see illustration)**.
10 Refit the sump as described in Section 14.
11 Refit the crankshaft sprocket with reference to Section 8, and the timing belt as described in Section 7.

Flywheel/driveplate end oil seal

Note: *The seal housing and crank sensor must be correctly positioned, and this requires a special tool.*
12 Remove the flywheel as described in Section 16.
13 Remove the sump as described in Section 14. This is necessary to ensure a satisfactory seal between the sump and oil seal housing on refitting.
14 Unbolt and remove the oil seal housing, complete with the oil seal.
15 The new oil seal will be supplied ready-fitted to a new oil seal housing.
16 Thoroughly clean the oil seal housing mating face on the cylinder block.
17 New oil seal/housing assemblies are supplied with a fitting tool to prevent damage to the oil seal as it is being fitted. Locate the tool over the end of the crankshaft **(see illustration)**.

17.19a Fit the oil seal/housing assembly over the end of the crankshaft...

17.19b ...then tighten the securing bolts to the specified torque

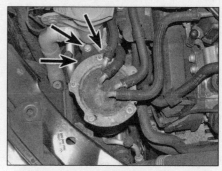

18.8 Undo the filter retaining bolts/nut

18 If the original oil seal housing was fitted using sealant, apply a thin bead of suitable sealant (VAG D 176 404 A2, or equivalent) to the cylinder block mating face of the oil seal housing. Note that the seal housing must be refitted within 5 minutes of applying the sealant.

Caution: DO NOT put excessive amounts of sealant onto the housing as it may get into the sump and block the oil pick-up pipe.

19 Carefully fit the oil seal/housing assembly over the end of the crankshaft, then refit the securing bolts and tighten the bolts progressively, in a diagonal sequence, to the specified torque **(see illustrations)**.

20 Remove the oil seal protector tool from the end of the crankshaft.

21 Refit the sump as described in Section 14.

22 Refit the flywheel as described in Section 16.

---

18 Engine/transmission mountings – inspection and renewal

Inspection

1 If improved access is required, jack up the front of the vehicle, and support it securely on axle stands (see *Jacking and vehicle support*). Remove the engine top cover, which also incorporates the air filter, and then remove the engine undershield(s).

2 Check the mounting rubbers to see if they

are cracked, hardened or separated from the metal at any point; renew the mounting if any such damage or deterioration is evident.

3 Check that all the mountings are securely tightened; use a torque wrench to check if possible.

4 Using a large screwdriver or a crowbar, check for wear in the mounting by carefully levering against it to check for free play. Where this is not possible, enlist the aid of an assistant to move the engine/transmission back-and-forth, or from side-to-side, whilst you observe the mounting. While some free play is to be expected, even from new components, excessive wear should be obvious. If excessive free play is found, check first that the fasteners are correctly secured, and then renew any worn components as described in the following paragraphs.

Renewal

Note: *New mounting securing bolts will be required on refitting.*

Right-hand mounting

5 Attach a hoist and lifting tackle to the engine lifting brackets on the cylinder head, and raise the hoist to just take the weight of the engine. Alternatively the engine can be supported on a trolley jack under the engine. Use a block of wood between the sump and the head of the jack, to prevent any damage to the sump.

6 For improved access, unscrew the coolant reservoir and move it to one side, leaving the coolant hoses connected.

7 Remove the air intake pipe.

8 Undo the retaining bolts and move the fuel filter to one side **(see illustration)**. Where applicable, move any wiring harnesses, pipes or hoses to one side to enable removal of the engine mounting.

9 Unscrew the bolts securing the mounting to the engine bracket, and then unscrew the bolts securing it to the body. Withdraw the mounting from the engine compartment **(see illustrations)**.

10 Refitting is a reversal of removal, bearing in mind the following points.

a) *Use new securing bolts.*

b) *There must be at least 10 mm between the engine mounting bracket and the right-hand side chassis member* **(see illustration)**.

c) *The side of the mounting support arm must be parallel to the side of the engine mounting bracket.*

d) *Tighten all fixings to the specified torque.*

Left-hand mounting

11 Remove the engine top cover.

12 Attach a hoist and lifting tackle to the engine lifting brackets on the cylinder head, and raise the hoist to just take the weight of the engine and transmission. Alternatively the engine can be supported on a trolley jack under the transmission. Use a block of wood between the transmission and the head of the jack, to prevent any damage to the transmission.

13 Remove the battery and battery tray, as described in Chapter 5A Section 3.

18.9a Remove the bracket

18.9b With the engine supported remove the mounting

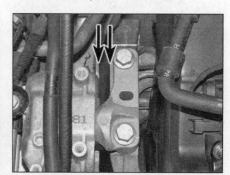

18.10 There must be at least 10 mm between the bracket and the chassis member

18.14 Remove the transmission to mounting bracket

18.19 Rear torque arm

14 Unscrew the bolts securing the mounting to the transmission, and the remaining bolts securing the mounting to the body, then lift the mounting from the engine compartment **(see illustration)**.

15 Refitting is a reversal of removal, bearing in mind the following points:
a) Use new mounting bolts.
b) The edges of the mounting support arm must be parallel to the edge of the mounting.
c) Tighten all fixings to the specified torque.

Rear mounting (torque arm)

16 Apply the handbrake, then jack up the front of the vehicle and support securely on axle stands (see *Jacking and vehicle support*). Remove the engine undershield(s) for access to the rear mounting (torque arm).

17 Support the rear of the transmission beneath the final drive housing. To do this, use a trolley jack and block of wood, or alternatively wedge a block of wood between the transmission and the subframe.

18 Working under the vehicle, unscrew and remove the bolt securing the mounting to the subframe.

19 Unscrew the two bolts securing the mounting to the transmission, then withdraw the mounting from under the vehicle **(see illustration)**.

20 Refitting is a reversal of removal, but use new mounting securing bolts, and tighten all fixings to the specified torque.

19 Engine oil cooler – removal and refitting

Note: *New sealing rings will be required on refitting.*

Removal

1 The oil cooler is mounted under the oil filter housing on the front of the cylinder block **(see illustration)**. Note that the oil cooler fitted to 2.0 litre models is slightly different.

2 Position a container beneath the oil filter to catch escaping oil and coolant.

3 Clamp the oil cooler coolant hoses to minimise coolant spillage, then remove the clips, and disconnect the hoses from the oil cooler. Be prepared for coolant spillage.

4 Unscrew the oil cooler securing plate from the bottom of the oil filter housing, then slide off the oil cooler. Recover the O-rings from the top and bottom of the oil cooler.

Refitting

5 Refitting is a reversal of removal, bearing in mind the following points:
a) Use new oil cooler O-rings.
b) Tighten the oil cooler securing plate securely.
c) On completion, check and if necessary top-up the oil and coolant levels.

20 Oil pressure warning light switch – removal and refitting

Removal

1 The oil pressure warning light switch is fitted to the oil filter housing **(see illustration 19.1)**. Remove the engine top cover to gain access to the switch (see Section 4).

2 Disconnect the wiring connector and wipe clean the area around the switch.

3 Unscrew the switch from the filter housing and remove it, along with its sealing washer. If the switch is to be left removed from the engine for any length of time, plug the oil filter housing aperture.

Refitting

4 Examine the sealing washer for signs of damage or deterioration and if necessary renew.

5 Refit the switch, complete with washer, and tighten it to the specified torque.

6 Securely reconnect the wiring connector then check and, if necessary, top-up the engine oil as described in Chapter 0 Section 5. On completion, refit the engine top cover.

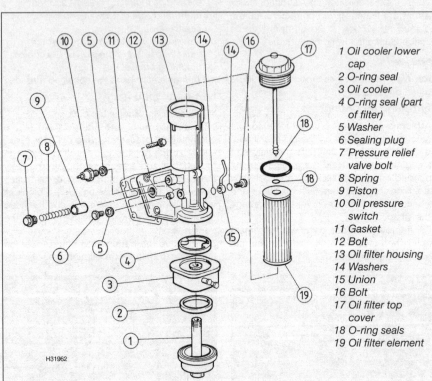

1 Oil cooler lower cap
2 O-ring seal
3 Oil cooler
4 O-ring seal (part of filter)
5 Washer
6 Sealing plug
7 Pressure relief valve bolt
8 Spring
9 Piston
10 Oil pressure switch
11 Gasket
12 Bolt
13 Oil filter housing
14 Washers
15 Union
16 Bolt
17 Oil filter top cover
18 O-ring seals
19 Oil filter element

H31962

19.1 Oil filter and oil cooler mounting details (1.9 litre engine)

Chapter 2 Part F
Engine removal and overhaul procedures

Contents

Degrees of difficulty

Easy, suitable for novice with little experience	**Fairly easy,** suitable for beginner with some experience	**Fairly difficult,** suitable for competent DIY mechanic	**Difficult,** suitable for experienced DIY mechanic	**Very difficult,** suitable for expert DIY or professional

Specifications

Engine codes*

1.6 litre petrol engine	BGU, BSE and BSF
2.0 litre petrol engine:	
Non-turbo	BLR, BVY, BLY and BVZ
Turbo	BWA, BWJ, CDLD and CDLA
Diesel engine:	
PD unit injector engine:	
1.9 litre, 8-valve, turbo, SOHC	BJB, BKC, BLS, BXE and BXF
2.0 litre:	
8-valve, turbo, SOHC	BMM
16-valve, turbo, DOHC	BKD and BMN
Common rail injection engine:	
1.6 litre engine	CAYC
2.0 litre engine	CFJA, CLCB, CEGA and CFHC

* **Note:** *See 'Vehicle identification' at the end of this manual for the location of engine code markings.*

Piston rings	**New**	**Wear limit**
End gaps:		
Petrol engines:		
1.6 engines:		
Compression rings	0.20 to 0.40 mm	0.80 mm
Oil scraper ring	0.25 to 0.50 mm	0.80 mm
2.0 litre:		
Compression rings	0.20 to 0.40 mm	0.80 mm
Oil scraper ring	0.25 to 0.50 mm	0.80 mm
Diesel engines:		
1.9 litre:		
Compression rings	0.20 to 0.40 mm	1.0 mm
Oil scraper ring	0.25 to 0.50 mm	1.0 mm
2.0 litre:		
Compression rings	0.25 to 0.50 mm	1.0 mm
Oil scraper ring	0.25 to 0.50 mm	1.0 mm
Ring-to-groove clearance:		
Petrol engines:		
1.6 litre:		
Compression rings	0.20 to 0.40 mm	0.80 mm
Oil scraper ring	0.25 to 0.50 mm	0.80 mm
2.0 litre:		
Compression rings	0.06 to 0.09 mm	0.20 mm
Oil scraper ring	0.03 to 0.06 mm	0.15 mm
Diesel engines		
1st compression ring	0.06 to 0.09 mm	0.25 mm
2nd compression ring.	0.05 to 0.08 mm	0.25 mm
Oil scraper ring	0.03 to 0.06 mm	0.15 mm

Crankshaft endfloat

	New	**Wear limit**
Petrol engines:		
1.6 litre	0.07 to 0.23 mm	0.30 mm
2.0 litre	0.07 to 0.23 mm	0.30 mm
Diesel engines	0.07 to 0.23 mm	0.30 mm

Cylinder head

Minimum permissible dimension between top of valve stem and top surface of cylinder head:	
Petrol engines:	
1.6 litre:	
Inlet valves	31.7 mm
Exhaust valves	31.7 mm
2.0 litre	No reworking permitted
Diesel engines	No reworking permitted
Minimum cylinder head height:	
Petrol engines:	
1.6 litre	132.9 mm
2.0 litre	No reworking permitted
Diesel engines	No reworking permitted
Maximum cylinder head gasket face distortion:	
All engines	0.1 mm

Valves

Valve stem diameter:	**Inlet valves**	**Exhaust valves**
Petrol engines:		
1.6 litre BGU, BSE and BSF	5.980 ± 0.007 mm	5.965 ± 0.007 mm
2.0 litre BLR, BVY, BLY and BVZ	5.98 mm	5.96 mm
2.0 litre BWA, BWJ, CDLD and CDLA	5.980 mm	5.95 mm
Diesel engines:		
1.6 litre CAYC	5.968 – 5.982 mm	5.958 – 5.972 mm
1.9 litre BJB, BKC, BLS, BXE and BXF	6.980 mm	6.956 mm
2.0 litre BMM	6.980 mm	6.98 mm
2.0 litre BKD and BMN	6.98 mm	6.956 mm
2.0 litre BKD and BMN	5.980 mm	5.965 mm
2.0 litre CFJA, CLCB, CEGA and CFHC.	5.980	5.965 mm
Valve seat angle (all engines)	45°	45°

Torque wrench settings

Refer to Chapter 2A, Chapter 2B, Chapter 2C, Chapter 2D, or Chapter 2E as applicable.

1 General Information

How to use this Chapter

1 Included in this Part of Chapter 2 are details of removing the engine from the car and general overhaul procedures for the cylinder head, cylinder block and all other engine internal components.

2 The information given ranges from advice concerning preparation for an overhaul and the purchase of new parts, to detailed step-by-step procedures covering removal, inspection, renovation and refitting of engine internal components.

3 After Section 6, all instructions are based on the assumption that the engine has been removed from the car. For information concerning in-car engine repair, as well as the removal and refitting of those external components necessary for full overhaul, refer to the relevant in-car repair procedure section and to Section 5 of this Chapter. Ignore any preliminary dismantling operations described in the relevant in-car repair sections that are no longer relevant once the engine has been removed from the car.

4 Apart from torque wrench settings, which are given at the beginning of the relevant in-car repair procedure in Chapter 2A, Chapter 2B, Chapter 2C, Chapter 2D or Chapter 2E (as applicable) all specifications relating to engine overhaul are given at the beginning of this Part of Chapter 2.

2 Engine overhaul –
general information

1 It is not always easy to determine when, or if, an engine should be completely overhauled, as a number of factors must be considered.

2 High mileage is not necessarily an indication that an overhaul is needed, while low mileage does not preclude the need for an overhaul. Frequency of servicing is probably the most important consideration. An engine which has had regular and frequent oil and filter changes, as well as other required maintenance, should give many thousands of miles of reliable service. Conversely, a neglected engine may require an overhaul very early in its life.

3 Excessive oil consumption is an indication that piston rings, valve seals and/or valve guides are in need of attention. Make sure that oil leaks are not responsible before deciding that the rings and/or guides are worn. Perform a compression (or leakdown) test, as described in Part A 2A Section 2 (1.6 litre petrol engines) B 2B Section 3 (2.0 litre petrol engines) C 2C Section 2 (1.6 litre common rail diesel engines) D 2D Section 2 (2.0 litre common rail diesel engines) or E 2E Section

2 (1.9 and 2.0 litre PD diesel engines) of this Chapter (as applicable), to determine the likely cause of the problem.

4 Check the oil pressure with a gauge fitted in place of the oil pressure switch, and compare it with that specified (see Specifications in Parts A, B, C, D or E of this Chapter). If it is extremely low, the main and big-end bearings, and/or the oil pump, are probably worn.

5 Loss of power, rough running, knocking or metallic engine noises, excessive valve gear noise, and high fuel consumption may also point to the need for an overhaul, especially if they are all present at the same time. If a complete service does not remedy the situation, major mechanical work is the only solution.

6 An engine overhaul involves restoring all internal parts to the specification of a new engine. During an overhaul, the pistons and the piston rings are renewed. New main and big-end bearings are generally fitted (where possible); if necessary, the crankshaft may be renewed to restore the journals. The valves are also serviced as well, since they are usually in less-than-perfect condition at this point. While the engine is being overhauled, other components, such as the starter and alternator, can be overhauled as well. The end result should be an as-new engine that will give many trouble-free miles. **Note:** *Critical cooling system components such as the hoses, thermostat and coolant pump should be renewed when an engine is overhauled. The radiator should be checked carefully, to ensure that it is not clogged or leaking. Also, it is a good idea to renew the oil pump whenever the engine is overhauled.*

7 Before beginning the engine overhaul, read through the entire procedure, to familiarise yourself with the scope and requirements of the job. Overhauling an engine is not difficult if you follow carefully all of the instructions, have the necessary tools and equipment, and pay close attention to all specifications. It can, however, be time-consuming. Plan on the car being off the road for a minimum of two weeks, especially if parts must be taken to an engineering works for repair or reconditioning. Check on the availability of parts and make sure that any necessary special tools and equipment are obtained in advance. Most work can be done with typical hand tools, although a number of precision measuring tools are required for inspecting parts to determine if they must be renewed. Often the engineering works will handle the inspection of parts and offer advice concerning reconditioning and renewal. **Note:** *Always wait until the engine has been completely dismantled, and until all components (especially the cylinder block and the crankshaft) have been inspected, before deciding what service and repair operations must be performed by an engineering works. The condition of these components will be the major factor to consider when determining whether to overhaul the original engine, or to buy a reconditioned unit. Do not, therefore,*

purchase parts or have overhaul work done on other components until they have been thoroughly inspected. As a general rule, time is the primary cost of an overhaul, so it does not pay to fit worn or sub-standard parts.

8 As a final note, to ensure maximum life and minimum trouble from a reconditioned engine, everything must be assembled with care, in a spotlessly-clean environment.

3 Engine/transmission removal
– preparation and precautions

1 If you have decided that the engine must be removed for overhaul or major repair work, several preliminary steps should be taken.

2 Locating a suitable place to work is extremely important. Adequate work space, along with storage space for the vehicle, will be needed. If a workshop or garage is not available, at the very least a solid, level, clean work surface is required.

3 If possible, clear some shelving close to the work area and use it to store the engine components and ancillaries as they are removed and dismantled. In this manner, the components stand a better chance of staying clean and undamaged during the overhaul. Laying out components in groups together with their fixings bolts, screws, etc, will save time and avoid confusion when the engine is refitted.

4 Clean the engine compartment and engine before beginning the removal procedure; this will help visibility and help to keep tools clean.

5 The help of an assistant is essential; there are certain instances when one person cannot safely perform all of the operations required to remove the engine from the vehicle. Safety is of primary importance, considering the potential hazards involved in this kind of operation. A second person should always be in attendance to offer help in an emergency. If this is the first time you have removed an engine, advice and aid from someone more experienced would also be beneficial.

6 Plan the operation ahead of time. Before starting work, obtain (or arrange for the hire of) all of the tools and equipment you will need. Access to the following items will allow the task of removing and refitting the engine to be completed safely and with relative ease: a hoist and lifting tackle – rated in excess of the weight of the engine, complete sets of spanners and sockets as described at the rear of this manual, wooden blocks, and plenty of rags and cleaning solvent for mopping-up spilled oil, coolant and fuel. A selection of different-sized plastic storage bins will also prove useful for keeping dismantled components grouped together. If any of the equipment must be hired, make sure that you arrange for it in advance, and perform all of the operations possible without it beforehand; this may save you time and money.

7 Plan on the vehicle being out of use for quite a while, especially if you intend to carry out an engine overhaul. Read through the whole of this Section and work out a strategy based on your own experience, and the tools, time and workspace available to you. Some of the overhaul processes may have to be carried out by a Seat dealer or an automotive machine shop – these establishments often have busy schedules, so it would be prudent to consult them before removing or dismantling the engine, to get an idea of the amount of time required to carry out the work.

8 When removing the engine from the vehicle, be methodical about the disconnection of external components. Labelling cables and hoses as they are removed will greatly assist the refitting process.

9 Always be extremely careful when lifting the engine from the engine compartment. Serious injury can result from careless actions. If help is required, it is better to wait until it is available rather than risk personal injury and/or damage components by continuing alone. By planning ahead and taking your time, a job of this nature, although major, can be accomplished successfully and without incident.

4 Engine and transmission – removal and refitting

Removal

1 For the home mechanic the easiest way to remove the engine and transmission assembly is by completely removing the front lock carrier panel and dragging the assembly out of the front of the engine bay. On models with air conditioning the AC system must be de-gassed (by an air conditioning specialist) before removing the panel. Alternatively the engine and transmission can be removed by lowering it from the engine bay. However this approach still requires the lock carrier moving to the service position (as described in Chapter 3 Section 3), so it may as well be removed completely.

2 Switch off the ignition and all electrical consumers, and remove the ignition key.

3 Where fitted, remove the engine top cover and air filter, as described in Chapter 4A Section 3 (petrol engines) or Chapter 4B Section 3 (diesel engines) and all associated air ducting **(see illustration)**.

4 Remove the battery as described in Chapter 5A Section 3, then remove the battery tray **(see illustrations)**.

5 Apply the handbrake, then jack up the front of the vehicle and support it on axle stands (see *Jacking and vehicle support*). Remove both front roadwheels.

6 Move the lock carrier to the service position as as described in Chapter 3 Section 3 or remove it completely. On models with AC the panel can only be removed after the AC has been de-gassed.

⚠ *Warning: The AC system must be evacuated using specialist equipment. It is a criminal offence to knowingly discharge the refrigerant to atmosphere.*

7 Drain the cooling system as described in Chapter 1A Section 33 (petrol engines) or Chapter 1B Section 33 (diesel engines).

8 Noting their locations, disconnect all wiring, coolant hoses, vacuum hoses and fuel lines from the engine/transmission, with reference to the relevant Chapters of this Manual. Alternatively, the engine wiring loom may remain on the engine by disconnecting it from the left-hand side of the engine compartment, and removing the engine management ECU as described in Chapter 4A Section 5 (petrol engines) or Chapter 4B Section 4 (diesel engines). Tape over or plug fuel lines to prevent entry of dust and dirt **(see illustrations)**.

4.3 Remove the air filter housing

4.4a Remove the battery…

4.4b …and tray

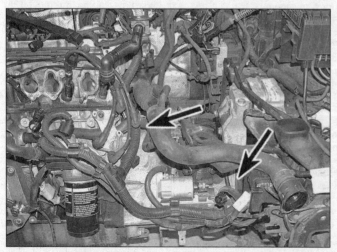

4.8a The wiring loom at the front of the engine

4.8b Disconnect and seal the fuel lines

9 Remove the front exhaust pipe **(see illustration)** as described in Chapter 4C Section 9 (petrol engines) or Chapter 4D Section 9 (diesel engines).

10 Unbolt the rear engine support/torque arm from the transmission **(see illustration)**.

11 Disconnect the gearchange mechanism **(see illustration)** with reference to Chapter 7A Section 2 (manual transmission) or Chapter 7B Section 4 (DSG transmission)

12 On manual transmission models, remove the clutch slave cylinder or disconnect the clutch hose (models with a concentric slave cylinder). **Note:** *Do not depress the clutch pedal once the slave cylinder has been removed.*

13 Whilst access is easy (assuming the lock carrier has been removed), refer to Chapter 3 Section 12 and unbolt the air conditioning compressor from the front of the engine. If the AC system has been de-gassed remove the compressor completely. If the AC system is still charged (because the lock carrier is only in the service position) suspend the compressor to one side of the engine compartment and do not remove the refrigerant lines.

14 Disconnect the driveshafts from the transmission drive flanges as described in Chapter 8 Section 2. Suspend them from the underbody **(see illustration)**.

15 Unbolt the washer fluid reservoir for access to the right-engine mounting. Also unbolt the coolant expansion tank and place to one side.

16 Connect a hoist and lifting tackle to the engine lifting brackets on the cylinder head, and raise the hoist to just take the weight of the engine/transmission **(see illustration)**.

17 Unbolt the right and left-hand engine mountings **(see illustrations)**.

18 Make a final check to ensure that all relevant wiring, hoses and pipes have been disconnected, then carefully swivel the engine/transmission assembly away from the sides of the engine compartment, lift it upwards or lower it (see paragraph 1), and withdraw forwards from the front of the car **(see illustration)**.

4.9 Disconnect the exhaust system from the manifold (or turbocharger)

4.10 Remove the rear engine torque arm

4.11 Disconnect the gearchange cables and unbolt the bracket

4.14 Use straps or cable ties to suspend the driveshafts from the underbody

4.16 Lift the engine/transmission with an engine crane

4.17a Remove the right-hand mounting…

4.17b …and the left-hand mounting

4.18 Removing the engine/transmission from the vehicle

Separation

Engine and transmission

19 Remove the starter motor **(see illustration)** as described in Chapter 5A Section 8.

20 Where applicable, unscrew the bolt securing the small engine-to-transmission plate to the transmission **(see illustration)**.

21 Ensure that both engine and transmission are adequately supported, then unscrew the remaining engine-to-transmission bolts, noting the location of each bolt, and the locations of any brackets secured by the bolts.

22 Carefully withdraw the transmission from the engine, ensuring that the weight of the transmission is not allowed to hang on the input shaft while it is engaged with the clutch friction disc. Recover the engine-to-transmission plate.

Reconnection and refitting

Engine and transmission

23 Reconnection and refitting are a reversal of removal, bearing in mind the following points:

a) *Ensure that any brackets noted before removal are in place on the engine-to-transmission bolts.*

b) *Tighten all fixings to the specified torque, where given.*

c) *Where applicable, have the air conditioning system recharged with refrigerant by a suitably-qualified professional.*

d) *Ensure that all wiring, hoses and pipes are correctly reconnected and routed as noted before removal.*

e) *Ensure that the fuel lines are correctly reconnected.*

f) *On completion, refill the cooling system as described in Chapter 1A Section 33 (petrol engines) or Chapter 1B Section 33 (diesel engines).*

**5 Engine overhaul –
preliminary information**

---

1 It is much easier to dismantle and work on the engine if it is mounted on a portable engine stand. These stands can often be hired from a tool hire shop. Before the engine is mounted on a stand, the flywheel should be removed, so that the stand bolts can be tightened into the end of the cylinder block/crankcase. **Note:** *Do not measure cylinder bore dimensions with the engine mounted on this type of stand.*

2 If a stand is not available, it is possible to dismantle the engine with it blocked up on a sturdy workbench, or on the floor. Be very careful not to tip or drop the engine when working without a stand.

3 If you intend to obtain a reconditioned engine, all ancillaries must be removed first, to be transferred to the new engine (just as they will if you are doing a complete engine overhaul yourself). These components include

4.19 Remove the starter motor

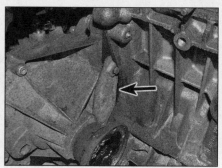
4.20 Where fitted remove the plate

the following (it may be necessary to transfer additional components, such as the oil level dipstick/tube assembly, oil filter housing, etc, depending on which components are supplied with the reconditioned engine:

Petrol engines

a) *Alternator (including mounting brackets) and starter motor (Chapter 5A Section 5).*

b) *The ignition system components including all sensors and spark plugs (Chapter 1A Section 29 and Chapter 5B Section 3).*

c) *The fuel injection system components (Chapter 4A Section 5).*

d) *All electrical switches, actuators and sensors, and the engine wiring harness.*

e) *Inlet and exhaust manifolds, and turbocharger (where applicable) (Chapter 4A Section 10, Chapter 4C Section 8 and Chapter 4C Section 6).*

f) *Engine mountings (see the relevant part of Section or Chapter 2B Section 17).*

g) *Clutch components (Chapter 6 Section 6).*

h) *Oil separator (where applicable).*

Diesel engines

a) *Alternator (including mounting brackets) and starter motor (Chapter 5A).*

b) *The glow plug/preheating system components (Chapter 5C Section 2).*

c) *All fuel system components, including fuel injectors, all sensors and actuators (Chapter 4B).*

d) *The brake vacuum pump (Chapter 9 Section 21).*

e) *All electrical switches, actuators and sensors, and the engine wiring harness.*

f) *Inlet and exhaust manifolds/turbocharger (where applicable) (Chapter 4B Section 6 and Chapter 4D Section 5).*

g) *Engine mountings (see Chapter 2C Section 16, Chapter 2D Section 16). or Chapter 2E Section 18.*

h) *Clutch components (Chapter 6 Section 6).*

All engines

Note: *When removing the external components from the engine, pay close attention to details that may be helpful or important during refitting. Note the fitted position of gaskets, seals, spacers, pins, washers, bolts, and other small components.*

4 If you are obtaining a short engine (the engine cylinder block/crankcase, crankshaft, pistons and connecting rods, all fully assembled), then the cylinder head, sump, oil pump, timing belt(s) and chain (as applicable – together with tensioner(s) and covers), auxiliary drivebelt (together with its tensioner), coolant pump, thermostat housing, coolant outlet elbows, oil filter housing and where applicable oil cooler will also have to be removed.

5 If you are planning a full overhaul, the engine can be dismantled in the order given below:

a) *Inlet and exhaust manifolds (see the relevant part of Chapter 4A, Chapter 4B, Chapter 4C and Chapter 4D).*

b) *Timing belt, sprockets and tensioner (see the relevant part of Chapter 2A, Chapter 2B, Chapter 2C, Chapter 2D or Chapter 2E).*

c) *Cylinder head (see the relevant part of Chapter 2A Section 12 (1.6 litre petrol), Chapter 2B Section 10 (2.0 litre petrol), Chapter 2C Section 11 (1.6 CR diesel), Chapter 2D Section 11 (2.0 litre CR diesel) or Chapter 2E Section 13 (1.9 and 2.0 litre PD diesel).*

d) *Flywheel/driveplate (see the relevant part of Chapter 2A, Chapter 2B, Chapter 2C, Chapter 2D or Chapter 2E).*

e) *Sump (see the relevant part of Chapter 2A, Chapter 2B, Chapter 2C, Chapter 2D or Chapter 2E).*

f) *Oil pump (see the relevant part of Chapter 2A, Chapter 2B, Chapter 2C, Chapter 2D or Chapter 2E).*

g) *Piston/connecting rod assemblies (see Section 9).*

h) *Crankshaft (see Section 10).*

---

6 Cylinder head – dismantling

Note: *A valve spring compressor tool will be required for this operation.*

1.6 litre petrol engines

1 With the cylinder head removed (as described in Chapter 2A Section 12), proceed as follows.

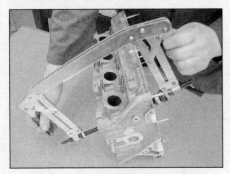

6.9a Compress a valve spring with a compressor tool

6.9b Remove the spring cap...

6.9c ...and valve spring

2 If not already removed, remove the the inlet and exhaust manifolds as described in Chapter 4A Section 10 and Chapter 4C Section 8 respectively.

3 Remove the camshaft and hydraulic tappets/roller rocker fingers, as described in Chapter 2A Section 9 and Chapter 2A Section 10.

4 If desired, unbolt the coolant housing from the rear of the cylinder head, and recover the seal.

5 If not already done, remove the camshaft position sensor.

6 Unscrew the securing nut, and recover the washer, and remove the timing belt tensioner pulley from the stud on the cylinder head.

7 Unbolt any remaining auxiliary brackets and/or engine lifting brackets from the cylinder head as necessary, noting their locations to aid refitting.

8 Turn the cylinder head over, and rest it on one side.

9 Using a valve spring compressor, compress each valve spring in turn until the split collets can be removed. Release the compressor, and lift off the spring cap and spring. If, when the valve spring compressor is screwed down, the spring cap refuses to free and expose the split collets, gently tap the top of the tool, directly over the spring cap, with a light hammer. This will free the retainer **(see illustrations)**.

10 Using a pair of pliers, or a removal tool, carefully extract the valve stem oil seal from the top of the valve guide **(see illustrations)**.

11 Withdraw the valve from the gasket side of the cylinder head **(see illustration)**.

12 It is essential that each valve is stored together with its collets, cap, spring and spring seat. The valves should be kept in their correct sequences, unless they are so badly worn that they are to be renewed.

2.0 litre petrol engines

13 With the cylinder head removed, proceed as follows.

14 Remove the inlet and exhaust manifolds as described in Chapter 4A Section 10 and Chapter 4C Section 8 respectively.

15 Unbolt any remaining auxiliary brackets and/or engine lifting brackets from the cylinder head as necessary, noting their locations to aid refitting.

16 Proceed as described in paragraphs 8 to 12, but when labelling the valve components, make sure that the valves are identified as inlet and exhaust, as well as numbered.

Diesel engines

17 With the cylinder head removed, proceed as follows.

18 Remove the inlet and exhaust manifolds as described in Chapter 4B Section 6 and Chapter 4D Section 5.

19 Remove the camshaft and hydraulic tappets, as described in Chapter 2C Section 9, Chapter 2D Section 9 or Chapter 2E Section 10.

20 Remove the glow plugs, with reference to Chapter 5C Section 2.

21 Remove the fuel injectors, with reference to Chapter 4B Section 5.

22 Unscrew the nut and remove the timing

belt tensioner pulley from the stud on the timing belt end of the cylinder head.

23 Unbolt any remaining auxiliary brackets and/or engine lifting brackets from the cylinder head as necessary, noting their locations to aid refitting.

24 Proceed as described in paragraphs 8 to 12.

---

7 Cylinder head and valves – cleaning and inspection

1 Thorough cleaning of the cylinder head and valve components, followed by a detailed inspection, will enable you to decide how much valve service work must be carried out during engine overhaul. **Note:** *If the engine has been severely overheated, it is best to assume that the cylinder head is warped – check carefully for signs of this.*

Cleaning

2 Using a suitable degreasing agent, remove all traces of oil deposits from the cylinder head, paying particular attention to the camshaft bearing surfaces, hydraulic tappet bores, valve guides and oilways. Scrape off any traces of old gasket from the mating surfaces, taking care not to score or gouge them. If using emery paper, do not use a grade of less than 100. Turn the head over and, using a blunt blade, scrape any carbon deposits from the combustion chambers and ports. Finally, wash the entire head

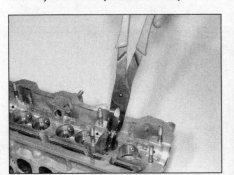

6.10a Use a removal tool...

6.10b ...to remove the valve stem oil seals

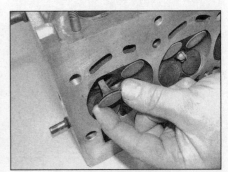

6.11 Removing a valve

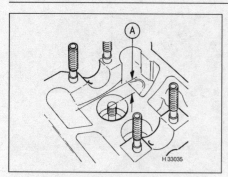

7.6 Measure the distance (A) between the top face of the valve stem and the top surface of the cylinder head

7.7 Measure the distortion of the cylinder head gasket surface

casting with a suitable solvent to remove the remaining debris.

3 Clean the valve heads and stems using a fine wire brush (or a power-operated wire brush). If the valve is covered with heavy carbon deposits, scrape off the majority of the deposits with a blunt blade first, then use the wire brush.

4 Thoroughly clean the remainder of the components using solvent and allow them to dry completely. Discard the oil seals, as new ones must be fitted when the cylinder head is reassembled.

Inspection

Cylinder head

Note: *If the valve seats are to be recut, ensure that the maximum permissible reworking dimension (where applicable) is not exceeded (the maximum dimension will only allow minimal reworking to produce a perfect seal between valve and seat). If the maximum dimension is exceeded, the function of the hydraulic tappets cannot be guaranteed, and the cylinder head must be renewed. Refer to paragraph 6 for details of how to calculate the maximum permissible reworking dimension. Reworking is not permitted on 2.0 litre petrol engines and all diesel engines.*

5 Examine the head casting closely to identify any damage or cracks that may have developed. Cracks can often be identified from evidence of coolant or oil leakage. Pay particular attention to the areas around the valve seats and spark plug/fuel injector holes. If cracking is discovered in this area, Seat state that on diesel engines and SOHC petrol engines, the cylinder head may be re-used, provided the cracks are no larger than 0.5 mm wide on diesel engines, or 0.3 mm wide on SOHC petrol engines. More serious damage will mean the renewal of the cylinder head casting.

6 Moderately pitted and scorched valve seats can be repaired by lapping the valves in during reassembly, as described later in this Chapter. Badly worn or damaged valve seats may be restored by recutting where permitted, however, the maximum permissible reworking dimension must not be exceeded, only minimal reworking being possible (see note at beginning of paragraph 5). To calculate the maximum permissible reworking dimension, proceed as follows **(see illustration)** :

a) *If a new valve is to be fitted, use the new valve for the following calculation.*

b) *Insert the valve into its guide in the cylinder head, and push the valve firmly on to its seat.*

c) *Using a flat edge placed across the top surface of the cylinder head, measure the distance between the top face of the valve stem, and the top surface of the cylinder head. Record the measurement obtained.*

d) *Consult the Specifications, and look up the value for the minimum permissible dimension between the top face of the valve stem and the top surface of the cylinder head.*

e) *Now take the measured distance and subtract the minimum permissible dimension, to give the maximum*

permissible reworking dimension; eg, Measured distance (34.4 mm) Minus Minimum permissible dimension (34.0 mm) = Maximum permissible reworking dimension (0.4 mm).

7 Measure any distortion of the gasket surfaces using a straight-edge and a set of feeler blades. Take one measurement longitudinally on the manifold mating surface(s). Take several measurements across the head gasket surface, to assess the level of distortion in all planes **(see illustration)**. Compare the measurements with the figures in the Specifications.

8 On 1.6 litre petrol engines, if the head is distorted beyond the specified limit, it may be possible to have it machined by an engineering works, provided that the minimum permissible cylinder head height is maintained.

9 On all other engines, if the head is distorted beyond the specified limit, the head must be renewed.

Camshaft

10 Inspection of the camshaft is covered in Parts A to F of this Chapter, as applicable.

Valves and associated components

11 Examine each valve closely for signs of wear. Inspect the valve stems for wear ridges, scoring or variations in diameter; measure their diameters at several points along their lengths with a micrometer, and compare with the figures given in the Specifications **(see illustration)**.

12 The valve heads should not be cracked, badly pitted or charred. Note that light pitting of the valve head can be rectified by lapping-in the valves during reassembly, as described in Section 8.

13 Check that the valve stem end face is free from excessive pitting or indentation; this could be caused by defective hydraulic tappets.

14 Using vernier calipers, measure the free length of each of the valve springs. As a manufacturer's figure is not quoted, the only way to check the length of the springs is by comparison with a new component. Note that valve springs are usually renewed during a major engine overhaul **(see illustration)**.

15 Stand each spring on its end on a flat surface, against an engineer's square **(see illustration)**. Check the squareness of the

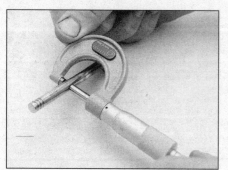

7.11 Measure the diameter of the valve stems using a micrometer

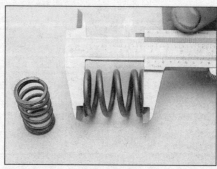

7.14 Measure the free length of each valve spring

7.15 Check the squareness of a valve spring

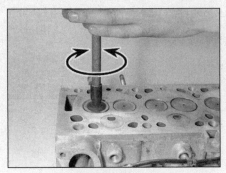

8.2 Grinding-in a valve

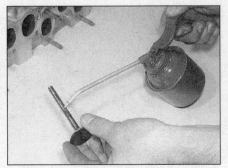

8.8a Lubricate the valve stem with clean engine oil

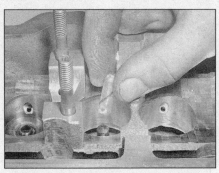

8.8b Fit a protective sleeve over the valve stem before fitting the stem seal

spring visually, and renew it if it appears distorted.

16 Renew the valve stem oil seals regardless of their apparent condition.

8 Cylinder head – reassembly

Note: *A valve spring compressor tool will be required for this operation.*

1.6 petrol engines (SOHC)

1 To achieve a gas-tight seal between the valves and their seats, it will be necessary to lap-in (or grind-in) the valves. To complete this process you will need a quantity of fine/coarse grinding paste and a grinding tool – this can either be of the rubber sucker type, or the automatic type which is driven by a rotary power tool.

2 Smear a small quantity of fine grinding paste on the sealing face of the valve head. Turn the cylinder head over so that the combustion chambers are facing upwards and insert the valve into the correct guide. Attach the grinding tool to the valve head and using a backward/forward rotary action, grind the valve head into its seat. Periodically lift the valve and rotate it to redistribute the grinding paste **(see illustration)**.

3 Continue this process until the contact between valve and seat produces an unbroken, matt grey ring of uniform width, on both faces. Repeat the operation on the remaining valves.

4 If the valves and seats are so badly pitted that coarse grinding paste must be used, bear in that there is a maximum permissible reworking dimension for the valves and seats. Refer to the Specifications at the beginning of this Chapter for the minimum dimension from the end of the valve stem to the top face of the cylinder head (see Section 7, paragraph 6). If this minimum dimension is exceeded due to excessive lapping-in, the hydraulic tappets may not operate correctly, and the cylinder head must be renewed.

5 Assuming the repair is feasible, work as described previously, but use coarse grinding paste initially, to achieve a dull finish on the

valve face and seat. Wash off the coarse paste with solvent and repeat the process using fine grinding paste to obtain the correct finish.

6 When all the valves have been ground in, remove all traces of grinding paste from the cylinder head and valves using solvent, and allow the head and valves to dry completely.

7 Turn the cylinder head on its side.

8 Working on one valve at a time, lubricate the valve stem with clean engine oil, and insert the valve into its guide. Fit one of the protective plastic sleeves supplied with the new valve stem oil seals over the end of the valve stem – this will protect the oil seal as it is being fitted **(see illustrations)**.

9 Dip a new valve stem seal in clean engine oil, and carefully push it over the valve stem and onto the top of the valve guide – take care not to damage the stem seal as it is fitted. Use a suitable long-reach socket or a valve stem seal fitting tool to press the seal firmly

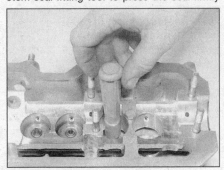

8.9 Use a special installer to fit a valve stem oil seal

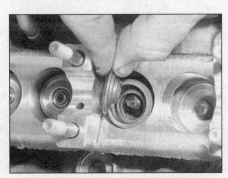

8.11a Fit the upper spring seat

into position **(see illustration)**. Remove the protective sleeve from the valve stem.

10 Locate the valve spring over the valve stem, ensuring that the lower end of the spring seats correctly on the cylinder head **(see illustration)**.

11 Fit the upper spring seat over the top of the spring, then using a valve spring compressor, compress the spring until the upper seat is pushed beyond the collet grooves in the valve stem. Refit the split collets. Gradually release the spring compressor, checking that the collets remain correctly seated as the spring extends. When correctly seated, the upper spring seat should force the collets securely into the grooves in the end of the valve stem **(see illustrations)**.

12 Repeat this process for the remaining sets of valve components, ensuring that all components are refitted to their original locations. The traditional method of settling

8.10 Fit a valve spring

8.11b Use grease to hold the split collets in the groove

8.19a Use a long-reach socket to fit a valve stem oil seal

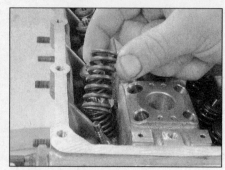

8.19b Fit a valve spring…

8.19c …and upper spring seat

the the collets by striking the head of the valve with a block of wood and a mallet after installation is not recommended. Instead slightly compress and release the valves several times with the spring compressor. Check before progressing any further that the split collets remain firmly seated in the grooves in the end of the valve stem.

13 Refit any auxiliary brackets and/or engine lifting brackets to their original locations, as noted before removal.

14 Refit the timing belt tensioner pulley, and secure with the nut and washer.

15 Where applicable, refit the camshaft position sensor.

16 Where applicable, refit the coolant housing to the rear of the cylinder head, using a new seal.

17 Refit the camshaft (Chapter 2A Section 9) and hydraulic tappets (Chapter 2A Section 10).

18 Refit the inlet and exhaust manifolds as described in Chapter 4A Section 10 and Chapter 4C Section 8.

2.0 petrol engines (DOHC)

19 Proceed as described in paragraphs 1 to 13 **(see illustrations)**.

20 Refit the inlet and exhaust manifolds

as described in Chapter 4A Section 10 and Chapter 4C Section 8.

Diesel engines

21 Proceed as described in paragraphs 1 to 13.

22 Refit the timing belt tensioner pulley to the stud on the cylinder head, and refit the securing nut.

23 Refit the fuel injectors, with reference to Chapter 4B Section 5.

24 Refit the glow plugs, with reference to Chapter 5C Section 2.

25 Refit the hydraulic tappets and camshaft, as described in Chapter 2C Section 9, Chapter 2D Section 9 or Chapter 2E Section 10.

26 Refit the inlet and exhaust manifolds, as described in Chapter 4B Section 6 and Chapter 4D Section 5.

9 Piston/connecting rod assemblies – removal

1 Proceed as follows according to engine type:

a) On 1.6 litre petrol engines, remove

the cylinder head, sump and oil baffle plate, and oil pump and pick-up pipe, as described in Chapter 2A.

b) On 2.0 litre engines, remove the cylinder head, sump and oil baffle plate, oil pump and pick-up pipe, and the balancer shaft assembly as described in Chapter 2B.

c) On diesel engines, remove the cylinder head, sump and oil baffle plate, and oil pump and pick-up pipe, as described in Chapter 2C, Chapter 2D or Chapter 2E.

2 Inspect the tops of the cylinder bores for ridges at the point where the pistons reach top dead centre. These must be removed otherwise the pistons may be damaged when they are pushed out of their bores. Use a scraper or ridge reamer to remove the ridges. Such a ridge indicates excessive wear of the cylinder bore.

3 Check the connecting rods and big-end caps for identification markings. Both connecting rods and caps should be marked with the cylinder number on one side of each assembly. Note that No 1 cylinder is at the timing belt end of the engine. If no marks are present, using a hammer and centre-punch, paint or similar, mark each connecting rod and big-end bearing cap with its respective cylinder number – note on which side of the

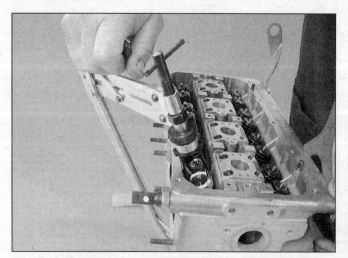

8.19d Compress a valve spring using a compressor tool

9.3 Mark the big-end caps and connecting rods with their cylinder numbers

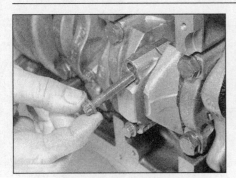
9.6a Unscrew the big-end bearing cap bolts...

9.6b ...and remove the cap

9.7 Wrap the threaded ends of the bolts with tape

connecting rods and caps the marks are made (see illustration).

4 Similarly, check the piston crowns for direction markings. An arrow on each piston crown should point towards the timing belt end of the engine. On some engines, this mark may be obscured by carbon build-up, in which case the piston crown should be cleaned to check for a mark. In some cases, the direction arrow may have worn off, in which case a suitable mark should be made on the piston crown using a scriber – do not deeply score the piston crown, but ensure that the mark is easily visible.

5 Turn the crankshaft to bring Nos 1 and 4 pistons to bottom dead centre.

6 Unscrew the bolts or nuts, as applicable, from No 1 piston big-end bearing cap. Lift off the cap, and recover the bottom half bearing shell. If the bearing shells are to be re-used, tape the cap and bearing shell together. Note that if the bearing shells are to be re-used, they must be fitted to the original connecting rod and cap (see illustrations).

7 Where the bearing caps are secured with nuts, wrap the threaded ends of the bolts with insulating tape to prevent them scratching the crankpins and bores when the pistons are removed (see illustration).

8 Using a hammer handle, push the piston up through the bore, and remove it from the top of the cylinder block. Where applicable, take care not to damage the piston cooling oil

spray jets in the cylinder block as the piston/ connecting rod assembly is removed. Recover the upper bearing shell, and tape it to the connecting rod for safe-keeping.

9 Loosely refit the big-end cap to the connecting rod, and secure with the bolts or nuts, as applicable – this will help to keep the components in their correct order.

10 Remove No 4 piston assembly in the same way.

11 Turn the crankshaft as necessary to bring Nos 2 and 3 pistons to bottom dead centre, and remove them in the same way.

12 Where applicable, remove the securing bolts, and withdraw the piston cooling oil spray jets from the bottom of the cylinder block (see illustrations).

10 Crankshaft – removal

Note: If no work is to be done on the pistons and connecting rods, there is no need to push the pistons out of the cylinder bores. The pistons should just be pushed far enough up the bores so that they are positioned clear of the crankshaft journals.

1 Proceed as follows according to engine type:

a) On 1.6 (SOHC) petrol engines, remove

the timing belt and crankshaft sprocket, sump and oil baffle plate, oil pump and pick-up pipe, flywheel/driveplate, and the crankshaft oil seal housings, as described in Chapter 2A.

b) On 2.0 (DOHC) petrol engines, remove the timing belt/chain, sump, oil pump and pick-up pipe, flywheel/driveplate, and the crankshaft oil seal housing, as described in Chapter 2B. **Note:** On 2.0 litre engines, it will necessary to remove the balancer shaft assembly.

c) On diesel engines, remove the timing belt and crankshaft sprocket, sump and oil baffle plate, oil pump and pick-up pipe, flywheel/driveplate, and the crankshaft oil seal housings, as described in Chapter 2C, Chapter 2D or Chapter 2E.

2 Remove the pistons and connecting rods, or disconnect them from the crankshaft, as described in Section 9.

3 Check the crankshaft endfloat as described in Section 13, then proceed as follows.

4 The main bearing caps should be numbered 1 to 5 from the timing belt end of the engine. If the bearing caps are not marked, mark them accordingly using a centre-punch. Note the orientation of the markings to ensure correct refitting.

5 Slacken and remove the main bearing cap bolts, and lift off each cap. If the caps appear to be stuck, tap them with a soft-faced mallet to free them from the cylinder block (see

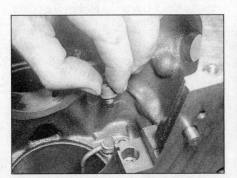
9.12a Remove the securing bolts...

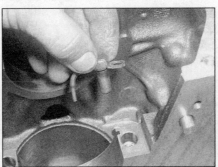
9.12b ...and withdraw the piston cooling oil spray jets

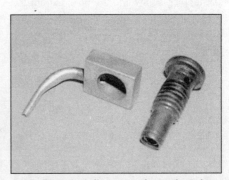
9.12c Piston cooling spray jet and retainer

10.5 Slacken and remove the main bearing cap bolts

10.7 Lift the crankshaft from the cylinder block

illustration). Recover the lower bearing shells, and tape them to their caps for safe-keeping.
6 Recover the lower crankshaft endfloat control thrustwasher halves from either side of the No 3 main bearing cap, noting their orientation.
7 Lift the crankshaft from the cylinder block **(see illustration).** Take care, as the crankshaft is heavy. On engines with a crankshaft speed/position sensor fitted, lay the crankshaft on wooden blocks – do not rest the crankshaft on the sensor wheel.
8 Recover the upper bearing shells from the cylinder block, and tape them to their respective caps for safe-keeping. Similarly, recover the upper crankshaft endfloat control thrustwasher halves, noting their orientation.
9 On engines with a crankshaft speed/position sensor wheel, unscrew the securing bolts, and remove the sensor wheel, noting which way round it is fitted.

11 Cylinder block/crankcase – cleaning and inspection

Cleaning

1 Remove all external components and electrical switches/sensors from the block, including mounting brackets, the coolant pump, the oil filter, and oil cooler **(see illustration)**, etc. For complete cleaning, the core plugs should ideally be removed. Drill a small hole in the plugs, then insert a self-tapping screw into the hole. Extract the

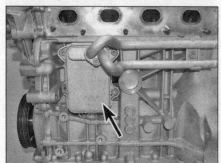

11.1 Oil cooler located on the front of the engine

plugs by pulling on the screw with a pair of grips, or by using a slide hammer.
2 Scrape all traces of gasket and sealant from the cylinder block/crankcase, taking care not to damage the sealing surfaces.
3 Remove all oil gallery plugs (where fitted). The plugs are usually very tight – they may have to be drilled out, and the holes retapped. Use new plugs when the engine is reassembled.
4 If the casting is extremely dirty, it should be steam-cleaned. After this, clean all oil holes and galleries one more time. Flush all internal passages with warm water until the water runs clear. Dry thoroughly, and apply a light film of oil to all mating surfaces and cylinder bores, to prevent rusting. If you have access to compressed air, use it to speed up the drying process, and to blow out all the oil holes and galleries.

⚠ *Warning: Wear eye protection when using compressed air.*

5 If the castings are not very dirty, you can do an adequate cleaning job with hot, soapy water and a stiff brush. Take plenty of time, and do a thorough job. Regardless of the cleaning method used, be sure to clean all oil holes and galleries very thoroughly, and to dry all components well. Protect the cylinder bores as described above, to prevent rusting.
6 Where applicable, check the piston cooling oil spray jets for damage, and renew if necessary. Check the oil spray hole and the oil passages for blockage.
7 All threaded holes must be clean, to ensure accurate torque readings during reassembly. To clean the threads, run the correct-size

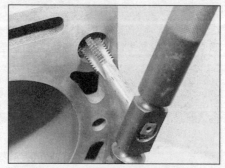

11.7 To clean the cylinder block threads, run a correct-size tap into the holes

tap into each of the holes to remove rust, corrosion, thread sealant or sludge, and to restore damaged threads **(see illustration)**. If possible, use compressed air to clear the holes free of debris produced by this operation. **Note:** *Take extra care to exclude all cleaning liquid from blind tapped holes, as the casting may be cracked by hydraulic action if a bolt is threaded into a hole containing liquid.*
8 After coating the mating surfaces of the new core plugs with suitable sealant, fit them to the cylinder block. Make sure that they are driven in straight and seated correctly, or leakage could result.
9 Apply suitable sealant to the new oil gallery plugs, and insert them into the holes in the block. Tighten them securely.
10 If the engine is not going to be reassembled immediately, cover it with a large plastic bag to keep it clean; protect all mating surfaces and the cylinder bores, to prevent rusting.

Inspection

11 Visually check the castings for cracks and corrosion. Look for stripped threads in the threaded holes. If there has been any history of internal coolant leakage, it may be worthwhile having an engine overhaul specialist check the cylinder block/crankcase for cracks with special equipment. If defects are found, have them repaired, if possible, or renew the assembly.
12 Check each cylinder bore for scuffing and scoring.
13 If in any doubt as the condition of the cylinder block have the block/bores inspected and measured by an engine reconditioning specialist. They will be able to advise on whether the block is serviceable, whether a rebore is necessary, and supply the appropriate pistons and rings.
14 If the bores are in reasonably good condition and not excessively worn, then it may only be necessary to renew the piston rings.
15 If this is the case, the bores should be honed, to allow the new rings to bed-in correctly and provide the best possible seal. Consult an engine reconditioning specialist
16 On diesel engines, if the oil/water pump housing was removed, it can be refitted at this stage if wished. Use a new gasket, and before fully tightening the bolts, align the housing faces with those of the engine block.
17 The cylinder block/crankcase should now be completely clean and dry, with all components checked for wear or damage, and repaired or overhauled as necessary.
18 Apply a light coating of engine oil to the mating surfaces and cylinder bores to prevent rust forming.
19 Refit as many ancillary components as possible, for safe-keeping. If reassembly is not to start immediately, cover the block with a large plastic bag to keep it clean, and protect the machined surfaces as described above to prevent rusting.

12 Piston/connecting rod assemblies – cleaning and inspection

Cleaning

1 Before the inspection process can begin, the piston/connecting rod assemblies must be cleaned, and the original piston rings removed from the pistons.

2 The rings should have smooth, polished working surfaces, with no dull or carbon-coated sections (showing that the ring is not sealing correctly against the bore wall, so allowing combustion gases to blow by) and no traces of wear on their top and bottom surfaces. The end gaps should be clear of carbon, but not polished (indicating a too-small end gap), and all the rings (including the elements of the oil control ring) should be free to rotate in their grooves, but without excessive up-and-down movement. If the rings appear to be in good condition, they are probably fit for further use; check the end gaps (in an unworn part of the bore) as described in Section 12.

3 If any of the rings appears to be worn or damaged, or has an end gap significantly different from the specified value, the usual course of action is to renew all of them as a set. **Note:** *While it is usual to renew piston rings when an engine is overhauled, they may be re-used if in acceptable condition. If re-using the rings, make sure that each ring is marked during removal to ensure that it is refitted correctly.*

4 Carefully expand the old rings over the top of the pistons. The use of two or three old feeler blades will be helpful in preventing the rings dropping into empty grooves **(see illustration)**. Be careful not to scratch the piston with the ends of the ring. The rings are brittle, and will snap if they are spread too far. They are also very sharp – protect your hands and fingers. Note that the third ring incorporates an expander. Keep each set of rings with its piston if the old rings are to be re-used. Note which way up each ring is fitted to ensure correct refitting.

5 Scrape away all traces of carbon from the top of the piston. A hand-held wire brush (or a piece of fine emery cloth) can be used, once the majority of the deposits have been scraped away.

6 Remove the carbon from the ring grooves in the piston, using an old ring. Break the ring in half to do this (be careful not to cut your fingers – piston rings are sharp). Be careful to remove only the carbon deposits – do not remove any metal, and do not nick or scratch the sides of the ring grooves.

7 Once the deposits have been removed, clean the piston/connecting rod assembly with paraffin or a suitable solvent, and dry thoroughly. Make sure that the oil return holes in the ring grooves are clear.

Inspection

8 If the pistons and cylinder bores are not damaged or worn excessively, and if the

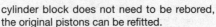

12.4 Old feeler blades can be used to prevent piston rings from dropping into empty grooves

cylinder block does not need to be rebored, the original pistons can be refitted.

9 Have the pistons and cylinder bore measure by an engine reconditioning specialist. They will be able to advise on possible repairs, and supply the correct replacement parts. **Note:** *If the cylinder block was rebored during a previous overhaul, oversize pistons may already have been fitted.*

10 Normal piston wear shows up as even vertical wear on the piston thrust surfaces, and slight looseness of the top ring in its groove. New piston rings should always be used when the engine is reassembled.

11 Carefully inspect each piston for cracks around the skirt, around the gudgeon pin holes, and at the piston ring 'lands' (between the ring grooves).

12 Look for scoring and scuffing on the piston skirt, holes in the piston crown, and burned areas at the edge of the crown. If the skirt is scored or scuffed, the engine may have been suffering from overheating, and/or abnormal combustion which caused excessively high operating temperatures. The cooling and lubrication systems should be checked thoroughly.

13 Scorch marks on the sides of the pistons show that blow-by has occurred.

14 A hole in the piston crown, or burned areas at the edge of the piston crown, indicates that abnormal combustion (pre-ignition, knocking, or detonation) has been occurring.

15 If any of the above problems exist, the causes must be investigated and corrected, or the damage will occur again. The causes

may include incorrect ignition/injection pump timing, inlet air leaks or incorrect air/fuel mixture (petrol engines), or a faulty fuel injector (diesel engines).

16 Corrosion of the piston, in the form of pitting, indicates that coolant has been leaking into the combustion chamber and/or the crankcase. Again, the cause must be corrected, or the problem may persist in the rebuilt engine.

17 Locate a new piston ring in the appropriate groove and measure the ring-to-groove clearance using a feeler blade **(see illustration)**. Note that the rings are of different widths, so use the correct ring for the groove. Compare the measurements with those listed; if the clearances are outside of the tolerance band, then the piston must be renewed. Confirm this by checking the width of the piston ring with a micrometer.

18 Examine each connecting rod carefully for signs of damage, such as cracks around the big-end and small-end bearings. Check that the rod is not bent or distorted. Damage is highly unlikely, unless the engine has been seized or badly overheated. Detailed checking of the connecting rod assembly can only be carried out by a Seat dealer or engine repair specialist with the necessary equipment.

19 The gudgeon pins are of the floating type, secured in position by two circlips. The pistons and connecting rods can be separated as follows.

20 Using a small flat-bladed screwdriver, prise out the circlips, and push out the gudgeon pin **(see illustrations)**. Hand

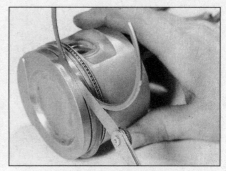

12.17 Measure the piston ring-to-groove clearance using a feeler blade

12.20a Use a small flat-bladed screwdriver to prise out the circlip…

12.20b …then push out the gudgeon pin and separate the piston and connecting rod

12.23a The piston crown is marked with an arrow which must point towards the timing belt/chain end of the engine

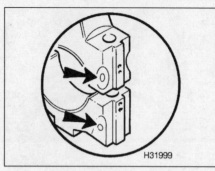

12.23b The recesses in the connecting rod and bearing cap must face the timing belt/chain end of the engine

pressure should be sufficient to remove the pin. Identify the piston and rod to ensure correct reassembly. Discard the circlips – new ones must be used on refitting. If the gudgeon pin proves difficult to remove, heat the piston to 60°C with hot water – the resulting expansion will then allow the two components to be separated.

21 Examine the gudgeon pin and connecting rod small-end bearing for signs of wear or damage. It should be possible to push the gudgeon pin through the connecting rod bush by hand, without noticeable play. Wear can be cured by renewing both the pin and bush. Bush renewal, however, is a specialist job – press facilities are required, and the new bush must be reamed accurately.

22 Examine all components, and obtain any new parts from your Seat dealer or engine reconditioning specialist. If new pistons are purchased, they will be supplied complete with gudgeon pins and circlips. Circlips can also be purchased individually.

23 The orientation of the piston with respect to the connecting rod must be correct when the two are reassembled. The piston crown is marked with an arrow (which may be obscured by carbon deposits); this must point towards the timing belt/chain end of the engine when the piston is installed. The connecting rod and its bearing cap both have recesses machined into them on one side, close to their mating surfaces – these recesses must both face the same way as the arrow on the piston crown (ie, towards the timing belt/chain end of the engine) when correctly installed.

Reassemble the two components to satisfy this requirement **(see illustrations)**.

24 Apply a smear of clean engine oil to the gudgeon pin. Slide it into the piston and through the connecting rod small-end. Check that the piston pivots freely on the rod, then secure the gudgeon pin in position with two new circlips. Ensure that each circlip is correctly located in its groove in the piston.

25 Repeat the cleaning and inspection process for the remaining pistons and connecting rods.

13 Crankshaft – checking endfloat and inspection

Checking endfloat

1 If the crankshaft endfloat is to be checked, this must be done when the crankshaft is still installed in the cylinder block/crankcase, but is free to move (see Section 10).

2 Check the endfloat using a dial gauge in contact with the end of the crankshaft. Push the crankshaft fully one way, and then zero the gauge. Push the crankshaft fully the other way, and check the endfloat. The result can be compared with the specified amount, and will give an indication as to whether new thrustwasher halves are required **(see illustration)**. Note that all thrustwashers must be of the same thickness.

3 If a dial gauge is not available, feeler blades can be used. First push the crankshaft fully

towards the flywheel end of the engine, then use feeler blades to measure the gap between the web of No 3 crankpin and the thrustwasher halves **(see illustration)**.

Inspection

4 Clean the crankshaft using paraffin or a suitable solvent, and dry it, preferably with compressed air if available. Be sure to clean the oil holes with a pipe cleaner or similar probe, to ensure that they are not obstructed.

⚠️ *Warning: Wear eye protection when using compressed air.*

5 Check the main and big-end bearing journals for uneven wear, scoring, pitting and cracking.

6 Big-end bearing wear is accompanied by distinct metallic knocking when the engine is running (particularly noticeable when the engine is pulling from low speed) and some loss of oil pressure.

7 Main bearing wear is accompanied by severe engine vibration and rumble – getting progressively worse as engine speed increases – and again by loss of oil pressure.

8 Check the bearing journal for roughness by running a finger lightly over the bearing surface. Any roughness (which will be accompanied by obvious bearing wear) indicates that the crankshaft requires regrinding (where possible) or renewal.

9 If the crankshaft has been reground, check for burrs around the crankshaft oil holes (the holes are usually chamfered, so burrs should not be a problem unless regrinding has been carried out carelessly). Remove any burrs with a fine file or scraper, and thoroughly clean the oil holes as described previously.

10 Have the crankshaft measured and inspected by an engine reconditioning specialist. They will be able to advise any possible repairs and supply the correct parts.

11 Check the oil seal contact surfaces at each end of the crankshaft for wear and damage. If the seal has worn a deep groove in the surface of the crankshaft, consult an engine overhaul specialist; repair may be possible, but otherwise a new crankshaft will be required.

12 If the crankshaft journals have not already been reground, it may be possible to have the crankshaft reconditioned, and to fit undersize shells (see Section). If no undersize shells are available and the crankshaft has worn beyond the specified limits, it will have to be renewed. Consult your VW dealer or engine specialist for further information on parts availability.

14 Main and big-end bearings – inspection and selection

Inspection

1 Even though the main and big-end bearings should be renewed during the engine overhaul, the old bearings should be retained

13.2 Measure crankshaft endfloat using a dial gauge

13.3 Measure crankshaft endfloat using feeler blades

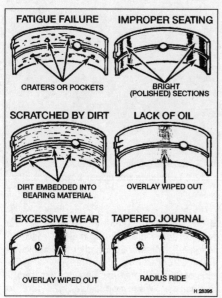

14.1 Typical bearing failures

FATIGUE FAILURE
CRATERS OR POCKETS

IMPROPER SEATING
BRIGHT (POLISHED) SECTIONS

SCRATCHED BY DIRT
DIRT EMBEDDED INTO BEARING MATERIAL

LACK OF OIL
OVERLAY WIPED OUT

EXCESSIVE WEAR
OVERLAY WIPED OUT

TAPERED JOURNAL
RADIUS RIDE

H 28395

for close examination, as they may reveal valuable information about the condition of the engine **(see illustration)**.

2 Bearing failure can occur due to lack of lubrication, the presence of dirt or other foreign particles, overloading the engine, or corrosion. Regardless of the cause of bearing failure, the cause must be corrected before the engine is reassembled, to prevent it from happening again.

3 When examining the bearing shells, remove them from the cylinder block/crankcase, the main bearing caps, the connecting rods and the connecting rod big-end bearing caps. Lay them out on a clean surface in the same general position as their location in the engine. This will enable you to match any bearing problems with the corresponding crankshaft journal. Do not touch any shell's internal bearing surface with your fingers while checking it, or the delicate surface may be scratched.

4 Dirt and other foreign matter gets into the engine in a variety of ways. It may be left in the engine during assembly, or it may pass through filters or the crankcase ventilation system. It may get into the oil, and from there into the bearings. Metal chips from machining operations and normal engine wear are often present. Abrasives are sometimes left in engine components after reconditioning, especially when parts are not thoroughly cleaned using the proper cleaning methods. Whatever the source, these foreign objects often end up embedded in the soft bearing material, and are easily recognised. Large particles will not embed in the bearing, but will score or gouge the bearing and journal. The best prevention for this cause of bearing failure is to clean all parts thoroughly, and keep everything spotlessly-clean during engine assembly. Frequent and regular engine oil and filter changes are also recommended.

5 Lack of lubrication (or lubrication breakdown) has a number of interrelated causes. Excessive heat (which thins the oil), overloading (which squeezes the oil from the bearing face) and oil leakage (from excessive bearing clearances, worn oil pump or high engine speeds) all contribute to lubrication breakdown. Blocked oil passages, which usually are the result of misaligned oil holes in a bearing shell, will also oil-starve a bearing, and destroy it. When lack of lubrication is the cause of bearing failure, the bearing material is wiped or extruded from the steel backing of the bearing. Temperatures may increase to the point where the steel backing turns blue from overheating.

6 Driving habits can have a definite effect on bearing life. Full-throttle, low-speed operation (labouring the engine) puts very high loads on bearings, tending to squeeze out the oil film. These loads cause the bearings to flex, which produces fine cracks in the bearing face (fatigue failure). Eventually, the bearing material will loosen in pieces, and tear away from the steel backing.

7 Short-distance driving leads to corrosion of bearings, because insufficient engine heat is produced to drive off the condensed water and corrosive gases. These products collect in the engine oil, forming acid and sludge. As the oil is carried to the engine bearings, the acid attacks and corrodes the bearing material.

8 Incorrect bearing installation during engine assembly will lead to bearing failure as well. Tight-fitting bearings leave insufficient bearing running clearance, and will result in oil starvation. Dirt or foreign particles trapped behind a bearing shell result in high spots on the bearing, which lead to failure.

9 Do not touch any shell's internal bearing surface with your fingers during reassembly as there is a risk of scratching the delicate surface, or of depositing particles of dirt on it.

10 As mentioned at the beginning of this Section, the bearing shells should be renewed as a matter of course during engine overhaul. To do otherwise is false economy.

Selection

11 Main and big-end bearings for the engines described in this Chapter are available in standard sizes and a range of undersizes to suit reground crankshafts.

12 Have the crankshaft measured by an engine reconditioning specialist. They will be able to supply the correctly sized bearings.

15 Engine overhaul – reassembly sequence

1 Before reassembly begins, ensure that all new parts have been obtained, and that all necessary tools are available. Read through the entire procedure to familiarise yourself with the work involved, and to ensure that all items necessary for reassembly of the engine

are at hand. In addition to all normal tools and materials, thread-locking compound will be needed. A suitable tube of liquid sealant will also be required for the joint faces that are fitted without gaskets.

2 In order to save time and avoid problems, engine reassembly can be carried out in the following order, referring to Part A, B, C, D or E of this Chapter unless otherwise stated. Where applicable, use new gaskets and seals when refitting the various components.

a) Crankshaft (Section 17).
b) Piston/connecting rod assemblies (Section 18).
c) Oil pump.
d) Sump.
e) Flywheel/driveplate.
f) Cylinder head.
g) Timing belt/chain, tensioner and sprockets.
h) Engine external components.

3 At this stage, all engine components should be absolutely clean and dry, with all faults repaired. The components should be laid out (or in individual containers) on a completely clean work surface.

16 Piston rings – refitting

1 Before fitting new piston rings, the ring end gaps must be checked as follows.

2 Lay out the piston/connecting rod assemblies and the new piston ring sets, so that the ring sets will be matched with the same piston and cylinder during the end gap measurement and subsequent engine reassembly.

3 Insert the top ring into the first cylinder, and push it down the bore using the top of the piston. This will ensure that the ring remains square with the cylinder walls. Position the ring approximately 15.0 mm the bottom of the cylinder bore, at the lower limit of ring travel. Note that the top and second compression rings are different.

4 Measure the end gap using feeler blades, and compare the measurements with the figures given in the Specifications **(see illustration)**.

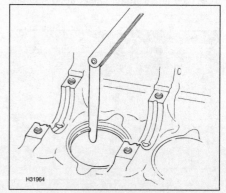

H31964

16.4 Check a piston ring end gap using a feeler blade

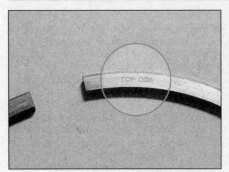

16.9 Piston ring TOP marking

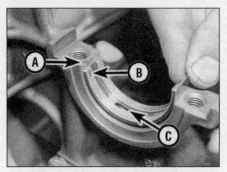

17.2 Bearing shell correctly refitted

A Recess in cylinder block
B Lug on bearing shell
C Oil hole

17.4 Lubricate the upper bearing shells

5 If the gap is too small (unlikely if genuine Seat parts are used), it must be enlarged, or the ring ends may contact each other during engine operation, causing serious damage. Ideally, new piston rings providing the correct end gap should be fitted. As a last resort, the end gap can be increased by filing the ring ends very carefully with a fine file. Mount the file in a vice equipped with soft jaws, slip the ring over the file with the ends contacting the file face, and slowly move the ring to remove material from the ends. Take care, as piston rings are sharp, and are easily broken.

6 With new piston rings, it is unlikely that the end gap will be too large. If the gaps are too large, check that you have the correct rings for your engine and for the particular cylinder bore size.

7 Repeat the checking procedure for each ring in the first cylinder, and then for the rings in the remaining cylinders. Remember to keep rings, pistons and cylinders matched up.

8 Once the ring end gaps have been checked and if necessary corrected, the rings can be fitted to the pistons.

9 Fit the piston rings using the same technique as for removal. Fit the bottom (oil control) ring first, and work up. Note that a two- or three-section oil control ring may be fitted; where a two-section ring is fitted, first insert the wire expander, then fit the ring. Ensure that the rings are fitted the correct way up – the top surface of the rings is normally marked TOP **(see illustration)**. Offset the piston ring

gaps by 120° from each other. **Note:** *Always follow any instructions supplied with the new piston ring sets ñ different manufacturers may specify different procedures. Do not mix up the top and second compression rings, as they have different cross-sections.*

17 Crankshaft – refitting

1 Wipe off the surfaces of the bearing shells in the crankcase and bearing caps.

2 Press the bearing shells into their locations, ensuring that the tab on each shell engages In the notch in the cylinder block or bearing cap, and that the oil holes In the cylinder block and bearing shell are aligned **(see illustration)**. Take care not to touch any shells bearing surface with your fingers.

3 Where applicable, refit the crankshaft speed/position sensor wheel, and tighten the securing bolts to the specified torque. Make sure that the sensor wheel is correctly orientated as noted before removal.

4 Liberally coat the bearing shells in the crankcase with clean engine oil of the appropriate grade **(see illustration)**. Make sure that the bearing shells are still correctly seated in their locations.

5 Lower the crankshaft into position so that No 1 cylinder crankpin is at BDC, ready for fitting No 1 piston. Ensure that the crankshaft endfloat control thrustwasher halves, either side of the No 3 main bearing location, remain in position. Where applicable, take care not to damage the crankshaft speed/position sensor wheel as the crankshaft is lowered into position.

6 Lubricate the lower bearing shells in the main bearing caps with clean engine oil. Make sure that the crankshaft endfloat control thrustwasher halves are still correctly seated either side of No 3 bearing cap **(see illustrations)**.

7 Fit the main bearing caps in the correct order and orientation – No 1 bearing cap must be at the timing belt end of the engine and the bearing shell tab locating recesses in the crankcase and bearing caps must be adjacent to each other **(see illustration)**. Insert the bearing cap bolts (using new bolts where necessary), and hand-tighten them only.

8 Working from the centre bearing cap outwards, tighten the bearing cap bolts to their specified torque. On engines where two Stages are given for the torque, tighten all bolts to the Stage 1 torque, then go round again, and tighten all bolts through the Stage 2 angle **(see illustrations)**.

9 Check that the crankshaft rotates freely by turning it by hand. If resistance is felt, recheck the bearing running clearances, as described previously.

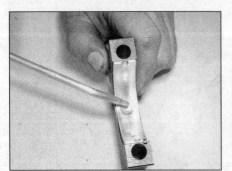

17.6a Lubricate the lower bearing shells...

17.6b ...and make sure that the thrustwashers are correctly seated

17.7 Fitting No 1 main bearing cap

17.8a Tighten the main bearing cap bolts to the specified torque...

17.8b ...then through the specified angle

10 Check the crankshaft endfloat as described at the beginning of Section 13. If the thrust surfaces of the crankshaft have been checked and new thrustwashers have been fitted, then the endfloat should be within specification.

11 Refit the pistons and connecting rods or reconnect them to the crankshaft as described in Section 18.

12 Proceed as follows according to engine type:

a) *On 1.6 (SOHC)) petrol engines, refit the crankshaft oil seal housings, flywheel/ driveplate, oil pump and pick-up pipe, sump and oil baffle plate, and the crankshaft sprocket and timing belt, as described in Chapter 2A.*

b) *On 2.0 (DOHC) petrol engines, refit the crankshaft oil seal housing, flywheel/ driveplate, oil pump and pick-up pipe, sump, and main timing belt, as described in Chapter 2B.* **Note:** *On 2.0 litre engines, it will necessary to refit the balancer shaft assembly.*

c) *On diesel engines, refit the crankshaft oil seal housings, flywheel/driveplate, oil pump and pick-up pipe, sump and oil baffle plate, and the crankshaft sprocket and timing belt, as described in Chapter 2C, Chapter 2D or Chapter 2E.*

18 Piston/connecting rod assemblies – refitting

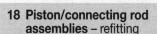

Note: *A piston ring compressor tool will be required for this operation.*

1 Note that the following procedure assumes that the crankshaft main bearing caps are in place.

2 Where applicable, refit the piston cooling oil spray jets to the bottom of the cylinder block, and tighten the securing bolts to the specified torque.

3 On engines where the big-end bearing caps are secured by nuts, fit new bolts to the connecting rods. Tap the old bolts out of the connecting rods using a soft-faced mallet, and tap the new bolts into position.

4 Ensure that the bearing shells are correctly fitted, as described at the beginning of this Section. If new shells are being fitted, ensure that all traces of the protective grease are cleaned off using paraffin. Wipe dry the shells and connecting rods with a lint-free cloth.

5 Lubricate the cylinder bores, the pistons, piston rings and upper bearing shells with clean engine oil **(see illustrations)**. Lay out each piston/connecting rod assembly in order on a clean work surface. Where the bearing caps are secured with nuts, pad the threaded ends of the bolts with insulating tape to prevent them scratching the crankpins and bores when the pistons are refitted.

6 Start with piston/connecting rod assembly No 1. Make sure that the piston rings are

still spaced as described in Section 16, then clamp them in position with a piston ring compressor tool.

7 Insert the piston/connecting rod assembly into the top of cylinder No 1. Lower the big-end in first, guiding it to protect the cylinder bores. Where oil jets are located at the bottoms of the bores, take particular care not to damage them when guiding the connecting rods onto the crankpins.

8 Ensure that the orientation of the piston in its cylinder is correct – the piston crown, connecting rod and big-end bearing cap have markings, which must point towards the timing belt end of the engine when the piston is installed in the bore – refer to Section 18 for details. On SOHC diesel engines, the piston crowns are specially shaped to improve the engine's combustion characteristics. Because of this, pistons 1 and 2 are different to pistons 3 and 4. When correctly fitted, the larger inlet valve chambers on pistons 1 and 2 must face the flywheel/driveplate end of the engine, and the larger inlet valve chambers on the remaining pistons must face the timing belt end of the engine. New pistons have number markings on their crowns to indicate their type – 1/2 denotes piston 1 or 2, 3/4 indicates piston 3 or 4. On DOHC diesel engines, make sure the arrows on the piston crowns point towards the timing belt end of the engine **(see illustrations)**.

9 Using a block of wood or hammer handle against the piston crown, tap the assembly into the cylinder until the piston crown is flush

18.5a Lubricate the pistons...

18.5b ...and big-end upper bearing shells with clean engine oil

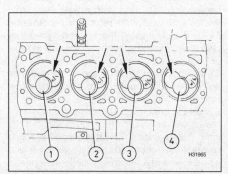

18.8a Piston orientation and coding on SOHC diesel engines...

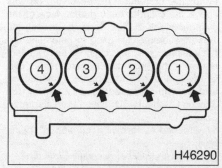

18.8b ...and on DOHC diesel engines

18.9 Using a hammer handle to tap the piston into its bore

18.13a Tighten the big-end bearing cap bolts/nuts to the specified torque…

18.13b …then through the specified angle

with the top of the cylinder **(see illustration)**.
10 Ensure that the bearing shell is still correctly installed in the connecting rod, then liberally lubricate the crankpin and both bearing shells with clean engine oil.
11 Taking care not to mark the cylinder bores, tap the piston/connecting rod assembly down the bore and onto the crankpin. On engines where the big-end caps are secured by nuts, remove the insulating tape from the threaded ends of the connecting rod bolts. Oil the bolt threads, and on engines where the big-end caps are secured by bolts, oil the undersides of the bolt heads.
12 Fit the big-end bearing cap, tightening its retaining nuts or bolts (as applicable) finger-tight at first. The connecting rod and its bearing cap both have recesses machined into them on one side, close to their mating surfaces – these recesses must both face the same way as the arrow on the piston crown (ie, towards the timing end of the engine) when correctly installed. Reassemble the two components to satisfy this requirement.
13 Tighten the retaining bolts or nuts (as applicable) to the specified torque and angle, in the two stages given in the Specifications **(see illustrations)**.
14 Refit the remaining three piston/connecting rod assemblies in the same way.
15 Rotate the crankshaft by hand. Check that it turns freely; some stiffness is to be expected if new parts have been fitted, but there should be no binding or tight spots.
16 On diesel engines, if new pistons have been fitted, or if a new short engine has been fitted, the projection of the piston crowns above the cylinder head mating face of the cylinder block at TDC must be measured. This measurement is used to determine the thickness of the new cylinder head gasket required. This procedure is described as part of the Cylinder head – removal, inspection and refitting procedure in Chapter 2C Section 11, Chapter 2D Section 11 or Chapter 2E Section 13.
17 Proceed as follows according to engine type:
a) On 1.6 (SOHC) petrol engines, refit the

oil pump and pick-up pipe, sump and oil baffle plate, and cylinder head, as described in Chapter 2A.
b) On 2.0 (DOHC) litre petrol engines, refit the oil pump and pick-up pipe, sump and oil baffle plate, cylinder head, and the balancer shaft assembly as described in Chapter 2B.
c) On diesel engines, refit the oil pump and pick-up pipe, sump and oil baffle plate, and cylinder head, as described in Chapter 2C, Chapter 2D or Chapter 2E.

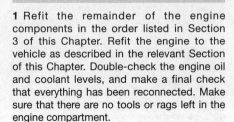

19 Engine – initial start-up after overhaul and reassembly

1 Refit the remainder of the engine components in the order listed in Section 3 of this Chapter. Refit the engine to the vehicle as described in the relevant Section of this Chapter. Double-check the engine oil and coolant levels, and make a final check that everything has been reconnected. Make sure that there are no tools or rags left in the engine compartment.
2 Where necessary, reconnect the battery leads with reference to Disconnecting the battery at the rear of this manual.

Petrol models

3 Remove the spark plugs, referring to Chapter 1A Section 29 for details.
4 The engine must be immobilised such that it can be turned over using the starter motor without starting – disable the fuel pump by unplugging the fuel pump power relay from the relay board with reference to Chapter 12, and also disable the ignition system by disconnecting the wiring from the DIS module or coils, as applicable.
Caution: To prevent damage to the catalytic converter, it is important to disable the fuel system.
5 Turn the engine using the starter motor until the oil pressure warning lamp goes out. If the lamp fails to extinguish after several seconds of cranking, check the engine oil level and oil filter security. Assuming these are correct,

check the security of the oil pressure switch wiring – do not progress any further until you are satisfied that oil is being pumped around the engine at sufficient pressure.
6 Refit the spark plugs, and reconnect the wiring to the fuel pump relay and DIS module or coils, as applicable.

Diesel models

7 Disconnect the injector harness wiring plug at the end of the cylinder head – refer to Chapter 4B Section 5 for details.
8 Turn the engine using the starter motor until the oil pressure warning lamp goes out.
9 If the lamp fails to extinguish after several seconds of cranking, check the engine oil level and oil filter security. Assuming these are correct, check the security of the oil pressure switch cabling – do not progress any further until you are satisfied that oil is being pumped around the engine at sufficient pressure.
10 Reconnect the injector wiring plug.

All models

11 Start the engine, but be aware that as fuel system components have been disturbed, the cranking time may be a little longer than usual.
12 While the engine is idling, check for fuel, water and oil leaks. Don't be alarmed if there are some odd smells and the occasional plume of smoke as components heat up and burn off oil deposits.
13 Assuming all is well, keep the engine idling until hot water is felt circulating through the top hose.
14 After a few minutes, recheck the oil and coolant levels, and top-up as necessary.
15 There is no need to retighten the cylinder head bolts once the engine has been run following reassembly.
16 If new pistons, rings or crankshaft bearings have been fitted, the engine must be treated as new, and run-in for the first 600 miles. Do not operate the engine at full-throttle, or allow it to labour at low engine speeds in any gear. It is recommended that the engine oil and filter are changed at the end of this period.

Chapter 3
Cooling, heating and air conditioning systems

Contents

Degrees of difficulty

Easy, suitable for novice with little experience	Fairly easy, suitable for beginner with some experience	Fairly difficult, suitable for competent DIY mechanic	Difficult, suitable for experienced DIY mechanic	Very difficult, suitable for expert DIY or professional

Specifications

Engine codes*

Petrol engines:
1.6 litre .... BGU, BSF and BSE
2.0 litre .... BLR, BVY, BLY, BVZ, BWA,BWJ, CDLD and CDLA
Diesel engines:
PD unit injector engine:
1.9 litre .... BJB, BKC, BLS, BXE and BXF
2.0 litre: .... BMM. BKD and BMN
Common rail injection engine
1.6 litre .... CAYC
2.0 litre .... CLCB, CEGA, CFJA and CFHC
* Note: *See 'Vehicle identification' at the end of this manual for the location of engine code markings.*

Cooling system pressure cap

Opening pressure.. 1.4 to 1.6 bar

Thermostat	Begins to open	Fully open
Petrol engines:		
1.6 litre	87°C	102°C
2.0 litre	87°C	Map-controlled
Diesel engines:		
1.6 litre	92°C	107°C
1.9 litre	85°C	105°C
2.0 litre (PD)	87°C	102°C
2.0 litre (CR)	87°C	102°C

Torque wrench settings

	Nm	lbf ft
Coolant pipe to map-controlled thermostat housing:		
2.0 litre non-turbo engine	10	7
2.0 litre turbo engine	5	4
Coolant pump:		
1.6 litre petrol engine	15	11
2.0 litre petrol engine	15	11
Diesel engines:		
PD unit injector engines	15	11
Common rail engines	9	6
Coolant pump pulley	20	15
Radiator	5	4
Radiator cooling fan shroud bolts	5	4
Thermostat cover bolts:		
1.6 litre petrol engine	15	11
2.0 litre petrol engine	Not applicable	
All diesel engines	15	11
Thermostat housing bolts:		
2.0 litre petrol engine	15	11
All diesel engines	Not applicable	

1 General information and precautions

1 A pressurised cooling system is used, with a pump, an aluminium crossflow radiator, two electric cooling fans, a thermostat and a heater matrix, as well as the interconnecting hoses. A secondary electric coolant pump and a temperature-sensitive by-pass thermostat are fitted in the coolant return from the turbocharger on 2.0 litre BWA and BWJ engines. The thermostat is located in the front, right-hand side of the cylinder block, in the coolant return from the radiator.

2 The system functions as follows. Coolant is circulated through the cylinder block and head passages by the coolant pump which is driven by the timing belt. On common rail diesel engines, an electrically operated coolant pump is also fitted. The coolant cools the cylinder bores, combustion surfaces and valve seats of the engine.

3 When the engine is cold, the thermostat is closed and the coolant only circulates around the engine and the heater matrix in the passenger compartment, however, when the engine reaches a predetermined temperature, the thermostat opens and the coolant passes through the radiator for additional cooling. The coolant enters the top of the radiator and is cooled, as it circulates down through the cooling tubes, by the inrush of air when the car is in forward motion. Airflow is supplemented by the action of the electric cooling fans when necessary. Upon leaving the bottom of the radiator, the coolant returns to the engine and the cycle is repeated. When it is closed, there is no circulation through the radiator, and when it is open, there is circulation. On some engines, the action of the thermostat is determined in conjunction with the engine management ECU.

4 Refer to Section 11 for information on the air conditioning system.

Precautions

⚠️ *Warning: Do not attempt to remove the expansion tank filler cap or disturb any part of the cooling system while the engine is hot, as there is a high risk of scalding. If the expansion tank filler cap must be removed before the engine and radiator have fully cooled (even though this is not recommended) the pressure in the cooling system must first be relieved. Cover the cap with a thick layer of cloth, to avoid scalding, and slowly unscrew the filler cap until a hissing sound can be heard. When the hissing has stopped, indicating that the pressure has reduced, slowly unscrew the filler cap until it can be removed; if more hissing sounds are heard, wait until they have stopped before unscrewing the cap completely. At all times keep well away from the filler cap opening.*

⚠️ *Warning: Do not allow antifreeze to come into contact with skin or painted surfaces of the vehicle. Rinse off spills immediately with plenty of water. Never leave antifreeze lying around in an open container or in a puddle in the driveway or on the garage floor. Children and pets are attracted by its sweet smell. Antifreeze can be fatal if ingested.*

2.4 Pull out the retaining clip

⚠️ *Warning: If the engine is hot, the electric cooling fan may start rotating even if the engine is not running, so be careful to keep hands, hair and loose clothing well clear when working in the engine compartment.*

⚠️ *Warning: Refer to Section 11 for additional precautions to be observed when working on models with air conditioning.*

2 Cooling system hoses – disconnection and renewal

Note: *Refer to the warnings given in Section 1 of this Chapter before proceeding.*

1 If the checks described in the relevant part of Chapter 1A Section 10 or Chapter 1B Section 10 reveal a faulty hose, it must be renewed as follows.

2 First drain the cooling system as described in Chapter 1A Section 33 or Chapter 1B Section 33. If the coolant is not due for renewal, it may be re-used if it is collected in a clean container.

3 To disconnect a hose, release its retaining clips, then move them along the hose, clear of the relevant inlet/outlet union. Carefully work the hose free.

4 In order to disconnect the radiator inlet and outlet hoses, apply pressure to hold the hose on to the relevant union, pull out the spring clip and pull the hose from the union **(see illustration)**. Note that the radiator inlet and outlet unions are fragile; do not use excessive force when attempting to remove the hoses. If a hose proves to be difficult to remove, try to release it by rotating the hose ends before attempting to free it.

5 When fitting a hose, first slide the clips onto the hose, then work the hose into position. If clamp type clips were originally fitted, it is a good idea to use screw type clips when refitting the hose. If the hose is stiff, use a little

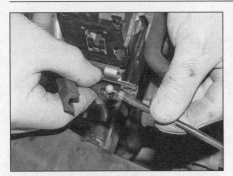

3.3a Disconnect the bonnet release cable…

3.3b …and disconnect the switch wiring plug

3.3c Disconnect the charge air hoses

3.3d Remove the upper bolts

3.3e …and remove the lower bolts. Substitute the lock carrier mounting bolts with one threaded rod on each side

3.3f Insert a block of wood between the lock carrier and the body

soapy water as a lubricant, or soften the hose by soaking it in hot water.

6 Work the hose into position, checking that it is correctly routed, then slide each clip along the hose until it passes over the flared end of the relevant union, before securing it in position with the retaining clip.

7 Prior to refitting a radiator inlet or outlet hose, renew the connection O-ring regardless of condition. The connections are a push-fit over the radiator unions.

8 Refill the cooling system as described in Chapter 1A Section 33 or Chapter 1B Section 33.

9 Check thoroughly for leaks as soon as possible after disturbing any part of the cooling system.

3 Radiator – removal, inspection and refitting

Removal

1 Switch off the ignition and all electrical consumers.

2 On 2.0 litre turbocharged petrol engines (engine codes BWA, BWJ, CDLA and CDLD) and 2.0 litre common rail diesel engines the lock carrier must be moved to the service position (as described below). On all 1.6 petrol engines and non-turbocharged 2.0 litre engines only the bumper cover requires removal. On all other engines the radiator can be removed after the cooling fan assembly has been removed.

3 Where required move the lock carrier to the service position as follows:

a) Remove the front bumper cover (Chapter 11 Section 6).

b) Disconnect the bonnet release cable and unplug the bonnet lock switch wiring connector (see illustration).

c) Disconnect the wiring plugs from both head-lights. Alternatively remove both headlights as described in Chapter 12 Section 8.

d) On models with a turbocharger, remove the air ducts.

e) Remove the air intake duct from the slam panel (see illustration).

f) On models fitted with AC remove the pipe support bracket from the right-hand upper chassi leg.

g) On models with an intercooler, disconnect the air hoses from the intercooler (see illustration).

h) Remove the slam panel upper mounting bolts (see illustration).

i) Disconnect the temperature sensor.

j) Remove a bolt from each side and insert a length of threaded rod (or the correct VAG tool T10093) into the bolt holes.

k) Remove the remaining mounting bolts and slide the lock carrier along 100 mm. Insert a block of wood between the lock carrier and the chassis leg to hold the carrier away from the body (see illustration).

4 Jack up and support the front of the vehicle (see Jacking and vehicle support) remove the undershield and then drain the cooling system as described in Chapter 1A Section 33 or Chapter 1B Section 33.

5 On all models remove the cooling fans and cowling, as described in Section 5.

6 On 1.6 litre petrol and non-turbocharged 2.0 litre petrol models remove the front bumper cover as described in Chapter 11 Section 6.

7 On all models disconnect the radiator hoses (see illustrations) – anticipate some coolant spillage as the hoses are removed.

3.7a Remove the lower hose…

3.7b …and the upper hose

3.8a Separate the condenser from the radiator...

3.8b ...and remove the radiator

3.14a Radiator lower mounting rubber grommets on the lock carrier...

3.14b ...and upper mounting rubbers on the radiator

8 On 1.6 petrol and 2.0 litre non-turbocharged models separate the radiator from the condenser (where fitted) and support panel **(see illustrations)**. Tip the top of the radiator forward and work the radiator up, then down and out from behind the bumper reinforcement.

9 On all other models remove the radiator from the lock carrier and then remove the radiator downwards (2.0 turbocharged petrol and 1.9 PD diesel models) or upwards (all other models).

Inspection

10 If the radiator has been removed due to suspected blockage, reverse flush it as described in the relevant part of Chapter 1A Section 33 or Chapter 1B Section 33.

11 Clean dirt and debris from the radiator fins, using an airline (in which case, wear eye protection) or a soft brush. Be careful, as the fins are sharp and easily damaged.

12 If necessary, a radiator specialist can perform a 'flow test' on the radiator, to establish whether an internal blockage exists.

13 A leaking radiator must be referred to a specialist for permanent repair. Do not attempt to weld or solder a leaking radiator, as damage may result.

14 Check the radiator mounting rubbers, and renew if necessary **(see illustrations)**.

Refitting

15 Refitting is a reversal of removal. On completion, refill the cooling system using the correct type of antifreeze as described in the relevant part of Chapter 1A Section 33 or Chapter 1B Section 33.

---

4 Thermostat – removal, testing and refitting

1.6 litre petrol engines

Removal

1 The thermostat is located behind a connection flange in the front right-hand side of the engine block. Remove the engine top cover.

2 Jack up and support the front of the vehicle (see *Jacking and vehicle support*) and then remove the engine undershield.

3 Drain the cooling system as described in Chapter 1A Section 33.

4 Disconnect the coolant hose from the thermostat cover/connection flange **(see illustration)**.

5 Unscrew the two securing bolts, and remove the thermostat cover/connection flange, noting the locations of any brackets secured by the bolts, then lift out the thermostat and O-ring seal **(see illustrations)**.

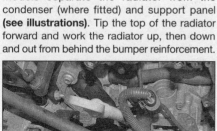

4.4 Remove the coolant hose

4.5a Remove the bolts and...

4.5b ...pull off the cover

4.5c Remove the seal and then...

4.5d ...the thermostat

2.0 litre petrol engines

Removal

6 Jack up and support the front of the vehicle (see *Jacking and vehicle support*) and then remove the engine undershield.

7 Drain the cooling system as described in Chapter 1A Section 33.

8 Remove the Alternator (Chapter 5A Section 5). On FSI models with AC, unbolt the AC compressor (Section 12) and secure it to one side. DO NOT disconnect the refrigerant lines.

9 On FSI models unbolt and remove the Alternator/AC compressor mounting bracket and then disconnect the wiring plug from the coolant base at the base of the thermostat housing.

10 Remove the coolant hose and unbolt the thermostat housing, Recover the seal. Note that FSI models with engine codes BLR and BLY use an ECU mapped thermostat.

1.9 litre and 2.0 litre 'PD' Diesel engines

Removal

11 Remove the alternator as described in Chapter 5A Section 5.

12 On 2.0 litre engines, disconnect the coolant temperatures sensor wiring plug and then remove the outlet hose from the thermostat.

13 Remove the housing bolts and then remove the cover. The thermostat arrangement is very similar to the 1.6 litre petrol engines described above.

1.6 litre 'CR' Diesel engines

Note: *Some engines have the same type of thermostat, as fitted to the 2.0 litre engine.*

Removal

14 Drain the coolant as described in Chapter 1B Section 33.

15 Unbolt and remove the charge air pipe from the front of the engine.

16 Release the lower end of the thermostat coolant pipe and then release the hose clip from the coolant hose at the thermostat housing. Remove the coolant hose.

17 Unscrew the two securing bolts, and remove the thermostat cover/housing complete with the thermostat. Note the locations of any brackets secured by the

4.17a Remove the housing…

4.17b and remove the seal

4.21 The thermostat housing bolts are behind the alternator

4.24 Remove the bolt

bolts. Recover the O-ring if it is loose (see illustrations).

18 Remove the thermostat by twisting the thermostat 90° clockwise, and then pull it from the cover.

2.0 litre 'CR' Diesel engines

Note: *All 2.0 litre engines have the three outlet type thermostat housing fitted, as do some 1.6 litre engines. Note also that some common rail engines have a thermostat housing that appears to have a removable thermostat (secured by screws at the base of the housing) but this is not (at the time of writing) available as a separate part.*

Removal

19 Remove the air filter housing as described in.

20 Remove the battery and then remove the battery support tray as described in Chapter 5A Section 3.

21 Remove the alternator (see illustration) as

described in Chapter 5A Section 5.

22 Disconnect the crankcase breather hose and the remove the air intake pipe from the turbocharger. Unclip the vacuum hoses as required.

23 Unclip the wiring loom from the front of the engine and then slacken the hose clips from the air intake duct. Unbolt the duct and remove it.

24 Release the spring clips (or worm drive clips) and then unbolt and remove the intercooler lower duct (see illustration).

25 Unbolt and then remove the dipstick guide tube.

26 Slacken the hose clips and then remove the hoses from the thermostat housing. Unbolt the coolant pipe from the top of the transmission to remove it. Release the main hose at the radiator outlet and remove it.

27 Remove the mounting bolts, pull off the housing and recover the seal (see illustrations).

4.27a Remove the bolts

4.27b Remove the thermostat

4.27c Check the condition of the seal

Testing – all models

28 On model fitted with a removable thermostat the traditional testing method is to place the thermostat in a pan of water with a thermometer. Heat the water and check the opening temperature of the thermostat. However, given the low cost of a new thermostat, if the thermostat is suspect then simply replace it.

29 On 2.0 litre FSI engines (BLR and BLY codes) additional testing is possible for the map-controlled thermostat – connect a 12 volt supply to the two terminals, and check that the minimum lift of the thermostat is 7.0 mm after 10 minutes, proving that the heating resistor is functioning correctly.

Refitting – all models

30 Refitting is a reversal of removal, bearing in mind the following points.

a) *Refit the thermostat using a new O-ring. Lubricate the seal with coolant.*

5.9 Disconnect the wiring plug

5.10b ...then remove the intake duct

b) *Where applicable, the thermostat should be fitted with the brace almost vertical.*
c) *Ensure that any brackets are in place on the thermostat cover bolts as noted before removal.*
d) *Refill the cooling system with the correct type and quantity of coolant as described in Chapter 1A Section 33 or Chapter 1B Section 33.*

---

5 Electric cooling fans – testing, removal and refitting

Testing

1 Two electric cooling fans are fitted to all models. The cooling fans are supplied with current through the ignition switch, cooling fan control unit (located on the motor),

5.10a Remove the cover and...

5.10c Unscrew the inner section

temperature sensor, the relays and fuses/fusible link (see Chapter 12 Section 3). The circuit is activated by the temperature sensor mounted in the outlet elbow at the bottom, left-hand side of the radiator, except on 2.0 litre turbo diesel and petrol engines, where the engine management ECU activates it according to the engine coolant temperature sensor. Testing of the cooling fan circuit is as follows.

2 If a fan does not appear to work, first check the fuses and fusible links. If they are good, run the engine until normal operating temperature is reached, then allow it to idle. If the fan does not cut-in within a few minutes, the cause may be the temperature sender (where applicable) which can be checked by a Seat dealer or suitably equipped garage using specialist diagnostic equipment.

3 The motors can be checked by disconnecting the motor wiring connector and connecting a 12 volt supply directly to the motor terminals. If the motor is faulty, it must be renewed, as no spares are available.

4 If the fan still fails to operate, check the cooling fan circuit wiring (Chapter 14). Check each wire for continuity and ensure that all connections are clean and free of corrosion.

5 On models with a cooling fan control unit, if no fault can be found, then it is likely that the cooling fan control unit is faulty. Testing of the unit should be entrusted to a Seat dealer or specialist; if the unit is faulty it must be renewed.

Removal

6 Switch off the ignition and all electrical consumers. Remove the engine top cover.

7 Jack up and support the front of the vehicle (See *Jacking and vehicle support*).

8 On models fitted with PD diesel engines move the lock carrier into the service position as described in Section 3.

9 Remove the engine undershield and disconnect the fan wiring plug at the base of the cowl **(see illustration)**.

10 Remove the air intake duct from the slam panel **(see illustrations)** and on PD diesel engine models remove the complete air filter assembly as described in Chapter 4A Section 3.

11 On models with an intercooler, remove the charge air inlet and outlet hoses.

12 Remove the upper and lower cowl/fan mounting screws and then lift the cowling and fan down (or up) and out of the engine bay **(see illustration)**.

13 To remove the fans and motors from the shroud, first disconnect and release the wiring plugs, then unscrew the nuts and remove the units **(see illustration)**.

Refitting

14 Refitting is a reversal of removal.

5.12 Remove the fans

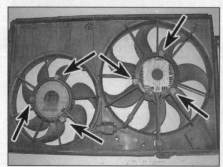

5.13 Remove the fan (or fans) from the shroud

6.6 Pull out the clip

6.9 Engine coolant temperature sensor on the 1.6 petrol engines

6 Cooling system electrical sensors – testing, removal and refitting

Cooling fan temperature sensor

Testing

1 Where fitted, the sensor is located in the outlet elbow at the bottom left-hand side of the radiator (see Section 5).
2 The sensor contains a thermistor, which consists of an electronic component whose electrical resistance decreases at a predetermined rate as its temperature rises. When the coolant is cold, the sensor resistance is high, current flow through the sensor is reduced. No resistance-to-temperature values are available, however a simple check can be made by checking that the resistance changes (with a multimeter) according to temperature – in a similar manner to checking a thermostat. The only method of accurately checking the sensor is with dedicated diagnostic equipment, but as the sensor is a low cost item, if the sensor is suspect it should be replaced.

Removal and refitting

3 The engine and radiator should be cold before removing the sensor. Switch off the ignition and all electrical consumers.
4 Either drain the cooling system (as described in Chapter 1A Section 33 or Chapter 1B Section 33), or have ready a

suitable plug which can be used to plug the sensor aperture whilst it is removed.
5 Disconnect the wiring plug from the sensor.
6 Pull out the clip and remove the sensor from the elbow (see illustration). Recover the O-ring seals from the elbow and sensor.
7 Refitting is a reversal of removal, but tighten the sensor securely. On completion, refill the cooling system with the correct type and quantity of coolant as described in Chapter 1A Section 33 or Chapter 1B Section 33 or top-up as described in *Weekly checks* Section 5.
8 Start the engine and run it until it reaches normal operating temperature, then continue to run the engine and check that the cooling fan cuts in and functions correctly.

Coolant temperature sensor

Testing

9 The sensor is located at the left-hand end of the cylinder head (see illustration).
10 The sensor contains a thermistor, which consists of an electronic component whose electrical resistance decreases at a predetermined rate as its temperature rises. When the coolant is cold, the sensor resistance is high, current flow through the sensor is reduced. The change in resistance and the voltage returned to the Engine Control Unit (ECU) is critical to the efficient operation of the engine. A faulty sensor may cause poor or inefficient running (as well as a temperature gauge that reads incorrectly). Diagnostic equipment can identify a faulty sensor, but

given the low cost of a replacement, if the sensor is suspect it should be replaced.

Removal and refitting

11 Where fitted, remove the engine top cover.
12 Disconnect the wiring from the sensor (see illustration), located on the thermostat housing at the left-hand end of the cylinder head. Partially drain the cooling system to below the level of the sensor (as described in Chapter 1A Section 33 or Chapter 1B Section 33).
13 Pull out the retaining clip and withdraw the sensor from the housing. Recover the O-ring (see illustrations).
14 Refitting is a reversal of removal. Bearing in mind the following points.
a) *Refit the sensor with a new O-ring.*
b) *Refill the cooling system as described in Chapter 1A Section 33 or Chapter 1B Section 33, or top-up as described in ' Weekly checks '.*

7 Coolant pump – removal and refitting

Note: *On all engines covered by this manual the coolant pump is driven by the timing belt (cambelt). If the coolant pump is being replaced the fitting of a new timing belt is recommended.*

Petrol engines

Removal

1 Drain the cooling system as described in Chapter 1A Section 33.
2 Remove the timing belt as described in Chapter 2A Section 7 (1.6 litre petrol engines) Chapter 2B Section 4 (2.0 litre petrol engines). If the timing belt is not being replaced, note the following points.
a) *The lower part of the timing belt guard need not be removed.*
b) *The timing belt should be left in position on the crankshaft sprocket.*
c) *Cover the timing belt with a cloth to protect it from coolant.*
3 Remove the two securing bolts, and remove the rear timing belt guard.

6.12 Disconnect the wiring...

6.13a ... then pull out the retaining clip...

6.13b ...and withdraw the coolant temperature sensor and O-ring seal from the thermostat housing

7.10a Undo the coolant pump bolts...

7.10b ...and remove the pump

7.10c Always replace the O-ring seal

4 Remove the remaining retaining bolts, and withdraw the coolant pump from the engine block. Recover the O-ring seal from the groove in the pump. If the pump is faulty, it must be renewed.

Refitting

5 Refitting is a reversal of removal, bearing in mind the following points.
a) Fit the coolant pump with a new O-ring.
b) Lubricate the O-ring with coolant.
c) Install the pump with the cast lug facing down.
d) Refill the cooling system as described in Chapter 1A Section 33.

Electric coolant pump

Note: *The additional coolant pump is only fitted to TFSI models.*
6 Removal and refitting is the same as the diesel engine version described below.

Diesel engines

Removal

7 Drain the cooling system as described in Chapter.
8 Remove the timing belt as described in Chapter 2C Section 7 (1.6 litre common rail engines), Chapter 2D Section 7 (2.0 litre common rail engines) or Chapter 2E Section 7 (PD diesel engines). If the timing belt is not being replaced, note the following points:
a) The lower part of the timing belt guard need not be removed.

b) The timing belt should be left in position on the crankshaft sprocket.
c) Cover the timing belt with a cloth to protect it from coolant.
9 Unscrew the timing belt idler pulley, and push the pulley downwards approximately 30 mm.
10 Unscrew the coolant pump retaining bolts, and remove the pump from the engine block. Recover the O-ring seal from the groove in the pump. If the pump is faulty, it must be renewed **(see illustrations)**.

Refitting

11 Refitting is a reversal of removal, bearing in mind the following points.
a) Fit the coolant pump with a new O-ring.
b) Lubricate the O-ring with coolant.
c) Install the pump with the cast lug facing down.
d) Refill the cooling system as described in Chapter.

Electric circulation pump

Note: *This pump is only fitted to some 2.0 litre common rail engines*
12 Raise the front of the vehicle and support is securely on axle stands (See *Jacking and vehicle support*).
13 Undo the fasteners and remove the engine undertray.
14 Drain the coolant (Chapter 1B Section 33) or fit hose clamps to the coolants hoses connected to the pump.Disconnect the wiring

plug, release the clips and disconnect the hoses from the pump **(see illustrations)**.
15 Undo the retaining bolt and remove the pump **(see illustration)**.
16 Refitting is a reversal of removal. Top up the coolant, and bleed the system as described in Chapter 1B Section 33.

8 Heating and ventilation system – general information

1 The heating/ventilation system consists of a four-speed blower motor (housed in the passenger compartment), face-level vents in the centre and at each end of the facia, and air ducts to the front and rear footwells.
2 The control unit is located in the facia, and the controls operate flap valves to deflect and mix the air flowing through the various parts of the heating/ventilation system. The flap valves are contained in the air distribution housing, which acts as a central distribution unit, passing air to the various ducts and vents.
3 Cold air enters the system through the grille at the rear of the engine compartment. A pollen filter is fitted to filter out dust, soot, pollen and spores from the air entering the vehicle.
4 The airflow, which can be boosted by the blower, flows through the various ducts, according to the settings of the controls. Stale air is expelled through ducts beneath the rear

7.14a Disconnect the wiring plug

7.14b Disconnect the hoses from the circulation pump

7.15 Circulation pump bracket retaining bolt (shown with the charge air pipe removed for clarity)

9.3a Remove the screws...

9.3b ...and release the complete panel

9.4a Disconnect the wiring plugs

9.4b The control unit can now be removed fromt the trim/vent panel

bumper. If warm air is required, the cold air is passed through the heater matrix, which is heated by the engine coolant.

5 If necessary, the outside air supply can be closed off, allowing the air inside the vehicle to be recirculated. This can be useful to prevent unpleasant odours entering from outside the vehicle, but should only be used briefly, as the recirculated air quality inside the vehicle will soon deteriorate.

9 Heating and ventilation system components – removal and refitting

Heater/ventilation control unit

1 Switch off the ignition and all electrical consumers, then set the heater controls to 'cold', blower to '0', and vent to 'footwell'.
2 On early models ((up to 02/09) remove the audio unit as described in Chapter 12 Section 22. On later models remove the audio unit trim, but leave the audio unit in place.
3 Remove the lower mounting screws and then prise the trim and centre vent panel from the facia **(see illustrations)**.

4 Disconnect the wiring plugs as the unit is removed and then remove the control unit **(see illustrations)**.
5 Refitting is a reversal of removal, but ensure the control knobs are positioned as previously noted. Check the operation of the controls.

Heater matrix

6 Remove the facia panel as described in Chapter 11 Section 27.
7 At the rear of the engine compartment, gain access to the coolant flow and return pipes. The number of components requiring removal

9.8a Clamp the coolant hoses

to access the pipes will depend on the type of engine fitted. Where fitted, remove the heat shield by releasing the single nut (some models) and the multiple spring clips.
8 Using hose clamps, clamp the heater matrix inlet and return hoses located on the bulkhead at the rear of the engine compartment. Place a container beneath the hoses, then remove them either at the quick release connectors or at the spring clips**(see illustrations)**. Note the location of the hoses for correct refitting.
9 With the hoses disconnected, remove the

9.8b Remove the coolant hoses

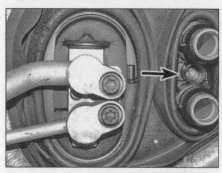

9.10 Slacken the bolt

9.11 Remove the vent duct

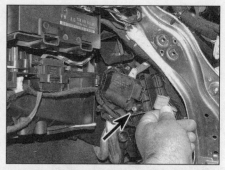

9.12a Disconnect the wiring plug and remove the screw

9.12b Remove the cover to expose...

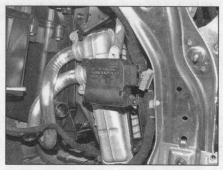

9.12c ... the heater matrix

9.13a Release the pipe clamps...

coolant from the matrix by blowing air into the upper tube, preferably using an airline.

10 Loosen (but do not remove) the bolt located between the matrix upper and lower tubes **(see illustration)**. This will make removal of the matrix easier.

11 Working inside the vehicle, remove the air distribution duct **(see illustration)**. On models fitted with an electric heater remove the heater.

12 Disconnect the wiring plug from the control motor and remove the motor fixing – there is no need to remove the motor completely. Undo the screws and remove the cover from the heater matrix **(see illustrations)**.

13 Place cloth rags or similar on the floor beneath the heater matrix, then release the pipe clamps and pull the coolant pipes from

the matrix **(see illustrations)**. Note that the seals may come out with the pipes or remain in the matrix.

14 With difficulty, remove the heater matrix from the heater unit **(see illustration)**.

15 Refitting is a reversal of removal, noting the following.

a) Make sure the seal is fitted correctly around the perimeter of the matrix.

b) Always fit new seals and clamps.

c) When reconnecting the coolant pipes, moisten the seals with coolant and make sure the conical ends locate in the matrix. After reconnecting the pipes, the clamps must turn easily before tightening them securely.

d) Top-up the coolant level with reference to 'Weekly checks' at the beginning of this Manual.

Heater unit

Note: *Removal of the complete heater box is a long and involved task. The main reason to remove it completely will be to replace the AC evaporator or some of the internal air distribution flaps. Most of the control motors and cables are accessible once the facia panel is removed (Chapter 11 Section 27). Note that some of the control motors are also accessible with the glovebox or steering column removed.*

16 Disconnect the battery (Chapter 5A Section 3) and on models fitted with AC have the system degassed by a suitably equipped garage or air conditioning specialist.

17 Disconnect the heater matrix coolant hoses **(see illustrations 9.8a and 9.8b)** and where available blow compressed air through the matrix to clear the coolant out.

9.13b ...and pull the coolant pipes from the matrix

9.13c Recover the seals

9.14 Removing the matrix from the heater unit

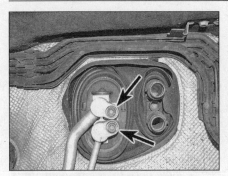

9.18a Unbolt the refrigerant lines

9.18b Seal the AC lines and the expansion valve as soon as they are disconnected

9.32 Disconnet the wiring plug

18 Where fitted remove the heat shield from the plenum chamber front panel. On models fitted with air conditioning disconnect the refrigerant lines at the plenum chamber **(see illustrations)**. Seal the open lines and the expansion valve.

19 Remove the bolt from the centre of the heater matrix pipes **(see illustration 9.10)** and then partially release the grommet from around the pipes and AC expansion valve (where AC is fitted).

20 Remove the facia panel as described in Chapter 11 Section 27 and then remove the crossmember (Chapter 11 Section 28).

21 Remove the accelerator pedal as described in Chapter 4A Section 5 (petrol engines) or Chapter 4B Section 4 and then fold back the carpet (at both footwells).

22 Remove the air distribution ducts (at both sides) for the rear footwells.

23 Remove the heater blower motor as described in this Section.

24 On models with AC disconnect the evaporator drain tube from the housing and then remove the metal support bracket from the left-hand rear of the housing.

25 Work around the housing and disconnect the wiring plugs as required. Release the wiring loom from the housing – again as required.

26 Gently pull the right-hand side of the housing forward until the bolt can be removed from the right-hand support bracket.

27 Carefully pull the heater unit from the bulkhead, taking care not to damage or bend

the matrix tubes on the bulkhead. The help of an assistant may be required. Be prepared for coolant spillage as the assembly is removed from inside the car.

28 If necessary, the heater may be further dismantled on the bench.

29 Refitting is a reversal of removal, but top-up the coolant with reference to *Weekly checks*, at the beginning of this Manual.

Heater blower motor

30 Switch off the ignition and all electrical consumers.

31 Remove the passenger side glovebox as described in Chapter 11 Section 25.

32 Where fitted remove the lower cover panel. Disconnect the wiring plug from the blower motor **(see illustration)**.

33 Where fitted, remove the single screw, then release the catch and turn the blower motor anti-clockwise to remove it from the heater housing **(see illustrations)**.

34 Refitting is a reversal of removal.

Heater blower motor series resistor

35 Remove the blower motor as described above.

36 Disconnect the wiring plug, remove the screws and pull out the resistor pack **(see illustration)**.

Caution: The resistor may be very hot if the heater has recently been in use.

37 Refitting is the reverse of removal.

Fresh/recirculating air flap positioning motor

Note: *All the various control motors for the heating and (where fitted) AC have a diagnostic capability and can output fault codes that can be read by a suitable code reader (scan tool). Consult a Seat dealer or suitably equipped garage before replacing any component part of the system.*

38 Turn on the ignition and select the 'fresh air' option on the heater control panel. Allow the motor to operate and park before turning off the ignition.

39 Remove the passenger side glovebox and fascia end panel as described in Chapter 11 Section 25.

40 Unclip the cover retaining the motor to the heater housing.

41 Disconnect the wiring, then remove the motor from its mountings and separate it from the air flap lever.

42 Refitting is a reversal of removal. Note that if the motor is renewed, its basic settings must be reprogrammed by a Seat dealer using specialist equipment.

Auxiliary heater

43 Some diesel models are equipped with an auxiliary electrically powered heating element within the heating unit.

44 Proceed as described in this Section for heater matrix removal. Undo the screws and remove the cover from the heater matrix. If

9.33a Remove the motor

9.33b The location of the screw fitting point. The screw is not fitted to all versions

9.36 Remove the wiring plug and screws

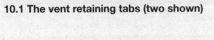

10.1 The vent retaining tabs (two shown)

11.6 Air conditioning compressor bolted to the front of the cylinder block

the upper screw is not accessible, turn the temperature flap control as necessary.
45 Note their fitted positions, then disconnect the wiring plugs from the element.
46 Pull the heating element from place.
47 Refitting is a reversal of removal.

10 Heating/ventilation system vents – removal and refitting

Side vents

1 In theory the vents can be prised out from the fascia using a small flat-bladed screwdriver. However this proved impossible due to the strong metal clips holding the vent in place **(see illustration)**. They can be removed if the fascia is removed (as described in Chapter 11 Section 27) as the tangs can then be compressed from the rear.
2 To refit, carefully push the vent into position until the locating clips engage.

Central facia vents

3 The central vents are removed to gain access to the heater control panel – as described in Section 9.

11 Air conditioning system – general information and precautions

General information

1 Air Conditioning (AC) is fitted as standard to most models, and is available as manually-operated system or automatically-operated climate control system (climatronic) with or without dual zone control. The Climatronic system works in conjunction with the heating and air conditioning systems to maintain a selected vehicle interior temperature fully automatically.
2 The air conditioning system enables the temperature of incoming air to be lowered, and dehumidifies the air, which makes for rapid demisting and increased comfort. The cooling side of the system works in the same way as a domestic refrigerator. Refrigerant gas is drawn into a belt-driven compressor

and passes into a condenser mounted in front of the radiator, where it loses heat and becomes liquid. The liquid passes through an expansion valve to an evaporator, where it changes from liquid under high pressure to gas under low pressure. This change is accompanied by a drop in temperature, which cools the evaporator. The refrigerant returns to the compressor and the cycle begins again.
3 Air blown through the evaporator passes to the air distribution unit, where it is mixed with hot air blown through the heater matrix to achieve the desired temperature in the passenger compartment.
4 The heating side of the system works in the same way as on models without air conditioning.
5 The operation of the system is controlled electronically by coolant temperature switches, and pressure switches which are screwed into the compressor high-pressure line. Any problems with the system should be referred to a Seat dealer or an air conditioning specialist.
6 The only operation which can be carried out easily without discharging the refrigerant is the renewal of the compressor drivebelt, which is covered in the relevant part of Chapter 1A Section 30 (petrol models) or Chapter 1B Section 30 (diesel models). Removal of the evaporator and condenser requires the evacuation of the refrigerant. If necessary the compressor can be unbolted and moved aside, without disconnecting its flexible hoses, after removing the drivebelt **(see illustration)**.

Precautions

• When an air conditioning system is fitted, it is necessary to observe special precautions whenever dealing with any part of the system, its associated components and any items which require disconnection of the system. If for any reason the system must be disconnected, entrust this task to your Seat dealer or an air conditioning specialist.

 Warning: The refrigeration circuit contains a refrigerant and it is therefore dangerous to disconnect any part of the system without specialised knowledge and equipment. The refrigerant is potentially dangerous and should only be handled by qualified persons. If it is splashed onto the skin it can cause frostbite. It is not itself poisonous, but in the presence of a naked flame (including a cigarette) it forms a poisonous gas. Uncontrolled discharging of the refrigerant is dangerous and potentially damaging to the environment.
• Do not operate the air conditioning system if it is known to be short of refrigerant, as this may damage the compressor.

12 Air conditioning system components – removal and refitting

Sunlight sensor

1 Switch off the ignition and all electrical consumers.
2 Using a small screwdriver, gently prise the cover off **(see illustration)**.
3 Disconnect the wiring plug and withdraw the sensor **(see illustration)**.
4 Refitting is a reversal of removal.

Outlet temperature sensor

5 Switch off the ignition and all electrical consumers.
6 Remove the lower trim panel from the drivers side as described in Chapter 11 Section 27.
7 The sensor is located in the air duct. Disconnect the wiring plug, rotate the sensor through 90 degrees and remove it.
8 Refitting is a reversal of removal.

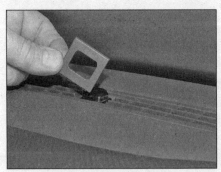

12.2 Remove the cover

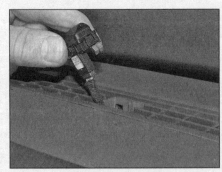

12.3 Remove the sensor

12.11a Disconnect the wiring plug

12.11b The sensor location – shown with the fascia removed for clarity

12.14a Disconnect the wiring plug

12.14b The sensor location in the duct (shown with the fascia removed for clarity)

12.16 Evaporator temperature sensor (position shown with evaporator removed)

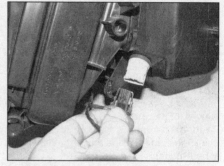

12.18 Disconnect the wiring connector from the sensor

Side vent temperature sensors

Note: *Sensors are fitted behind both fascia side vents.*

9 Switch off the ignition and all electrical consumers.

10 Remove the glovebox as described in Chapter 11 Section 25.

11 Disconnect the wiring plug from the sensor **(see illustrations)**.

12 Turn the sender through 90°, and withdraw it from the housing.

13 To remove the right-hand sensor, remove the cover panel from the diagnostic socket and then remove the lighting switch as described in Chapter 12 Section 5.

14 Disconnect the wiring plug and remove the sensor **(see illustrations)**.

15 Refitting is a reversal of removal.

Evaporator temperature sensor

16 The sensor is located to the right-hand side of the heater housing, behind the evaporator **(see illustration)**, to the left of the clutch pedal.

17 Working in the driver's side foot well, undo the retaining screw and unclip the panel from the front of the centre console.

18 Disconnect the wiring plug connector, then rotate the sensor 90° and pull it from the housing **(see illustration)**.

19 Refitting is a reversal of the removal procedure.

Ambient temperature sensor

20 On some models remove unclip the centre grille to access the sensor. On other models

remove the front bumper cover as described in Chapter 11 Section 6.

21 Unclip the sensor from its retainer and disconnect the wiring plug **(see illustration)**.

22 Refitting is a reversal of removal. Make sure the wiring is fully connected to prevent entry of water.

Evaporator

23 Remove the heater matrix and heater housing, as described earlier in Section 9.

24 Remove the rubber gaiter and foam cushion from around the coolant and refrigerant pipes **(see illustration)**.

25 Release any retaining clips and remove the coolant pipes from the housing **(see illustration)**.

26 Noting the routing and their fitted

12.21 Ambient temperature sender location at the front of the radiator/condenser

12.24 Remove the rubber/foam grommet from the pipes

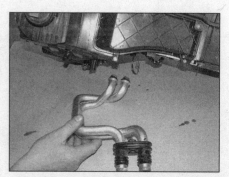

12.25 Remove the heater pipes from the housing

12.26a Release the wiring loom from the retaining clips …

12.26b … and locating pegs on the heater housing

12.27 Unscrew the bolts securing the two sections of the heater housing

connections remove the wiring loom from around the heater housing **(see illustrations)**.
27 Working your way around the housing, undo the retaining bolts and split the housing into two parts **(see illustration)**.
28 To remove the evaporator, undo the retaining bolts and split the upper and lower halves of the housing, then withdraw the evaporator from the housing together with the refrigerant lines and rubber grommet **(see illustrations)**.
29 Refitting is a reversal of removal, fit new seals and have the system recharged by a Seat dealer or refrigeration specialist.

Condenser

Note: *It is possible to remove the cooling*
fans, radiator, intercooler (where fitted) and condenser as a single unit. If this approach is adopted the aid of an assistant is essential.

⚠️ *Warning: It is a criminal offence to knowingly discharge refrigerant to the atmosphere.*

30 Have the refrigerant evacuated from the air conditioning system by a Seat dealer or automotive refrigeration specialist. Many garages will have the equipment and mobile services are available at a reasonable cost.
31 Move the lock carrier to the service position (as described in Section 3) and then remove then remove the radiator as described in Section 3 of this Chapter.
32 Where fitted, remove the intercooler as

described in Chapter 4C Section 7 or Chapter 4D Section 7.
33 Undo the screws and disconnect the refrigerant lines from the condenser **(see illustration)**. Recover the seals and plug the lines and condenser openings to prevent entry of foreign matter and water vapour.
34 Unbolt the condenser from the radiator and then lift out the condenser **(see illustrations)**.
35 Refitting is a reversal of removal, but fit new seals, and have the system recharged by a Seat dealer or refrigeration specialist.

Compressor

36 Have the refrigerant evacuated from the air conditioning system by a Seat dealer or refrigeration specialist. Note that the compressor can be moved forward and secured to the front panel if removal is only required for access to other components. There is no need to have the system evacuated if this is the case.
37 Where fitted, remove the intercooler pipes (Chapter 4C Section 7 -petrol engines or Chapter 4D Section 7 – diesel engines) and then remove the auxiliary drivebelt as described in Chapter 1A Section 30 (petrol engines) or Chapter 1B Section 30 (diesel engines).
38 Unscrew the retaining bolts and disconnect the refrigeant lines from the

12.28a Unscrew the bolts and separate the evaporator housing …

12.28b …then withdraw the evaporator

12.33 Disconnect the refrigerant lines from the condenser

12.34a Remove the fixings (one shown, but four in total)

12.34b Lift out the condenser

12.38 Undo the refrigerant line securing bolts on the compressor

12.39 Disconnect the wiring connector

12.40a Remove the bolts and…

compressor **(see illustration)**. Remove the O-ring seals and renew them if necessary. Plug the open pipes and ports to prevent the ingress of moisture.

39 Disconnect the wiring plug connector from the rear of the compressor **(see illustration)**.

40 Undo the retaining bolts **(see illustrations)** and withdraw the compressor from the mounting bracket.

41 Refitting is a reversal of the removal procedure, ensure that all the fixings are tightened to the specified torque settings, where given. On completion, have new O-rings fitted to the refrigerant lines, and then have the system recharged by a Seat dealer or refrigeration specialist.

Receiver/drier

Note: *Check for the availability of parts before removal, there are different makes of receiver/drier and some cannot be renewed separately from the condenser.*

42 The receiver/drier is attached to the left-hand side of the condenser.

43 Remove the condenser, as described in paragraphs 9 to 14.

Modine type

44 Unscrew the cap (with a T50 torx socket) from the top of the receiver/drier **(see illustration)**. Peel back the warning label, where fitted.

Caution: DO NOT remove this cap if the system has not been evacuated, as it will be under high pressure.

45 Insert a bolt into the receiver/drier sealing cover, push downwards slightly and extract the retaining circlip. Pull the receiver/drier element from place.

Showa type

46 The receiver/drier is bolted to the side of the condenser, undo the retaining bolt(s) and withdraw it from the condenser.

12.40b …pull the compressor off the mounting bracket (shown with the AC lines attached)

47 Refitting is a reversal of the removal procedure, ensure that all the fixings are tightened to the specified torque settings, where given. On completion, fit new O-rings and then have the system recharged by a Seat dealer or refrigeration specialist.

Refrigerant pressure sensor

48 The sensor is mounted low down in front of the compressor.

12.49a The pressure sensor

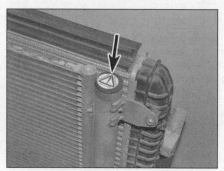

12.44 Unscrew the cap to access the receiver/drier

49 Disconnect the wiring plug and unscrew the sensor **(see illustrations)** – note there is a schraeder valve below the sensor and this will seal as the sensor is removed. However caution will be required, because if the valve is faulty refrigerant will escape. If there is any sound of escaping refrigerant, immediately refit the sensor and have the system professionally evacuated before removing the sensor.

50 Refitting is a reversal of removal.

12.49b The sensor removed showing the schrader valve

Chapter 4 Part A
Petrol engine fuel systems

Contents

Degrees of difficulty

Easy, suitable for novice with little experience	Fairly easy, suitable for beginner with some experience	Fairly difficult, suitable for competent DIY mechanic	Difficult, suitable for experienced DIY mechanic	Very difficult, suitable for expert DIY or professional

Specifications

Engine codes*

1.6 litre	BGU, BSF and BSE
2.0 litre	
Non-turbo.	BLR, BVY, BLY and BVZ
Turbo	BWA, BWJ, CDLD and CDLA

* Note: *See 'Vehicle identification' at the end of this manual for the location of engine code markings.*

System type

1.6 litre engines:
Engine codes BSE and BSF. . . . . Siemens Simos 7.1
Engine codes BGU. . . . . Siemens Simos 3.3
2.0 litre engines:
Engine code BLR, BVY, BVZ and BLY . . . . Bosch Motronic MED 9.5
Engine code BWA, BWJ, CDLD and CDLA . . . . Bosch Motronic MED 9.1
* Note: *See 'Vehicle identification' at the end of this manual for the location of engine code markings.*

Fuel system data

Fuel pump type . . . . Electric, immersed in fuel tank
Fuel pump delivery rate:
1.6 litre engines . . . . 580 cc/min (battery voltage of 10.5 V)
2.0 litre FSI engines . . . . 275 cc/min (battery voltage of 10.5 V)
2.0 litre TFSI . . . . 580 cc/min (battery voltage of 10.5 V)
Regulated fuel pressure:
1.6 litre engines . . . . 3.8 to 4.2 bar
2.0 litre FSI engines . . . . 4 to 4.8 bar
2.0 litre TFSI engines . . . . 4.0 to 7.0 bar
Engine idle speed. . . . . non-adjustable, electronically controlled
Idle CO content (non-adjustable, electronically-controlled). . . . . 0.5% max
Injector electrical resistance (typical) . . . . 12 to 17 ohms

Torque wrench settings

	Nm	lbf ft
All models		
Accelerator pedal to bulkhead.	10	7
Fuel lift pump/gauge sender retaining ring.	110	81
Fuel tank strap retaining bolt	25	18
Knock sensor(s)	20	15
Oxygen sensor(s)	50	37
1.6 litre engine		
Fuel rail to inlet manifold	8	6
Inlet manifold support:		
To cylinder head	15	11
To inlet manifold	8	6
Inlet manifold to cylinder head.	25	18
Inlet manifold upper part-to-lower part screws	3	2
Throttle housing/module mounting bolts	8	6
2.0 litre engines		
Fuel pressure sensor	20	15
Fuel rail mounting bolts	10	7
High-pressure fuel line to pump:		
Engine codes BWJ and BWA:		
Union nut	25	19
Banjo bolt	17	13
Engine codes BLR, BLY, BVZ and BLY:		
Low-pressure union bolt.	15	11
High-pressure union bolt	15	11
High-pressure fuel pump	10	7
Inlet manifold-to-cylinder head nuts/bolts	10	7
Inlet manifold upper part to lower part.	10	7
Throttle housing mounting bolts.	8	6

1 General information and precautions

General information

1 The systems described in this Chapter are all self-contained engine management systems, which control both the fuel injection and ignition. This Chapter deals with the fuel system components only – see Chapter 4C for information on the turbocharger, exhaust and emission control systems, and to Chapter 5B for details of the ignition system.

2 The fuel injection system consists of a fuel tank, an electric fuel lift pump/level sender unit, a fuel filter, fuel supply and return lines, a throttle housing/module, four electronic fuel injectors, and an Electronic Control Unit (ECU) together with its associated sensors, actuators and wiring. Two basic fuel injection systems are fitted – an indirect (low pressure) injection system where the injectors inject fuel into the inlet manifold upstream of the inlet valves, and a direct (high pressure) injection system where the injectors inject fuel directly into the combustion chambers.

3 The indirect injection system uses a low pressure fuel rail and injectors fitted to the inlet manifold.

4 The direct injection system (fitted to FSi (Fuel Stratified injection) and TFSi (Turbocharged Fuel Stratified injection) engines) uses a high-pressure fuel pump mounted on top of the camshaft housing, actuated by a plunger in contact with the inlet camshaft. Fuel under high pressure is fed into a fuel rail incorporated into the lower inlet manifold. The injectors are located between the lower inlet manifold and cylinder head, and inject fuel directly into the combustion chambers.

5 The two basic fuel systems function in a very similar way, but there are significant detail differences, particularly in the sensors used and in the inlet manifold arrangements.

6 The fuel lift pump is immersed in the fuel inside the tank, and delivers a constant supply of fuel through a cartridge filter to the fuel rail or high-pressure fuel pump (according to engine). The fuel pressure regulator maintains a constant fuel pressure to the fuel injectors. The location of the regulator varies according to the engine fitted. On 1.6 and later 2.0 litre engines the pressure regulator is located on the fuel rail or high-pressure pump. On early 2.0 litre engines the regulator is either fitted to the rear of the fuel filter or integrated into the fuel filter.

7 The fuel injectors are opened and closed by an Electronic Control Unit (ECU), which calculates the injection timing and duration according to engine speed, crankshaft/camshaft position, throttle position and rate of opening, inlet manifold depression, inlet air temperature, coolant temperature, roadspeed and exhaust gas oxygen content information, received from sensors mounted on and around the engine.

8 Inlet air is drawn into the engine through the air cleaner, which contains a renewable paper filter element. On some non-turbo models, the inlet air temperature is regulated by a valve mounted in the air cleaner inlet trunking, which blends air at ambient temperature with hot air, drawn from over the exhaust manifold.

9 The temperature and pressure of the air entering the throttle housing is measured by a sensor mounted on the inlet manifold. This information is used by the ECU to fine-tune the fuelling requirements for different operating conditions. Turbocharged engines have an additional air temperature sensor mounted downstream of the throttle housing, which monitors the (compressed) air temperature after it has been through the turbocharger and intercooler.

10 1.6 litre engines have a variable-length inlet manifold. A vacuum-controlled flap is used to divert the inlet air into one of two paths through the manifold, the paths being of different lengths. Controlling the inlet air in this way has the effect of altering the engine's torque characteristics at different engine speeds and loads.

11 Idle speed control is achieved partly by an electronic throttle valve positioning module, which is part of the throttle housing, and partly by the ignition system, which gives fine control of the idle speed by altering the ignition timing. As a result, manual adjustment of the engine idle speed is not necessary or possible.

12 The exhaust gas oxygen content is constantly monitored by the ECU by oxygen sensors (also known as lambda sensors), one before the catalytic converter, and one after – this improves sensor response time and accuracy, and the ECU compares the signals from each sensor to confirm that the converter is working correctly. The ECU uses the information from the sensors to modify the injection timing and duration to maintain the optimum air/fuel ratio.

13 The ECU also controls the operation of the activated charcoal filter evaporative loss system – refer to Chapter 4C Section 2 for further details.

14 It should be noted that fault diagnosis of all the engine management systems described in this Chapter is only possible with dedicated electronic test equipment. Problems with the systems operation should therefore be referred to a Seat dealer or suitably equipped garage for assessment. Note however that low cost diagnostic equipment is now available to the home mechanic and whilst it will not offer the depth of coverage of professional equipment it may help to guide the home mechanic in the right direction. Once the fault has been identified, the removal/refitting sequences detailed in the following Sections will then allow the appropriate component(s) to be renewed as required.

Precautions

Warning: Petrol is extremely flammable – great care must be taken when working on any part of the fuel system.

• Do not smoke, or allow any naked flames or uncovered light bulbs near the work area. Note that gas-powered domestic appliances with pilot flames, such as heaters boilers and tumble-dryers, also present a fire hazard – bear this in mind if you are working in an area where such appliances are present. Always keep a suitable fire extinguisher close to the work area, and familiarise yourself with its operation before starting work. Wear eye protection when working on fuel systems, and wash off any fuel spilt on bare skin immediately with soap and water. Note that fuel vapour is just as dangerous as liquid fuel – possibly more so; a vessel that has been emptied of liquid fuel will still contain vapour, and can be potentially explosive.

• Many of the operations described in this Chapter involve the disconnection of fuel lines, which may cause an amount of fuel spillage. Before commencing work, refer to the above 'Warning' and the information in 'Safety first!' at the beginning of this manual.

• Residual fuel pressure always remains in the fuel system, long after the engine has been switched off. This pressure must be relieved in a controlled manner before work can commence on any component in the fuel system – refer to Section 9 for details.

• When working with fuel system components, pay particular attention to cleanliness – dirt entering the fuel system may cause blockages, which will lead to poor running.

• In the interests of personal safety and equipment protection, many of the procedures in this Chapter suggest that the negative lead be removed from the battery terminal. This firstly eliminates the possibility of accidental short-circuits being caused as the vehicle is being worked upon, and secondly prevents damage to electronic components (eg, sensors, actuators, ECUs) which are particularly sensitive to the power surges caused by disconnection or reconnection of the wiring harness whilst they are still 'live'. Refer to Chapter 5A Section 3.

2 Fuel pipes and connections

1 Disconnect the cable from the negative battery terminal (see Chapter 5A Section 3) before proceeding.

2 The fuel supply pipe connects the fuel pump in the fuel tank to the fuel rail on the engine.

3 Whenever you're working under the vehicle, be sure to inspect all fuel and evaporative emission pipes for leaks, kinks, dents and other damage. Always replace a damaged fuel pipe immediately.

4 If you find signs of dirt in the pipes during disassembly, disconnect all pipes and blow them out with compressed air. Inspect the fuel strainer on the fuel pump pick-up unit for damage and deterioration.

Steel tubing

5 It is critical that the fuel pipes be replaced with pipes of equivalent type and specification.

6 Some steel fuel pipes have threaded fittings. When loosening these fittings, hold the stationary fitting with a spanner while turning the union nut.

Plastic tubing

Warning: When removing or installing plastic fuel tubing, be careful not to bend or twist it too much, which can damage it. Also, plastic fuel tubing is NOT heat resistant, so keep it away from excessive heat.

7 When replacing fuel system plastic tubing, use only original equipment replacement plastic tubing.

Flexible hoses

8 When replacing fuel system flexible hoses, use original equipment replacements, or hose to the same specification.

9 Don't route fuel hoses (or metal pipes) within 100 mm of the exhaust system or within 280 mm of the catalytic converter. Make sure that no rubber hoses are installed directly against the vehicle, particularly in places where there is any vibration. If allowed to touch some vibrating part of the vehicle, a hose can easily become chafed and it might start leaking. A good rule of thumb is to maintain a minimum of 8.0 mm clearance around a hose (or metal pipe) to prevent contact with the vehicle underbody.

Disconnecting Fuel pipe Fittings

10 Typical fuel pipe fittings:

2.10a Two-tab type fitting; depress both tabs with your fingers, then pull the fuel pipe and the fitting apart

2.10b On this type of fitting, depress the two buttons on opposite sides of the fitting, then pull it off the fuel pipe

2.10c Threaded fuel pipe fitting; hold the stationary portion of the pipe or component (A) while loosening the union nut (B) with a flare-nut spanner

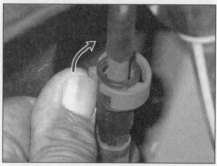

2.10d Plastic collar-type fitting; rotate the outer part of the fitting

2.10e Metal collar quick-connect fitting; pull the end of the retainer off the fuel pipe and disengage the other end from the female side of the fitting...

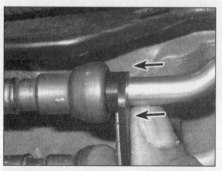

2.10f ...insert a fuel pipe separator tool into the female side of the fitting, push it into the fitting and pull the fuel pipe off the pipe

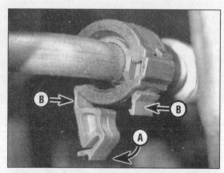

2.10g Some fittings are secured by lock tabs. Release the lock tab (A) and rotate it to the fully-opened position, squeeze the two smaller lock tabs (B)...

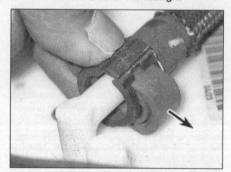

2.10h ...then push the retainer out and pull the fuel pipe off the pipe

2.10i Spring-lock coupling; remove the safety cover, install a coupling release tool and close the tool around the coupling...

2.10j ...push the tool into the fitting, then pull the two pipes apart

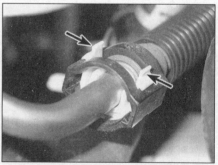

2.10k Hairpin clip type fitting: push the legs of the retainer clip together, then push the clip down all the way until it stops and pull the fuel pipe off the pipe

3.1 Release the inlet duct

3.2a Disconnect the wiring plug...

3.2b ...and release the spring clips

3 Air filter housing and inlet system – removal and refitting

Removal

Engine codes BWA, BWJ, CDLA and CDLD

1 Release the spring type hose clip from the inlet air duct (see illustration).
2 Disconnect the wiring plug from the airflow sensor and then release the spring clips from the inlet manifold (see illustrations).
3 Unclip the crankcase breather hose from the housing and then carefully pull up the housing from the engine (see illustration).

Engine codes BGU, BSE, BSF BLR, BVY, BLY and BVZ

4 The air cleaner is located in front of the battery

3.3a Unclip the breather hose...

3.3b ...and remove the cover

on the left-hand side of the engine compartment. Where fitted remove the engine cover.
5 Release the outlet duct spring clip (see illustration).
6 Remove the fixing and then lift out the filter housing (see illustrations).

7 If required the inlet duct/damping chamber can now be removed (see illustration).
8 Finally remove the outlet ducting by releasing it from the throttle body and breather hose (see illustration).

3.5 Release the spring type hose clip

3.6a Remove the fixing

3.6b Release the ducting and then...

3.6c ...pull up the complete housing to remove it

3.7 Remove the inlet duct

3.8 Release the ducting (1.6 engine shown)

Refitting

9 Refitting is a reversal of removal, noting the following points:

a) *If removed, ensure that the air filter element is correctly refitted, referring to Chapter 1A Section 28 if necessary.*

b) *It is most important that an airtight seal is made between the air cleaner and the throttle housing/module.*

4 Inlet air temperature control system – general information

Note: *This system is not fitted to all models.*

1 Where fitted, the inlet air temperature control system consists of a temperature-controlled flap valve, mounted in its own housing in the air cleaner inlet trunking or in the air cleaner lid, and a duct to the warm-air collector plate over the exhaust manifold.

2 The temperature sensor in the flap valve housing senses the temperature of the inlet air, and opens the valve when a preset lower limit is reached. As the flap valve opens, warm air drawn from around the exhaust manifold blends with the inlet air.

3 As the temperature of the inlet air rises, the sensor closes the flap progressively, until the warm-air supply from the exhaust manifold is completely closed off, and only air at ambient temperature is admitted to the air cleaner.

4 With the ducting removed from the valve housing, the sensor is visible. If a hairdryer and suitable freeze spray is available, the action of the sensor can be tested.

5 Fuel system components – removal and refitting

Note: *Observe the precautions in Section 1 before working on any component in the fuel system. Information on the engine management system sensors which are more directly related to the ignition system will be found in Chapter 5B.*

Throttle body

1.6 and 2.0 litre (normally aspirated engines)

1 Remove the air filter housing and ducting as described in Section 3.

2 On 2.0 litre models with a heated throttle body, clamp the flow and return pipes and then (anticipating some coolant loss) remove the coolant pipes.

3 Disconnect the wiring from the throttle body **(see illustration)**.

4 Unscrew and remove the through-bolts, then lift the throttle housing/module away from the inlet manifold. Recover the O-ring seal **(see illustrations)**.

2.0 litre turbocharged engines

5 Remove the engine cover/air filter housing as described in Section 3.

6 Jack up and supprt the front of the vehicle – see *Jacking and vehicle support* and then remove the engine undershield.

7 Remove the alternator as described in Chapter 5A Section 5.

8 On models fitted with a recirculation valve (engine codes BWJ, CDLD and CDLA) disconnect the wiring plug and then remove the by-pass air hose.

9 Unbolt the charge air pipe from the engine and remove it.

10 Disconnect the wiring plug, remove the bolts and lower the throttle body from the inlet manifold. Recover the sealing gasket **(see illustration 10.1)**.

11 Refitting is a reversal of removal, noting the following:

a) *Use a new throttle housing-to-inlet manifold seal.*

b) *Tighten the throttle housing through-bolts evenly to the specified torque.*

c) *Ensure that all hoses and electrical connectors are refitted securely.*

Fuel injectors (and fuel rail where applicable)

Note: *Observe the precautions in Section 1 before working on any component in the fuel system. If a faulty injector is suspected, before removing the injectors, it is worth trying the effect of one of the proprietary injector-cleaning treatments. These can be added to the petrol in the tank, and are intended to clean the injectors as you drive. Note that Seat technicians use tool T10133/2A to remove the injectors and fit the new injector seals – although the tool may not be required to remove the injectors, tools T10133/5 and T10133/6 (part of the kit) it will be required to fit the new seals.*

1.6 litre engines

12 On the right-hand side of the engine compartment, near the coolant expansion tank, disconnect the fuel supply line. Plug the end of the line to prevent entry of dust and dirt.

13 Disconnect the wiring from the injectors.

14 Remove the lower part of the inlet manifold complete with the fuel rail, as described in Section.

15 Pull out the clips and remove the injectors from the fuel rail **(see illustration)**.

2.0 litre engine codes BWA, BWJ, CDLA and CDLD

⚠️ **Warning:** *The fuel injection system operates at high pressure and must be de-pressurised before starting work as described in Section 9.*

Note: *Two versions of the fuel rail are fitted. Removal and refitting is essentially the same for both versions.*

16 Remove the inlet manifold complete with fuel rail as described in Section 10, then disconnect all hoses and fuel lines, and unbolt the fuel rail from the inlet manifold. Tape over or seal the intake ports in the cylinder head to prevent entry of dust and dirt.

17 Using a screwdriver, bend the lugs on the injector support rings to one side, then remove the rings. Note that the lugs will most likely break, and new rings will be required for refitting.

5.3 Disconnect the wiring plug

5.4a Remove the throttle body

5.4b ...and recover the O-ring seal

5.15 Remove the injector retaining clips

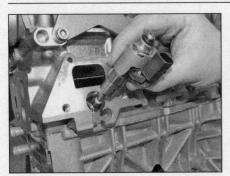

5.19 Removing the fuel injectors from the cylinder head

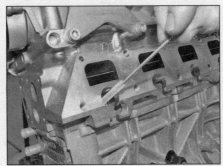

5.29a Thoroughly clean the injector seatings in the cylinder head

5.29b Fitting new injector seals

18 At this stage Seat technicians use a special slide-hammer tool to remove the injectors. This is VAG tool T10133/2 (early version) or T10133/2A (later version). Where the earlier version is available it will require a slight modification to fit around the injector body. The modification requires the filing of a 6mm diameter semi-circle into the closed end of the tool.

19 Where the correct tool is not available is also possible to remove the injectors by carefully prising them from the cylinder head or fabricating a tool **(see illustration)**. Where this is not possible the fuel injectors must be removed with the correct tool.

2.0 litre engine codes BVY, BVZ, BLY, BLR

 Warning: The fuel injection system operates at high pressure and must be de-pressurised before starting work as described in Section 9.

20 Remove the inlet manifold as described in Section.

21 On the right-hand side of the engine compartment, near the coolant expansion tank, disconnect the fuel supply line and, where applicable, the return line. Plug the ends of the lines to prevent entry of dust and dirt.

22 Disconnect the wiring from the injectors.

23 Disconnect the crankcase ventilation hose, and pull off the engine oil dipstick together with the guide tube.

24 Unbolt the fuel pipes.

25 Unbolt and remove the fuel rail together with the manifold flaps from the injectors.

26 Tape over or seal the intake ports in the cylinder head to prevent entry of dust and dirt.

27 Using a screwdriver, bend the lugs on the injector support rings to one side, then remove the rings. Note that the lugs will most likely break, and new rings will be required for refitting.

28 At this stage Seat technicians use a special slide-hammer tool to remove the injectors. This is VAG tool T10133/2 (early version) or T10133/2A (later version). Where the earlier version is available it will require a slight modification to fit around the injector body. The modification requires the filing of

a 6mm diameter semi-circle into the closed end of the tool. Where the correct tool is not available is also possible to remove the injectors by carefully prising them from the cylinder head or fabricating a tool. Where this is not possible the fuel injectors must be removed with the correct tool.

All engine codes

29 Refitting is a reversal of removal, but thoroughly clean the cylinder head seatings, and fit new injector seals **(see illustrations)**. On all engine codes except BSE, BSF and BSF, the seals are made of Teflon, and the special VAG tool will be required to compress the seals before fitting the injectors to the cylinder head. Do not grease or oil the seals on these engines. On engine codes BSE, BGU and BSF (ie, low-pressure injection system), lightly moisten the O-ring seals with clean engine oil.

Fuel pressure regulator

Note: *Observe the precautions in Section before working on any component in the fuel system.*

30 The fuel pressure regulator is located on the fuel filter beneath the rear of the car on 1.6 litre engines and 2.0 litre FSI engines (engine codes BLR, BVY, BLY and BVZ). On some variants it is removable, on others it is integrated into the filter assembly and is replaced with the filter **(see illustration)**. On all other models the regulator is part of the high pressure pump mounted on the cylinder

head/inlet manifold. On these models if the regulator is faulty it must be replaced along with the high pressure pump as described later in this Section.

31 First, chock the front wheels, then jack up the rear of the car and support on axle stands (see *Jacking and vehicle support*).

32 Anticipating some fuel spillage, place a suitable container beneath the filter, then disconnect the hoses.

33 Unscrew the clamp bolt and remove the filter from its mounting.

34 Pull out the clip and remove the pressure regulator from the rear of the filter. Recover the seal and O-ring and discard them as new ones must be used on refitting.

35 Refitting is a reversal of removal, but use a new seal and O-ring.

Throttle valve positioner

36 The positioner is matched to the throttle housing/module during manufacture, and is not available separately – if defective, a complete throttle housing/module will be required.

Throttle pedal/position sensor

37 All models are fitted with a 'fly-by-wire' throttle where the position sensor is integral with the accelerator pedal. Working inside the car, remove the footrest panel from below the accelerator and brake pedals **(see illustration)**.

38 Prise off the cap, and undo the screw

5.30 Fuel filter with integrated pressure regulator

5.37 Remove the footrest panel...

5.38 ...prise off the cap and undo the screw...

5.39 ...release the clips (sensor removed for clarity)...

5.40 ...then remove the throttle pedal/ sensor and disconnect the wiring

securing the throttle pedal to the bulkhead **(see illustration)**.

39 The throttle pedal is clipped to the floor. A pair of screwdrivers (or the VAG tool T10238) will be required to release the pedal **(see illustration)**.

40 Withdraw the throttle pedal and disconnect the wiring and support **(see illustration)**.

41 Refitting is a reversal of removal.

Inlet air temperature/ pressure sender

1.6 litre engines

42 The combined air temperature and air pressure sensor is located on the top of the upper inlet manifold. First, disconnect the wiring **(see illustration)**.

43 Undo the two screws and remove the sensor from the inlet manifold **(see illustration)**. Recover the seal and discard, as a new one must be used on refitting.

44 Refitting is a reversal of removal, but fit a new seal.

2.0 litre FSi engines (BLR, BVY, BLY and BVS)

45 The air temperature sensor is located on the side of the inlet manifold **(see illustration 10.1b)**. First, disconnect the wiring.

46 Undo the screws and remove the sensor. Recover the seal and discard, as a new one must be used on refitting.

47 Refitting is a reversal of removal, but fit a new seal.

2.0 litre TFSi engines (BWA, BWJ, CDLD and CDLA)

48 The air temperature and pressure sensor is located on the bottom right-hand of the inlet manifold **(see illustration 10.1c)**. Note that access is limited.

49 Disconnect the wiring from the sensor.

50 Undo the two screws and remove the sensor from the inlet manifold. Recover the seal and discard it. A new should be used on refitting.

51 Refitting is a reversal of removal, but fit a new seal.

Coolant temperature sensor

52 Refer to Chapter.

Oxygen (lambda) sensors

⚠️ *Warning: Working on the sensors is only advisable with the engine (and therefore the exhaust system) completely cold. The catalytic converter in particular will be very hot for some time after the engine has been switched off.*

53 All models have one sensor threaded into the exhaust manifold or at the top of the exhaust downpipe, ahead of the catalytic converter, and a second sensor mounted in the exhaust front pipe or intermediate pipe, downstream of the converter. Refer to Chapter 4C Section 9 for more details **(see illustration)**.

54 Working from the sensor, trace the wiring harness from the oxygen sensor back to the connector, and disconnect it. Typically, the wiring plug is coloured black for the upstream sensor, and brown for the downstream sensor – do not confuse exhaust gas temperature sensors fitted to the exhaust system – mark them for identification prior to disconnection. Access to downstream sensor connector is gained by undo the 4 retaining nuts and removing the plastic cover **(see illustration)**. Unclip the sensor wiring from any retaining clips, noting how it is routed.

55 Access to the upstream sensor is possible on some models from above, while

5.42 Disconnect the wiring plug

5.43 Remove the screws

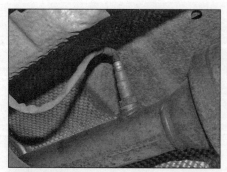

5.53 The pre converter Oxygen sensor on the 2.0 litre model (engine codes BLY and BVZ)

5.54 The downstream sensor connector is accessed once the plastic cover is removed

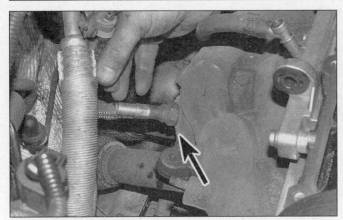

5.55 Oxygen sensor located on the exhaust manifold (1.6 litre engine). Shown with the inlet manifold removed

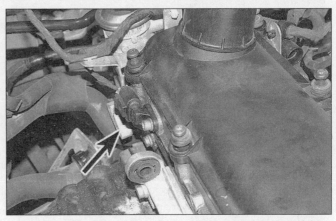

5.63 The camshaft position sensor (shown with inlet manifold removed)

the downstream sensor (where fitted) is only accessible from below **(see illustration)**.

56 Unscrew and remove the sensor, taking care to avoid damaging the sensor probe as it is removed. **Note:** *As a flying lead remains connected to the sensor after it has been disconnected, if the correct-size spanner is not available, a slotted socket will be required to remove the sensor.*

57 Apply a little high-temperature anti-seize grease to the sensor threads – avoid contaminating the probe tip.

58 Refit the sensor, tightening it to the correct torque. Reconnect the wiring.

Crankshaft position sensor

59 On all engines, the engine speed sensor is mounted on the front, left-hand side of the cylinder block, adjacent to the mating surface

of the block and transmission bellhousing, next to the oil filter. If necessary, drain the engine oil and remove the oil filter and cooler to improve access, with reference to Chapter 1A Section 6.

60 Trace the wiring back from the sensor, and unplug the harness connector.

61 Unscrew the retaining bolt and withdraw the sensor from the cylinder block.

62 Refitting is a reversal of removal.

Camshaft position sensor

1.6 litre engines

63 The camshaft position sensor is located on the left-hand rear face of the cylinder head **(see illustration)**. Where fitted, remove the engine cover and then remove the throttle body as described in this Section.

64 Disconnect the wiring **(see illustration)**.

65 Undo the screw and remove the sensor. Recover the O-ring seal.

66 Refitting is a reversal of removal, but fit a new O-ring seal.

2.0 litre engine

67 The camshaft position sensor is located on the right-hand end of the cylinder head, near the front.

68 Remove the engine cover/air filter.

69 Disconnect the wiring plug. Undo the screw and remove the sensor. Recover the O-ring seal.

70 Refitting is a reversal of removal, but fit a new O-ring seal.

Clutch pedal switch

71 The clutch pedal switch is clipped to the clutch master cylinder on the pedal bracket. Remove the master cylinder as described in Section.

72 Unclip the pedal switch from the bottom of the master cylinder.

73 Refitting is a reversal of removal.

Electronic control unit (ECU)

Caution: Always wait at least 30 seconds after switching off the ignition before disconnecting the wiring from the ECU. When the wiring is disconnected, all the learned values are erased. Note also that if the ECU is renewed, the identification of the new ECU must be transferred to the immobiliser control unit by a Seat dealer.

74 The ECU is located centrally behind the engine compartment bulkhead, under one of the windscreen cowl panels. Remove the wiper arms and cowl panel as for windscreen wiper motor removal and refitting, described in Chapter 12 Section 19.

75 The ECU is either secured by shear bolts or blind (pop) rivets. Where the security cover is secured with rivets, drill them out **(see illustration)**. Where the ECU is secured with shear bolts it may be possible to grip the remains of the bolt and unscrew them, if not they will need to be drilled out.

76 Remove the security cover then disconnect the wiring plugs. To disconnect the front plug, use a screwdriver to lever up the catch **(see illustrations)**.

5.64 Disconnect the camshaft position sensor

5.75 Drill out the pop rivets…

5.76a …then remove the security cover…

5.76b …and disconnect the front wiring plug by levering up the catch

5.77a Withdraw the electronic control unit...

5.77b ...and disconnect the rear wiring plug

77 Withdraw the electronic control unit from its location and disconnect the rear plug **(see illustrations)**.

78 Refitting is a reversal of removal. Bear in mind the comments made in the Caution above.

High-pressure fuel pump

Caution: The system must be depressurised before the pump is removed – see Section 9.

79 The high-pressure fuel pump is located on top of the camshaft housing. First, remove the engine top cover/air cleaner.

80 Set the engine to TDC as described in Chapter 2B Section 2.

81 Position cloth rags around the high-pressure fuel pump, then disconnect the

fuel lines as follows. Note that there are two versions of the fuel line connections, removal of either is similar.

82 Disconnect the wiring plugs and remove the wiring guide where fitted **(see illustrations)**.

83 Unscrew the mounting bolts, then remove the high-pressure fuel pump from the top of the camshaft housing **(see illustration)**. Recover the O-ring seal and discard it, as a new one must used on refitting.

84 Extract the bucket tappet/push rod from the housing.

85 Refitting is a reversal of removal, noting the following:

a) Clean the mating faces of the pump and camshaft housing.

b) Lubricate the bucket tappet/push rod

5.82a Disconnect the fuel pressure regulator wiring plug...

5.82b ...and the fuel pressure wiring plug

5.83 Remove the high-pressure fuel pump mounting bolts (one hidden)

6.1 Fuel filter location

with clean engine oil, then insert it in the camshaft housing.

c) Smear clean engine oil on the new O-ring seal, then refit the fuel pump together with the O-ring seal. Insert the mounting bolts and tighten to the specified torque.

d) Reconnect the wiring, then reconnect the fuel lines and tighten the union nut(s) and (where used) the banjo bolts to the specified torque. To ensure correct seating, the nut(s) and bolts must be progressively tightened to their specified torque.

e) Refit the engine top cover/air cleaner.

Fuel pressure sensors

86 Low and high fuel pressure sensors are fitted to all FSi and TSi engines. On all engines the low pressure sensor is fitted to the fuel supply line (before the pump) and the high pressure sensor is fitted to the fuel rail. Note that all versions also have a fuel pressure limiting valve fitted to the fuel line at the fuel rail.

87 Depressurise the fuel injection system as described in Section 9.

88 Disconnect the wiring from the sensor, then unscrew and remove it. Be prepared for some loss of fuel by positioning a suitable container beneath the sensor.

89 Refitting is a reversal of removal, but tighten the sensor to the specified torque.

---

6   Fuel filter – renewal

Note: *Observe the precautions in Section 1 before working on any component in the fuel system.*

Note: *On models fitted with a pressure regulator, the filter maybe supplied complete with a pressure regulator.*

⚠ *Warning: The fuel system must be de-pressurised before starting work as described in Section 9.*

1 The fuel filter is located in front of the fuel tank, on the right-hand underside of the car **(see illustration)**. Anticipate some fuel spillage before starting work and have a container and shop towels to hand.

2 Jack up the right-hand rear of the car, and support it on axle stands (see *Jacking and vehicle support*). When positioning the axle stand, ensure that it will not inhibit access to the filter.

3 To further improve access, the handbrake cable can be unhooked from the adjacent wire clip.

4 Disconnect the fuel hoses at each end of the filter, noting their locations for refitting. The connections are of quick-release type, disconnected by squeezing the catch on each. Where the new fuel filter was supplied with sealing plugs, fit these to the old filter as soon as each fuel line is removed **(see**

6.4 Disconnect the fuel hoses and seal the fuel filter

6.6a Remove the clamp screw...

6.6b ...and remove the fuel filter

illustration). This will minimise the loss of fuel.

5 Before removing the filter, look for an arrow marking, which points in the direction of fuel flow – in this case, towards the front of the car. The new filter must be fitted the same way round.

6 Remove the retaining clamp screw, and slide the filter out of position **(see illustrations)**. Dispose of the old filter carefully – even if the fuel inside is tipped out, the filter element will still be soaked in fuel, and will be highly flammable.

7 On models with an integral pressure regulator, unclip the regulator frpm the filter body and fit it to the new fuel filter.

8 Offer the new filter into position, ensuring that the direction-of-flow arrow is pointing towards the front of the car. Refit the clamp screw and tighten it securely.

9 Connect the fuel hoses to each end of the filter, in the same positions as noted on removal. Push the hoses fully onto the filter stubs, and if necessary, clip them back to the underside of the car. Hook the handbrake cable back in place, if it was disturbed.

10 Lower the car to the ground, then start the engine and check for signs of fuel leakage at both ends of the filter.

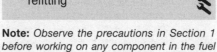

7 Fuel lift pump and gauge sender unit – removal and refitting

Note: *Observe the precautions in Section 1 before working on any component in the fuel system.*

⚠️ *Warning: Avoid direct skin contact with fuel – wear protective clothing and gloves when handling fuel system components. Ensure that the work area is well-ventilated to prevent the build-up of fuel vapour.*

General information

1 The fuel lift pump and gauge sender unit are combined in one assembly, which is mounted in the top of the fuel tank. The unit protrudes into the fuel tank, and its removal involves exposing the Specifications of the tank to the atmosphere.

Removal

2 Ensure that the vehicle is parked on a level surface, then disconnect the battery negative lead and position it away from the terminal as described in Chapter 5A Section 3.

3 Remove the rear seat cushion (Chapter 11

Section 22), and lift the carpet from the load space floor.

4 Unclip and disconnect the wiring connector block, then prise up the cover and disconnect the wiring from the top of the lift pump/gauge sender unit **(see illustrations)**.

5 Pad the area around the supply and return fuel hoses with rags to absorb any spilt fuel from the fuel lines, then squeeze the catches to release the hose clips and disconnect them **(see illustration)**. Observe the supply and return arrow markings on the ports – label the fuel hoses accordingly to ensure correct

7.4a Prise up the cover and...

7.4b ... disconnect the wiring from the pump/gauge unit

7.5 Fuel supply and return hoses

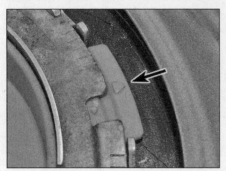

7.6a Note the alignment marks...

7.6b ...then use a suitable tool to unscrew...

7.6c ...and remove the securing ring

refitting later. The supply pipe is black, and may have white markings, while the return pipe is blue, or has blue markings.

6 Note the position of the alignment marks,

7.7 Removing the lift pump/gauge sender unit from the fuel tank

then unscrew and remove the securing ring. Use a pair of water pump pliers (or home-made tool) to grip and rotate the securing ring **(see illustrations)**.

7 Lift out the lift pump/gauge sender unit, holding it above the level of the fuel in the tank until the excess fuel has drained out. Recover the flange and seal **(see illustration)**.

8 With the pump/sender unit removed from the car, lay it on an absorbent card or rag **(see illustration)**. Inspect the float at the end of the sender unit swinging arm for punctures and fuel ingress – renew the unit if it appears damaged.

9 The fuel pick-up incorporated in the assembly is spring-loaded to ensure that it always draws fuel from the lowest part of the tank. Check that the pick-up is free to move under spring tension with respect to the sender unit body.

10 Inspect the rubber seal from the fuel tank aperture for signs of fatigue – renew it if necessary **(see illustration)**.

11 Inspect the sender unit wiper and track; clean off any dirt and debris that may have accumulated, and look for breaks in the track.

12 If required, the sender unit can be separated from the assembly, as follows. Disconnect the two small wires (note their positions), then remove the four screws and slide the unit downwards to remove **(see illustrations)**.

13 The unit top plate can be removed by releasing the plastic tags at either side; recover the large spring which fits onto a peg on the plate underside **(see illustrations)**.

Refitting

14 Refit the lift pump/sender unit by following

7.8 Lift pump/gauge sender unit removed from the car

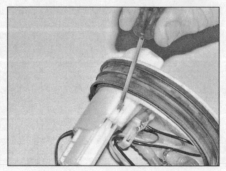

7.10 If not removed with the unit, recover the rubber seal and check its condition

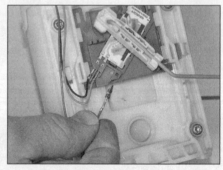

7.12a Disconnect the small wires...

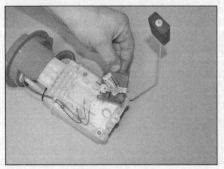

7.12b ...then undo the screws and slide out the sender unit

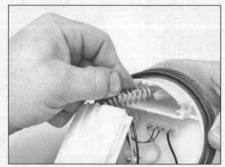

7.13a Prise up the retaining tags...

7.13b ...and recover the spring fitted under the top plate

7.14a Locate the seal in the tank aperture...

7.14b ...then fit the lift pump/sender unit

the removal procedure in reverse, noting the following points:

a) Take care not to bend the float arm as the unit is refitted.

b) Smear the outside of tank aperture rubber seal with clean fuel or lubricating spray, to ease fitting. Locate the seal in the tank aperture before fitting the lift pump/sender unit **(see illustrations)**.

c) The arrow markings on the sender unit body and the access aperture must be aligned.

d) Reconnect the fuel hoses to the correct ports – observe the direction-of-flow arrow markings, and refer to paragraph 5. Ensure that the fuel hose fittings click fully into place.

e) On completion, check that all associated pipes are securely clipped to the tank, then run the engine and check for fuel leaks.

8 Fuel tank – removal and refitting

Note: *Observe the precautions in Section 1 before working on any component in the fuel system.*

Removal

1 Before the tank can be removed, it must be drained of as much fuel as possible. As no drain plug is provided, it is preferable to carry out this operation with the tank almost empty.

2 Open the fuel filler flap, and unscrew the

8.10 Disconnect the fuel line at the filter

fuel filler cap – leave the cap loosely in place.

3 Disconnect the battery negative lead and position it away from the terminal as described in Chapter 5A Section 3. Using a hand pump or syphon, remove any remaining fuel from the bottom of the tank.

4 Loosen the right-hand rear wheel bolts, then jack up the rear of the car and remove the right-hand rear wheel.

5 Remove the right-hand rear wheel arch liner.

6 Gain access to the top of the fuel pump/sender unit as described in Section 7, and disconnect the wiring harness from the top of the pump/sender unit at the wiring plug.

7 Unscrew the fuel filler flap unit retaining screw (on the side opposite the flap hinge), and ease the flap unit out of position. Recover the rubber seal which fits around the filler neck.

8 Unbolt the filler pipe from the body.

9 Release the mounting rubbers and support the rear of the exhaust system to allow removal of the fuel tank.

10 Disconnect the fuel and charcoal canister breather lines as necessary **(see illustration)**.

11 Position a trolley jack under the centre of the tank. Insert a block of wood between the jack head and the tank to prevent damage to the tank surface. Raise the jack until it just takes the weight of the tank.

12 Unscrew the mounting bolt and detach the tank strap **(see illustration)**.

13 Lower the jack and tank away from the underside of the vehicle. If necessary,

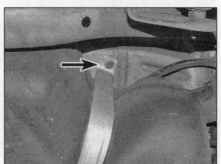

8.12 Remove the bolt

unscrew the nuts and remove the heat shield from the tank.

14 If the tank is contaminated with sediment or water, remove the fuel pump/sender unit (see Section 7) and swill the tank out with clean fuel. The tank is injection-moulded from a synthetic material, and if damaged, it should be renewed. However, in certain cases it may be possible to have small leaks or minor damage repaired. Seek the advice of a suitable specialist before attempting to repair the fuel tank.

Refitting

15 Refitting is the reverse of the removal procedure, noting the following points:

a) When lifting the tank back into position, make sure the mounting rubbers are correctly positioned, and take care to ensure none of the hoses get trapped between the tank and vehicle body.

b) Ensure that all pipes and hoses are correctly routed, are not kinked, and are securely held in position with their retaining clips.

c) Tighten the tank strap retaining bolt to the specified torque.

d) On completion, refill the tank with fuel, and exhaustively check for signs of leakage prior to taking the vehicle out on the road.

9 Fuel injection system – depressurisation

Note: *Observe the precautions in Section 1 before working on any component in the fuel system.*

⚠ **Warning: The following procedure will merely relieve the pressure in the fuel system – remember that fuel will still be present in the system components and take precautions accordingly before disconnecting any of them.**

⚠ **Warning: Models fitted with the high pressure fuel injection systems (TFSi and FSi engines) can run up to 120 bar on the high side of the system and up to 6 bar on the low side. Exercise extreme caution when working with these systems.**

Warning: Do not work on any aspect of the fuel system with a hot engine.

1 The fuel system referred to in this Section is defined as the tank-mounted fuel pump, the fuel filter, the fuel injectors, the fuel pressure regulator, and the metal pipes and flexible hoses of the fuel lines between these components. All these contain fuel, which will be under pressure while the engine is running and/or while the ignition is switched on. The pressure will remain for some time after the ignition has been switched off, and must be relieved before any of these components are disturbed for servicing work. Ideally, the engine should be allowed to cool completely before work commences.

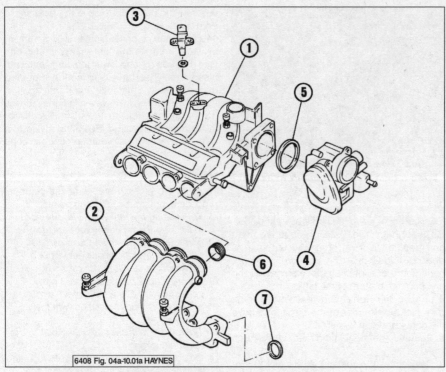

10.1a The two part 1.6 litre inlet manifold

1 Inlet manifold upper section	3 Pressure sensor (with air temperature sensor)	5 Seal (replace if damaged)
2 Inlet manifold lower section	4 Throttle body	6 Seal (replaced if damaged)
		7 Seal (replace)

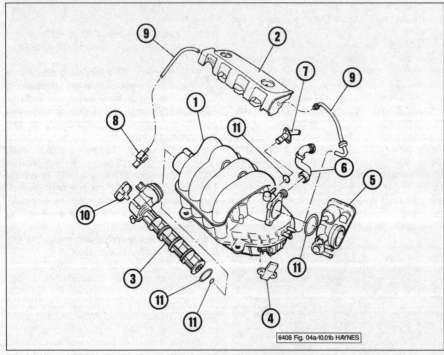

10.1b The 2.0 litre FSi manifold

1 Inlet manifold	4 Pressure sensor (with air temperature sensor)	7 EGR pipe
2 Vacuum reservoir	5 Throttle body	8 Control valve
3 Variable inlet manifold runner	6 Crankcase breather hose	9 Vacuum hose
		10 End cover
		11 Seal (replace if damaged)

2 The fuel pump will run and prime the system as soon as a door is opened. To prevent this, on all engines, identify and remove the fuel pump fuse (or relay). Refer to Chapter 12 Section 3.

3 On all engines with the Bosch high-pressure system (FSi and TFSi engines) disconnect the wiring from the fuel pressure regulating valve on the high-pressure pump, then run the engine at idling speed for 10 seconds. **Note:** *With the engine switched off, disconnect the fuel lines as quickly as possible, as the internal fuel pressure will rise due to heat from the engine.*

4 On 1.6 litre engines (with the fuel pump disabled) crank the engine for about ten seconds. The engine may fire and run for a while, but let it continue running until it stops. The fuel injectors should have opened enough times during cranking to considerably reduce the line fuel pressure, and reduce the risk of fuel spraying out when a fuel line is disturbed.

5 Disconnect the battery negative lead and position it away from the terminal as described in Chapter 5A Section 3.

6 Place a suitable container beneath the relevant connection/union to be disconnected, and have a large rag ready to soak up any escaping fuel not being caught by the container.

7 Slowly open the connection to avoid a sudden release of pressure, and position the rag around the connection to catch any fuel spray which may be expelled. Once the pressure has been released, disconnect the fuel line. Insert plugs to minimise fuel loss and prevent the entry of dirt into the fuel system.

8 After working on the fuel system, reconnect the wiring/fuse/relay as applicable.

10 Inlet manifold and associated components – removal and refitting

Note: *Observe the precautions in Section 1 before working on any component in the fuel system.*

1 The design of the inlet manifold varies considerably depending on engine type **(see illustrations)**.

1.6 litre engine (two-piece manifold – upper section)

Note: *It is possible to remove the manifold as a complete assembly if required.*

Removal

2 Where fitted, remove the engine cover and then disconnect the battery negative lead. Position it away from the terminal as described in Chapter 5A Section 3.

3 Disconnect the air inlet duct and remove it as described in Section 3.

4 With reference to Section 5 remove the

10.4 Remove the throttle body

10.5 Unclip and disconnect the EVAP control solenoid and hoses

throttle body from the inlet manifold (see illustration).

5 Release the hose clip and disconnect the brake servo vacuum hose and the EVAP lines from the manifold (see illustration).

6 On models fitted with a secondary air injection system, disconnect the wiring plug and hoses. If the upper section is being removed prior to removing the lower section, remove the air pump as described in Chapter 4C Section 5.

7 Disconnect and unclip all wiring, coolant and vacuum hoses from the manifold (as applicable), noting their location for correct refitting (see illustrations). Use clamps on the coolant hoses to minimise spillage.

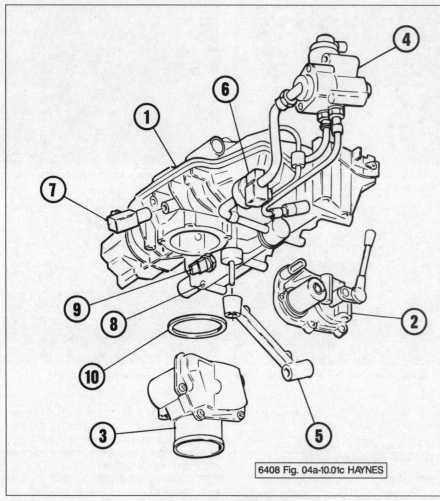

6408 Fig. 04a-10.01c HAYNES

10.1c The 2.0 litre TFSi manifold

1 Inlet manifold
2 Inlet manifold flap control motor
3 Throttle body
4 High pressure pump (with regulator)
5 Support strut
6 Control valve (charcoal filter)
7 Air temperature sensor
8 Fuel rail
9 Fuel pressure sensor
10 Seal

10.7a Disconnect the wiring plug and vacuum lines from the manifold runner control

10.7b Disconnect the wiring plug from the manifold pressure sensor

10.9a Remove the manifold to support bracket screws (at both sides)

10.9b The upper manifold mounting bracket (shown with manifold removed)

10.9c Remove the screws (at both ends)...

10.9d ...and at the centre

10.10 Removing the upper part of the inlet manifold

10.14 Remove the fuel line

8 As applicable, unbolt the support brackets from the manifold.

9 Remove the screws securing the upper part of the manifold to the lower part **(see illustrations)**. Where necessary, prise out the plastic flanges.

10 Lift the upper part of the manifold off the lower part, and remove it from the engine compartment. Recover the seals **(see illustration)**.

Refitting

11 Refitting is a reversal of removal. Use new seals, gaskets or ducts as necessary, and tighten the upper-to-lower part bolts to

the specified torque. It is most important that there are no air leaks at the joint.

1.6 litre engine (two-piece manifold – lower part)

Removal

Note: *The lower section is removed complete with the fuel rail and injectors.*

12 Remove the upper part of the manifold as described previously in this Section.

13 Where fitted, remove the secondary air injection pump as described in Chapter 4C Section 5.

14 Disconnect the wiring from the fuel pressure sensor and fuel injectors. Remove the fuel line(s)

(see illustration). Note that the fuel line can remain connected if the manifold is only being removed for access to the spark plugs.

15 Unclip the wiring loom conduit from the manifold **(see illustration)**.

16 Check around the manifold, and unclip any hoses or wiring which may still be attached.

17 Progressively loosen the nuts and bolts and withdraw the manifold from the cylinder head. With the manifold partially removed, disconnect the wiring plugs from the injectors and move the wiring loom to the side. Recover the seals/gasket(s) – all should be renewed when refitting the manifold **(see illustration)**.

10.15 Release the wiring loom

10.17 Removing the lower part of inlet manifold

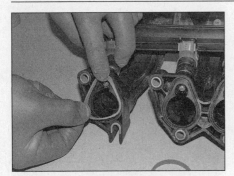

10.18a Fit new seals to the cylinder head mounting face...

10.18b ...and to the upper section mounting face

Refitting

18 Refitting is a reversal of removal. Use new seals or gaskets as necessary **(see illustrations)** and tighten the manifold-to-head nuts and bolts to the specified torque. It is most important that there are no air leaks at the joint.

One-piece manifold (2.0 litre FSi models)

Removal

19 where fitted, remove the engine cover and then disconnect the battery negative lead. Position it away from the terminal (see Chapter 5A Section 3).
20 Disconnect and remove the vacuum reservoir from the top of the manifold.
21 Unplug and disconnect the wiring plugs from the inlet manifold runner control valve, the air temperature sensor, the manifold pressure sensor and the throttle body.
22 Remove the air filter housing as described in Section 3.
23 Where fitted disconnect the EGR hose and on models with engine codes BLR and BLY, fit hose clamps and then remove the coolant hoses.
24 With reference to Section 5, remove the throttle housing/module from the inlet manifold.
25 Disconnect the vacuum hoses for the fuel pressure regulator, and (if not already removed) for the brake servo. Note how the hoses are routed, for use when refitting.
26 Remove the dipstick and then disconnect the crankcase breather hose.
27 Release the spring clips at the cylinder head and then remove the lower mounting bolts from the support bracket.
28 Lift out the manifold (towards the battery) and remove it.

Refitting

29 Refitting is a reversal of removal.

One piece manifold (2.0 litre TFSi models)

Removal

30 Depressurise the fuel system (Section 9) and disconnect the battery (Chapter 5A Section 3).
31 Disconnect and remove the fuel lines form the high pressure pump. Anticipate some fluid spillage.
32 Disconnect the breather hose **(see illustration)**.
33 Disconnect the wiring plugs from the fuel pressure sensor, the high pressure pump, the throttle body and the charge air pipe pressure sensor **(see illustration)**.
34 Partially drain the coolant and then disconnect the coolant hose.
35 Remove the dipstick and then unbolt the guide tube from the manifold.
36 Where fitted remove the turbocharger control valve and then remove the throttle body inlet duct.
37 Remove the dipstick guide tube lower mounting and then remove the tube. Unbolt and remove the throttle body (Section 5). Recover the O-ring seal.
38 Unbolt and remove the support strut from the bottom of the manifold and then remove the manifold to head bolts. Lift of the manifold and recover the seals.

Refitting

39 Refitting is a reversal of removal. Use a new gasket or seals, as applicable, and tighten the retaining bolts/nuts to the specified torque. It is most important that there are no air leaks at the joint.

11 Fuel injection system – testing and adjustment

1 If a fault appears in the fuel injection system, first ensure that all the system wiring connectors are securely connected and free of corrosion. Then ensure that the fault is not due to poor maintenance; ie, check that the air cleaner filter element is clean, the spark plugs are in good condition and correctly gapped, the cylinder compression pressures are correct, the ignition system wiring is in good condition and securely connected, and the engine breather hoses are clear and undamaged, referring to Chapter 1A, Chapter 2A, Chapter 2B, Chapter 5B or Chapter 4C.
2 If these checks fail to reveal the cause of the problem, the vehicle should be taken

10.32 Remove the breather hose

10.33 Disconnect the throttle body wiring plug

11.2 The diagnostic socket (DLC – Data Link Connector

to a Seat dealer or suitably equipped garage for testing. A diagnostic connector is incorporated in the engine management system wiring harness, into which dedicated electronic test equipment can be plugged **(see illustration)**. The test equipment is capable of 'interrogating' the engine management system ECU electronically and accessing its internal fault log (reading fault codes).

3 Fault codes can only be extracted from the ECU using a dedicated fault code reader. A Seat dealer will obviously have the VAG approved tester, but many independent garages will have invested in the equipment (or at least near dealer level test equipment). Low cost diagnostic testers are also available in the aftermarket and whilst these will not have the depth of coverage of the factory tester (or other professional equipment) they will at least display the mandatory EOBD (European On Board Diagnostic) emissions related fault codes. Some aftermarket equipment may also be capable of interrogating other on board systems, such as the ABS for example.

4 Using this equipment, faults can be isolated. However just because a fault code for a misfire (for example) has been logged, some basic system testing will still be required to isolate the reason for the misfire.

Chapter 4 Part B
Diesel engine fuel systems

Contents

Degrees of difficulty

| **Easy,** suitable for novice with little experience | **Fairly easy,** suitable for beginner with some experience | **Fairly difficult,** suitable for competent DIY mechanic | **Difficult,** suitable for experienced DIY mechanic | **Very difficult,** suitable for expert DIY or professional |

Specifications

Engine codes*

1.9 litre – 'PD' (unit injector)	BJB, BKC, BLS, BXE and BXF
2.0 litre – 'PD' (unit injector)	BMM, BKD and BMN
1.6 litre – common rail injection	CAYC
2.0 litre – common rail injection	CFJA, CLCB, CEGA and CFHC

See 'Vehicle identification' at the end of the manual for the location of the engine code markings

General

Fuel injection system	Electronic, direct, unit injectors or common rail injection
Firing order...	1-3-4-2
Maximum engine speed......................................	N/A (ECU controlled)
Engine fast idle speed	N/A (ECU controlled)

Tandem pump (PD Unit injector engines only)

Fuel pressure at 1500 rpm	3.5 bar

Turbocharger

Type	Garrett or KKK

Torque wrench settings

	Nm	lbf ft
Camshaft position sensor	10	7
Common fuel rail	15	11
EGR pipe flange bolts	25	18
EGR valve mounting bolt:		
CR engines	20	15
PD engines except BDJ and BST engine codes	22	16
PD engines BDJ and BST	20	15
Engine speed/TDC sender	5	4
Flap motor housing	10	7
Injector clamp/mounting:		
1.9 litre PD unit injector engines*:		
Stage 1	12	9
Stage 2	Angle-tighten a further 270°	
2.0 litre PD unit injector engines*:		
Stage 1	3	2
Stage 2	Angle-tighten a further 90°	
Stage 3	Angle-tighten a further 180°	
Common rail injector engines:		
Clamp nut	10	7
Cover bolt	5	4
Inlet manifold to cylinder head:		
PD engines*	22	16
CR engines	8	6
Pump injector rocker shaft bolts*:		
Stage 1	20	15
Stage 2	Angle-tighten a further 90°	
Tandem pump bolts:		
Upper	20	15
Lower	10	7

* Use new fixings

1 General information and precautions

General information

1 All engines covered by this Manual are fitted with a direct-injection fuelling system, incorporating a fuel tank, an engine-bay mounted fuel filter with an integral water separator, fuel supply and return lines and four fuel injectors. Models with the 1.9 or 2.0 litre PD Unit injector engines, also have a fuel cooler mounted beneath the right-hand side of the car, on the underbody.

2 However, two different types of fuel injection system are fitted. The first is first is known as the 'PD' (Pumpe Duse) or Unit injector system, where fuel is delivered by a camshaft driven 'tandem pump' at low pressure to the injectors (known as 'Unit injectors'). A 'roller rocker' assembly, mounted above the camshaft bearing caps, uses an extra set of camshaft lobes to compress the top of each injector once per firing cycle. This arrangement creates high injection pressures. The precise timing of the pre-injection and main injection is controlled by the engine management ECM and a solenoid on each injector.

3 The second type is the familiar Common Rail (CR) system, where fuel is supplied from a timing belt-driven high-pressure pump to a common fuel rail (or reservoir). The four injectors are fitted into the cylinder head and are connected to the fuel rail by rigid metal pipes. The precise timing of the pre-, main, and post-injections are controlled by the engine management ECM and an electrically operated Piezo crystal incorporated into the injector design. All engines are fitted with a turbocharger.

4 The direct-injection fuelling system is controlled electronically by a diesel engine management system, comprising an Electronic Control Module (ECM) and its associated sensors, actuators and wiring. In addition, the ECM manages the operation of the Exhaust Gas Recirculation (EGR) emission control system (Chapter 4D Section 3), the turbocharger boost pressure control system (on models fitted with a turbocharger) and the glow plug control system (Chapter 5C Section 1).

5 A flap valve/throttle valve module fitted to the intake manifold is closed by the ECM for 3 seconds as the engine is switched off, to minimise the air intake as the engine shuts down. This minimises the vibration felt as the pistons come up against the volume of highly compressed air present in the combustion chambers.

6 It should be noted that fault diagnosis of the diesel engine management system is only possible with dedicated electronic test equipment. Problems with the system's operation should therefore be referred to a Seat dealer or suitably-equipped specialist for assessment. Once the fault has been identified, the removal/refitting sequences detailed in the following Sections will then allow the appropriate component(s) to be renewed as required.

7 The EOBD diagnostic connector is located under the drivers side of the facia **(see illustration)**.

Precautions

8 Many of the operations described in this Chapter involve the disconnection of fuel lines, which may cause an amount of fuel spillage. Before commencing work, refer to the warnings below and the information in *Safety First!* at the beginning of this manual.

⚠ *Warning: When working on any part of the fuel system, avoid direct contact skin contact with diesel fuel – wear protective clothing and gloves when handling fuel system components. Ensure that the work area is well-ventilated to prevent the build-up of diesel fuel vapour.*

• **Fuel injectors operate at extremely high**

1.7 The EOBD diagnostic connector is located under the drivers side of the facia

pressures and the jet of fuel produced at the nozzle is capable of piercing skin, with potentially fatal results. When working with pressurised injectors, take care to avoid exposing any part of the body to the fuel spray. It is recommended that a diesel fuel systems specialist should carry out any pressure testing of the fuel system components.

• Under no circumstances should diesel fuel be allowed to come into contact with coolant hoses – wipe off accidental spillage immediately. Hoses that have been contaminated with fuel for an extended period should be renewed.

• Diesel fuel systems are particularly sensitive to contamination from dirt, air and water. Pay particular attention to cleanliness when working on any part of the fuel system, to prevent the ingress of dirt. Thoroughly clean the area around fuel unions before disconnecting them. Only use lint-free cloths and clean fuel for component cleansing.

• Store dismantled components in sealed containers to prevent contamination and the formation of condensation.

2 Fuel pipe and connectors

1 Disconnect the cable from the negative battery terminal (see Chapter 5A Section 3) before proceeding.
2 The fuel supply pipe connects the fuel pump in the fuel tank to the fuel filter on the engine.
3 Whenever you're working under the vehicle, be sure to inspect all fuel and evaporative emission pipes for leaks, kinks, dents and other damage. Always replace a damaged fuel pipe immediately.
4 If you find signs of dirt in the pipes during disassembly, disconnect all pipes and blow them out with compressed air. Inspect the fuel strainer on the fuel pump pick-up unit for damage and deterioration.

Steel tubing

5 It is critical that the fuel pipes be replaced with pipes of equivalent type and specification.
6 Some steel fuel pipes have threaded fittings. When loosening these fittings, hold the stationary fitting with a spanner while turning the union nut.

Plastic tubing

⚠️ **Warning: When removing or installing plastic fuel tubing, be careful not to bend or twist it too much, which can damage it. Also, plastic fuel tubing is NOT heat resistant, so keep it away from excessive heat.**

7 When replacing fuel system plastic tubing, use only original equipment replacement plastic tubing.

Flexible hoses

8 When replacing fuel system flexible hoses, use original equipment replacements, or hose to the same specification.

9 Don't route fuel hoses (or metal pipes) within 100 mm of the exhaust system or within 280 mm of the catalytic converter. Make sure that no rubber hoses are installed directly against the vehicle, particularly in places where there is any vibration. If allowed to touch some vibrating part of the vehicle, a hose can easily become chafed and it might start leaking. A good rule of thumb is to maintain a minimum of 8.0 mm clearance around a hose (or metal pipe) to prevent contact with the vehicle underbody.

Disconnecting Fuel pipe Fittings

10 Typical fuel pipe fittings see Chapter 4A Section 2.

3 Air filter assembly – removal and refitting

Note: There are variations in the procedure for removing the air filter assembly, however the removal procedure is essentially the same for all models.

Removal

1 Unclip the plastic cover from the air intake ducting, then release the two securing clips (one at each side) and remove the ducting from the front crossmember (see illustrations). Note, on some models there may be a retaining screw securing the inner part of the lower ducting to the crossmember.
2 Disconnect the wiring plug from the air mass meter (see illustration).
3 Disconnect the vacuum pipe from the air intake hose (see illustrations).

3.1a Unclip the cover from the air intake

3.1b Release the securing clips …

3.1c …and remove the air ducting

3.2 Disconnect the air mass meter wiring plug (common rail shown)

3.3a Disconnecting the vacuum pipe on common rail engines

3.3b The vacuum pipe on PD engines

3.4 Disconnect the air intake hose

3.5a Undo the retaining screw …

3.5b … and remove the air filter assembly out of the engine compartment

3.5c Release the drain tube as it is removed

3.6 Make sure the assembly is located correctly on the mounting pegs

4 Loosen the clip and disconnect the air duct from the air mass meter **(see illustration)**.

5 Undo the retaining bolt and lift the air cleaner assembly upwards to release it from the locating pegs **(see illustrations)**. As the air cleaner is removed, withdraw the water drain hose, under the air filter assembly out from the inner wing panel.

Refitting

6 Refit the air filter assembly by following the removal procedure in reverse. Make sure the assembly is located correctly on the lower mounting pegs **(see illustration)**.

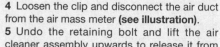

4 Diesel engine management system – component removal and refitting

Throttle pedal/position sensor

1 Remove the footrest panel by removing the plastic screws. Lift out the panel and then. Release the securing clip and disconnect the wiring connector from the top of the accelerator pedal **(see illustrations)**.

2 Prise off the cap, and undo the screw securing the throttle pedal to the bulkhead **(see illustration)**.

3 The throttle pedal is clipped to two metal pegs in the floor panel, using a lever and thin screwdriver, release the securing clips **(see illustration)**.

4.1a Remove the footrest panel

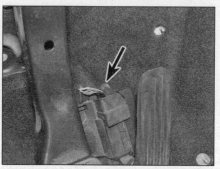

4.1b Release the wiring connector locking clip

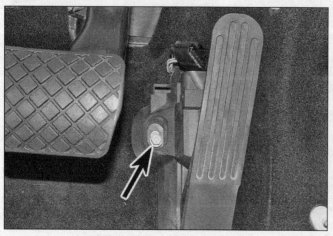

4.2 Undo the pedal retaining bolt

4.3 Throttle pedal securing pegs in the floor panel

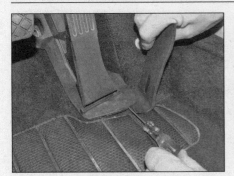

4.4a Release the outer clip by pressing the screwdriver forwards

4.4b Release the inner clip by pushing the screwdriver inwards, and then to the left to release the clip

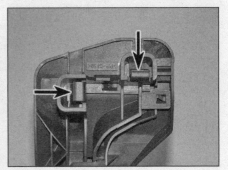

4.4c Direction to release the two securing clips

4 The outer clip can be released by pressing the screwdriver inwards to release the securing clip, whilst levering the pedal upwards. Then insert the screwdriver into the inner slot and push it to the left to release the securing clip **(see illustrations)**. Withdraw the throttle pedal from the floor panel and out from the vehicle.

5 Refitting is a reversal of removal.

Throttle valve housing/module (Common rail engines)

6 Where fitted, remove the engine cover **(see illustration)**.

7 Slacken the retaining clips and remove the air intake rubber hose from the throttle housing **(see illustration)**. If required, undo the bolts securing the air intake plastic hose from the intercooler to the throttle housing, this will allow easier removal of the intake rubber hose.

8 Disconnect the wiring plug connector, from the throttle housing/module **(see illustration)**.

9 Undo the retaining bolt and disconnect the dipstick guide tube from the throttle housing **(see illustration)**.

10 Unscrew and remove the retaining bolts, then lift the throttle housing/module away from the inlet manifold **(see illustrations)**. Recover the O-ring seal; a new one will be required for refitting.

11 Refitting is a reversal of removal, noting the following:

a) Use a new throttle housing-to-inlet manifold seal **(see illustration)**.

4.6 Remove the engine cover

4.7 Slacken the two retaining clips

4.8 Disconnect the wiring plug

4.9 Undo the dipstick tube retaining bolt

b) Tighten the throttle housing bolts evenly to the specified torque.

c) Ensure that all hoses and electrical connectors are refitted securely.

Inlet manifold flap motor/ housing (PD engines)

Note: On some models the inlet manifold flap housing is vacuum-operated and is

4.10a Undo the three retaining bolts ...

4.10b ... and remove the throttle housing module

4.11 Fit a new seal to the housing on refitting

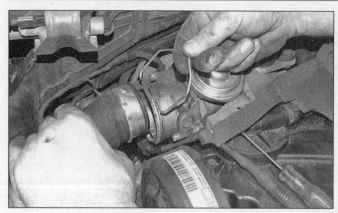

4.13a Release the spring clip and...

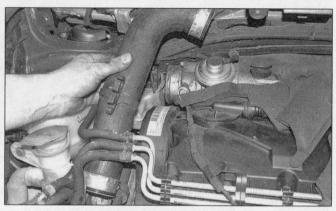

4.13b ...remove the ducting

incorporated into the EGR valve, whereas other engine codes have a separate electrically-operated flap motor.

Removal

12 Remove the engine cover.
13 Loosen the clip (or release the spring clip) and disconnect the air trunking from the flap housing **(see illustrations)**.
14 Where fitted disconnect the flap control motor wiring plug from the housing.
15 Disconnect the EGR vacuum hose from the top of the housing and where required unbolt the EGR valve pipe and recover the gasket.
16 Unscrew the bolts **(see illustration)** securing the flap housing to the EGR valve/housing/manifold, remove the flap housing and recover the O-ring seal.

4.16 Remove the bolts

4.25 The charge air pressure sensor (common rail engine shown)

Refitting

17 Refitting is a reversal of removal. Renew the O-ring seal if it appears damaged.

Coolant temperature sensors

18 Refer to Chapter 3 Section 6.

Fuel temperature sensor

Removal

19 The fuel temperature sensor is located in the fuel return line to the fuel filter in the engine compartment on PD unit injection engines or in the fuel supply line at the top of the high-pressure fuel pump on common rail injection engines. First, disconnect the wiring.
20 On PD unit injection engines, remove the securing clip, then extract the sensor from

4.21 Fuel temperature sensor – common rail engines

4.27 Crankshaft position speed sensor

the fuel line housing and recover the O-ring seal.
21 On common rail injection engines, release the clamps and disconnect the hoses from the sensor assembly **(see illustration)**.

Refitting

22 Refit the fuel temperature sensor by reversing the removal procedure, using a new O-ring seal (where applicable).

Inlet air temperature sensor
PD unit injection engines only

23 All models have an air temperature sensor built into the air mass meter – this sensor is an integral part of the air mass meter, and cannot be renewed separately.

Charge air pressure/temperature sensor
Removal

24 The charge air pressure and temperature sensor is fitted either on the intercooler inlet elbow (engine codes BMM and BLS), or on the air hose from the intercooler to the inlet manifold (engine codes BKC and BJB). On these engines access to the sensor is from above. On all common rail engines the sensor is fitted to the lower hose. On these engines, jack up the front of the vehicle and support it on axle stands (see *Jacking and vehicle support*), then remove the engine undertray.
25 Disconnect the wiring then undo the screw and remove the sensor from its location **(see illustration)**.

Refitting

26 Refit the sensor by reversing the removal procedure, using a new O-ring seal.

Engine speed/TDC sensor

Note: *Access to the sensor is limited. Removing the oil cooler hoses and coolant hoses improves access. Note that the illustrations show the sensor position with the transmission removed for clarity.*
27 The engine speed/TDC sensor is mounted on the front cylinder block, adjacent to the mating surface of the block and transmission bellhousing **(see illustration)**.

4.29a Undo the retaining bolt ...

4.29b ... and withdraw the sensor

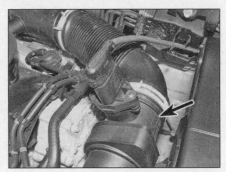

4.31 Air mass meter location on engine code BKC

28 Access is from beneath the engine compartment. Apply the handbrake, and then jack up the front of the vehicle and support it on axle stands (see *Jacking and vehicle support*). Remove the engine undertray.
29 Remove the retaining screw and withdraw the sensor from the cylinder block **(see illustrations)**.
30 Refit the sensor by reversing the removal procedure.

Air mass meter

Removal
31 On all models the air mass meter is located in the outlet from the air cleaner assembly on the left-hand side of the engine compartment **(see illustration)**.
32 Disconnect the wiring plug, loosen the clips and disconnect the air ducting from the air mass meter **(see illustrations)**.
33 On common rail engines remove the screws and remove the meter and housing from the air filter housing. On PD engines remove the screws and lift the meter from the housing. Note however that replacement air mass meters are often supplied complete with the housing to avoid damage in transit.
Caution: Handle the air mass meter carefully – its internal components are easily damaged.

Refitting
34 Refitting is a reversal of removal. Renew the O-ring seal if it appears damaged.

Absolute pressure (altitude) sensor
35 The absolute pressure sensor is an

integral part of the ECU, and hence cannot be renewed separately.

Clutch pedal switch

Removal
36 The clutch pedal switch is clipped to the clutch master cylinder on the pedal bracket. Remove the master cylinder as described in Chapter 6 Section 4.
37 Unclip the pedal switch from the bottom of the master cylinder.

Refitting
38 Refitting is a reversal of removal.

Electronic control unit (ECU)
Caution: Always wait at least 30 seconds after switching off the ignition before disconnecting the wiring from the ECU. When the wiring is disconnected, all the learned values are erased, however any

4.32a Disconnecting the wiring plug on PD engines and..

Specifications of the fault memory are retained. Note that if the ECU is renewed, the identification of the new ECU must be transferred to the immobiliser control unit by a VW dealer or suitably equipped garage.

Removal
39 The ECU is located centrally behind the engine compartment bulkhead, in the plenum chamber. Remove the wiper arms and cowl panel and front section of the plenum chamber as for windscreen wiper motor removal and refitting (as described in Chapter 12 Section 19).
40 Where the cover is secured with pop-rivets or shear bolts, drill them out **(see illustration)**. Where it is secured with bolts, unscrew and remove them.
41 Remove the security cover then unbolt the ECU **(see illustration)**.

4.32b ...on common rail engines

4.32c Release the spring clip

4.40 Sometimes it is possible to remove the shear bolts with pliers

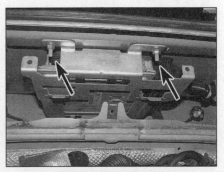

4.41 Remove the bolts

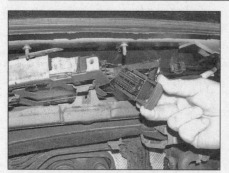

4.42 Disconnect the wiring plugs by levering up the catch

4.44 The rear locating tab

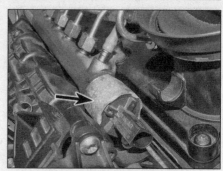

4.45 Fuel pressure regulator valve

42 Pull the ECU forward, rotate it slightly and disconnect the wiring plugs **(see illustration)**.
43 Remove the ECU.

Refitting

44 Refitting is a reversal of removal. Make sure that the locating tab is correctly engaged **(see illustration)** before refitting the mounting bolts. Fit new shear bolts

Fuel pressure regulating valve (common rail engines)

45 The fuel pressure regulating valve is fitted to the left-hand end of the fuel rail **(see illustration)**. If the valve is removed from the fuel rail, then it will need to be renewed, as it has a deformable sealing lip as part of the valve.
46 To check the operation of the regulating valve, first disconnect the fuel return hose from the fuel rail and plug the end **(see illustration)**. Then fit a piece of hose to the

fuel rail and the other end into a container. There are three checks that can be made, the first two with the engine running and the third if the vehicle will not start:

a) *Start the engine and run at idle for 30 seconds, there should be approx. 75ml of fuel (1.6 litre engines) or 100ml of fuel (2.0 litre engines) in the container.*
b) *Start engine and increase engine speed to 2000rpm, there should be 0ml of fuel in the container (allow for a few droplets of fuel).*
c) *On vehicles that will not run, turn the ignition key and crank the engine, there should be 0ml of fuel in the container (allow for a few droplets of fuel).*

47 If any of these readings are not attained, renew the regulating valve.
48 To renew the valve, remove the fuel rail as described in Section 12.
49 Clean around the valve, then slacken the valve from the end of the fuel rail **(see**

illustration) ; counterhold the fuel rail using the flats on the housing. Plug the end of the rail to prevent dirt from entering.
50 Fit the new valve by reversing the removal procedure, making sure that the threads are all clean before fitting. Check the deformable seal on the new valve, before fitting, to check it is not damaged. Apply a small amount of Molybdenum grease to seal and threads.

Fuel pressure sender (common rail engines)

51 The fuel pressure sender is fitted to the right-hand end of the fuel rail. If the engine will not start, disconnect the fuel pressure sender wiring connector and see if the engine will start. If the engine starts, the fuel pressure sender is faulty. With the connector removed a value is taken from the control unit, so that the engine will start, in this mode the maximum engine speed is limited to 3000rpm.
52 To renew the sender, first disconnect the wiring plug connector **(see illustration)**.
53 Clean around the sender, then slacken it from the end of the fuel rail **(see illustration)** ; plug the end of the rail to prevent dirt from entering.
54 Refit the pressure sender by reversing the removal procedure, making sure that the threads are all clean before refitting. The sender has a deformable seal, check for damage. Keep the threads free of oil and grease.

Camshaft position sensor

55 The camshaft position sensor (Hall sender) is located behind the timing belt cover, below the camshaft sprocket **(see illustration)**.

4.46 Disconnect the fuel hose

4.49 Pressure regulator valve fitted to the left-hand end of the fuel rail

4.52 Disconnect the wiring connector from the fuel pressure sender

4.53 The fuel pressure sender is fitted to the right-hand end of the fuel rail

4.55 Camshaft position sensor

4.57 Disconnect the sensor wiring connector

4.58 Unbolt the Idler pulley

4.59 Prise up the aperture cover

56 Remove the timing belt, as described in Chapter 2C Section 7 (1.6 litre common rail engine) Chapter 2D Section 7 (2.0 litre common rail engine) or Chapter 2E Section 7 (1.9 and 2.0 litre PD engines).

57 Disconnect the sensor wiring plug connector, located at the rear of the oil filter housing against the cylinder block **(see illustration)**.

58 To make access easier undo the retaining bolt and remove the timing belt idler pulley **(see illustration)**.

59 Using a screwdriver prise out the aperture cover in the rear plastic cover, then withdraw the wiring plug through the cover, unhooking it from the rear cover **(see illustration)**.

60 Undo the retaining bolt and remove the camshaft sensor from the cylinder head.

61 Refitting is a reversal of removal, but tighten the bolt to the specified torque setting and fit rubber plugs to the aperture for the wiring in the rear plastic cover.

---

5 Injectors – general information, removal and refitting

⚠ *Warning: Exercise extreme caution when working on the fuel injectors. Never expose the hands or any part of the body to injector spray, as the high pressure can cause the*

fuel to penetrate the skin, with possibly fatal results. You are strongly advised to have any work which involves testing the injectors under pressure carried out by a dealer or fuel injection specialist. Refer to the precautions given in Section 1 of this Chapter before proceeding.

General information

1 Injectors do deteriorate with prolonged use, and it is reasonable to expect them to need reconditioning or renewal after 60 000 miles (100 000 km) or so. Accurate testing, overhaul and calibration of the injectors must be left to a specialist.

Note: *Take care not to allow dirt into the injectors or fuel pipes during this procedure. Do not drop the injectors or allow the needles at their tips to become damaged. The injectors are precision-made to fine limits, and must not be handled roughly. Keep the injectors identified for position to ensure correct refitting. Always keep the injectors upright at all times where possible. Store them securely (see illustration)*

PD Engines

Removal

2 With reference to Chapter 2E Section 6 and Chapter 2E Section 4, remove the upper timing belt cover and camshaft cover.

3 Using a spanner or socket, turn the crankshaft pulley until the rocker arm for the injector which is to be removed, is at its highest, ie, the injector plunger spring is under the least amount of tension.

4 Slacken the locknut of the adjustment screw on the end of the rocker arm above the injector, and undo the adjustment screw until the rocker arm lies against the plunger pin of the injector **(see illustration)**.

5 Starting at the outside and working in, gradually and evenly slacken and remove the rocker shaft retaining bolts. Lift off the rocker shaft. Check the contact face of each adjustment screw, and renew any that show signs of wear.

6 Undo the clamping block securing bolt and remove the block from the side of the injector **(see illustration)**.

7 Using a small screwdriver, carefully prise the wiring connector from the injector.

8 Seat technicians use a slide hammer (VAG special tool T10055) to pull the injector from the cylinder head. This is a slide hammer which engages in the side of the injector. If this tool is not available, it is possible to fabricate an equivalent using a short section of angle-iron, a length of threaded rod, a cylindrical weight, and two locknuts. Weld/braze the rod to the angle-iron, slide the weight over the rod, and lock the two nuts together at the end of the rod to provide the

5.1 A storage rack is simple to construct

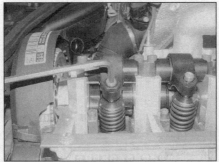

5.4 Undo the adjustment screw until the rocker arm lies against the plunger pin of the injector

5.6 Remove the clamping block securing bolt

a 5 mm
b 15 mm
c 25 mm
d Weld/braze
   the rod to the
   angle-iron
e Threaded rod
f Cylindrical
   weight
g Locknut

H32626

5.8a Unit injector removal tool

5.8b Seat the slide hammer/tool in the slot on the side on the injector, and pull the injector out

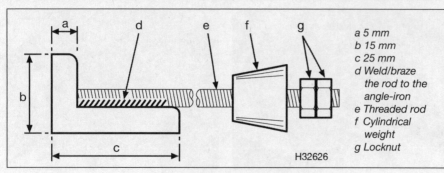

5.9a Undo the two nuts at the back of the head and slide the injector loom/rail out

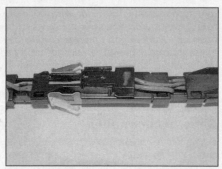

5.9b The injector connectors will slide into the loom/rail to prevent them from being damaged as the assembly iswithdrawn/ inserted into the cylinder head

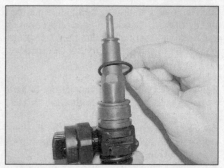

5.10 Care must be used to ensure that the injector O-rings are fitted without being twisted

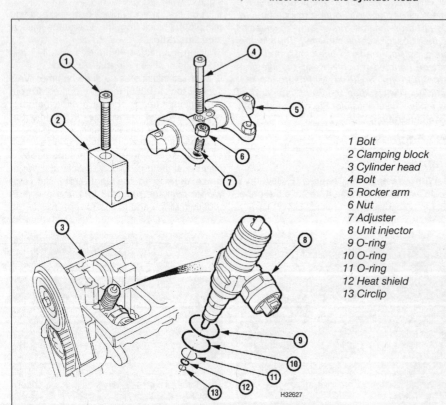

1 Bolt
2 Clamping block
3 Cylinder head
4 Bolt
5 Rocker arm
6 Nut
7 Adjuster
8 Unit injector
9 O-ring
10 O-ring
11 O-ring
12 Heat shield
13 Circlip

H32627

5.11 Unit injector

stop for the weight **(see illustration)**. Seat the slide hammer/tool in the slot on the side on the injector, and pull the injector out using a few gently taps. Recover circlip, the heat shield and O-rings and discard. New ones must be used for refitting **(see illustration)**.

9 If required, the injector wiring loom/rail can be removed from the cylinder head by undoing the two retaining nuts/bolts at the back of the head. To prevent the wiring connectors fouling the cylinder head casting as the assembly is withdrawn, insert the connectors into the storage slots in the plastic wiring rail. Carefully push the assembly to the rear, and out of the casting **(see illustrations)**.

Refitting

10 Prior to refitting the injectors, the three O-rings, heat insulation washer and clip must be renewed. Due to the high injection pressures, it is essential that the O-rings are fitted without being twisted. Seat recommend the use of three special assembly sleeves to install the O-rings squarely. It may be prudent to entrust O-ring renewal to a Seat dealer or suitably-equipped injection specialist, rather than risk subsequent leaks **(see illustration)**.

11 After renewing the O-rings, fit the heat shield and secure it in place with the circlip **(see illustration)**.

12 Smear clean engine oil onto the O-rings, and push the injector evenly down into the cylinder head onto its stop.

13 Fit the clamping block alongside the injector, but only hand-tighten the new retaining bolt at this stage.

14 It is essential that the injectors are fitted at right-angles to the clamping block. In order to achieve this, measure the distance from the rear face of the cylinder head to the rounded section of the injector **(see illustrations)**. The dimensions (a) are as follows:

    Cylinder 1 = 333.0 ± 0.8 mm
    Cylinder 2 = 245.0 ± 0.8 mm
    Cylinder 3 = 153.6 ± 0.8 mm
    Cylinder 4 = 65.6 ± 0.8 mm

15 Once the injector(s) are aligned correctly, tighten the clamping bolt to the specified Stage 1 torque setting, and the Stage 2 angle tightening setting. **Note:** *If an injector has been renewed, it is essential that the adjustment screw, locknut of the corresponding rocker and ball-pin are renewed at the same time. The ball-pins simply pull out of the injector spring cap. There is an O-ring in each spring cap to stop the ball-pins from falling out.*

16 Smear some grease (VW No G000 100) onto the contact face of each rocker arm adjustment screw, and refit the rocker shaft assembly to the camshaft bearing caps, tightening the retaining bolts as follows. Starting from the inside out, hand-tighten the bolts. Again, from the inside out, tighten the bolts to the Stage 1 torque setting. Finally, from the inside out, tighten the bolts to the Stage 2 angle tightening setting.

17 The following procedure is only necessary if an injector has been removed and refitted/renewed. Attach a DTI (Dial Test Indicator) gauge to the cylinder head upper surface, and position the DTI probe against the top of the adjustment screw **(see illustration)**. Turn the crankshaft until the rocker arm roller is on the highest point of its corresponding camshaft lobe, and the adjustment screw is at its lowest. Once this position has been established, remove the DTI gauge, screw the adjustment screw in until firm resistance is felt, and the injector spring cannot be compressed further. Turn the adjustment screw anti-clockwise 180°, and tighten the locknut to the specified torque. Repeat this procedure for any other injectors that have been refitted.

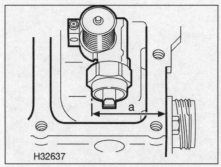

5.14a Measure the distance (a) from the rear of the cylinder head to the rounded section of the injector (see text)

5.14b Use a set square against the edge of the injector…

5.14c …and measure the distance to the rear of the cylinder head

5.17 Attach a DTI (Dial Test Indicator) gauge to the cylinder head upper surface, and position the DTI probe againstthe top of the adjustment screw

18 Reconnect the wiring plug to the injector.

19 Refit the camshaft cover and upper timing belt cover, as described in Chapter 2E Section 4 and Chapter 2E Section 6.

20 Bleed the fuel system as described in Section 11.

Common rail engines

Removal

Note: *Take care not to allow dirt into the injectors or fuel pipes during this procedure. Do not drop the injectors or allow the needles at their tips to become damaged. The injectors are precision-made to fine limits, and must not be handled roughly. Keep the injectors identified for position to ensure correct refitting.*

21 Pull the plastic cover over the engine upwards from its mountings. Where fitted, remove the foam insulation over the injectors.

22 Ensure the area around the injectors and the pipes/return hoses is clean and free from debris. The use of a vacuum cleaner is recommended. Plug all fuel lines when they have been disconnected to prevent any dirt ingress **(see illustration)**.

1.6 litre engine

23 Disconnect the injector wiring plug connectors **(see illustration)**.

5.22 Fit sealing caps to prevent dirt ingress

5.23 Disconnect the wiring plugs from the injectors

5.24a Remove the retaining clips ...

5.24b ... and disconnect the fuel return pipes

5.25a Use two spanners to counterhold the fuel pipe to the injector ...

5.25b ...and remove the high pressure fuel pipes

5.26 Remove the injector clamp retaining bolts

24 Using a pair of long-nose pliers, withdraw the retaining clip from the side of the injector and remove the fuel return hose **(see illustrations)**. Discard the o-ring seals, as new ones will be required for refitting.

25 Counterhold the injector with an open-ended spanner when releasing the pipe union. Undo the unions and remove the high-pressure pipes from between the fuel rail and the injectors **(see illustrations)**. Plug the openings to prevent contamination.

26 Undo the bolt securing the injector clamp **(see illustration)**, note that one clamp secures two injectors in place.

27 Seat technicians use a slide hammer (tool T10055) and adapter (T10402) to pull the injector from the cylinder head. If this tool is

not available, it may be possible to fabricate an equivalent tool to pull the injector out of the cylinder head.

28 Two injectors will need to be removed together, as the clamping piece is slotted into both injectors. Recover the copper seal and O-rings and discard. New ones must be used for refitting **(see illustration)**. **Note:** *The injectors can only be refitted to their original positions. Mark the injectors to avoid confusion if refitting the original injectors.*

2.0 litre engines

29 Disconnect the injector wiring plug connectors.

30 Push the return hose connector

downwards at its outer tabs, then pull up the centre piece and disconnect them from the top of the injectors **(see illustrations)**. Plug the openings to prevent contamination.

31 Undo the unions and remove the high-pressure pipes from between the fuel rail and the injectors. Counterhold the injector with an open-ended spanner when releasing the pipe union.

32 Undo the bolt securing the injector clamp between the two injectors, note that one clamp secures two injectors in place.

33 Seat technicians use a slide hammer (VAG tool T10055) and adapter (T10415) to pull the injector from the cylinder head. If this tool is not available, it may be possible to fabricate an equivalent tool to pull the injector out of the cylinder head.

34 Two injectors will need to be removed together, as the clamping piece is slotted into both injectors. Recover the copper seal and O-rings and discard. New ones must be used for refitting. **Note:** *The injectors can only be refitted to their original positions. Mark the injectors to avoid confusion if refitting the original injectors.*

Refitting

35 If required, renew the injector seals in the top of the camshaft cover. Using a screwdriver, prise the seal out from the cover; the new seal can then be pressed firmly

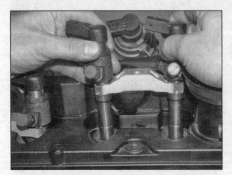

5.28 Remove two injectors at a time with retaining clamp

5.30a Hold down the outer tabs, and prise up the centre piece...

5.30b ...then pull the return hose connector upwards from the injector

5.35a Carefully prise out the seal ...

5.35b ...making sure the spring does not fall into the cover

5.37a Clean off the carbon around the end of the injector ...

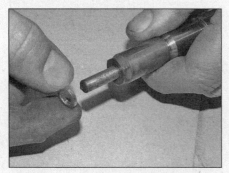

5.37b ... and fit a new sealing washer

5.38 Fit new O-ring seals to the fuel return lines

5.39 Fit the new return pipe O-ring to the top of the injector

into the cover **(see illustrations)**. There are different size seals depending on engine code, make sure the correct seals are supplied. Also make sure the spring on the inside lip of the seal does not drop into the camshaft cover.

36 Ensure the area around the injector locations in the cylinder head are clean and free from debris. Use a vacuum cleaner if available. Clean any carbon deposits from the injector and sealing surfaces with a cloth soaked in clean engine oil or rust-releasing spray.

37 To remove the copper sealing washer, spray rust-releasing spray around the injector nozzle, then clamp the seal in a vice, and use

a twisting motion to pull the injector from the seal. Push the new copper seal into place **(see illustrations)**. Do not touch the very end of the injector, or you could block up the nozzle.

38 On 1.6 litre engines, apply a little clean engine oil to the return pipe connection, and fit the new O-ring **(see illustration)**.

39 On 2.0 litre engines, apply a little clean engine oil to the return pipe connection on the injector, and fit the new O-ring **(see illustration)**.

40 To renew the main injector O-ring seal, Seat specify the use of tool VAG special tool T10377. This tool allows the O-ring to slide over the end of the injector without twisting.

With care, the seals can be fitted without the tool **(see illustrations)**.

41 Apply a smear of clean engine oil to the main O-ring seal, and insert the injector into place in the cylinder head **(see illustrations)**. Note that if the original injectors are being refitted, they must go into their original positions. Tighten the injector clamping bolt/ nut to the specified torque.

42 Refit the high-pressure fuel pipes and tighten the unions to the specified torque. Note that the pipes may be re-used providing the tapered seats are undamaged and the pipes are not deformed, constricted or corroded. Counterhold the injector with an open-ended spanner when tightening the pipe union.

5.40 Fit a new main O-ring seal without twisting it

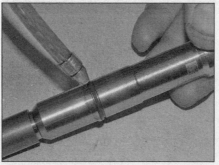

5.41a Lubricate the seal with some clean oil ...

5.41b ...then slide the injectors down into the cylinder head

5.43 Fit new clips to secure the fuel return lines

43 On 1.6 litre engines, when fitting the return hoses to the sides of the injector, use new securing clips **(see illustration)**.

44 The remainder of refitting is a reversal of removal, noting the following:

a) If one or more injectors have been renewed, the 'injector delivery calibration values' and 'injector voltage calibration values' must be entered into the ECM using VAG group diagnostic equipment. Entrust this task to a Seat dealer or suitably equipped specialist.

b) After completion of the work, the fuel system must be bled as described in Section 11.

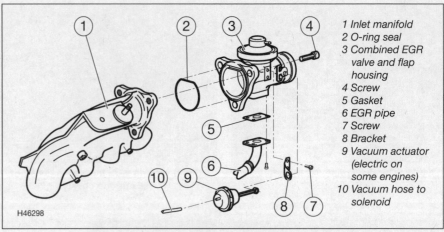

1 Inlet manifold
2 O-ring seal
3 Combined EGR valve and flap housing
4 Screw
5 Gasket
6 EGR pipe
7 Screw
8 Bracket
9 Vacuum actuator (electric on some engines)
10 Vacuum hose to solenoid

6.1a Typical inlet manifold details (1.9 litre PD engines)

6 Inlet manifold – removal and refitting

PD Engines

Removal

1 All PD engines have a conventional inlet manifold. 1.9 litre engines have a simple manifold fitted with an EGR control and throttle valve/shut down valve mounted at one end. 2.0 litre engines have an inlet manifold with a centrally mounted EGR control and throttle valve/shut down valve **(see illustrations)**. Note the illustrations show typical manifolds – there are variations, but all are similar.

2 Where fitted, remove the engine cover by pulling it up.

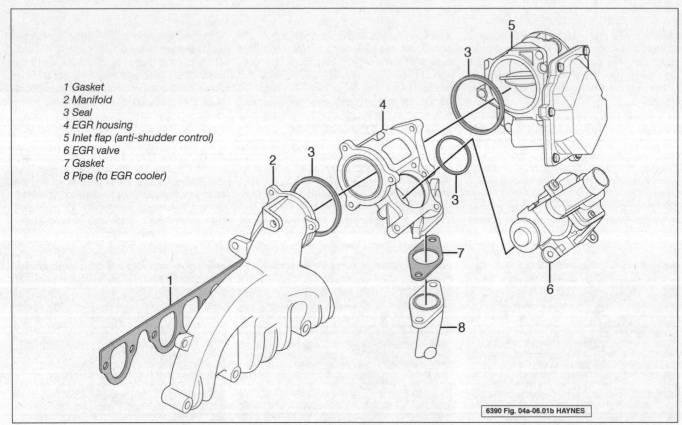

1 Gasket
2 Manifold
3 Seal
4 EGR housing
5 Inlet flap (anti-shudder control)
6 EGR valve
7 Gasket
8 Pipe (to EGR cooler)

6390 Fig. 04a-06.01b HAYNES

6.1b Inlet manifold 2.0 litre PD engine (BMM)

6.9 Undo the coolant pipe retaining screws

6.10 Undo the fuel line retaining screws

6.11 Disconnect the wiring connector

3 Remove the flap motor and EGR valve with reference to Section 6 of this Chapter, and to Chapter 4D Section 3.

4 Access can be improved where required if the front section of the plenum chamber is removed as described in Chapter 12 Section 19.

5 Where applicable, remove the heat shield from the manifold, then unscrew the mounting bolts and remove the inlet manifold from the cylinder head. Recover the gasket and discard, as a new one must be used on refitting.

Refitting

6 Refitting is a reversal of removal, using new manifold, EGR pipe and manifold flap assembly gaskets/O-rings. Tighten the mounting bolts to the specified torque.

Common rail engines

Removal

Plastic manifold

7 Remove the throttle housing, as described in Section 4 of this Chapter.

8 Remove the fuel rail from the top of the inlet manifold, as described in Section 12.

9 Undo the two retaining screws and move the coolant return pipe to one side **(see illustration)**.

10 Undo the two retaining screws and move the fuel return pipe to one side **(see illustration)**.

11 Where fitted, disconnect the wiring plug connector from the manifold flap motor **(see illustration)**.

12 Undo the retaining screw and remove

the EGR cooler changeover valve from the manifold **(see illustration)**.

13 Slacken the retaining clamp and disconnect the EGR pipe **(see illustration)**.

14 Undo the manifold retaining bolts, starting from the outside and working inwards in a diagonal sequence **(see illustrations)**. Lift the manifold from the cylinder head and retrieve the gasket seals; discard, as new ones will be required for refitting.

Refitting

15 Refitting is a reversal of removal, using new seals and gaskets **(see illustration)**. Remember to renew any self-locking nuts. Tighten the manifold retaining bolts to the specified torque setting, starting from the inside and working outwards in a diagonal sequence.

6.12 Undo the solenoid valve mounting bracket securing bolt

6.13 Slacken the EGR pipe retaining clamp

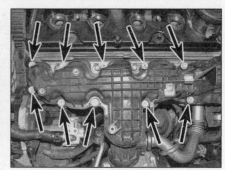

6.14a Undo the retaining bolts...

6.14b ...and remove the inlet manifold

6.14c Cover the inlet manifold recess with tape

6.15 Renew the inlet manifold seals

7 Fuel filter – renewal

1 Refer to Chapter 1B Section 29.

8 Fuel lift pump and gauge sender unit – removal and refitting

Note: *Observe the precautions in Section 1 before working on any component in the fuel system.*

⚠ **Warning: Avoid direct skin contact with fuel – wear protective clothing and gloves when handling fuel system components. Ensure that the work area is well ventilated to prevent the build-up of fuel vapour.**

General information

1 The fuel lift pump and gauge sender unit are combined in one assembly, which is mounted in the top of the fuel tank. The unit protrudes into the fuel tank, and its removal involves exposing the contents of the tank to the atmosphere.

Removal

2 Ensure that the vehicle is parked on a level surface, then disconnect the battery negative lead and position it away from the terminal. **Note:** *Refer to Chapter 5A Section 3.*
3 Remove the rear seat squab as described in Chapter 11 Section 22 and then prise off the cover. Release the locking tab and disconnect the wiring plug **(see illustration)**.
4 Disconnect the fuel lines and seal them. Note the position of the flow and return lines **(see illustration)**.
5 Note the position of the alignment marks, then unscrew and remove the securing ring. Use a pair of water pump pliers (or home-made tool) to grip and rotate the securing ring **(see illustrations)**.
6 Lift out the lift pump/gauge sender unit, holding it above the level of the fuel in the tank until the excess fuel has drained out. Recover the flange and seal **(see illustration)**.

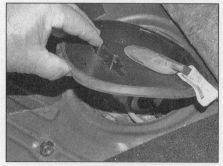

8.3a Lift off the cover and...

8.4 The fuel lines are marked

8.5b ...then use a suitable tool to unscrew...

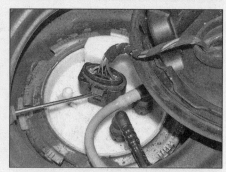

8.3b ...disconnect the wiring plug

8.5a Note the alignment marks...

8.5c ...and remove the securing ring

7 With the pump/sender unit removed from the car, lay it on an absorbent card or rag **(see illustration)**. Inspect the float at the end of the sender unit swinging arm for punctures and fuel ingress – renew the unit if it appears damaged.
8 The fuel pick-up incorporated in the assembly is spring-loaded to ensure that it

always draws fuel from the lowest part of the tank. Check that the pick-up is free to move under spring tension with respect to the sender unit body.
9 Inspect the rubber seal from the fuel tank aperture for signs of fatigue – renew it if necessary **(see illustration)**.
10 Inspect the sender unit wiper and track;

8.6 Removing the lift pump/gauge sender unit from the fuel tank

8.7 Lift pump/gauge sender unit removed from the car

8.9 If not removed with the unit, recover the rubber seal and check its condition

8.12a Disconnect the small wires...

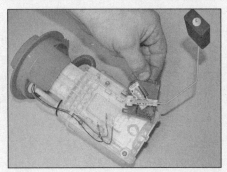

8.12b ...then undo the screws and slide out the sender unit

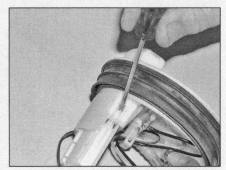

8.13a Prise up the retaining tags ...

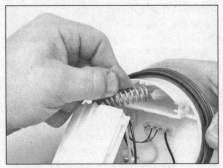

8.13b ...and recover the spring fitted under the top plate

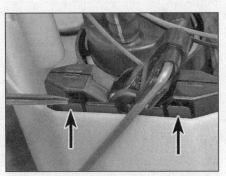

8.14a Press the clips to the outside...

8.14b ...and slide the sender unit upwards

clean off any dirt and debris that may have accumulated, and look for breaks in the track.
11 If required, the sender unit can be separated from the assembly. Two versions are fitted.

Version One

12 Disconnect the two small wires (note their positions), then remove the four screws and slide the unit downwards to remove **(see illustrations)**.
13 The unit top plate can be removed by releasing the plastic tags at either side; recover the large spring that fits onto a peg on the plate underside **(see illustrations)**.

Version Two

14 Gently pull the sender unit to the side and

upwards at the same time. If the sender unit is reluctant to release, gently push the retaining tabs outwards at the same time as pull the sender unit upwards **(see illustrations)**.
15 Note their fitted positions, release the clips and disconnect the sender unit wiring plugs **(see illustration)**. **Note:** *On later sender units, the plugs are released by depressing the clips on the front and rear of the plastic housing.*

Refitting

16 Refit the lift pump/sender unit by following the removal procedure in reverse, noting the following points:
a) *Take care not to bend the float arm as the unit is refitted.*

b) *Smear the outside of tank aperture rubber seal with clean fuel or lubricating spray, to ease fitting. Locate the seal in the tank aperture before fitting the lift pump/sender unit* **(see illustrations)**.
c) *Tighten the securing ring to the specified torque setting.*
d) *Reconnect the fuel hoses to the correct ports – observe the direction-of-flow arrow markings, and refer to paragraph 6. Ensure that the fuel hose fittings click fully into place.*
e) *On completion, check that all associated pipes are securely clipped to the tank, then run the engine and check for fuel leaks.*

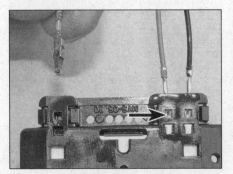

8.15 Depress the clips and pull the wiring plugs from the sender

8.16a Locate the seal in the tank aperture...

8.16b ...then fit the lift pump/sender unit

9.9 Disconnect the fuel lines

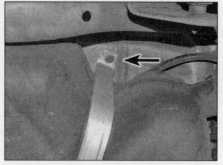

9.11 Remove the tank straps

b) *Ensure that all pipes and hoses are correctly routed, are not kinked, and are securely held in position with their retaining clips.*
c) *Tighten the tank strap retaining bolts to the specified torque.*
d) *On completion, refill the tank with fuel, and exhaustively check for signs of leakage prior to taking the vehicle out on the road.*

10 Fuel pump – removal and refitting

Tandem fuel pump (PD unit injection engines only)

Note: *Disconnecting the central connector for the unit injectors may cause a fault code to be logged by the engine management ECU. This code can only be erased by a Seat dealer or suitably-equipped garage.*

Removal

1 The tandem fuel pump is located on the left-hand end of the cylinder head. It is effectively a vacuum pump (for the braking system) and a fuel lift pump (for the fuel injection system). First, remove the engine top cover.
2 Disconnect the fuel supply hose (marked white) and the return hose (marked blue) from the fuel filter and drain all fuel from the hoses into a suitable container.
3 Remove the air cleaner assembly together with the air mass meter and air ducting, with reference to Section 3.
4 Release the retaining clip (where fitted) and disconnect the brake servo pipe from the tandem pump **(see illustration)**.
5 Disconnect the fuel supply hose (marked white) and the return hose (marked blue) from the tandem pump.
6 Unscrew the four retaining bolts and remove the tandem pump from the cylinder head. Recover the gasket and discard, as a new one must be used on refitting. There are no serviceable parts within the tandem pump. If the pump is faulty, it must be renewed.

Refitting

7 Refitting is a reversal of removal, but use a new gasket and tighten the mounting bolts to the specified torque. Ensure the pump pinion engages correctly with the drive slot in the camshaft **(see illustration)**. It is recommended that the tandem pump is primed with fuel as follows. Disconnect the fuel filter return hose (marked blue), and connect the hose to a hand vacuum pump. Operate the vacuum pump until fuel comes out of the return hose. Take care not to suck any fuel into the vacuum pump. Reconnect the return hose to the fuel filter.
8 Have the engine management ECU's fault memory interrogated and erased by a Seat dealer or suitably-equipped garage.

9 Fuel tank – removal and refitting

Note: *Observe the precautions in Section 1 before working on any component in the fuel system.*

Removal

1 Before the tank can be removed, it must be drained of as much fuel as possible. As no drain plug is provided, it is preferable to carry out this operation with the tank almost empty.
2 Disconnect the battery negative lead and position it away from the terminal (Chapter 5A Section 3), then using a hand pump or syphon, remove any remaining fuel from the tank.
3 Lift up the rear seat squab and disconnect the wiring plug from the pump/sender unit – as described in Section 8.
4 Open the fuel filler flap, and unscrew the fuel filler cap – leave the cap loosely in place.
5 Loosen the left-hand rear wheel bolts, then jack up the rear of the car (see *Jacking and vehicle support*). Remove the relevant rear roadwheel. and remove the left-hand rear wheel.
6 Undo the retaining screws, release the retaining clips and remove the left-hand rear wheel arch liner.
7 Open the fuel filler flap, and unscrew the fuel cap from the top of the filler neck.
8 Unscrew the bolts at the filler neck.

9 Disconnect the fuel lines from the connections at the front of the fuel tank **(see illustration)**.
10 Position a trolley jack under the centre of the tank. Insert a block of wood between the jack head and the tank to prevent damage to the tank surface. Raise the jack until it just takes the weight of the tank.
11 Unscrew the mounting bolts and detach the tank straps **(see illustration)**.
12 Lower the jack and tank away from the underside of the vehicle.
13 Continue to lower the tank, whilst guiding the filler neck through the suspension arm. Remove the tank from the vehicle.
14 If the tank is contaminated with sediment or water, remove the fuel pump/sender unit (see Section 8) and swill the tank out with clean fuel. The tank is injection-moulded from a synthetic material, and if damaged, it should be renewed. However, in certain cases it may be possible to have small leaks or minor damage repaired. Seek the advice of a suitable specialist before attempting to repair the fuel tank.

Refitting

15 Refitting is the reverse of the removal procedure, noting the following points:
a) *When lifting the tank back into position, make sure the mounting rubbers are correctly positioned, and take care to ensure none of the hoses get trapped between the tank and vehicle body.*

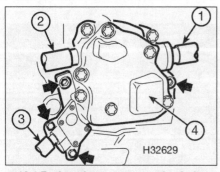

10.4 Fuel tandem pump securing bolts

1 Brake servo hose   3 Fuel return hose
2 Fuel supply hose   4 Tandem pump

10.7 Ensure that the tandem pump pinion engages correctly with the drive slot in the camshaft

10.10 Disconnect the pump wiring connector

10.13 Remove the lifting bracket from above the fuel pump

10.14 Remove the high pressure fuel line from the pump

High pressure fuel pump (common rail engines)

Removal

1.6 litre engines

9 Remove the timing belt and pump sprocket as described in Chapter 2C Section 7 and Chapter 2C Section 8.

10 Disconnect the wiring connector from the fuel metering valve on top of the fuel pump **(see illustration)**.

11 Undo the two retaining screws and move the coolant return pipe to one side **(see illustration 6.9)**.

12 Undo the two retaining screws and move the fuel return pipe to one side **(see illustration 6.10)**.

13 Undo the pipe retaining bracket screw and the two retaining bolts, then remove the lifting eye from the front of the engine **(see illustration)**.

14 Slacken the fuel pipe unions and remove the high-pressure pipe from the pump to the fuel rail **(see illustration)**.

15 Release the retaining clips and disconnect the two fuel hoses from the top of the fuel pump **(see illustration)**.

16 Counterhold the pump hub using VAG tool No. T10051, and undo the pump hub nut. In the absence of this special tool, counterhold the hub using something to lock the sprocket, whilst the centre nut is slackened **(see illustration)**.

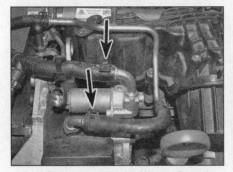

10.15 Disconnect the fuel hoses from the pump

10.16 Remove the fuel pump centre hub nut

17 Using a suitable two-legged puller (VAG tool No. T40064) and two bolts, remove the hub complete with sprocket from the pump shaft.

18 Undo the 3 retaining bolts and remove the pump **(see illustration)**.

2.0 litre engines

19 Remove the pump sprocket as described in Chapter 2D Section 8.

20 Disconnect the fuel supply hose from the pump **(see illustration)**. Plug all openings to prevent contamination.

21 Disconnect the wiring plug from the sensor on the pump.

22 Slacken the unions and disconnect the high-pressure fuel pipe between the pump and common fuel rail. Undo the retaining

screws to release the fuel pipe from the support brackets.

23 On some models it may be necessary to disconnect he wiring connectors from the glow plugs. Pulling only under the ridge at the top of the connectors, disconnect the wiring connectors. **Note:** *Take care not to damage the connectors or wiring, as the glow plug wiring loom/connectors are only available as a complete assembly.*

24 Undo the bolts securing the coolant pipe to the intake manifold, and position the pipe to one side.

25 Disconnect the fuel return pipes from the pump and the common fuel rail **(see illustration)**.

26 Counterhold the pump hub using VAG tool

10.18 Fuel pump mounting bolts

10.20 Release the clip and disconnect the fuel supply hose

10.25 Fuel pump return hose

10.26 Counterhold the pump hub with a C-spanner, and undo the nut

10.27a Use a two-legged puller and 8 mm bolts to pull the hub from the pump shaft

10.27b Note the locating peg in the pump shaft

No. T10051, and undo the pump hub nut. In the absence of this special tool, counterhold the hub using a suitable C-spanner **(see illustration)**.

27 Using a suitable two-legged puller and two 8 mm bolts, remove the hub from the pump shaft **(see illustrations)**.

28 Undo the 3 retaining bolts and remove the pump.

Refitting

29 Refitting is a reversal of removal, noting the following points:

a) *Ensure all fuel pipes/hose connections are clean and free from debris.*

b) *The high-pressure fuel pipe from the pump to the common rail maybe re-used providing it's not been damaged.*

c) *Tighten all fasteners to their specified torque where given.*

d) *Fill the pump with clean fuel through the*

fuel supply pipe aperture prior to starting **(see illustration)**.

e) *Bleed the fuel system as described in Section 11.*

Supplementary fuel pump (common rail engines only)

Removal

30 The supplementary fuel pump is not fitted to all models, where fitted, it is located on a bracket bolted to the top of the right-hand engine mounting **(see illustration)**.

31 Disconnect the wiring plug from the exhaust gas pressure sensor, then undo the retaining bolt and move the sensor and bracket to one side **(see illustration)**.

32 Undo the retaining bolts and lift the supplementary fuel pump upwards to access the fuel pipes and wiring plug **(see illustration)**.

33 Release the fuel hoses from the retaining clips, and disconnect the fuel pump wiring plug.

34 Release the clamps and disconnect the fuel supply pipe from the fuel filter. Plug/cover the openings to prevent contamination.

35 Disconnect the fuel temperature sensor wiring plug, release the clamp and disconnect the fuel supply pipe from the high-pressure fuel pump. Plug/cover the openings to prevent contamination.

Refitting

36 Refitting is a reversal of removal, noting the following points:

a) *Ensure the hoses are not kinked.*

b) *Ensure all connections are clean and free from debris.*

c) *Note the hose connections: The return hose is blue (or blue markings), and the supply hose is white.*

d) *Bleed the system as described in Section 11.*

11 Fuel system bleeding

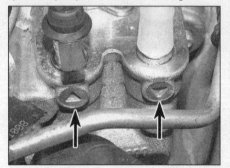

10.29 The triangles adjacent to the pump apertures indicate fuel flow

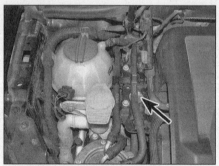

10.30 Supplementary fuel pump location

10.31 Undo the retaining screw

10.32 Undo the bolts and remove the pump bracket

PD engines

1 Attach a hand-held vacuum pump to the return outlet (coloured blue) from the tandem fuel pump, and operate the pump until a steady stream of fuel appears from the outlet. Start the engine and check for leaks. See also Chapter 1B Section 29 (Fuel filter).

Common rail engines

2 Prime the high pressure fuel pump by filling it with clean diesel through the fuel supply aperture **(see illustration 10.29a)**, then operate the starter for shorts bursts (no more than 10 seconds at a time) until the engine starts. Operate the engine at a fast idle (approx 2000 rpm) for several minutes before allowing it to return to its normal idle speed.

3 If the engine fails to start, it must be filled/bled using Seat diagnostic equipment (VAS5051 etc.). Using this equipment operates the electric fuel pumps for 3 minutes. Alternatively remove the access cover from the in tank fuel pump, identify the pump

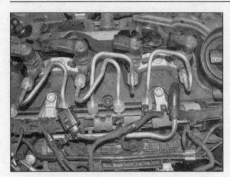

12.3a Remove the high pressure pipes from the rail and injectors …

12.3b …and fit sealing caps to prevent dirt ingress

12.4a Disconnect the hose from the return pipe …

supply wiring and supply the pump directly with 12 volts (see the wiring diagrams at the end of this manual for details).

4 Once the engine has been started, test drive the vehicle over a distance of at least 15 miles with at least one period of full acceleration. If there is any air left in the fuel system, the engine management ECM may switch to 'limp home' mode, and store a fault code. Have the fault code cleared and road test the vehicle again.

12 Fuel rail –
removal and refitting

Note: *Observe the precautions in Section 1 before working on any component in the fuel system.*

Removal

1 Remove the engine cover by pulling it upwards and then remove the foam insulation (where fitted). Ensure the area around the fuel rail and pipes is clean and free from debris. If available, use a vacuum cleaner.

2 Disconnect the wiring connectors from the fuel injectors.

3 Counterhold the injector with an open-ended spanner when releasing the

12.4b …and the return pipe from the fuel rail

pipe union. Undo the unions and remove the high-pressure pipes from between the fuel rail and the injectors **(see illustrations)**. Plug the openings to prevent contamination.

4 Release the retaining clips and disconnect the fuel return hose from the fuel rail **(see illustrations)**

5 Undo the retaining bolts, disconnect the coolant return hose from the expansion tank, and move the coolant pipe/hose to one side.

6 Slacken the fuel pipe unions and remove the high-pressure pipe from the pump to the fuel rail.

7 Disconnect the wiring plugs from the glow plugs, fuel pressure regulating valve, and the

12.7 Lift the wiring loom over and place to one side

fuel pressure sensor at each end of the fuel rail. Unclip the wiring loom retaining bracket from the top of the fuel rail and move it to one side **(see illustration)**.

8 Undo the multi-spline retaining bolts and remove the fuel rail **(see illustrations)**.

9 If required, note their fitted positions, then unscrew the fuel pressure sensor and pressure regulating valve from the fuel rail, as described in Section 4. **Note:** *VW insist that once removed, the pressure regulating valve cannot be re-used.*

Refitting

10 Where applicable, refit the fuel pressure sensor and the new regulating valve to the fuel rail, and tighten them to the specified torque. Note that the threads of the sensors must be clean and free from oil and grease.

11 The remainder of refitting is a reversal of removal, noting the following points:

a) *Refit the high-pressure fuel pipes and tighten the unions to the specified torque. Note that the high-pressure fuel pipes may be re-used providing the tapered seats are undamaged and the pipes are not deformed, constricted or corroded.*

b) *After completion of the work, the fuel system must be bled as described in Section 11.*

12.8a Undo the fuel rail mounting bolts – 1.6 litre engines

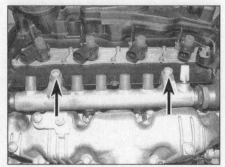

12.8b Undo the fuel rail mounting bolts – 2.0 litre engines

Notes

Chapter 4 Part C
Emission control and exhaust systems – petrol engines

Contents

Degrees of difficulty

Easy, suitable for novice with little experience	**Fairly easy,** suitable for beginner with some experience	**Fairly difficult,** suitable for competent DIY mechanic	**Difficult,** suitable for experienced DIY mechanic	**Very difficult,** suitable for expert DIY or professional

Specifications

Engine codes*

1.6 litre	BGU, BSF and BSE
2.0 litre:	
FSi (non-turbo)	BLR, BVY, BLY and BVZ
TFSi (turbo)	BWA, BWJ, CDLD and CDLA

* **Note:** *See 'Vehicle identification' at the end of this manual for the location of engine code markings.*

Emission control applications

1.6 litre engine codes BGU, BSE and BSF	One cat in intermediate pipe. Pre-cat oxygen sensor in exhaust manifold, after-cat sensor behind cat in intermediate pipe, EGR system with potentiometer control on engine code BGU, secondary air injection on engine codes BGU and BSE. Engine code BSF has no secondary air and no EGR
2.0 litre engine codes BLR, and BVY	Two catalytic converters and two pre-catalytic convertors. Two pre-cat oxygen sensors in exhaust manifold and two post cat sensors. Engine code BLR has an additional sensor (5 in total) Variable valve timing and EGR fitted
2.0 litre engine codes BLY and BVZ.	Two catalytic converters, one pre-cat oxygen sensor and one post cat sensor. Variable valve timing, but no EGR system
2.0 litre engine codes BWJ, BWA, CDLA and CDLD	One catalytic converter, one pre-cat oxygen sensor and one post-cat sensor. Variable valve timing, but no EGR system

Torque wrench settings

	Nm	lbf ft
Coolant pipe banjo bolts	35	26
EGR pipe flange bolts to throttle housing	10	7
EGR pipe flange nuts/bolts:		
1.6 litre engine codes BGU, BSE and BSF:	25	18
To EGR valve	8	6
To cylinder head	18	13
2.0 litre engine codes BLR, BVY BL and BVZ:		
To EGR valve	10	7
To exhaust manifold	15	11
To inlet manifold	10	7
EGR pipe mounting bolts:		
1.6 litre engine codes BGU, BSE and BSF	25	18
All other engines	10	7
EGR pipe union nut:		
1.6 litre engine codes BGU, BSE and BSF	60	44
EGR valve mounting/through-bolts:		
1.6 litre engine codes BAG, BLF and BLP	10	7
2.0 litre engine codes BLR, BVX and BVY	25	18
Exhaust clamp nuts	40	30
Exhaust manifold nuts:		
2.0 litre models*	25	18
Exhaust manifold support bracket to engine	25	18
Exhaust manifold-to-downpipe nuts*	40	30
Exhaust mounting bracket nuts and bolts	25	18
Intercooler mounting bolts	10	7
Oil supply pipe banjo bolts*	30	22
Oil return pipe bolts	8	6
Oxygen sensor	50	37
Secondary air adapter plate mounting bolts	10	7
Secondary air combi-valve mounting bolts	10	7
Secondary air pipe union nuts	25	18
Turbocharger support bracket to engine	25	18
Turbocharger-to-downpipe nuts**	40	30
Turbocharger-to-manifold bolts**	30	22
Turbocharger-to-support bracket bolt	30	22

* Do not re-use
** Use thread-locking compound

1 General Information

Emission control systems

1 All petrol models are designed to use unleaded petrol, and are controlled by engine management systems that are programmed to give the best compromise between driveability, fuel consumption and exhaust emission production. In addition, a number of systems are fitted that help to minimise other harmful emissions. A crankcase emission control system is fitted, which reduces the release of pollutants from the engine's lubrication system, and a catalytic converter is fitted which reduces exhaust gas pollutant. An evaporative loss emission control system is fitted which reduces the release of gaseous hydrocarbons from the fuel tank.

Crankcase emission control

2 To reduce the emission of unburned hydrocarbons from the crankcase into the atmosphere, the engine is sealed and the blow-by gases and oil vapour are drawn from inside the crankcase, through a wire-mesh oil separator, into the inlet tract to be burned by the engine during normal combustion.

3 Under conditions of high manifold depression, the gases will be sucked positively out of the crankcase. Under conditions of low manifold depression, the gases are forced out of the crankcase by the (relatively) higher crankcase pressure. If the engine is worn, the raised crankcase pressure (due to increased blow-by) will cause some of the flow to return under all manifold conditions.

Exhaust emission control

4 To minimise the amount of pollutants which escape into the atmosphere, all petrol models are fitted with one or two catalytic converters in the exhaust system. The fuelling system is of the closed-loop type, in which an oxygen (lambda) sensor in the exhaust system provides the engine management system ECU with constant feedback, enabling the ECU to adjust the air/fuel mixture to optimise combustion.

5 The oxygen sensor has a built-in heating element, controlled by the ECU through the oxygen sensor relay, to quickly bring the sensor's tip to its optimum operating temperature. The sensor's tip is sensitive to oxygen, and sends a voltage signal to the ECU that varies according to the amount of oxygen in the exhaust gas. If the inlet air/fuel mixture is too rich, the exhaust gases are low in oxygen so the sensor sends a low-voltage signal, the voltage rising as the mixture weakens and the amount of oxygen rises in the exhaust gases. Peak conversion efficiency of all major pollutants occurs if the inlet air/fuel mixture is maintained at the chemically-correct ratio for the complete combustion of petrol of 14.7 parts (by weight) of air to 1 part of fuel (the stoichiometric ratio). The sensor output voltage alters in a large step at this point, the ECU using the signal change as a reference point and correcting the inlet air/fuel mixture accordingly by altering the fuel injector pulse width.

6 Most models have two sensors, one before and one after the main catalytic converter, however, some models have four sensors, two before and two after the catalytic converter. This enables more efficient monitoring of the exhaust gas, allowing a faster response time.

1.12 Typical exhaust pipe mounting

2.2 Charcoal canister location on right-hand inner wing

2.3a Remove the spring clip and pull off the hose

2.3b Disconnect inlet line at the quick release behind the coolant reservoir...

2.3c ...or at the check valve

The overall efficiency of the converter itself can also be checked.

7 An Exhaust Gas Recirculation (EGR) system is also fitted to some models. This reduces the level of nitrogen oxides produced during combustion by introducing a proportion of the exhaust gas back into the inlet manifold, under certain engine operating conditions, via a plunger valve. The system is controlled electronically by the engine management ECU.

8 A secondary air system is fitted to 1.6 litre engine codes BGU and BSE, to reduce cold-start emissions when the catalytic converter is still warming-up. The system is an electric air pump, fed with air from the air cleaner, and a system of valves. When the engine is cold, air is pumped into the exhaust manifold, and mixes with the exhaust gas, in order to prolong combustion of any unburnt exhaust gases. The process also helps to bring the catalytic converter to its working temperature more quickly. When the engine coolant temperature is high enough, and the converter is operating normally, the system is switched off by the engine management ECU.

Evaporative emission control

9 To minimise the escape of unburned hydrocarbons into the atmosphere, an evaporative loss emission control system is fitted to all petrol models. The fuel tank filler cap is sealed and a charcoal canister is mounted underneath the right-hand wing to collect the petrol vapours released from the fuel contained in the fuel tank. It stores them until they can be drawn from the canister (under the control of the fuel injection/ignition

system ECU) via the purge valve(s) into the inlet tract, where they are then burned by the engine during normal combustion.

10 To ensure that the engine runs correctly when it is cold and/or idling and to protect the catalytic converter from the effects of an over-rich mixture, the purge control valve(s) are not opened by the ECU until the engine has warmed-up, and the engine is under load; the valve solenoid is then modulated on and off to allow the stored vapour to pass into the inlet tract.

Exhaust systems

11 On most models, the exhaust system is the exhaust manifold, front pipe(s), intermediate pipe and silencer, and tailpipe and silencer. The systems fitted differ in detail depending on the engine fitted – for example, in how the catalytic converter is incorporated into the system. On some

models the converter is incorporated in the exhaust manifold, while on others it is part of the exhaust front pipe. Where two converters are fitted, an additional exhaust section with catalytic converter is fitted behind the front pipe and catalytic converter. On turbocharged models, the turbocharger is integral with the exhaust manifold, and is driven by the exhaust gases.

12 The system is supported by various metal brackets screwed to the vehicle floor, with rubber vibration dampers fitted to suppress noise **(see illustration)**.

2 Evaporative loss emission control system – information and component renewal

1 The evaporative loss emission control system consists of the purge valve, the activated charcoal filter canister and a series of connecting vacuum hoses.

2 The canister is located on the right-hand side of the engine compartment, in front of the windscreen washer reservoir filler **(see illustration)**

3 Pull off the larger hose (which leads to the control solenoid) If required, the check valve valve can now be prised out from the top of the canister **(see illustrations)**.

4 Unclip the canister form the mounting bracket **(see illustration)**.

5 If required the mounting bracket can now be unbolted from the engine mounting **(see illustration)**.

2.4 Removing the charcoal canister

2.5 Removing the mounting bracket

2.6 The purge valve at the rear of the engine (1.6 litre shown)

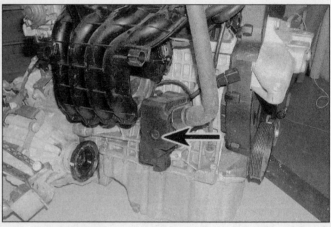

3.1 Oil separator box on rear of engine (seen with engine removed)

6 The control valve is mounted away from the canister (the exact location depends on the engine fitted). Disconnect the wiring plug and remove the hoses **(see illustration)**.

7 Refitting is a reversal of removal.

3 Crankcase emission system – general information

1 The crankcase emission control system consists of hoses connecting the crankcase to the air cleaner or inlet manifold. Oil separator units are fitted to some petrol engines, usually at the back of the engine **(see illustration)**.

2 The system requires no attention other than to check at regular intervals that the hoses, valve and oil separator are free of blockages and in good condition.

4 Exhaust Gas Recirculation (EGR) system – component removal and refitting

Note: *When an EGR control valve is replaced it must be adapted to the vehicle with a suitable diagnostic tool.*

1 The EGR system consists of the EGR valve, the modulator (solenoid) valve and a series of connecting vacuum hoses.

2 The EGR valve is mounted as follows:

a) *On engine codes BGU and BSE it is located on the left-hand end of the cylinder head, with connecting hoses from the air cleaner, throttle housing and exhaust manifold.*

b) *On engine codes BLR, BVY, BLY and BVZ it is located on the right-hand rear of the cylinder head, with connecting pipes to the inlet manifold on the front of the head, and to the exhaust manifold on the rear of the head. The exhaust gases are cooled by coolant from the engine cooling system.*

3 To improve access, remove the engine top cover. Removal details vary according to model.

EGR valve

1.6 litre engines

Note: *Only fitted to engine code BGU.*

4 Jack up and support the front of the vehicle (See, *Jacking and vehicle support*) and remove the engine undershield.

5 Remove the heat shield from the driveshaft and then unbolt the EGR pipe from the exhaust manifold.

6 Remove the engine top cover and the air filter ducting from the throttle body.

7 Unbolt the EGR pipe from the throttle body and then disconnect the EGR pipe and wiring plug from the EGR valve **(see illustration)**.

8 Unscrew the two nuts from the EGR pipe flange at the EGR valve.

9 Move the EGR pipe flange at the EGR valve aside, off the two studs; recover the gasket. Slide the EGR valve off the mounting studs, and remove it.

10 Refitting is a reversal of removal. Use new gaskets, and tighten the EGR pipe mountings to the specified torque.

2.0 litre FSi (non-turbo)

11 Remove the engine top cover.

12 Disconnect the wiring from the EGR valve located on the right-hand rear of the cylinder head.

13 Fit hose clamps to the hoses connected to the EGR valve, then loosen the clips and disconnect them.

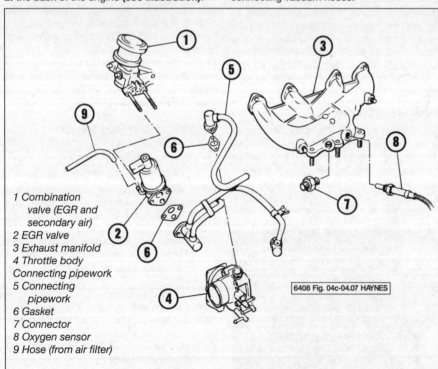

1 Combination valve (EGR and secondary air)
2 EGR valve
3 Exhaust manifold
4 Throttle body Connecting pipework
5 Connecting pipework
6 Gasket
7 Connector
8 Oxygen sensor
9 Hose (from air filter)

6408 Fig. 04c-04.07 HAYNES

4.7 EGR valve and associated components

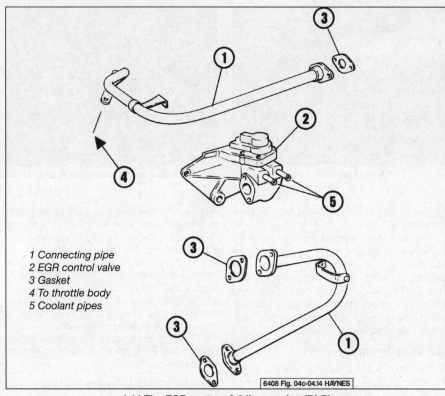

1 Connecting pipe
2 EGR control valve
3 Gasket
4 To throttle body
5 Coolant pipes

6408 Fig. 04c-04.14 HAYNES

4.14 The EGR system 2.0 litre engine (BLR)

14 Unscrew the flange and mounting bolts and disconnect the connecting pipe from the inlet manifold and EGR valve. Recover the gasket **(see illustration)**.
15 Unscrew the flange and mounting bolts and disconnect the connecting pipe from the exhaust manifold and EGR valve. Recover the gaskets.
16 Unscrew the mounting bolts and remove the EGR valve from the cylinder head. Note the location of the bracket.
17 Refitting is a reversal of removal but use new gaskets and a new O-ring seal. Tighten the bolts to the specified torque where given. Check and if necessary top-up the cooling system.

5 Secondary air injection system – information and component renewal

Note: *Secondary air injection is only fitted to 1.6 litre engine codes BGU and BSE.*
1 The secondary air injection system (also known as a 'pulse-air' system) consists of an electrically-operated air pump (fed with air from the air cleaner), a relay for the air pump, a vacuum-operated air supply combi-valve, a solenoid valve to regulate the vacuum supply, and pipework to feed the air into the exhaust manifold. For more information on the principles of operation, refer to Section 1 **(see illustration)**.

5.1 Secondary air injection system components – 1.6 litre BGU engine (BSE engine code is similar)

1 Air supply combi-valve
2 Vacuum hose
3 Wiring plug
4 Secondary air inlet valve
5 To connection on brake servo pipe
6 Air pump relay
7 O-ring
8 Pressure hose
9 Inlet hose
10 From air cleaner
11 Secondary air pump
12 Wiring plug
13 Pump mounting nut
14 Mounting bracket bolt
15 Mounting bracket bolt
16 Pump mounting bracket
17 Inlet manifold lower section
18 Adapter plate bolts
19 Adapter plate
20 Gasket
21 Mounting nut
22 EGR valve
23 Union adapter
24 Secondary air pipe union nut
25 Secondary air pipe
26 To union on exhaust manifold
27 Adapter plate bolts

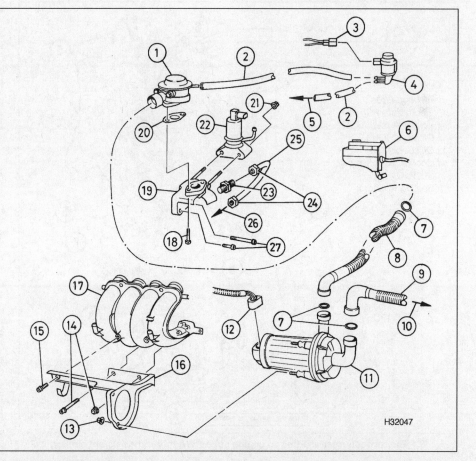

H32047

5.3 Remove the hose

5.5 Remove the pump

5.6 Remove the mounting bracket

Air pump

2 The air pump is mounted on a bracket attached to the lower section of the inlet manifold. To improve access to the pump, remove the engine top cover.

3 Disconnect the air hoses on top of the pump by squeezing together the lugs on the hose end fittings, and pulling the hoses upwards **(see illustration)**. Recover the O-ring seal from each hose – new seals should be used when refitting. The hoses are of different sizes, and so cannot be refitted incorrectly.

4 Disconnect the wiring plug from the air pump.

5 Unscrew the pump-to-mounting bracket nuts/nuts and bolts (as applicable) and slide the pump out of the mounting bracket **(see illustration)**.

6 The mounting bracket can be now be removed from the inlet manifold **(see illustration)**. This is required for access to the ignition coil. The bracket is secured by two bolts and a nut, or by three bolts, depending on model. Where three bolts are used, they are of different lengths, so note their locations.

7 Refitting is a reversal of removal.

Air supply combi-valve

8 The combi-valve is mounted on top of the exhaust manifold. To improve access to the valve, remove the engine top cover.

9 Where necessary, loosen and remove the nuts securing the heat shield over the exhaust manifold; this will allow the shield to be moved for access to the combi-valve mounting bolts.

10 Disconnect the vacuum hose and the large-diameter air hose from the valve – the air hose is released by squeezing together the lugs on the hose end fitting.

11 Remove the two valve mounting bolts from below the valve, and lift the valve off its mounting flange. Recover the gasket.

12 Refitting is a reversal of removal. Use a new gasket, and tighten the mounting bolts to the specified torque.

Combi-valve adapter elbow

13 The combi-valve is mounted on an adapter elbow, which is screwed onto the exhaust manifold. This elbow can be removed if required (after removing the combi-valve as described above and the EGR valve) by unscrewing the adapter mounting bolts; recover the gasket. When refitting, use a new gasket, and tighten the adapter elbow mounting bolts to the specified torque.

---

6 Turbocharger – general information, precautions, removal and refitting

General information

1 The turbocharger is integral with the exhaust manifold and cannot be removed separately. Lubrication is provided by an oil supply pipe that runs from the engine oil filter mounting. Oil is returned to the sump through a return pipe that connects to the sump **(see illustration)**. The turbocharger unit has a separate wastegate valve and vacuum actuator diaphragm, which is used to control the boost pressure applied to the inlet manifold.

2 The turbocharger's internal components rotate at a very high speed, and as such are sensitive to contamination; a great deal of damage can be caused by small particles of dirt, particularly if they strike the delicate turbine blades.

Precautions

3 The turbocharger operates at extremely high speeds and temperatures. Certain

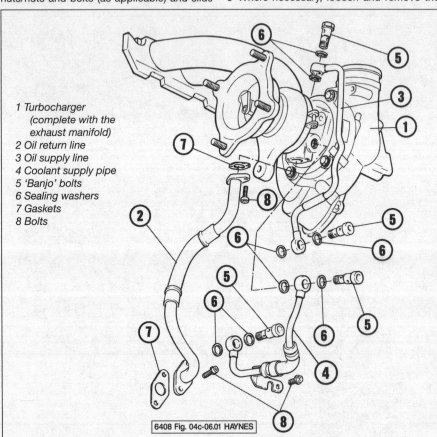

1 Turbocharger
  (complete with the
  exhaust manifold)
2 Oil return line
3 Oil supply line
4 Coolant supply pipe
5 'Banjo' bolts
6 Sealing washers
7 Gaskets
8 Bolts

6408 Fig. 04c-06.01 HAYNES

6.1 The turbocharger as fitted to engine code BWA (other models are similar)

precautions must be observed to avoid premature failure of the turbo, or injury to the operator.

● Do not operate the turbo with any parts exposed. Foreign objects falling onto the rotating vanes could cause excessive damage and (if ejected) personal injury.

● Do not operate the turbo with any parts exposed. Foreign objects falling onto the rotating vanes could cause excessive damage and (if ejected) personal injury.

● Cover the turbocharger air inlet ducts to prevent debris entering, and clean using lint-free cloths only.

● Do not race the engine immediately after start-up, especially if it is cold. Give the oil a few seconds to circulate.

● Observe the recommended intervals for oil and filter changing, and use a reputable oil of the specified quality. Neglect of oil changing, or use of inferior oil, can cause carbon formation on the turbo shaft and subsequent failure. Thoroughly clean the area around all oil pipe unions before disconnecting them, to prevent the ingress of dirt. Store dismantled components in a sealed container to prevent contamination.

Caution: Thoroughly clean the area around all oil pipe unions before disconnecting them, to prevent the ingress of dirt. Store dismantled components in a sealed container to prevent contamination. Cover the turbocharger air inlet ducts to prevent debris entering, and clean using lint-free cloths only.

Removal

Note: *The turbocharger is removed upwards from the rear of the engine compartment.*

4 Apply the handbrake, then jack up the front of the vehicle and support it on axle stands (see *Jacking and vehicle support*). Remove the right-hand front roadwheel, the wheel arch liner, and the engine compartment undertray.

5 Remove the engine top cover/air filter and the air inlet duct as described in.

6 Drain the cooling system as described in Chapter 1B Section 33.

7 Remove the windscreen grille panel and plenum chamber front panel as described in Chapter 12 Section 19.

8 Disconnect the charge air duct from the right-hand bottom of the intercooler. Unscrew the mounting bolts beneath the crankshaft pulley and remove the charge air duct from the turbocharger.

9 Disconnect the air duct from the right-hand bottom of the intercooler, then unscrew the mounting bolts beneath the crankshaft pulley and remove the duct from the turbocharger.

10 Disconnect the wiring from the turbocharger and unbolt the wiring loom bracket.

11 Unbolt the air duct adapter plate from the turbocharger.

12 Disconnect the oxygen sensor wiring at the connector on the bulkhead. Also remove the wiring from the spark plug HT leads, and place the wiring loom to one side.

13 Disconnect the hose from the coolant expansion tank, then disconnect the coolant hoses located at the left-hand end of the cylinder head.

14 At the rear of the engine compartment, disconnect the upper hose from the heater matrix. Unbolt the protective plate and coolant pipe.

15 Unbolt the crankcase breather line and heat shield from the turbocharger, disconnect the line from the camshaft cover and remove.

16 Remove the charcoal canister hose to the turbocharger from the camshaft cover.

17 Loosen the union bolt from the oil supply pipe to the top of the turbocharger.

18 Unscrew and remove the upper nuts securing the exhaust front pipe to the turbocharger **(see illustration)**.

19 Working under the car, unbolt and remove the right-hand driveshaft (Chapter 8 Section 2).

20 Remove the heat shield from above the removed driveshaft and then unscrew and remove the lower nuts securing the exhaust front pipe to the turbocharger.

21 Unbolt the exhaust mounting bracket from the crossmember, then loosen the clamp bolts and separate the front pipe from the centre section. Pull the front pipe off of the turbocharger, move it slightly to the rear and secure to one side. The front pipe is not removed completely, nor is the oxygen sensor wiring completely removed.

22 Unbolt the oil supply line and coolant supply pipe from the turbocharger.

23 Unbolt the oil return line from the turbocharger.

24 Unbolt the turbocharger support bracket.

25 To provide additional working room, unscrew the bolts securing the rear engine torque arm to the bottom of the transmission, then pull the engine approximately 20 mm to the rear and secure in this position with a ratchet-type strap.

26 As applicable, detach the coolant pipe located on the right-hand side of the engine, near the engine mounting. The retaining nuts only require loosening.

27 Unscrew the nuts securing the turbocharger/exhaust manifold to the cylinder head, noting the location of the coolant pipe bracket on the right-hand stud (where

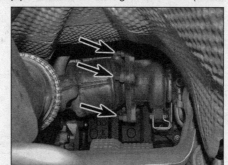

6.18 Remove the mounting nuts (one hidden)

applicable). Slide the manifold from the studs and withdraw the assembly upwards from the engine compartment.

28 With the turbocharger on the bench, the divert air valve may be removed by unscrewing the mounting bolts. Recover the O-ring seal and discard as a new one must be used on refitting.

Refitting

29 Refit the turbocharger by following the removal procedure in reverse, noting the following points:

a) *Renew all gaskets, sealing washers and O-rings.*

b) *When refitting the diverter air valve, make sure the location dowel enters the hole in the manifold before tightening the bolts.*

c) *Before reconnecting the oil supply pipe, fill the turbocharger with fresh oil using an oil can.*

d) *Tighten all nuts and bolts to the specified torque, where given.*

e) *Ensure that the air hose clips are securely tightened, to prevent air leaks.*

f) *Disable the ignition system and crank the engine over until the oil pressure warning light extinguishes.*

g) *When the engine is started after refitting, allow it to idle for approximately one minute to give the oil time to circulate around the turbine shaft bearings. Check for signs of oil or coolant leakage from the relevant unions.*

---

7 Intercooler – general information, removal and refitting

1 The intercooler is effectively an 'air radiator', used to cool the pressurised inlet air before it enters the engine.

2 When the turbocharger compresses the inlet air, one side-effect is that the air is heated, causing the air to expand. If the inlet air can be cooled, a greater effective volume of air will be inducted, and the engine will produce more power.

3 The compressed air from the turbocharger, which would normally be fed straight into the inlet manifold, is instead ducted around the engine to the base of the intercooler. The intercooler is mounted at the front of the car, in the airflow. The heated air entering the base of the unit rises upwards, and is cooled by the airflow over the intercooler fins, much as with the radiator. When it reaches the top of the intercooler, the cooled air is then ducted into the throttle housing.

Removal

4 The intercooler is located in front of the radiator on the lock carrier (the crossmember incorporating the headlights and radiator grille). Remove the radiator as described in Chapter 3 Section 3 for access to the intercooler. This procedure includes moving the lock carrier to its Service position.

7.5 Remove the charge air hoses by releasing the spring clips

8.4 Remove the bracket

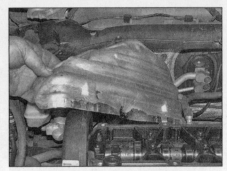

8.6 Remove the heat shield

5 Release the air ducts from the inlet and outlet elbows at the bottom of the intercooler **(see illustration)**.

6 Undo the screws securing the upper mounting rubbers to the lock carrier, and lift the intercooler upwards from the lower mountings, taking care not to damage the cooling fins.

7 Examine the mounting rubbers and renew them if necessary. Also, examine the intercooler for any damage, and check the air hoses for splits.

Refitting

8 Refitting is a reversal of removal, referring to Chapter 3 Section 3 when refitting the radiator. Ensure that the air hose clips are securely refitted, to prevent air leaks.

---

8 Exhaust manifold –
removal and refitting

Note: *This Section only covers removal of the exhaust manifold on non-turbo models. It is integral with the turbocharger on turbo models.*

Removal

1.6 litre engine

Note: *It is possible to remove the manifold*

down past the subframe and out of the vehicle, however for the home mechanic removing the manifold up and out of the engine bay is the easiest method, even though this approach requires removal of the inlet manifold.

1 Where fitted, remove the engine top cover.

2 Remove the plenum chamber front panel as described in Chapter 12 Section 19.

3 Remove the inlet manifold as described in Chapter 4A Section 10.

4 On the bulkhead at the rear of the engine compartment, disconnect the oxygen sensor wiring at the connector and then remove the wiring loom support bracket **(see illustration)**.

5 On engine code BGU, remove the EGR connecting pipe at the rear of the cylinder head.

6 Unscrew the nuts and remove the heat shield from the exhaust manifold **(see illustration)**.

7 Unbolt the front exhaust pipe together (with the catalytic converter) from the manifold, as described in Section 9. Note that with a good selection of tools, it is possible to access all the nuts from the top – there is no need to jack up the front of the vehicle.

8 Unscrew the nuts and withdraw the exhaust manifold from the cylinder head. Recover the gasket and discard, as a new one must be used on refitting **(see illustrations)**.

2.0 litre non-turbo

9 Apply the handbrake, then jack up the front of the vehicle and support it on axle stands (see *Jacking and vehicle support*). Remove the engine undershield.

10 Where fitted, remove the engine top cover.

11 Unscrew the nuts securing the exhaust twin-branch front pipe to the exhaust manifold, then unbolt the exhaust front mounting from the subframe, and the support bracket from the transmission. Lower the front pipe from the manifold and support it on an axle stand. Recover the gasket and discard it, as a new one must be used on refitting.

12 At the rear of the engine compartment, disconnect the oxygen sensor wiring on the bulkhead (BLR and BVY engines).

13 The EGR valve must now be removed and placed to one side, leaving the coolant hoses still attached. To do this, detach the EGR connecting pipe and release all wiring and hose supports from the valve, then unbolt and remove the valve.

14 Unbolt the heat shield from the exhaust manifold.

15 Unscrew the nuts and withdraw the exhaust manifold from the cylinder head (see

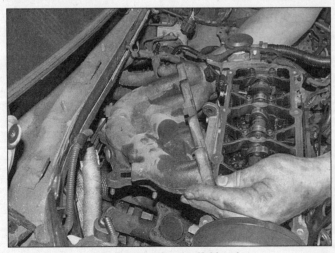

8.8a Remove the manifold and...

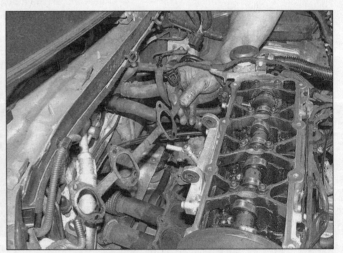

8.8b ...recover the gasket

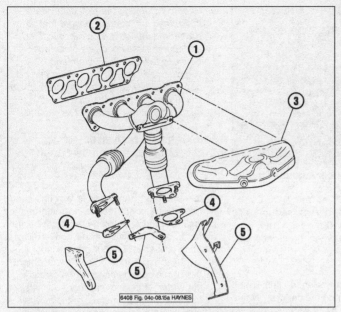

8.15a The 2.0 litre exhaust manifold (BLY and BVZ engines)

1 Exhaust manifold	4 Gaskets
2 Manifold gasket	5 Support brackets
3 Heat shield	

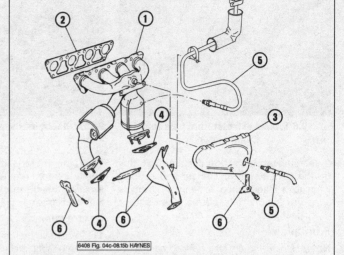

8.15b The 2.0 litre exhaust manifold (BLR and BVY engines)

1 Manifold and catalytic converter	4 Gaskets
2 Manifold gasket	5 O2 Sensors
3 Heat shield	6 Support brackets

illustrations). Discard the nuts as new ones must be used on refitting. Also, recover the gasket and discard.

2.0 litre turbo

16 The exhaust manifold is integral with the turbocharger. Refer to Section 6 for the removal and refitting procedure.

Refitting

17 Refitting is a reversal of the removal procedure, but fit new gaskets and tighten all nuts and bolts to the specified torque where given.

9 Exhaust system – component renewal

⚠️ *Warning: Allow ample time for the exhaust system to cool before starting work. In particular, note that the catalytic converter runs at very high temperatures. If there is any chance that the system may still be hot, wear suitable gloves. When removing the exhaust sections, take care not to damage the oxygen sensors if they are not removed from their locations.*

Removal

1 The original Seat system fitted in the factory is in three sections: the front section, the centre section, and the rear section and silencer(s) **(see illustration)** A catalytic converter is fitted to all models, and is variously located in the exhaust manifold, front pipe, or centre section. Some models have two catalytic converters fitted.

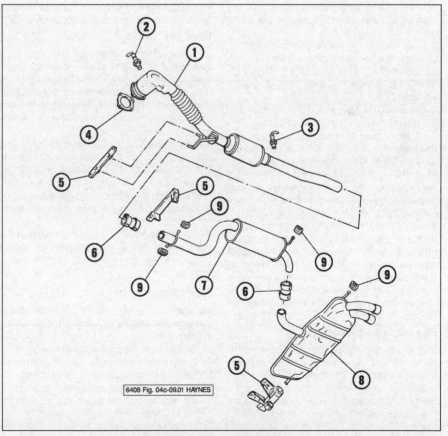

9.1 Typical exhaust system (TFSi shown)

1 Catalytic convertor and primary silencer	4 Gasket	7 Middle silencer
2 Oxygen sensor (pre-catalytic convertor)	5 Support bracket	8 Rear silencer
3 Oxygen sensor (post-catalytic convertor)	6 Connector sleeve	9 Rubber mounting

9.6 Typical exhaust front pipe rear connector

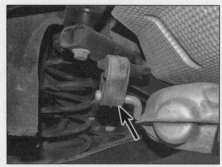

9.13 Exhaust rear silencer mounting rubber

2 To remove part of the system, first jack up the front or rear of the car and support it on axle stands (see *Jacking and vehicle support*). Alternatively, position the car over an inspection pit or on car ramps.

Front pipe

Note: *Handle the flexible, braided section of the front pipe carefully, and do not bend it excessively.*

3 Before removing the front section of the exhaust, establish how many oxygen sensors are fitted. Trace the wiring back from each sensor, and disconnect the wiring connector.
4 Unclip the oxygen sensor wiring from any clips or brackets, noting how it is routed for refitting.
5 If a new front section is being fitted, unscrew the oxygen sensors from the pipe, as applicable. If two sensors are fitted, note which fits where, as they should not be interchanged.
6 Unscrew the nuts or loosen the clamp at the rear of the pipe **(see illustration)**.
7 Unscrew and remove the nuts securing the front flange to the exhaust manifold or turbocharger. On some models, the shield over the right-hand driveshaft inner CV joint must be removed to improve access. Separate the front joint, and move it down sufficiently to clear the mounting studs.
8 Support the front of the pipe, then detach from the centre section. Where a clamp is fitted, slide the clamp either forwards or backwards to separate the joint. Twist the front pipe slightly from side-to-side, while pulling towards the front to release it from the rear section. When the pipe is free, lower it to the ground and remove it from under the car.

Rear pipe and silencers

9 If the factory-fitted Seat rear section is being worked on, examine the pipe between the two silencers for three pairs of punch marks, or three line markings. The centre marking indicates the point at which to cut the pipe, while the outer marks indicate the position for the ends of the new clamp required when refitting. Cut through the pipe using the centre mark as a guide, making the cut as square to the pipe as possible if either section is to be re-used.

10 If the factory-fitted rear section has already been renewed, loosen the nuts securing the clamp between the silencers so that the clamp can be moved.

Centre silencer

11 To remove the centre silencer, first unbolt the clamp or flange behind the catalyst. Where necessary, remove the bolts securing the two mounting brackets to the centre cradle under the car. To improve access, also remove the nuts securing the cradle to the underside of the car, and lower the cradle completely.
12 As necessary, slide the clamps at either end of the silencer section to release the pipe ends, and lower the silencer out of position.

Rear silencer

13 Depending on model, the rear silencer is supported either just at the very back, or in front and behind, by a rubber mounting which is bolted to the underside of the car **(see illustration)**. The silencer is attached to these mountings by metal pegs which push into the rubber section of each mounting.
14 Unscrew the bolts, and release the mounting(s) from the underside of the car. On models with two silencer mountings, it may prove sufficient to unbolt only one, and to prise the silencer from the remaining mounting, but for preference, both should be removed.
15 Where applicable, slide the clamp at the front end of the silencer section to release the pipe ends, and lower the silencer out of position.

Refitting

16 Each section is refitted by a reversal of the removal sequence, noting the following points:
a) Ensure that all traces of corrosion have been removed from the flanges or pipe ends, and renew all necessary gaskets.
b) If necessary, renew the clamps, and use the markings on the pipes as a guide to the clamp's correct fitted position.
c) Inspect the mountings for signs of damage or deterioration, and renew as necessary.
d) If using exhaust assembly paste, make sure this is only applied to joints downstream of the catalyst.

e) Prior to tightening the exhaust system mountings and clamps, ensure that all rubber mountings are correctly located and that there is adequate clearance between the exhaust system and vehicle underbody. Try to ensure that no unnecessary twisting stresses are applied to the pipes – move the pipes relative to each other at the clamps to relieve this.

10 Catalytic converter – general information and precautions

1 The catalytic converter is a reliable and simple device which needs no maintenance in itself, but there are some facts of which an owner should be aware if the converter is to function properly for its full service life:
a) DO NOT use leaded or lead-replacement petrol in a car equipped with a catalytic converter – the lead (or other additives) will coat the precious metals, reducing their converting efficiency and will eventually destroy the converter.
b) Always keep the ignition and fuel systems well-maintained in accordance with the manufacturer's schedule (see Chapter 1A).
c) If the engine develops a misfire, do not drive the car at all (or at least as little as possible) until the fault is cured.
d) DO NOT push- or tow-start the car – this will soak the catalytic converter in unburned fuel, causing it to overheat when the engine does start.
e) DO NOT switch off the ignition at high engine speeds – ie, do not 'blip' the throttle immediately before switching off the engine.
f) DO NOT use fuel or engine oil additives – these may contain substances harmful to the catalytic converter.
g) DO NOT continue to use the car if the engine burns oil to the extent of leaving a visible trail of blue smoke.
h) Remember that the catalytic converter operates at very high temperatures. DO NOT, therefore, park the car in dry undergrowth, over long grass or piles of dead leaves after a long run.
i) Remember that the catalytic converter is FRAGILE – do not strike it with tools during servicing work, and take care handling it when removing it from the car for any reason.
j) In some cases, a sulphurous smell (like that of rotten eggs) may be noticed from the exhaust. This is common to many catalytic converter-equipped cars, and has more to do with the sulphur content of the brand of fuel being used than the converter itself.
k) The catalytic converter, used on a well-maintained and well-driven car, should last for between 50 000 and 100 000 miles – if the converter is no longer effective, it must be renewed.

Chapter 4 Part D
Emission control and exhaust systems – diesel engines

Contents

Degrees of difficulty

Easy, suitable for novice with little experience	**Fairly easy,** suitable for beginner with some experience	**Fairly difficult,** suitable for competent DIY mechanic	**Difficult,** suitable for experienced DIY mechanic	**Very difficult,** suitable for expert DIY or professional

Specifications

Engine codes*

1.9 litre – 'PD' (unit injector)	BJB, BKC, BLS, BXE and BXF
2.0 litre – 'PD' (unit injector)	BKD, BMN, and BMM
1.6 litre – common rail injection	CAYC
2.0 litre – common rail injection	CFJA, CLCB, CEGA and CFHC

*See 'Vehicle identification' at the end of the manual for the location of the engine code markings

Emission control applications

1.6 litre CR	EGR and combined catalytic converter and particulate filter. EU5 emissions standard.
1.9 litre PD	EGR and combined catalytic converter. BLS and BXJ engines have a particulate filter. EU4 emissions standard.
2.0 litre PD (BMM 8 valve)	EGR and combined catalytic converter and particulate filter. EU4 emissions standard.
2.0 litre PD (16 valve)	EGR and catalytic converter. (Engine BMN has a particulate filter). EU4 emissions standard.
2.0 litre CR	EGR and combined catalytic converter and particulate filter. EU5 emissions standard.

Torque wrench settings

	Nm	lbf ft
1.9/2.0 litre PD engines		
EGR cooler/valve to bracket/cylinder head	10	7
EGR coolant outlet pipe	22	16
EGR recirculation pipe (to manifold)	25	18
EGR recirculation pipe (to valve)	22	16
EGR valve to cooler	10	7
EGR valve to inlet manifold (BMM engine)	10	7
Exhaust gas temperature sensor	45	33
Exhaust system clamp:		
Sleeve with individual clamps	25	18
Sleeve with single clamp	35	26
Catalytic converter support bracket	40	30
Lambda (O2) sensor	50	37
Particulate filter/catalytic converter clamp (to turbocharger)	7	5
Particulate filter support bracket (on subframe)	25	18
Particulate filter pressure sensor line (at filter)	45	33
Turbocharger/manifold heat shield nuts*	25	18
Turbocharger/manifold to cylinder head nuts/bolts*	25	18
Turbocharger oil drain bolts	20	15
Turbocharger oil drain connector (to block)	40	30
Turbocharger oil drain pipe union nut	22	16
Turbocharger oil feed supply connector	30	22
Turbocharger oil feed pipe union nut	22	16
1.6/2.0 litre common rail engines		
Catalytic converter/particulate filter mounting bracket nuts/bolts . . .	25	18
Particulate filter/catalytic converter clamp (to turbocharger)	7	5
EGR mounting bolts	9	6
EGR recirculation pipe bolts	9	6
EGR recirculation pipe nuts	24	18
Exhaust gas temperature probe	60	44
Exhaust system clamp:		
Sleeve with individual clamps	25	18
Sleeve with single clamp	35	26
Lambda (O2) sensor	50	37
Particulate filter pressure sensor line (at filter)	45	33
Particulate filter support bracket (on subframe)	25	18
Turbocharger/manifold to cylinder head nuts*	24	18
Turbocharger oil drain banjo bolt*	30	22
Turbocharger oil feed pipe	22	16
Turbocharger oil drain (at turbocharger) bolts*	17	13

*Use new fixings

1 General Information

Emission control systems

1 All diesel-engined models have a crankcase emission control system, and in addition, are fitted with a catalytic converter. All diesel engines are fitted with an Exhaust Gas Recirculation (EGR) system to reduce exhaust emissions.

Crankcase emission control

2 To reduce the emission of unburned hydrocarbons from the crankcase into the atmosphere, the engine is sealed and the blow-by gases and oil vapour are drawn from inside the crankcase, through a wire mesh oil separator, into the inlet tract to be burned by the engine during normal combustion.

3 Under conditions of high manifold depression, the gases will be sucked positively out of the crankcase. Under conditions of low manifold depression, the gases are forced out of the crankcase by the (relatively) higher crankcase pressure. If the engine is worn, the raised crankcase pressure (due to increased blow-by) will cause some of the flow to return under all manifold conditions.

Exhaust emission control

4 An oxidation catalyst is fitted in the exhaust system of all diesel-engined models. This has the effect of removing a large proportion of the gaseous hydrocarbons, carbon monoxide and particulates present in the exhaust gas. Some models covered in this manual, have a particulate filter, which is fitted behind the catalytic converter in order to filter out soot particles.

5 An Exhaust Gas Recirculation (EGR) system is fitted to all diesel-engined models. This reduces the level of nitrogen oxides produced during combustion by introducing a proportion of the exhaust gas back into the inlet manifold under certain engine operating conditions. The system is controlled electronically by the diesel engine management ECU.

Exhaust systems

6 The exhaust system consists of the exhaust manifold, front flexible pipe with catalytic converter/particulate filter and rear pipe section including silencer. The turbocharger is integral with the exhaust manifold, and is driven by the exhaust gases.

7 The system is supported by various metal brackets screwed to the vehicle floor, with rubber vibration dampers fitted to suppress noise.

2 Crankcase emission system – general information

1 The crankcase emission control system consists of hoses connecting the crankcase to the air cleaner, cylinder head cover and/or inlet manifold **(see illustration)**.

2 The system requires no attention other than to check at regular intervals that the hoses and pressure-regulating valve are free of blockages and in good condition.

3 Exhaust Gas Recirculation (EGR) system – component removal and refitting

1 The EGR system consists of the EGR valve, which directs exhaust gas from the exhaust manifold to the inlet manifold, and a control (changeover) valve that actuates the system according to engine load and speed **(see illustration)**. There is a vacuum-operated changeover unit, which is fitted on the EGR cooler for recirculating the exhaust gas. All this is connected by a series of vacuum hoses, and metal corrugated cooler pipes. There is also a recirculation potentiometer fitted to the EGR valve, this is controlled directly from the engine management ECU, which then activates the control motor.

2 A wide range of EGR control valves, coolers, sensors and actuators are fitted to the Leon range. The exact detail of removal (and refitting) varies between models, but the principle of removal is the same for all models **(see illustrations)**.

2.1 Disconnecting the breather hose

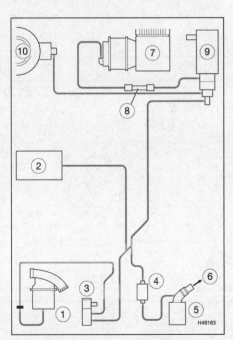

3.1 Typical vacuum hose layout

1 EGR cooler changeover vacuum unit
2 Cylinder head cover
3 EGR cooler changeover valve
4 Non-return valve
5 Vacuum pump
6 To brake servo connection
7 Air filter
8 Silencer
9 Charge pressure control solenoid valve
10 EGR valve

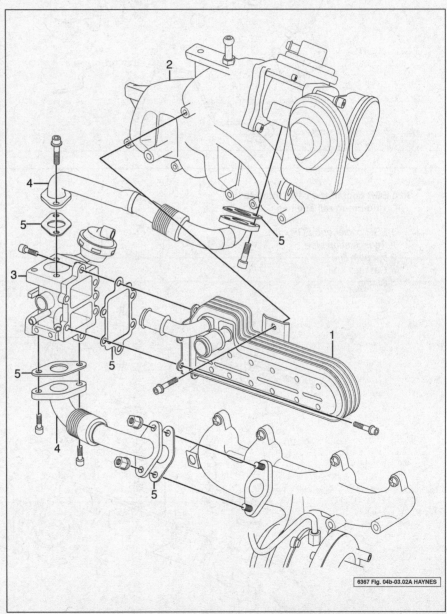

6367 Fig. 04b-03.02A HAYNES

3.2a Typical EGR and cooler fitted to most 1.9 litre PD engines

1 EGR cooler
2 Inlet manifold with control valve
3 By-pass (change-over) valve
4 Connector pipes
5 Gaskets

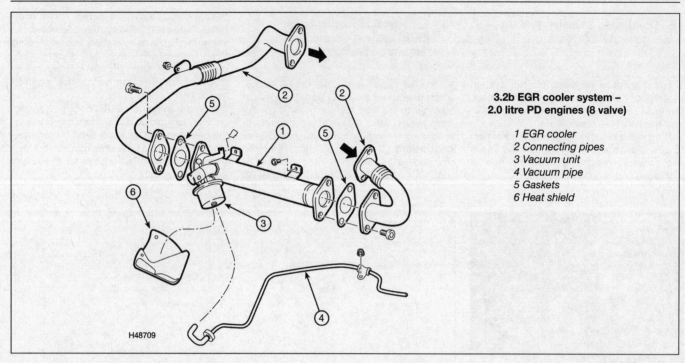

3.2b EGR cooler system –
2.0 litre PD engines (8 valve)

1 EGR cooler
2 Connecting pipes
3 Vacuum unit
4 Vacuum pipe
5 Gaskets
6 Heat shield

H48709

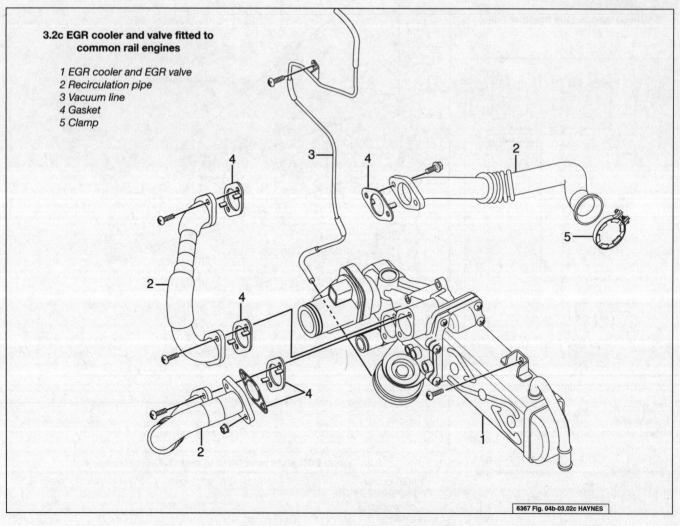

3.2c EGR cooler and valve fitted to
common rail engines

1 EGR cooler and EGR valve
2 Recirculation pipe
3 Vacuum line
4 Gasket
5 Clamp

6367 Fig. 04b-03.02c HAYNES

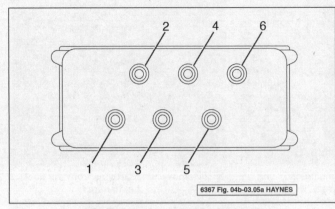

3.5a The six pin solenoid valve block

1 Vaccum supply
2 EGR valve
3 By-pass (change-over) flap of
  EGR cooling
4 Air filter
5 Vacuum reservoir
6 Turbocharger vacuum control

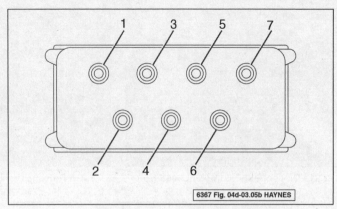

3.5b The seven pin valve block

1 Vacant
2 Vacuum supply
3 EGR valve
4 By-pass (change-over) flap of
  EGR cooling
5 Air filter
6 Vacuum reservoir
7 Turbocharger vacuum control

3 The EGR valve and cooler are either integrated as a single assembly or fitted as separate assemblies. The gas cooler uses coolant from the engine cooling system to cool the exhaust gases before passing them to the inlet manifold. The system is activated by a solenoid valve mounted at the front of the inlet manifold, which is controlled by the ECU.

4 Where an integrated cooler/EGR valve is fitted, it is joined to the exhaust manifold and cylinder head by flanged corrugated metal pipes. The system is activated by a solenoid valve mounted at the front of the inlet manifold, which is controlled by the ECU.

EGR cooler 'change-over' solenoid valve

5 The EGR vacuum-solenoid valve is either mounted on the front of the inlet manifold or integrated into the vacuum solenoid valve block **(see illustrations)**. Do not confuse the EGR solenoid with the turbo charge pressure control solenoid, which is mounted on the bulkhead.

6 Disconnect the wiring plug from the solenoid valve **(see illustration)**.

7 Unclip the valve from the mounting bracket **(see illustration)**, then identify the vacuum

hoses for refitting and disconnect them from the valve.

8 Refitting is a reversal of removal. Ensure that the hoses and wiring plug are reconnected securely and correctly.

EGR valve

Engine code BMM

9 Remove the throttle valve housing from the inlet manifold, as described in.

10 Undo the bolts securing the pipe to the left-hand side of the EGR valve **(see illustration)**.

11 Disconnect the wiring plug connector from the right-hand side of the EGR valve.

12 Undo the bolts and detach the valve from the inlet manifold. Discard the O-ring seal – a new one must be fitted.

13 Refitting is a reversal of removal, using a new EGR valve O-ring seal, and tightening the retaining bolts to the specified torque.

All other engine codes

14 The EGR valve is incorporated into the EGR cooler, remove the cooler as described in paragraphs 20 to 26. At the time of writing, it was not possible to separate the EGR valve from the cooler.

3.5c EGR cooler solenoid valve

EGR cooler

15 Remove the exhaust front pipe/catalytic converter/ particulate filter as described in Section 9.

16 Drain the cooling system as described in Chapter 1B Section 33. If required the coolant hoses can be clamped before removal, to save draining the complete system.

Engine codes BMM

17 Release the retaining clips and disconnect the coolant hoses from the EGR cooler. Be prepared for coolant spillage.

3.6 Disconnect the wiring connector …

3.7 …then unclip the valve from the bracket

3.10 EGR pipe-to-valve bolts

3.21a Disconnect the coolant hose from the cooler

3.21b Release the retaining clips and remove hose

3.22 Remove the EGR pipe from the cooler to the manifold

3.23a Undo the EGR pipe bolts to the cooler

3.23b Undo the EGR pipe bolts to the cylinder head

18 Undo the retaining bolts and disconnect the EGR metal connecting pipes from each end of the cooler.

19 Undo the upper bolts and lower banjo bolt and remove the turbocharger oil return pipe/support from the rear of the engine.

20 Disconnect the vacuum hose from vacuum unit on the lower part of the cooler, then undo the retaining bolts, and remove the cooler from the rear of the engine.

All other engine codes

21 Release the retaining clips and disconnect the coolant hoses from each end of the EGR cooler **(see illustrations)**. Be prepared for coolant spillage.

22 Undo the two retaining screws from the cooler, the two retaining nuts from the exhaust manifold and remove the EGR metal connecting pipe **(see illustration)**.

23 Undo the two retaining screws from the cooler, the two retaining screws from the cylinder head and remove the EGR metal connecting pipe **(see illustrations)**.

24 Undo the upper bolts and lower banjo bolt and remove the turbocharger oil return pipe/support from the rear of the engine **(see illustrations)**.

25 Disconnect the wiring plug connector from the EGR valve **(see illustration)**.

26 Disconnect the vacuum hose from vacuum unit on the lower part of the cooler **(see illustration)**, then undo the four retaining bolts, and remove the cooler from the rear of the engine.

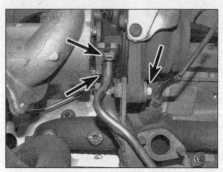

3.24a Undo the upper bolts...

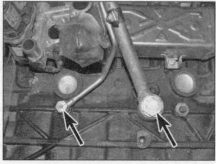

3.24b ...and lower banjo bolts ...

3.24c ...then remove the pipes and support bracket

3.25 Disconnect the wiring connector from the EGR valve

3.26 EGR cooler securing bolts

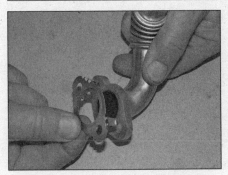

3.27 Fit new gaskets to the EGR pipes

5.6a Disconnecting the breather hose...

5.6b ...undo the retaining screw...

27 Refitting is a reversal of removal, using new gaskets **(see illustration)**, and tightening the retaining bolts to the specified torque.

4 Turbocharger – general information and precautions

General information

1 A turbocharger is fitted to all engines, and is integral with the exhaust manifold.
2 The turbocharger increases engine efficiency by raising the pressure in the inlet manifold above atmospheric pressure. Instead of the air simply being sucked into the cylinders, it is forced in.
3 Energy for the operation of the turbocharger comes from the exhaust gas. The gas flows through a specially shaped housing (the turbine housing) and in so doing, spins the turbine wheel. The turbine wheel is attached to a shaft, at the end of which is another vaned wheel, known as the compressor wheel. The compressor wheel spins in its own housing, and compresses the inducted air on the way to the inlet manifold.
4 Between the turbocharger and the inlet manifold, the compressed air passes through an intercooler (see Section 7 for details). The purpose of the intercooler is to remove from the inducted air some of the heat gained in being compressed. Because cooler air is denser, removal of this heat further increases engine efficiency.
5 Boost pressure (the pressure in the inlet manifold) is limited by a wastegate, which diverts the exhaust gas away from the turbine wheel in response to a pressure-sensitive actuator.
6 The turbo shaft is pressure-lubricated by an oil feed pipe from the engine oil filter mounting. The shaft 'floats' on a cushion of oil. Oil is returned to the sump through a return pipe that connects to the rear of the cylinder block.

Precautions

The turbocharger operates at extremely high speeds and temperatures. Certain precautions must be observed to avoid premature failure of the turbo, or injury to the operator.

• *Do not operate the turbo with any parts exposed – foreign objects falling onto the rotating vanes could cause excessive damage and (if ejected) personal injury.*
• *Cover the turbocharger air inlet ducts to prevent debris entering, and clean using lint-free cloths only.*
• *Do not race the engine immediately after start-up, especially if it is cold. Give the oil a few seconds to circulate.*
• *Observe the recommended intervals for oil and filter changing, and use a reputable oil of the specified quality. Neglect of oil changing, or use of inferior oil, can cause carbon formation on the turbo shaft and subsequent failure. Thoroughly clean the area around all oil pipe unions before disconnecting them, to prevent the ingress of dirt. Store dismantled components in a sealed container to prevent contamination.*

5 Turbocharger and exhaust manifold – removal and refitting

Note: *This Section describes removal of the turbocharger together with the exhaust manifold. On all diesel engines covered in this manual, the turbocharger cannot be removed from the exhaust manifold.*

Removal

1 Apply the handbrake, then jack up the front of the vehicle and support it on axle stands

5.6c ...and disconnect the air intake hose

(see *Jacking and vehicle support*). Remove the engine compartment undertray and the engine top cover.
2 Remove the plenum chamber front panel as described in Chapter 12 Section 19.
3 Remove the air filter assembly and ducting as described in.
4 On 1.9 and 2.0 litre PD engines, remove the subframe (Chapter 10 Section 22) and the right-hand driveshaft (Chapter 8 Section 2).
5 Remove the exhaust front pipe/catalytic converter/ particulate filter as described in Section 9.
6 Undo the retaining bolt, unclip the breather pipe and disconnect the air intake hose from the turbocharger **(see illustrations)**.
7 Disconnect the wiring connector and vacuum pipe from the vacuum unit on top of the turbocharger **(see illustrations)**.
8 Trace the wiring from the exhaust gas

5.7a Disconnect the wiring connector...

5.7b ...and the vacuum hose from the vacuum pump

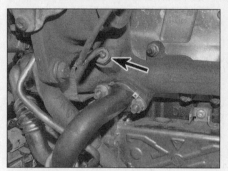

5.8a Locate the exhaust gas temperature sensor…

5.8b …and disconnect the wiring connector on the bulkhead

5.9 Disconnect the charge air pipe/ pulsation damper from the turbo

temperature sensor and then disconnect the sensor wiring plug. Detach the connector from the bracket **(see illustrations)**.

9 Slacken the retaining clips, undo the two retaining bolts and withdraw the pulsation damper from the charge air ducting to the turbo **(see illustration)**.

10 Undo the retaining screws and remove the EGR metal connecting pipes, with reference to Section 3.

11 Undo the two upper securing bolts and the lower banjo bolts for the oil supply and return for the turbocharger. Then remove the oil return pipe/support from the rear of the engine **(see illustrations 3.24a, 3.24b & 3.24c)**.

12 Undo the two retaining screws and remove the heat shield from the manifold **(see illustration)**.

13 Check around the turbocharger/exhaust manifold and unclip any wiring or hoses still connected, noting their fitted position and the routing of all cables.

14 Undo the retaining nuts and manoeuvre the exhaust manifold/turbocharger assembly out from the engine compartment **(see illustration)**.

Refitting

15 Refit the turbocharger by following the removal procedure in reverse, noting the following points:

a) Renew exhaust manifold gaskets, sealing washers and O-rings **(see illustration)**.

b) Fit a new gasket to the oil return pipe upper flange **(see illustration)**.

c) Fit a new banjo bolt to the oil return pipe **(see illustration)**.

d) Fit new sealing washers to the oil supply pipe **(see illustration)**.

e) Before reconnecting the oil supply pipe, fill the turbocharger with fresh oil using an oil can.

f) Tighten all nuts and bolts to the specified torque, where given.

g) Ensure that the air hose clips are securely tightened, to prevent air leaks.

h) When the engine is started after refitting, allow it to idle for approximately one minute to give the oil time to circulate around the turbine shaft bearings. Check for signs of oil or coolant leakage from the relevant unions.

5.12 Undo the two retaining screws from the heat shield

5.14 Remove the turbocharger/ manifold from the engine compartment

5.15a Fit a new exhaust manifold gasket…

5.15b …oil return/supply pipe gasket…

5.15c …oil return banjo bolt…

5.15d …and oil supply sealing washers

6.2 Charge pressure solenoid valve

6.3 Disconnect the wiring connector …

6.4 … unclip the vacuum pipes …

6.5 … and disconnect them from the solenoid valve

6 Turbocharger charge control system components – description, removal and refitting

Description

1 The turbocharger pressure valve is operated by vacuum supplied by the brake vacuum pump mounted on the left-hand end of the cylinder head. The vacuum supply is controlled by an electrically operated solenoid valve activated by the engine management ECU. The solenoid valve is mounted on the bulkhead at the rear of the engine compartment. The valve has three hoses connecting it to the air cleaner, turbocharger and vacuum pump. It is important that the hoses are connected correctly to the solenoid valve (see illustration 3.1).

Charge pressure control solenoid valve

2 The charge pressure solenoid valve is mounted at the rear of the engine compartment (see illustration) on the bulkhead.

3 Disconnect the wiring from the valve (see illustration).
4 Undo the two retaining nuts and withdraw the solenoid valve away from the bulkhead, unclipping the vacuum hoses from their retaining clips (see illustration).
5 Remove the vacuum hoses from the ports on the control solenoid valve (see illustration), noting their order of connection carefully to aid correct refitting.
6 Refitting is a reversal of removal.

Charge pressure valve (wastegate)

7 The charge pressure valve is an integral part of the turbocharger, and cannot be renewed separately.

7 Intercooler – general information, removal and refitting

General information

1 The intercooler is effectively an 'air radiator', used to cool the pressurised inlet air before it enters the engine.

2 When the turbocharger compresses the inlet air, one side-effect is that the air is heated, causing it to expand. If the inlet air can be cooled, a greater effective volume of air will be inducted, and the engine will produce more power.
3 The compressed air from the turbocharger, which would normally be fed straight into the inlet manifold, is instead ducted around the engine to the base of the intercooler. The intercooler is mounted at the front of the car, in the airflow. The heated air entering the base of the unit rises upwards, and is cooled by the airflow over the intercooler fins, much as with the radiator. When it reaches the top of the intercooler, the cooled air is then ducted into the throttle housing.
4 There is a charge air pressure/temperature sensor fitted to the air inlet ducting from the intercooler to the inlet manifold. See, for removal and refitting procedure.

Removal

5 The intercooler is located in front of the radiator on the lock carrier (the crossmember incorporating the headlights and radiator

7.6a Remove the charge air hoses …

7.6b …from both sides of the intercooler

7.6c Where fitted release the spring clips…

7.6d …and remove the hoses

7.7a Undo the retaining screws …

the spring type clips on some models) and release the air ducts from the inlet and outlet elbows at the bottom of the intercooler **(see illustrations)**.

7 Working from the front of the vehicle, undo the two retaining screws (one at each side) to release the intercooler side mountings **(see illustrations)**.

8 Tilt the intercooler and air-conditioning condenser backwards slightly, taking care not to damage the refrigerant pipes.

Warning: Refer to the precautions given in Chapter 3 Section 1, before working on the air-conditioning system.

9 With the intercooler tilted back, undo the screw at the right-hand lower side of the intercooler to release the refrigerant pipe mounting bracket **(see illustration)**.

10 Working at the front of the vehicle, undo the four retaining screws (two at each side) and pull the intercooler upwards to release it from the lower crossmember, then withdraw it from out of the engine compartment **(see illustrations)**. Make sure the condenser is secured safely to the front crossmember, to prevent any damage to the fins or the refrigerant pipes.

7.7b …to release the intercooler mountings

7.9 Undo the retaining screw

Refitting

11 Refitting is a reversal of removal, noting the following points:
a) *Make sure the mountings on the side of the intercooler are not damaged* **(see illustration)**.

grille). Remove the radiator as described in Chapter 3 Section 3, for access to the intercooler. This procedure may include

moving the lock carrier to its Service position (Chapter 3 Section 3).

6 Slacken the retaining clips (or release

7.10a Undo the two retaining screws…

7.10b …and the two from the other side of the condenser

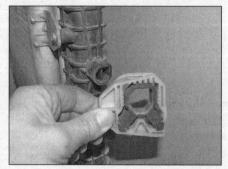

7.11a Check the condition of the mounting …

7.11b ...and the rubber grommets in the crossmember

8.3 Unclip the wiring loom

8.4 Undo the exhaust/turbo securing clamp

b) *Check the rubber grommets are still located in the lower crossmember* **(see illustration)**.
c) *Refer to Chapter 3 Section 3, when refitting the radiator.*
d) *Ensure that the air hose clips are securely refitted, to prevent air leaks.*

8.5 Unclip the hoses from the retaining clips on the timing cover

8.6 Undo the retaining bolt

8 Catalytic converter/particulate filter – removal and refitting

⚠️ *Warning: Allow ample time for the exhaust system to cool before starting work. In particular, note that the catalytic converter runs at very high temperatures. If there is any chance that the system may still be hot, wear suitable gloves.*

Note: *This Section describes removal of the exhaust front flexible pipe complete with the catalytic converter or the combined catalytic and particulate filter (depending on the model). On all engines removal of the subframe allows easy access to all the fixings, and whilst this is not strictly necessary it may be a better option if corroded fasteners are encountered.*

1 Disconnect the battery negative lead and position it away from the terminal – see 'disconnecting the battery' in Chapter 5A Section 3.
2 Working at the engine compartment bulkhead, disconnect the wiring plugs for the oxygen sensor and exhaust gas temperature sensors.
3 Trace the wiring back from the sensors, noting their routing and release the wiring harness from the retaining clips **(see illustration)**. Note the sensors can stay fitted to the exhaust at this point, if a new catalytic converter/particulate is fitted, then the sensors can be swapped over with the system of the vehicle.
4 Slacken the retaining bolt to release the clamp between the turbocharger and the catalytic converter/particulate filter **(see illustration)**.
5 Release the hoses from the timing cover **(see illustration)**, release the retaining clip and disconnect the pressure hose from

the exhaust to the differential pressure sender.
6 Working at the right-hand rear corner of the cylinder head, slacken and remove the upper mounting bolt **(see illustration)**.
7 Apply the handbrake, then jack up the front of the vehicle and support it on axle stands (see *Jacking and vehicle support*). Remove the engine compartment undertray.
8 If required the front subframe should now be removed, as described in Chapter 10 Section 22. On all engines remove the pendulum mount as described in Chapter 2C Section 16 (1.6 litre CR engines) Chapter 2D Section 16 (2.0 litre CR engines) or Chapter 2E Section 18 (PD engines).
9 Where fitted, undo the two retaining bolts and remove the heat shield from above the right-hand drive shaft. Note that removal of

the right-hand driveshaft will ease access considerably (Chapter 8 Section 2).
10 To prevent any damage to the front flexible pipe use a couple of pieces of metal bar and some retaining clips to prevent the flexible pipe from bending to far and getting damaged **(see illustration)**.
Caution: Handle the flexible, braided section of the front pipe carefully, and do not bend it excessively, if it is bent more than 10° it could get damaged.
11 Slacken the retaining bolts and slide the exhaust clamp to the rear of the vehicle, to disengage the exhaust front pipe from the rear section. Where required remove the transmission tunnel support struts **(see illustration)**.
12 Undo the nuts securing the catalytic converter/particulate filter lower support

8.10 Support the flexible pipe with flat metal bar and securing clips

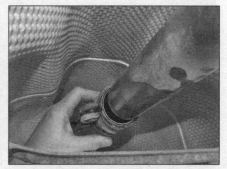

8.11 Slide the exhaust clamp along the exhaust

8.12a Undo the upper securing nuts ...

8.12b ...and lower securing nuts ...

9 Exhaust system – component renewal

⚠️ **Warning: Allow ample time for the exhaust system to cool before starting work. In particular, note that the catalytic converter runs at very high temperatures. If there is any chance that the system may still be hot, wear suitable gloves.**

Removal

1 A variety of exhaust systems are fitted to the Leon range of vehicles. The original Seat system fitted in the factory is in two sections. The front section includes the catalytic converter (or 'catalyst'), and particulate filter. The rear section, which connects to the front section under the centre of the vehicle has a silencer fitted. On some models the silencer is alongside the fuel tank, on other models the silencer is at the rear of the vehicle **(see illustrations)**.

bracket to the cylinder block **(see illustrations)**.

13 Carefully manoeuvre the catalytic converter/particulate filter downwards from place, rotating the assembly as it is removed. Take care not to damage any wiring, making sure that all wiring has been disconnected and release from their retaining clips. Also

take care not to bend the flexible section of the exhaust pipe at an angle of more than 10°.

Refitting

14 Refit the exhaust catalytic converter/ particulate filter assembly by following the removal procedure in reverse, but renew all gaskets **(see illustration)** and tighten all nuts and bolts to the specified torque, where given.

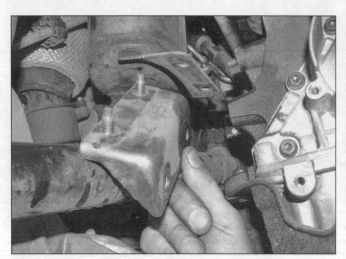

8.12c ...then remove the lower mounting bracket

8.14 Renew the gasket between the particulate filter and the turbo

9.1a Rear exhaust system – 2.0 litre CR engines

9.1b Rear exhaust system – 1.9 litre PD engines

9.4 Slacken the bolts on the exhaust pipe clamp

9.5a Rear silencer exhaust rubber mounting

9.5b Rubber mounting on centre section of exhaust

9.6 Mounting bracket across exhaust tunnel

2 To remove part of the system, first jack up the front or rear of the car and support it on axle stands (see *Jacking and vehicle support*). Alternatively, position the car over an inspection pit or on car ramps.

Front pipe (catalytic converter/particulate filter)

Caution: Handle the flexible, braided section of the front pipe carefully, and do not bend it excessively, if it is bent more than 10° it could get damaged.

3 The front flexible pipe section is part of the catalytic converter/particulate filter, remove the front pipe assembly as described in Section 8.

Rear pipe and single silencer

4 Unscrew the clamp bolt securing the rear section of exhaust to the front section **(see illustration)**.

5 Release the silencer from the rubber mountings, then twist it from the centre section and withdraw it from under the car **(see illustrations)**.

6 If required, undo the retaining bolts and remove the mounting brackets from across the exhaust tunnel, to make removal easier **(see illustration)**.

Rear pipe with two silencers

7 If the factory-fitted Seat rear section is being worked on, examine the pipe between the two silencers for three pairs of punch marks, or three line markings. The centre marking indicates the point at which to cut the pipe, while the outer marks indicate the position for the ends of the new clamp required when refitting. Cut through the pipe using the centre mark as a guide, making the cut as square to the pipe as possible if either resulting section is to be re-used.

8 If the factory-fitted rear section has already been renewed, unscrew the clamp bolts between the two sections, to remove individual silencers.

Refitting

9 Each section is refitted by a reversal of the removal sequence, noting the following points:

a) *Ensure that all traces of corrosion have been removed from the flanges or pipe ends, and renew all necessary gaskets.*

b) *The design of the clamps used between the exhaust sections means that they play a greater role in ensuring a gas-tight seal – fit new clamps if they are in less than perfect condition.*

c) *When fitting the clamps, use the markings on the pipes as a guide to the clamp's correct fitted position.*

d) *Inspect the mountings for signs of damage or deterioration, and renew as necessary.*

e) *If using exhaust assembly paste, make sure this is only applied to joints downstream of the catalyst.*

f) *Prior to tightening the exhaust system mountings and clamps, ensure that all rubber mountings are correctly located and that there is adequate clearance between the exhaust system and vehicle underbody.*

10 Catalytic converter – general information and precautions

1 The catalytic converter fitted to diesel models is simpler than that fitted to petrol models, but it still needs to be treated with respect to avoid problems. The converter is a reliable and simple device which needs no maintenance in itself, but there are some facts of which an owner should be aware if the converter is to function properly for its full service life:

a) *DO NOT use fuel or engine oil additives – these may contain substances harmful to the catalytic converter.*

b) *DO NOT continue to use the car if the engine burns (engine) oil to the extent of leaving a visible trail of blue smoke.*

c) *Remember that the catalytic converter is FRAGILE – do not strike it with tools during servicing work, and take care handling it when removing it from the car for any reason.*

d) *The catalytic converter, used on a well-maintained and well-driven car, should last for between 50,000 and 100,000 miles – if the converter is no longer effective, it must be renewed.*

Chapter 5 Part A
Starting and charging systems

Contents

Degrees of difficulty

Easy, suitable for novice with little experience 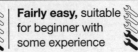	**Fairly easy,** suitable for beginner with some experience	**Fairly difficult,** suitable for competent DIY mechanic	**Difficult,** suitable for experienced DIY mechanic	**Very difficult,** suitable for expert DIY or professional

Specifications

General
System type . . . . 12 volt, negative earth

Starter motor
Rating:
  Petrol engines . . . . 12V, 1.1 kW
  Diesel engines . . . . 12V, 2.0 kW

Battery
Ratings . . . . 36 to 72 Ah (depending on model and market)

Alternator
Rating . . . . 55, 60, 70 or 90 amp
Minimum brush length . . . . 5.0 mm

Torque wrench settings

	Nm	lbf ft
Alternator mounting bolts	25	18
Alternator mounting bracket:		
Petrol engines	45	33
Diesel engines	45	33
Battery clamping plate bolt	22	16
Starter mounting bolts:		
M10	40	30
M12	80	60

1 General information and precautions

General information

1 The engine electrical system consists of the charging and starting systems. Because of their engine-related functions, these are covered separately from the body electrical devices such as the lights, instruments, etc (which are covered in Chapter 12). On petrol engine models refer to Chapter 5B for information on the ignition system, and on diesel models refer to Chapter 5C for the preheating system.

2 The electrical system is of the 12 volt negative earth type.

3 The battery is a maintenance-free (sealed for life) type and is charged by the alternator, which is belt-driven from the crankshaft pulley.

4 The starter motor is of the pre-engaged type, with an integral solenoid. On starting, the solenoid moves the drive pinion into engagement with the flywheel ring gear before the starter motor is energised. Once the engine has started, a one-way clutch prevents the motor armature being driven by the engine until the pinion disengages from the flywheel.

5 Further details of the various systems are given in the relevant Sections of this Chapter. While some repair procedures are given, the usual course of action is to renew the component concerned.

Precautions

 Warning: It is necessary to take extra care when working on the electrical system to avoid damage to semi-conductor devices (diodes and transistors), and to avoid the risk of personal injury. In addition to the precautions given in ' Safety first 0 Section 2 !', observe the following when working on the system:

 Warning: Always remove rings, watches, etc, before working on the electrical system. Even with the battery disconnected, capacitive discharge could occur if a component's live terminal is earthed through a metal object. This could cause a shock or nasty burn.

Warning: Do not reverse the battery connections. Components such as the alternator, electronic control units, or any other components having semi-conductor circuitry could be irreparably damaged.

Warning: Never disconnect the battery terminals, the alternator, any electrical wiring or any test instruments when the engine is running.

Warning: Do not allow the engine to turn the alternator when the alternator is not connected.

Warning: Never test for alternator output by flashing the output lead to earth.

Warning: Always ensure that the battery negative lead is disconnected when working on the electrical system.

Warning: If the engine is being started using jump leads and a slave battery, connect the batteries positive-to-positive and negative-to-negative (see Jump starting at the beginning of the manual). This also applies when connecting a battery charger.

Warning: Before using electric-arc welding equipment on the car, disconnect the battery, alternator and components such as the electronic control units (where applicable) to protect them from the risk of damage.

Caution: All audio units fitted as standard equipment by Seat have a built-in security code to deter thieves. Under normal circumstances, if the power source to the unit is cut, the audio unit will work as soon as the power is reconnected. If the audio unit is replaced the anti-theft code of the new unit must be entered. Always ensure that the code is available.

2 Battery – testing and charging

Testing

1 All original equipment batteries are sealed for life maintenance-free batteries. These may be one of three types:

- Standard lead acid. Sealed for life with a visual charge indicator fitted.
- Absorbent Glass Mat (AGM) Lead acid battery with the electrolyte absorbed in a glass matrix. No visual indicator fitted.
- Enhanced Flooded Battery (EFB). Visual indicator fitted. Used on some stop-start models.

2 Topping-up and testing of the electrolyte in each cell is not possible. The condition of the battery can therefore only be tested using a battery condition indicator or a voltmeter.

3 All models (except those with AGM batteries) are fitted with a maintenance-free battery with a built-in charge condition indicator. The indicator is located in the top of the battery casing, and indicates the condition of the battery from its colour. If the indicator shows green, then the battery is in a good state of charge. If the indicator turns darker, eventually to black, then the battery requires charging, as described later in this Section. If the indicator shows clear/yellow, then the electrolyte level in the battery is too low to allow further use, and the battery should be renewed. Do not attempt to charge, load or jump-start a battery when the indicator shows clear/yellow.

Note: *From 2009 Seat introduced batteries that only have a two colour visual indicator. If the indicator shows black there is sufficient electrolyte in the cell, if the indicator shows light yellow the electrolyte level is low and the battery must be replaced. Note that the indicator is only a guide to the electrolyte level on one cell and is not a guide to the state of charge.*

4 If testing the battery using a voltmeter, connect the voltmeter across the battery and note the voltage. The test is only accurate if the battery has not been subjected to any kind of charge for the previous six hours. If this is not the case, switch on the headlights for 30 seconds, then wait four to five minutes before testing the battery after switching off the headlights. All other electrical circuits must be switched off, so check that the doors and tailgate are fully shut when making the test.

5 If the voltage reading is less than 12.2 volts, then the battery is discharged, whilst a reading of 12.2 to 12.4 volts indicates a partially discharged condition. The battery should be recharged as described later in this Section.

6 The preferable method of testing the battery is to use a digital battery analyser. Most garages (and battery supply specialists) will have one. These machines are capable of assessing the condition and performance of the battery without removing it from the vehicle or disconnecting the battery terminals. Note also that many modern battery chargers will also have some ability to test the battery **(see illustration)**.

Charging

Note: *The following is intended as a guide only. Always refer to the manufacturer's recommendations (often printed on a label attached to the battery) before charging a battery.*

7 If the battery is to be recharged, we recommend that you use an 'intelligent'

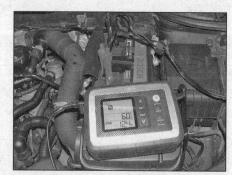

2.6 Modern battery chargers can often test battery capacity

2.7 Battery charging using a modern battery charger

3.2 Open the cover

3.3a Unbolt and then...

charger **(see illustration)**. Where an AGM or EFB battery is fitted it is essential that the charger is specifically capable of charging these types of battery. Whilst some chargers are capable of safely charging the battery with the battery connected to the vehicle, if you are unsure, always disconnect the battery. If the battery is disconnected (e.g. if it is to be removed and recharged on the bench), note that certain 'learned' values will be lost from the engine management ECU memory, requiring the car to be driven over a short distance after refitting the battery. Also, when the battery is reconnected, the warning lights for the ESP and electro-mechanical steering will light up and stay on. They will extinguish if you drive briefly in a straight line at a speed of 9 to 13 mph.

---

3 Battery – disconnecting, reconnecting, removal and refitting

Disconnecting the battery

Note: *If the vehicle has a security-coded audio unit, make sure that you have the code number before disconnecting the battery.*

1 Lower the drivers window and remove the key from the ignition.
2 The battery is located on the left-hand side of the engine compartment. Open the cover to

gain access to the battery **(see illustration)**. Note that some models have an insulated cover.
3 Loosen the clamp nut and disconnect the battery negative lead (-) **(see illustrations)**.
4 Move the negative (earth) lead away from the terminal. Cover the lead (or battery terminal) with a suitable insulator (a plastic bag is ideal) or simply secure it to the side with a cable tie.

Reconnecting the battery

5 Reconnect the negative lead. Push it down until it is flush with the battery terminal post and then tighten the nut to the specified torque.
6 Refit the battery cover and close the bonnet.
7 Reach through the open drivers window and turn on the sidelights. Wait a minute or so to allow the on board computer systems to boot up and for the battery voltage to stabilise.
8 Start the vehicle (from outside where possible) and then open and close all the power windows. Adjust the clock time and re-activate the audio unit by inserting the security code (where applicable).
9 To restore the electric window automatic opening/closing function (where fitted) ensure all windows and doors are closed, and then lock the vehicle via the driver's door. Now

3.3b ...disconnect the battery negative terminal

unlock the driver's door, and then lock it again – hold the key in the locking position for at least one second.
10 After reconnecting the battery and starting the vehicle the ESP (Electronic Stability Programme) the TCS (Traction Control System) and the power steering warning light will remain on until the vehicle is driven in a straight line at a speed of 10 mph or more.

Removal

11 Disconnect the battery negative cable as described above and then disconnect the battery positive cable **(see illustrations)**.
12 Remove the cover and then unscrew the

3.11a Remove the cover

3.11b Unbolt and then...

3.11c ...disconnect the positive lead

3.12a Remove the surround

3.12b Remove the retaining clamp bolt...

3.12c ...and remove the clamp

3.12d Lift out the battery

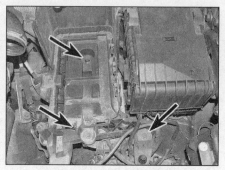

3.13a ...remove the bolts...

3.13b ...and lift out the support tray

retaining clamp bolt. Lift the battery from the engine compartment. **(see illustrations).**

13 If necessary the battery tray can now be removed. Remove the air filter housing as described in Chapter 4A Section 3 (petrol engines) or (diesel engines). Unbolt and remove the battery tray **(see illustrations)**.

Refitting

14 Refit the battery by following the removal procedure in reverse.

4 Alternator/charging system – testing in vehicle

Note: *Refer to Section 1 of this Chapter before starting work.*

1 If the charge warning light fails to illuminate when the ignition is switched on, first check the alternator wiring connections for security. If the light still fails to illuminate, check the continuity of the warning light feed wire from the alternator to the bulbholder. If all is satisfactory, the alternator is at fault and should be renewed or taken to an auto-electrician for testing and repair.

2 Similarly, if the charge warning light comes on with the ignition, but is then slow to go out when the engine is started, this may indicate an impending alternator problem. Check all the items listed in the preceding paragraph, and refer to an auto-electrical specialist if no obvious faults are found.

3 If the charge warning light illuminates when the engine is running, stop the engine and check that the auxiliary drivebelt is not broken (see Chapter 1A Section 11 or Chapter 1B Section 30) and that the alternator connections are secure. If all is so far satisfactory, check the alternator brushes and slip-rings as described in Section 6. If the fault persists, the alternator should be renewed, or taken to an auto-electrician for testing and repair.

4 If the alternator output is suspect even though the warning light functions correctly, the regulated voltage may be checked as follows.

5 Connect a voltmeter across the battery terminals, and start the engine.

6 Increase the engine speed until the voltmeter reading remains steady; the reading should be approximately 12 to 13 volts, and no more than 14 volts.

7 Switch on as many electrical accessories (eg, the headlights, heated rear window and heater blower) as possible, and check that the alternator maintains the regulated voltage at around 13 to 14 volts.

8 If the regulated voltage is not as stated, this may be due to worn brushes, weak brush springs, a faulty voltage regulator, a faulty diode, a severed phase winding or worn or damaged slip-rings. The brushes and slip-rings may be checked, but if the fault persists, the alternator should be renewed or taken to an auto-electrician.

5 Alternator – removal and refitting

Removal

1 Disconnect the battery negative lead and position it away from the terminal as described in Section 3.

2 Where fitted, remove the engine cover from the top of the engine. Also (where fitted) disconnect the vacuum hose, undo the retaining bolts and remove the vacuum reservoir **(see illustration)**.

3 Remove the auxiliary drivebelt from the alternator pulley as described in Chapter 1A Section 30 (petrol engines) or Chapter 1B

5.2 Remove the engine cover – common rail diesel model shown

5.8 Move the fuel filter to the side

5.9a Common rail alternator mounting bolts

5.9b Disconnect the control wire

Section 30 (diesel engines). Mark the drivebelt for direction to ensure it is refitted in the same position.

1.6 and 2.0 litre common rail diesel models

4 On 2.0 litre engines only, jack up and support the front of the vehicle (see *Jacking and vehicle support*) and then remove the engine undershield

5 Remove the right-hand wing liner and then remove the radiator cooling fan(s) as described in Chapter 3 Section 5.

6 Disconnect the wiring plug from the AC compressor and then unbolt the compressor from the mounting bracket. DO NOT disconnect the refrigerant lines. Secure the compressor to the side with cable ties or similar.

7 Disconnect the wiring plug from the boost pressure sensor and then unbolt and remove the charge air pipe.

8 Unbolt the fuel filter and secure it to the side **(see illustration)** – DO NOT disconnect the fuel lines.

9 Remove the upper alternator mounting bolt and then slacken the lower bolt. Rotate the alternator towards the front of the engine and disconnect the control wire plug **(see illustrations)**.

10 Remove the protective cap and unbolt the main output cable and then unbolt the cable clamp form the back of the alternator.

11 Remove the remaining bolt and then on 2.0 litre engines lower the alternator from the engine by manoeuvering it past the AC compressor. On 1.6 litre engines remove the alternator up and out of the engine bay.

PD diesel models

12 Unclip the fuel filter from the mounting bracket and secure it to the side – DO NOT disconnect the fuel lines.

13 Unbolt the charge air pipe from the engine. Release the hose clips and manoeuvre the hose out and past the fuel lines **(see illustration)**.

14 If not already done so remove the auxiliary drivebelt.

15 Remove the protective cap (where fitted), unscrew and remove the nut and washers, then disconnect the battery positive cable from the alternator terminal. Where applicable, unscrew the nut and remove the cable guide. Disconnect the control wire plug **(see illustration)**.

16 Remove the mounting bolts **(see illustration)** and lift the alternator up and out from the engine bay.

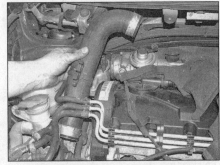

5.13 Remove the charge air pipe

Petrol engines

17 Unclip and move the EVAP charcoal canister to the side and if not already done so, remove the auxiliary drivebelt.

18 On 2.0 litre FSI engines disconnect the wiring plug and vacuum hose from the inlet manifold flap control solenoid. Remove the fixings and pull the control valve from the inlet manifold.

19 Remove the protective cap (where fitted), unscrew and remove the nut and washers, then disconnect the main output cable from the alternator terminal. Disconnect the control wire plug and then where applicable, unscrew

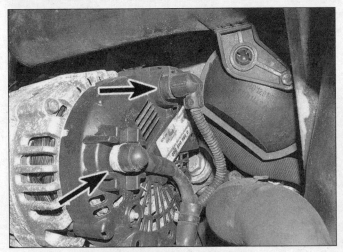

5.15 The alternator wiring plugs

5.16 Alternator mounting bolts

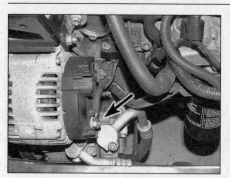

5.19 Remove the nut

5.20a Remove the mounting bolts (1.6 litre)

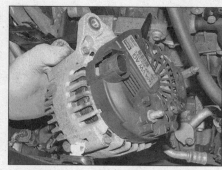

5.20b Lift out the alternator (1.6 litre)

the nut and remove the cable guide **(see illustration)**.

20 On 2.0 litre FSI engines move the small coolant hose to the side (there is no need to disconnect the hose). Remove the mounting bolts and lift out the alternator **(see illustrations)**.

Refitting

21 Refitting is a reversal of removal. Refer to Chapter 1A Section 30 (petrol engines) or Chapter 1B Section 30 (diesel engines) as applicable for details of refitting and tensioning the auxiliary drivebelt. Tighten the alternator mounting bolts to the specified torque.

6 Alternator – brush holder/ regulator module renewal

Removal

1 Remove the alternator, as described in Section 5.

2 Place the alternator on a clean work surface, with the pulley facing down.

Bosch

3 Undo the screw and the two retaining nuts, and lift away the outer plastic cover **(see illustration)**.

4 Unscrew the three securing screws, and remove the voltage regulator **(see illustrations)**.

Valeo

5 Prise off the spring clips, and remove the outer plastic cover **(see illustration)**.

6 Undo the two screws and single nut, and remove the voltage regulator **(see illustrations)**.

7 Slide off the brush cover by depressing the lugs on each side.

Inspection

8 Measure the free length of the brush

6.3 On the Bosch type, remove the outer cover...

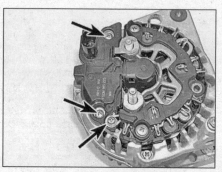

6.4a ...undo the screws...

6.4b ...and remove the brush holder/ regulator

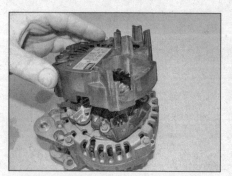

6.5 On the Valeo type, remove the outer plastic cover...

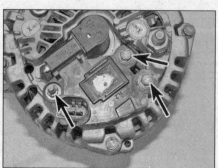

6.6a ...undo the two screws and single nut...

6.6b ...and remove the voltage regulator

contacts **(see illustration)**. Check the measurement with the Specifications; renew the module if the brushes are worn below the minimum limit.

9 Clean and inspect the surfaces of the slip-rings **(see illustration)**, at the end of the alternator shaft. If they are excessively worn, or damaged, the alternator must be renewed.

Refitting

Bosch

10 Refit the voltage regulator using a reversal of the removal procedure, tightening the screws securely. On completion, refer to Section 5 and refit the alternator.

Valeo

11 Depress the carbon brushes into the housing, then refit the voltage regulator and tighten the screws and nut securely. Slide on the brush cover until it is heard to engage. On completion, refer to Section 5 and refit the alternator.

---

7 Starting system – testing

Note: *Refer to Section 1 of this Chapter before starting work.*

1 If the starter motor fails to operate when the ignition key is turned to the appropriate position, the following possible causes may be to blame:

a) *The battery is faulty.*

b) *The electrical connections between the switch, solenoid, battery and starter motor are somewhere failing to pass the necessary current from the battery through the starter to earth.*

c) *The solenoid is faulty.*

d) *The starter motor is mechanically or electrically defective.*

2 To check the battery, switch on the headlights. If they dim after a few seconds, this indicates that the battery is discharged – recharge (see Section 2) or renew the battery. If the headlights glow brightly, operate the ignition switch and observe the lights. If they dim, then this indicates that current is reaching the starter motor, therefore the fault must lie in

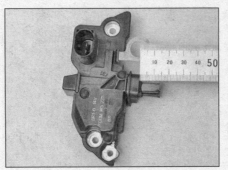

6.8 Measure the brush length

the starter motor. If the lights continue to glow brightly (and no clicking sound can be heard from the starter motor solenoid), this indicates that there is a fault in the circuit or solenoid – see following paragraphs. If the starter motor turns slowly when operated, but the battery is in good condition, then this indicates that either the starter motor is faulty, or there is considerable resistance somewhere in the circuit.

3 If a fault in the circuit is suspected, disconnect the battery leads (including the earth connection to the body), the starter/solenoid wiring and the engine/transmission earth strap. Thoroughly clean the connections, and reconnect the leads and wiring, then use a voltmeter or test light to check that full battery voltage is available at the battery positive lead connection to the solenoid, and that the earth is sound.

4 If the battery and all connections are in good condition, check the circuit by disconnecting the wire from the solenoid blade terminal. Connect a voltmeter or test light between the wire end and a good earth (such as the battery negative terminal), and check that the wire is live when the ignition switch is turned to the start position. If it is, then the circuit is sound – if not the circuit wiring can be checked as described in Chapter 12 Section 2.

5 The solenoid contacts can be checked by connecting a voltmeter or test light between the battery positive feed connection on the starter side of the solenoid, and earth. When the ignition switch is turned to the start position, there should be a reading or lighted

6.9 Clean and inspect the surfaces of the slip-rings

bulb, as applicable. If there is no reading or lighted bulb, the solenoid is faulty and should be renewed.

6 If the circuit and solenoid are proved sound, the fault must lie in the starter motor. It may be possible to have the starter motor overhauled by a specialist, but check on the availability and cost of spares before proceeding, as it may prove more economical to obtain a new or exchange motor.

---

8 Starter motor – removal and refitting

Removal

1 Disconnect the battery negative lead – see Section 3.

2 Apply the handbrake, then jack up the front of the vehicle and support it on axle stands (see *Jacking and vehicle support*). Remove the engine undertray.

3 Where fitted remove the engine cover from the top of the engine **(see illustration 5.2)**.

4 Remove the air filter housing as described in.

5 On some models it's necessary to remove the charge air pipe from the front of the engine.

6 Remove the protective cover and then unscrew the nut and disconnect the battery positive lead from the starter motor. Also disconnect the wire from the solenoid **(see illustrations)**.

8.6a Remove the protective cover

8.6b Unplug the solenoid wiring plug...

8.6c ...and then unbolt the main supply cable from the starter motor

8.7 Unbolt the earth cable

8.9 Removing the wiring loom support bracket

8.10a Unscrew the upper mounting bolt...

8.10b ...then guide the starter motor out of the bellhousing

7 Unscrew the nut and disconnect the earth cable **(see illustration)**.

8 Remove the engine undershield.

9 Working form below, unscrew the nut from the lower starter mounting bolt and remove the wiring loom support bracket **(see illustration)**.

10 Unscrew the upper mounting bolt noting and then guide the starter motor out of the bellhousing aperture and downwards out of the engine compartment **(see illustrations)**. Note that on some engines the starter motor is best removed after removing the left-hand wing liner.

Refitting

11 Refit the starter motor by following the removal procedure in reverse. Tighten the mounting bolts to the specified torque.

---

9 Starter motor –
   testing and overhaul

---

1 If the starter motor is thought to be defective, it should be removed from the vehicle and taken to an auto-electrician for assessment. In the majority of cases, new starter motor brushes can be fitted at a reasonable cost. However, check the cost of repairs first as it may prove more economical to purchase a new or exchange motor.

Chapter 5 Part B
Ignition system – petrol engines

Contents

Degrees of difficulty

| **Easy,** suitable for novice with little experience | **Fairly easy,** suitable for beginner with some experience | **Fairly difficult,** suitable for competent DIY mechanic | **Difficult,** suitable for experienced DIY mechanic | **Very difficult,** suitable for expert DIY or professional |

Specifications

System type

1.6 litre engines:	
Engine codes BGU, BSE and BSF	Siemens Simos 7
2.0 litre engines:	
Engine code BVY, BVZ and BWA	Bosch Motronic MED9.5.10
Engine code BLY and BLR	Bosch Motronic MED9.5
Engine code BWA, CDLA and CDLD	Bosch Motronic MED9.1

Ignition coil

Type:	
Engine codes BGU, BSE, and BSF	Single DIS coil with four HT lead outputs
All other engine codes	One coil per spark plug
Secondary winding resistance (DIS).	4000 to 6000 ohms

Spark plugs . . . . *See Chapter,Specifications*

Torque wrench settings

	Nm	lbf ft
Ignition coil mounting bolts (single DIS coil)	10	7
Knock sensor mounting bolt	20	15
Spark plugs	25	18

1 General Information

1 The Bosch and Simos systems are self-contained engine management systems, which control both the fuel injection and ignition. This Chapter deals with the ignition system components only – refer to Chapter for details of the fuel system components.

2 The ignition systems fitted are of the 'distributorless' (DIS – Distributorless Ignition System) type. Despite the many different system names and designations, as far as the ignition systems fitted to the Leon are concerned, there are essentially only two types of system used – BGU, BSE, and BSF engine codes have a single ignition coil unit with four HT lead terminals, while all other engines have four separate coils, one fitted to each spark plug.

3 The ignition timing cannot be adjusted by conventional means, and the advance and retard functions are carried out by the Electronic Control Unit (ECU).

4 The ignition system consists of the spark plugs, HT leads (where applicable), electronic ignition coil unit (or four separate coils), crankshaft position sensor, camshaft position sensor and the ECU together with its associated sensors and wiring.

5 The component layout varies from system to system, but the basic operation is the same for all models: the ECU supplies a voltage to the input stage of the ignition coil, which causes the primary windings in the coil to be energised. The supply voltage is periodically interrupted by the ECU and this results in the collapse of primary magnetic field, which then induces a much larger voltage in the secondary coil, called the HT voltage. This voltage is directed to the spark plug in the cylinder currently on its ignition stroke. The spark plug electrodes form a gap small enough for the HT voltage to arc across, and the resulting spark ignites the fuel/air mixture in the cylinder. The timing of this sequence of events is critical, and is regulated solely by the ECU.

6 The ECU calculates and controls the ignition timing primarily according to engine speed, crankshaft position, camshaft position, and inlet airflow rate information, received from sensors mounted on and around the engine. Other parameters that affect ignition timing are throttle position and rate of opening, inlet air temperature, coolant temperature and engine knock, monitored by sensors mounted on the engine. Note that most of these sensors have a dual role, in that the information they provide is equally useful in determining the fuelling requirements as in deciding the optimum ignition or firing point – therefore, removal of some of the sensors mentioned below is described in Chapter 4A.

7 The ECU computes engine speed and crankshaft position from toothed impulse ring attached to the engine crankshaft, with an engine speed sensor whose inductive head runs just above ring. As the crankshaft rotates, the ring 'teeth' pass the engine speed sensor, which transmits a pulse to the ECU every time a tooth passes it. At the top dead centre (TDC) position, there is one missing tooth in the ring periphery, which results in a longer pause between signals from the sensor. The ECU recognises the absence of a pulse from the engine speed sensor at this point, and uses it to establish the TDC position for No 1 piston. The time interval between pulses, and the location of the missing pulse, allow the ECU to accurately determine the position of the crankshaft and its speed. The camshaft position sensor enhances this information by detecting whether a particular piston is on an inlet or an exhaust cycle.

8 Information on engine load is supplied to the ECU by the inlet manifold pressure sensor, and from the throttle position sensor. The engine load is determined by computation based on the quantity of air being drawn into the engine. Further engine load information is sent to the ECU from the knock sensor(s). These sensors are sensitive to vibration, and detect the knocking which occurs when the engine starts to 'pink' (pre-ignite). If pre-ignition occurs, the ECU retards the ignition timing of the cylinder that is pre-igniting in steps until the pre-ignition ceases. The ECU then advances the ignition timing of that cylinder in steps until it is restored to normal, or until pre-ignition occurs again.

9 Sensors monitoring coolant temperature, throttle position, camshaft position, roadspeed, and (where applicable) air conditioning system operation. From all this constantly changing data, the ECU selects, and if necessary modifies, a particular ignition advance setting from a map of ignition characteristics stored in its memory.

10 The ECU also uses the ignition timing to finely adjust the engine idle speed, in response to signals from the air conditioning switch (to prevent stalling), or if the alternator output voltage falls too low.

11 In the event of a fault in the system due to loss of a signal from one of the sensors, the ECU reverts to an emergency ('limp-home') program. This will allow the car to be driven, although engine operation and performance will be limited. A warning light on the instrument panel will illuminate if the fault is likely to cause an increase in harmful exhaust emissions.

12 It should be noted that comprehensive fault diagnosis of all the engine management systems described in this Chapter is only possible with dedicated electronic test equipment. In the event of a sensor failing or other fault occurring, a fault code will be stored in the ECU's fault log, which can only be extracted from the ECU using a dedicated fault code reader. A Seat dealer will obviously have such a reader, as will most garages. Low cost diagnostic tools are available in the aftermarket, but at the lower end of the market these may be limited to only displaying

1.12 Checking for fault codes with a middle of the range code reader

the mandatory emissions related fault codes (see illustration). Once the fault has been identified, the removal/refitting sequences detailed in the following Sections will then allow the appropriate component(s) to be renewed as required.

Ignition coil(s)

13 The single coil fitted to BGU, BSE and BSF engine codes operates on the 'wasted spark' principle. The coil unit in fact contains two separate coils – one for cylinders 1 and 4, the other for cylinders 2 and 3. Each of the two coils produces an HT voltage at both outputs every time its primary coil voltage is interrupted – ie, cylinders 1 and 4 always 'fire' together, then 2 and 3 'fire' together. When this happens, one of the two cylinders concerned will be on the compression stroke (and will ignite the fuel/air mixture), while the other one is on the exhaust stroke – because the spark on the exhaust stroke has no effect, it is effectively wasted, hence the term 'wasted spark'.

14 On engine codes other than BGU, BSE and BSF, each spark plug has its own dedicated 'plug-top' HT coil which fits directly onto the spark plug (no HT leads are therefore needed). Unlike the 'wasted spark' system, on these models a spark is only generated at each plug once every engine cycle.

2 Ignition system – testing

⚠ **Warning: Extreme care must be taken when working on the system with the ignition switched on; it is possible to get a substantial electric shock from a vehicle's ignition system. Persons with cardiac pacemaker devices should keep well clear of the ignition circuits, components and test equipment. Always switch off the ignition before disconnecting or connecting any component and when using a multimeter to check resistances.**

Engines with one coil per plug

1 If a fault appears in the engine management (fuel injection/ignition) system which is

thought to ignition related, first ensure that the fault is not due to a poor electrical connection or poor maintenance; ie, check that the air cleaner filter element is clean, the spark plugs are in good condition and correctly gapped, that the engine breather hoses are clear and undamaged, referring to Chapter 1A for further information. If the engine is running very roughly, check the compression pressures as described in Chapter 2A Section 2 (1.6 litre engines) or Chapter 2B Section 3 (Chapter).

2 If these checks fail to reveal the cause of the problem, the vehicle should be taken to a Seat dealer or well equipped garage for testing. A diagnostic connector is incorporated in the engine management circuit into which a special electronic diagnostic tester can be plugged **(see illustration 1.12)**. The tester will locate the fault quickly and simply, alleviating the need to test all the system components individually which is a time-consuming operation that carries a high risk of damaging the ECU.

3 The only ignition system checks which can be carried out by the home mechanic are those described in Chapter 1A Section 29, relating to the spark plugs. If necessary, the system wiring and wiring connectors can be checked as described in Chapter 12 Section 2 ensuring that the ECU wiring connector(s) have first been disconnected.

Engines with single DIS coil

4 Refer to the information given in paragraphs 1 to 3. The only other likely cause of ignition trouble is the HT leads, linking the HT coil to the spark plugs. Check the leads as follows. Never disconnect more than one HT lead at a time to avoid possible confusion.

5 Pull the first lead from the plug by gripping the end fitting, not the lead, otherwise the lead connection may be fractured. Check inside the end fitting for signs of corrosion, which will look like a white crusty powder. Push the end fitting back onto the spark plug, ensuring that it is a tight fit on the plug. If not, remove the lead again and use pliers to carefully crimp the metal connector inside the end fitting until it fits securely on the end of the spark plug.

6 Using a clean rag, wipe the entire length of the lead to remove any built-up dirt and

3.1 The DIS coil is at the front of the engine, above the oil filter housing

grease. Once the lead is clean, check for burns, cracks and other damage. Do not bend the lead excessively, nor pull the lead lengthwise – the conductor inside is quite fragile, and might break.

7 Disconnect the other end of the lead from the HT coil. Again, pull only on the end fitting. Check for corrosion and a tight fit in the same manner as the spark plug end.

8 If an ohmmeter is available, check for continuity between the HT lead terminals. If there is no continuity the lead is faulty and must be renewed (as a guide, the resistance of each lead should be in the region of 4 to 8 k ohms).

9 Refit the lead securely on completion of the check then check the remaining leads one at a time, in the same way. If there is any doubt about the condition of any HT leads, renew them as a complete set.

3 HT coil(s) – removal and refitting

Removal

Engines with single DIS coil

1 On all models, the ignition coil unit is mounted on the front of the engine **(see illustration)**. On models fitted with a secondary air pump the pump must be removed first as described in Chapter 4C Section 5.

2 Make sure the ignition is switched off (take

3.5 Remove the HT leads

out the key).

3 Where applicable and/or necessary for access, remove the engine top cover(s). Removal details vary according to model, but the cover retaining nuts are concealed under circular covers, which are prised out of the main cover. Where plastic screws or turn-fasteners are used, these can be removed using a wide-bladed screwdriver. Remove the nuts or screws, and lift the cover from the engine, releasing any wiring or hoses attached.

4 The original HT leads should be marked from 1 to 4, corresponding to the cylinder/spark plug they serve (No 1 is at the timing belt end of the engine). Some leads are also marked from A to D, and corresponding markings are found on the ignition coil HT terminals – in this case, cylinder A corresponds to No 1, B to No 2, and so on. If there are no markings present, label the HT leads before disconnecting, and either paint a marking on the ignition coil terminals or make a sketch of the lead positions for use when reconnecting.

5 Disconnect the HT leads from the ignition coil terminals **(see illustration)**.

6 Unplug the main wiring plug (LT connector) at the base (or side) of the ignition coil, then unscrew the three mounting bolts and remove the coil unit from the engine **(see illustrations)**.

Engines with one coil per spark plug

7 Removal of the ignition coils is covered in the spark plug renewal procedure in

3.6a Unplug the connector....

3.6b ...remove the screws and...

3.6c ...remove the coil pack

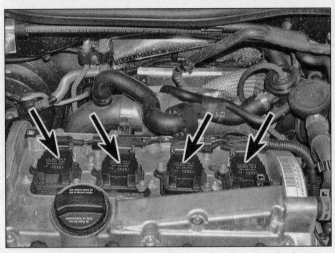

3.7 Ignition coils on engines with one coil per spark plug

5.3 The knock sensor (1.6 litre with secondary air pump removed)

Chapter 1A Section 29, since the coils must be removed for access to the plugs **(see illustration)**.

Refitting

8 Refitting is a reversal of the relevant removal procedure.

9 Securely tighten the coil mounting bolts. Use the marks noted before disconnecting when refitting the HT leads – if wished, spray a little water-dispersant (such as WD-40) onto each connector as it is refitted (this can also be used on the LT wiring connector).

4 Ignition timing – checking and adjusting

1 The ignition timing is under the control of the engine management system ECU and is not manually adjustable without access to dedicated electronic test equipment. A basic setting cannot be quoted because the ignition timing is constantly being altered to control engine idle speed (see Section 1 for details).

5 Knock sensor(s) – removal and refitting

Removal

1 The knock sensor(s) is/are located on the inlet manifold side of the cylinder block. **Note:** *Some models, have two knock sensors.*
2 Remove the engine top cover (where fitted)) to gain access to the sensor from above. On models fitted with a secondary air system remove the air pump to improve access (Chapter 4C Section 5).
3 Disconnect the wiring connector from the sensor or trace the wiring back from the sensor and disconnect its wiring connector (as applicable) **(see illustration)**.
4 Unscrew the mounting bolt and remove the sensor from the cylinder block.

Refitting

5 Refitting is the reverse of removal. Ensure the mating surfaces of the sensor and cylinder block are clean and dry and ensure the mounting bolt is tightened to the specified torque to ensure correct operation.

Chapter 5 Part C
Preheating system

Contents

Degrees of difficulty

Easy, suitable for novice with little experience	Fairly easy, suitable for beginner with some experience	Fairly difficult, suitable for competent DIY mechanic	Difficult, suitable for experienced DIY mechanic	Very difficult, suitable for expert DIY or professional

Specifications

Glow plugs
Electrical resistance . . . . 1.0 ohm
Current consumption (typical – no value quoted by Seat) . . . . 8 amps (per plug)

Torque wrench setting	Nm	lbf ft
Glow plug to cylinder head:		
PD engines	15	11
Common rail engines	18	13

1 General Information

1 To assist cold starting and to control emissions during the warm up phase, diesel engine models are fitted with a preheating system, which consists of four glow plugs, a glow plug control unit, a facia-mounted warning light and the associated electrical wiring. The glow plug control unit on some models, is incorporated in the ECU, other models have a control unit/relay located under the fusebox in the engine compartment (see Section 3).

2 The glow plugs are miniature electric heating elements, encapsulated in a metal case with a probe at one end and electrical connection at the other. Each inlet tract has a glow plug threaded into it, which is positioned directly in line with the incoming spray of fuel. When the glow plug is energised, the fuel passing over it is heated, allowing its optimum combustion temperature to be achieved more readily in the combustion chamber.

3 The duration of the preheating period is governed by the ECU, which monitors the temperature of the engine through the coolant temperature sensor and alters the preheating time to suit the conditions. Pre-heating only takes place at coolant temperature below 9°C.

4 A facia-mounted warning light informs the driver that preheating is taking place. The light extinguishes when sufficient preheating has taken place to allow the engine to be started, but power will still be supplied to the glow plugs for a further period until the engine is started. If no attempt is made to start the engine, the power supply to the glow plugs is switched off to prevent battery drain and glow plug burn-out. If the warning light flashes, or comes on during normal driving, this indicates a fault with the diesel engine management system, which should be investigated by a Seat dealer or suitably equipped garage as soon as possible.

5 After the engine has been started, the glow plugs continue to operate for a further period of time. This helps to improve fuel combustion whilst the engine is warming-up, resulting in quieter, smoother running and reduced exhaust emissions. On models fitted with a DPF (Diesel Particulate Filter) the glow plugs are used to assist with filter cleaning when regeneration of the filter takes place.

2 Glow plugs – testing, removal and refitting

Caution: Some models use a ceramic type glow plug (denoted by a white or silver wiring plug). These plugs are fragile and should be handled accordingly. If they are fractured and any part of the plug remains in the combustion chamber, then all particles must be removed or engine damage will occur. Only remove the glow plugs if they are confirmed as faulty and do not remove them for compression or cylinder leakage testing.

⚠️ *Warning: A correctly functioning glow plug will become red-hot in a very short time. This should be in mind when removing the glow plugs, if they have recently been in use. If a*

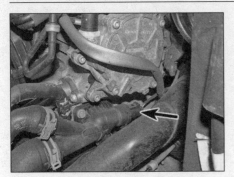

2.4 Temperature sensor location

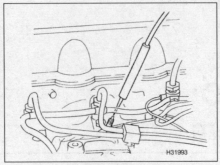

2.10 Testing the glow plugs using a multimeter

2.16a Glow plug wiring connector for No 1 injector

glow plug is dropped, it may be damaged internally, which could result in ceramic fragments entering the engine causing extensive damage. Do not fit a glow plug that has been dropped.

Testing

1 If the system malfunctions, testing is ultimately by substitution of known good units, but some preliminary checks may be made as described in the following paragraphs.

2 Before testing the system, use a multi-meter to check that the battery voltage is at least 11.5. Switch off the ignition.

3 Remove the engine top cover. **Note:** *On DOHC engines, access to the glow plugs involves removing the camshaft cover. Removal details vary according to model, but the cover retaining nuts are concealed under circular covers, which are prised out of the main cover. Remove the nuts, and lift the cover from the engine, releasing any wiring or hoses attached.*

4 Disconnect the wiring plug from the coolant temperature sender at the left-hand end of the engine (left as seen from the driver's seat) – refer to Chapter 3 Section 6 **(see illustration)**. Disconnecting the sender in this way simulates a cold engine, which is a requirement for the glow plug system to activate.

5 Disconnect the wiring connector from the most convenient glow plug, and connect a suitable multimeter between the wiring connector and a good earth.

6 Have an assistant switch on the ignition for approximately 20 seconds.

7 Battery voltage should be displayed – note that the voltage will drop to zero when the preheating period ends.

8 If no supply voltage can be detected at the glow plug, then either the glow plug relay (where applicable) or the supply wiring must be faulty. Also check that the glow plug fuse or fusible link has not blown – if it has, this may indicate a serious wiring fault. Investigate the cause or consult a Seat dealer (or suitably equipped garage) for advice.

9 To locate a faulty glow plug, first disconnect the battery negative cable and position it away from the terminal.

10 Disconnect the wiring plug from the glow plug terminal. Measure the electrical resistance between the glow plug terminal and the engine earth **(see illustration)**. At the time of writing, this information is not available – as a guide, a resistance of more than a few ohms indicates that the plug is defective.

11 If a suitable ammeter is available, connect it between the glow plug and its wiring connector, and measure the steady-state current consumption (ignore the initial current surge, which will be about 50% higher). As a guide, high current consumption (or no current draw at all) indicates a faulty glow plug.

12 As a final check, remove the glow plugs and inspect them visually, as described in the next sub-Section. A badly burned or charred stem may be an indication of a faulty fuel injector.

Removal

Note: *Refer to the Warning at the start of this Section before proceeding.*

PD unit injection engines

13 Disconnect the battery as described in Chapter 5A Section 3.

14 Pull the plastic cover on the top of the engine upwards from its mountings. Where fitted, remove the noise insulation from above the injectors.

15 On the DOHC engines (code BKD) remove the camshaft cover as described in Chapter 2E Section 4.

16 Disconnect the wiring connectors from the glow plugs. Label the connectors where necessary to make refitting easier **(see illustrations)**. On some models the glow plug wiring is clipped to the injector leak-off hoses – make sure that the clips are not lost as the wiring is removed.

17 Unscrew and remove the glow plug(s).

18 Inspect the glow plug stems for signs of damage. A badly burned or charred stem may be an indication of a faulty fuel injector.

Common rail injection engines

19 Remove the engine cover and (where fitted) remove the noise insulation.

20 Disconnect the wiring plugs from the injectors, exhaust gas pressure sensor and fuel rail pressure sensor **(see illustration)**.

21 Undo the retaining bolts and detach the coolant pipe from the intake manifold. Move the pipe to the front of the manifold.

2.16b Removing the wiring loom/rail from the glow plugs

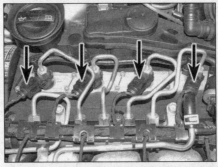

2.20 Disconnect the wiring plugs from the injectors

2.23 Disconnect the return hose from the fuel rail

2.23a Using long nose pliers …

2.23b … pull the connectors from the top of the glow plugs

22 Release the retaining clip and disconnect the fuel return hose from the fuel rail **(see illustration)**. Be prepared for fuel spillage.

23 Pull the connectors from the top of the glow plugs and move the wiring loom to one side. Be sure to only pull on the underside of the ridge at the top of the connectors **(see illustrations)**.

24 Clean the area around the glow plugs; use a vacuum cleaner if possible. Spray brake cleaner (or similar) around the glow plug opening, letting it penetrate briefly, and then blow out with compressed air.

Caution: Always wear protective goggles to protect your eyes, when using compressed air.

25 Using a universal joint, extension and a deep 10 mm socket, unscrew and remove the glow plug(s) from the cylinder head. Note that the plug must be kept 'straight' when being removed, as it can be easily damaged.

Refitting

26 Refitting is a reversal of removal, but tighten the glow plugs to the specified torque.

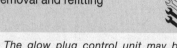

3 Glow plug control unit –
removal and refitting

Note: *The glow plug control unit may be incorporated inside the engine management ECU. Follow the procedure below for models with separate glow plug control unit.*

Removal

1 The glow plug control unit (where applicable) is located under the fuse box in the engine compartment **(see illustration)**.

2 To gain access to the control unit, remove the battery and battery tray, as described in Chapter 5A Section 3.

3 Release the retaining clip and slide the control unit mounting bracket from the underside of the fusebox housing **(see illustration)**.

4 Disconnect the wiring connector and remove the control unit

Refitting

5 Refitting is a reversal of removal.

3.1 Glow plug control unit location

3.3 Slide the control unit out and disconnect the wiring connector

Notes

Chapter 6
Clutch

Contents

Degrees of difficulty

Easy, suitable for novice with little experience	**Fairly easy,** suitable for beginner with some experience	**Fairly difficult,** suitable for competent DIY mechanic	**Difficult,** suitable for experienced DIY mechanic	**Very difficult,** suitable for expert DIY or professional

Specifications

General

Type:

Luk clutch . . . . Single dry friction disc, diaphragm spring with spring-loaded hub, self-adjusting pressure plate (SAC)

Sachs clutch . . . . Single dry friction disc, diaphragm spring with spring-loaded hub

Clutch operation . . . . Hydraulic with master and slave cylinders

Torque wrench settings

	Nm	lbf ft
Clutch pedal crash bar bolts*:		
1 bolt fitment	20	15
2 bolt fitment	10	7
Clutch pedal mounting bracket nuts*	25	18
Clutch pedal pivot nut*	25	18
Clutch pressure plate-to-flywheel bolts*:		
M7 bolt	20	15
M6 bolt	13	10
Clutch release bearing guide sleeve bolts:		
0A4 and 0AF transmissions*:		
Stage 1	5	4
Stage 2	Angle-tighten a further 90º	
02S transmissions	20	15
Concentric release bearing bolts*:		
Metal housing, with locking fluid	12	9
Plastic housing, with locking fluid	15	11
Clutch slave cylinder bolts	20	15

*Use new bolts/nuts

1 General Information

1 The clutch is of single dry plate type, incorporating a diaphragm spring pressure plate, and is hydraulically operated.

2 The pressure plate is bolted to the rear face of the flywheel, and the friction disc is located between the pressure plate and the flywheel friction surface. The friction disc hub is splined to the transmission input shaft and is free to slide along the splines. Friction lining material is riveted to each side of the disc, and the disc hub incorporates cushioning springs to absorb transmission shocks and ensure a smooth take-up of drive.

3 On all transmissions except 02Q, when the clutch pedal is depressed, the slave cylinder pushrod moves the release lever forwards. On 02Q transmissions, the slave cylinder is fitted concentrically around the transmission input shaft within the bellhousing. The release bearing is forced onto the pressure plate diaphragm spring fingers. As the centre of the diaphragm spring is pushed in, the outer part of the spring moves out and releases the pressure plate from the friction disc. Drive then ceases to be transmitted to the transmission.

4 When the clutch pedal is released, the diaphragm spring forces the pressure plate into contact with the linings on the friction disc, and at the same time pushes the disc slightly forward along the input shaft splines into engagement with the flywheel. The friction

2.2 Preparing to bleed the clutch on 02S transmissions

disc is now firmly sandwiched between the pressure plate and flywheel. This causes drive to be taken up.

5 As the linings wear on the friction disc, the pressure plate rest position moves closer to the flywheel resulting in the 'rest' position of the diaphragm spring fingers being raised. The hydraulic system requires no adjustment since the quantity of hydraulic fluid in the circuit automatically compensates for wear every time the clutch pedal is operated.

2 Hydraulic system – bleeding

⚠ *Warning: Hydraulic fluid is poisonous, thoroughly wash off spills from bare skin without delay. Seek immediate medical advice if any fluid is swallowed or gets into the eyes. Certain types of hydraulic fluid are inflammable and may ignite when brought into contact with hot components. Hydraulic fluid is also an effective paint stripper. If spillage occurs onto painted bodywork or fittings, it should be washed off immediately, using copious quantities of cold water. It is also hygroscopic (i.e. it can absorb moisture from the air) which then renders it useless. Old fluid may have suffered contamination, and should never be re-used.*

1 If any part of the hydraulic system is dismantled, or if air has accidentally entered the system, the system will need to be bled. The presence of air is characterised by the pedal having a spongy feel and it results in difficulty in changing gear.

2 On 02Q transmissions (fitted with a concentric clutch slave cylinder) the design of the clutch hydraulic system makes bleeding using the conventional method of pumping the clutch pedal hard to achieve. In order to remove all air present in the system, the best approach is to use pressure bleeding equipment. This is available from auto accessory shops at relatively low cost. An alternative method is to reverse bleed the clutch by pushing fluid back towards the master cylinder using a large syringe filled with brake/clutch fluid. Conventional bleeding is

possible on models fitted with an external slave cylinder **(see illustration)**.

3 When pressure bleeding equipment is used, should be connected to the brake/clutch hydraulic fluid reservoir in accordance with the manufacturer's instructions. The system is bled through the bleed screw of the clutch slave cylinder (all transmissions except 02Q), which is located on the top of the transmission housing. On 02Q transmissions, the slave cylinder bleed screw is located at the front of the transmission, above the starter motor **(see illustrations)** and can be accessed by removing the air filter housing as described in Chapter 4A Section 3 or Chapter 4B Section 3

4 Bleed the system until the fluid being ejected is free from air bubbles. Close the bleed screw, then disconnect and remove the bleeding equipment.

5 Check the operation of the clutch to see that it is satisfactory. If air still remains in the system, repeat the bleeding operation.

6 Discard any fluid that is bled from the system, even if it looks clean. Hydraulic fluid absorbs water and its re-use can cause internal corrosion of the master and slave cylinders, leading to excessive wear and failure of the seals.

3 Clutch pedal – removal and refitting

Removal

Note: *The complete pedal assembly and clutch master cylinder can be removed as a single item if the clutch master cylinder fluid lines are disconnected – see Section 4.*

1 Move the driver's seat fully to the rear, and adjust the steering column to its highest position.

2 Disconnect the battery negative lead and position it away from the terminal – see Chapter 5A Section 3.

3 Where fitted, remove the driver's side lower facia trim panel. To further improve access, remove the steering column shrouds as described in Chapter 11 Section 25.

4 Release the retaining clip and unclip the wiring loom guide from under the steering column **(see illustration)**.

2.3a Clutch slave cylinder bleed screw – 02S transmissions

2.3b Clutch bleed screw – 02Q transmissions

3.4 Unclip the wiring bracket from the steering column

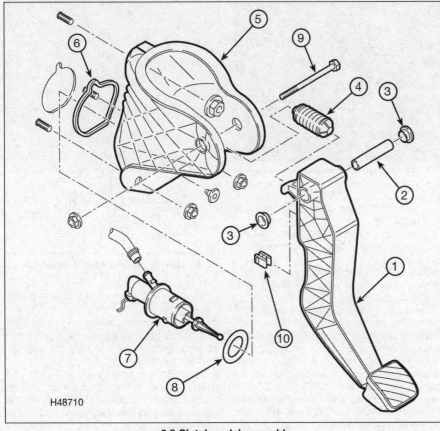

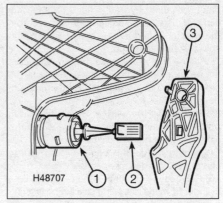

3.9 Master cylinder pushrod to pedal securing clip

1 Master cylinder pushrod
2 Securing clip
3 Clutch pedal

H48710

3.8 Clutch pedal assembly

1 Clutch pedal	5 Mounting bracket	8 Seal
2 Pivot pin	6 Seal	9 Pivot bolt
3 Bearing bushes	7 Master cylinder	10 Pedal securing clip
4 Over centre spring		

5 Undo the retaining bolt(s) and remove the crash bar from above the clutch pedal. There are two types of crash bar fitted; depending on vehicle, there could be either one or two bolts securing the crash bar.

6 Where fitted, release the securing clips and remove the insulation padding from around the clutch pedal mounting bracket.

7 Where fitted disconnect the wiring plug and then rotate the clutch pedal position switch through 45 degrees anti-clockwise to remove it. **Note:** *On later models this switch is fitted to the master cylinder.*

8 Undo the retaining nut and remove the pedal pivot bolt, move the pedal downwards and withdraw the over-centre spring from inside the top of the mounting bracket **(see illustration)**. **Note:** *The clutch pedal will still be attached to the master cylinder pushrod.*

9 Working at each side of the clutch pedal, release the retaining clip to disengage the master cylinder pushrod from the pedal **(see illustration)**. Note that a special tool is available to release the push rod locking clip (T20059B). Slide out the pivot pin and the pedal can now be removed from the mounting bracket.

Refitting

10 Refitting is a reversal of removal, bearing in mind the following points:
a) *Press the pushrod retaining clip firmly into the pedal until it is heard to engage.*
b) *Two types of clutch pedal switch are fitted to early models. One has a coloured plunger tip – on these the tip must NOT be extended before installation. The other version has a plain coloured plunger – on these the plunger must be fully extended before installation.*
c) *Tighten all fixings to the specified torque, where given.*
d) *On completion, check the brake/clutch fluid level, and top-up if necessary.*

4 Master cylinder – removal and refitting

Note: *Refer to the warning at the beginning of Section 2 regarding the hazards of working with hydraulic fluid.*

Removal

Note: *No spare parts are available from VAG/ Seat for the master cylinder. If the master*

cylinder is faulty or worn, the complete assembly must be renewed.
Note: *Access can be improved if the windscreen cowl panel and plenum chamber front panel are removed first, as described in Chapter 12 Section 19.*

1 The clutch master cylinder is located through the bulkhead and is attached to the clutch pedal mounting bracket **(see illustration)**. Hydraulic fluid for the unit is supplied from the brake master cylinder reservoir.

2 Before proceeding, place cloth rags on the carpet inside the car to prevent damage from spilt hydraulic fluid.

3 Remove the engine top cover/air filter assembly.

4 Working down the back of the engine compartment, clamp the hydraulic fluid hose (if it's rubber) leading from the brake fluid reservoir to the clutch master cylinder using a brake hose clamp. If the hose is plastic, then the end will need to be plugged to prevent fluid loss.

5 Similarly (where possible), clamp the rubber section of the hydraulic hose leading from the master cylinder to the slave cylinder using a brake hose clamp, to prevent loss of hydraulic fluid. The rubber section of pipe may be further down towards the slave cylinder.

6 Position a suitable container, or a wad of clean cloth, beneath the master cylinder to

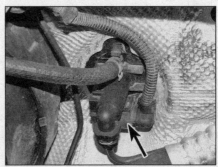

4.1 Location of master cylinder on bulkhead

catch escaping hydraulic fluid. Release the clip and disconnect the fluid supply hose from the master cylinder – be prepared for fluid spillage **(see illustration)**.

7 Pull the fluid outlet hose retaining clip from the union on the master cylinder, then pull the pipe from the union **(see illustration)**. Again, be prepared for fluid spillage.

8 Disconnect the wiring from the clutch position sender on the side of the master cylinder – later models only.

9 Working inside the vehicle, remove the clutch pedal as described in Section 3.

10 Release the clip and withdraw the master cylinder from the mounting bracket by twisting it anti-clockwise.

Refitting

11 Refitting is a reversal of removal, but bleed the clutch hydraulic system as described in Section 2.

5 Slave cylinder – removal and refitting

Removal

Note: *No spare parts are available from VAG/Seat for the slave cylinder. If the slave cylinder is faulty or worn, the complete assembly must be renewed.*

02S, 0AF and 0A4 transmissions

1 The slave cylinder is located on the top of the transmission casing **(see illustration)**.

2 Remove the air filter housing, as described

4.6 Disconnect the fluid supply hose

in Chapter 4A Section 3 (petrol engines) or (diesel engines).

3 Access can be further improved if the the battery and battery tray with reference to Chapter 5A Section 3.

4 Disconnect the gear selector cables from the gear selector levers, as described in Chapter 7A Section 2.

5 Prise out the retaining clips and remove the outer cables from the mounting bracket, then unbolt the mounting bracket from the top of the transmission.

6 Where required, undo the bolts and remove the gearbox support bracket from above the slave cylinder **(see illustration)**.

Caution: Refer to the warning at the beginning of Section 2 regarding the hazards of working with hydraulic fluid.

7 Place a wad of clean rag beneath the fluid line connection on the slave cylinder to catch escaping fluid.

8 Release the fluid pipe retaining clip from the

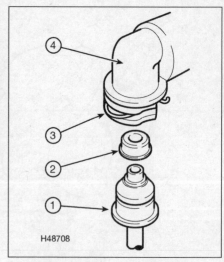

4.7 Master cylinder hose connection

1 Fluid pipe connection   3 Securing clip
2 Seal   4 Master cylinder

union on the slave cylinder, then pull the pipe from the union. Position the fluid pipe clear of the slave cylinder, and plug the end. Be prepared for some fluid spillage.

9 Unscrew the two bolts securing the slave cylinder to the transmission casing, and withdraw the slave cylinder from the transmission **(see illustration)**.

02Q transmission

Note: *The slave cylinder and release bearing are a single item, often referred to as a concentric release bearing.*

Note: *As the transmission must be removed to access the bearing, always check the clutch friction material, the clutch cover and the condition of the flywheel. This is particularly important on models fitted with a dual mass flywheel. Always replace the slave cylinder/release bearing with the clutch and clutch cover.*

10 Remove the transmission as described in Chapter 7A Section 3.

11 Release the retaining clip and pull the fluid bleeder connection from the outside of the transmission casing **(see illustration)**.

12 Undo the bolts and remove the slave cylinder/release bearing assembly **(see illustration)**.

5.1 Clutch slave cylinder

5.6 Remove the support bracket

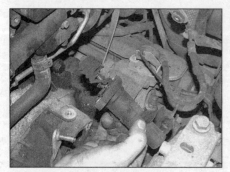

5.9 Removing the clutch slave cylinder from the transmission

5.11 Prise up the clip and pull the connection from place

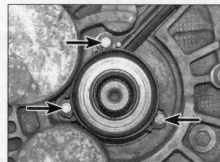

5.12 Slave cylinder/release bearing retaining bolts

Refitting

13 Refitting is a reversal of removal, bearing in mind the following points:

a) *Tighten all fixings to the specified torque where given.*

b) *On completion, bleed the clutch hydraulic system as described in Section 2.*

---

6 Clutch friction disc and pressure plate – removal, inspection and refitting

---

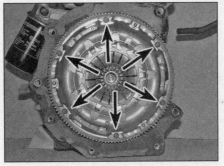

6.3a Undo the pressure plate retaining screws

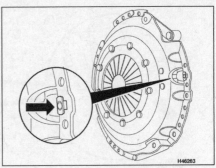

6.3b Ensure the stop pin is free to move

⚠ *Warning: Dust created by clutch wear and deposited on the clutch components may contain asbestos, which is a health hazard. DO NOT blow it out with compressed air or inhale any of it. DO NOT use petrol or petroleum-based solvents to clean off the dust. Brake system cleaner or methylated spirit should be used to flush the dust into a suitable receptacle. After the clutch components are wiped clean with clean rags, dispose of the contaminated rags and cleaner in a sealed container.*

Note: *Always inspect the dual mass flywheel (as described in the relevant engine Chapter) and replace it if necessary.*

Removal

1 Access to the clutch is obtained by removing the transmission as described in Chapter 7A Section 3.

2 If the clutch is to be refitted, mark the clutch pressure plate and flywheel in relation to each other.

3 Hold the flywheel stationary, and then unscrew the clutch pressure plate bolts ¼ of a turn at a time **(see illustration)**. With the bolts unscrewed two or three turns, check that the pressure plate is not binding on the dowel pins. If necessary, use a screwdriver to release the pressure plate. On models with the Sachs clutch, as the bolts are removed the stop pin must slacken. If it doesn't, press the pin towards the flywheel **(see illustration)**.

4 Remove all the bolts, then lift the clutch pressure plate and friction disc from the flywheel.

Inspection

Note: *Due to the amount of work necessary*

to remove and refit clutch components, it is usually considered good practice to renew the clutch friction disc, pressure plate assembly and release bearing as a matched set, even if only one of these is actually worn enough to require renewal. It is also worth considering the renewal of the clutch components on a preventative basis if the engine and/or transmission have been removed for some other reason.

5 Clean the pressure plate friction surface, clutch friction disc and flywheel. Do not inhale the dust, as it may contain asbestos which is dangerous to health.

6 Examine the fingers of the diaphragm spring for wear or scoring. If the depth of wear exceeds half the thickness of the fingers, a new pressure plate assembly must be fitted.

7 Examine the pressure plate for scoring, cracking, distortion and discoloration. Light scoring is acceptable, but if excessive, a new pressure plate assembly must be fitted **(see illustration)**. If the distortion of the friction surface exceeds 1.0 mm, renew it.

8 Examine the friction disc linings for wear and cracking, and for contamination with oil or grease. The linings are worn excessively if they are worn down to, or near, the rivets. Check the disc hub and splines for wear by temporarily fitting it on the transmission input shaft. Renew the friction disc as necessary.

9 Examine the flywheel friction surface for scoring, cracking and discoloration (caused by overheating). If excessive, it may be possible to have the flywheel machined by an engineering works, otherwise it should be renewed.

10 Ensure that all parts are clean, and free of oil or grease, before reassembling. Apply just a small amount of lithium-based grease (VAG No. G000100) to the splines of the friction disc hub. Do not use copper-based grease. Note that new pressure plates and clutch covers may be coated with protective grease. It is only permissible to clean the grease away from the friction disc lining contact area. Removal of the grease from other areas will shorten the service life of the clutch.

Refitting

11 Commence reassembly by locating the friction disc on the flywheel, with the raised side of the hub facing outwards (normally marked 'Getriebeseite' or 'Gearbox side'). If possible, the centralising tool (see paragraph 20) should be used to hold the disc on the flywheel at this stage **(see illustration)**.

Models with self-adjusting clutch (SAC)

12 On models with a Self-adjusting clutch (SAC), where a new friction disc is fitted, but the pressure plate is to be re-used, it is necessary to reset the pressure plate adjusting ring prior to assembly as follows.

13 Insert three 8 mm bolts into the pressure plate mounting holes at intervals of 120°. The bolts should be inserted from the flywheel side, and retained by nuts **(see illustration)**.

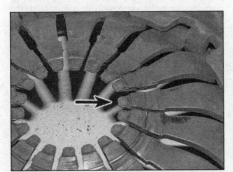

6.7 A worn and damaged clutch cover should always be replaced

6.11 The friction disc should be marked 'Getriebeseite' or 'Gearbox side'

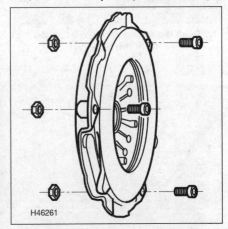

6.13 Insert three 8 mm bolts from the flywheel side, and secure with nuts

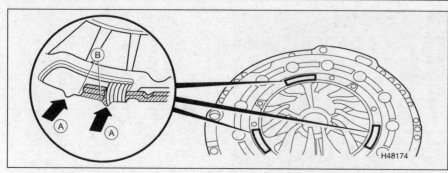

6.15 The edges of the adjuster ring (B) must be between the notches (A)

6.17 Fit the pressure plate over the locating dowel pins

6.20 With the pressure plate screws tightened, remove the centralising tool

14 Place the pressure plate face down on the bed of an hydraulic press so that only the heads of the bolts make contact with the press bed, then place a circular spacer over the ends of the diaphragm springs fingers.

15 Use 2 screwdrivers to attempt to rotate the adjuster ring anti-clockwise. Apply just enough pressure with the hydraulic press until it's just possible to move the adjuster ring **(see illustration)**.

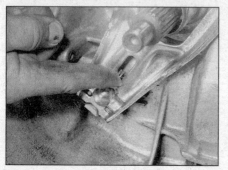

7.2 Push the spring clip to release the arm from the ball-stud

7.4a Use a screwdriver to depress the retaining tags…

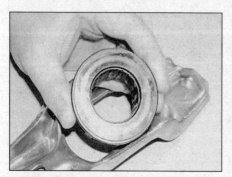

7.4b …then remove the release bearing from the arm

7.5 Guide sleeve on the transmission

16 Once the adjuster ring edges are between the notches, relieve the pressure. The ring is now reset. **Note:** *New pressure plates are supplied in this reset position.*

All models

17 Locate the clutch pressure plate on the disc, and fit it onto the location dowels **(see illustration)**. If refitting the original pressure plate, make sure that the previously-made marks are aligned.

18 Insert the bolts finger-tight to hold the pressure plate in position.

19 The friction disc must now be centralised, to ensure correct alignment of the transmission input shaft with the disc centre. To do this, a proprietary tool may be used, or alternatively, use a wooden mandrel made to fit inside the friction disc and the hole in the centre of the crankshaft. Insert the tool through the friction disc into the crankshaft, and make sure that it is central.

20 Tighten the pressure plate bolts progressively and in diagonal sequence, until the specified torque setting is achieved, then remove the centralising tool **(see illustration)**.

21 Check the release bearing in the transmission bellhousing for smooth operation – if the clutch is to be replaced always renew the release bearing (the bearing is normally part of the clutch kit). Renew it with reference to Section 7.

22 Refit the transmission with reference to Chapter 7A Section 3.

7 Release bearing and lever
– removal, inspection and refitting

Removal

Note: *On 02Q transmissions, the release bearing is part of the clutch slave cylinder (see Section 5).*

1 Remove the transmission as described in Chapter 7A Section 3.

2 Using a screwdriver, prise the release lever from the ball-stud on the transmission housing. If this proves difficult, push the retaining spring from the release lever first **(see illustration)**. Where applicable, remove the plastic pad from the stud.

3 Slide the release bearing, together with the lever, from the guide sleeve, and withdraw it over the transmission input shaft.

4 Separate the release bearing from the lever **(see illustrations)**.

5 If the guide sleeve is worn excessively, unbolt it and remove the O-ring seal **(see illustration)**. Note that a new guide bush may be included with the new clutch assembly.

Inspection

6 Spin the release bearing by hand, and check it for smooth running **(see illustration)**. Any tendency to seize or run rough will necessitate renewal of the bearing. If the bearing is to be re-used, wipe it clean with a

7.6 Typical release bearing failure

7.8 Fit the new guide sleeve

7.10 Fit the arm onto the stud

7.11a Slide on the bearing...

7.11b ...and clip it onto the arm

dry cloth; the bearing should not be washed in a liquid solvent, as this will remove the internal grease. Note that a new replacement release bearing is always included in a new clutch kit.

7 Clean the release lever, ball-stud and guide sleeve.

Refitting

8 If the guide sleeve was removed fit the new guide sleeve and tighten the bolts to the specified torque **(see illustration)**. If preferred, the guide sleeve may be assembled to the release bearing and lever, and the components fitted over the input shaft as one unit.

9 Lubricate the ball-stud in the transmission bellhousing with molybdenum sulphide-based grease.

10 Push the operating arm onto the pivot ball stud **(see illustration)**.

11 Slide the bearing over the guide sleeve and clip it into the operating arm **(see illustrations)**.

12 Replacement clutch kits are usually supplied with a lubricant suitable for the guide sleeve and the splines of the input shaft. Where no lubricant is supplied lubricate the transmission input shaft with light oil – DO NOT use grease.

13 Check that the bearing slides smoothly up and down the guide sleeve and then refit the transmission as described in Chapter 7A Section 3.

Notes

Chapter 7 Part A
Manual transmission

Contents

Degrees of difficulty

Easy, suitable for novice with little experience	**Fairly easy,** suitable for beginner with some experience	**Fairly difficult,** suitable for competent DIY mechanic	**Difficult,** suitable for experienced DIY mechanic	**Very difficult,** suitable for expert DIY or professional

Specifications

General

Type .... Five or six speed, transversely-mounted, front-wheel-drive layout with integral transaxle differential/final drive and 1 reverse gear

Code:
  5-speed 0A4/0AF
  6-speed 02Q/02S

Oil capacities

0A4 transmission*....................................... 1.7 litres
0AF transmission* .... 2.0 litres
02Q transmission* .... 2.3 litres
02S transmission*....................................... 1.9 litres
* Capacities are the refill amounts for a previously fitted transmission`

Torque wrench settings

	Nm	lbf ft
Driveshaft flange bolt	30	22
Flywheel cover plate (where fitted)......................	10	7
Gearchange cable support bracket	20	15
Gear lever housing to body:		
M6 nuts...	8	6
M8 nuts...	25	18
Left-hand mounting bracket bolts: *		
To transmission: *		
Stage 1	40	30
Stage 2	Angle-tighten a further 90°	
Transmission bracket to wing bracket: *		
Stage 1	60	45
Stage 2	Angle-tighten a further 90°	
Bracket to body:		
Stage 1	60	45
Stage 2	Angle-tighten a further 90°	
Transmission minor support bracket bolts: *		
Stage 1	20	15
Stage 2	Angle-tighten a further 90°	

Torque wrench settings (continued)

	Nm	lbf ft
Pendulum mount: *		
M10 X 30 mm bolt		
Stage 1	40	30
Stage 2	Angle-tighten a further 90°	
M10 X 35 mm bolt		
Stage 1	50	37
Stage 2	Angle-tighten a further 90°	
M14 X 70 mm bolt		
Stage 1	100	74
Stage 2	Angle-tighten a further 90°	
Reversing light switch	20	15
Transmission to engine:		
M12 bolts	80	59
M10 bolt	40	30
Oil filler/level plug		
Plug with 12 point socket	45	33
Plug with hex head	30	22

Do not re-use

1 General Information

1 The manual transmission is bolted directly to the left-hand end of the engine. This layout has the advantage of providing the shortest possible drive path to the front wheels, as well as locating the transmission in the airflow through engine bay, optimising cooling. The unit is cased in aluminium alloy.

2 Drive from the crankshaft is transmitted through the clutch to the gearbox input shaft, which is splined to accept the clutch friction disc.

3 All forward gears are fitted with synchromesh. The floor-mounted gear lever is connected to the gearbox by shift cables.

4 Levers on the transmission actuate internal selector forks, which are connected to the synchromesh sleeves. The sleeves are locked to the gearbox shafts but can slide axially by means of splined hubs, and they press baulk rings into contact with the respective gear/ pinion. The coned surfaces between the baulk rings and the pinion/gear act as a friction clutch, that progressively matches the speed of the synchromesh sleeve (and hence the gearbox shaft) with that of the gear/pinion. This allows gearchanges to be carried out smoothly.

5 Drive is transmitted to the differential crownwheel, which rotates the differential case and planetary gears, thus driving the sun gears and driveshafts. The rotation of the differential planetary gears on their shaft allows the inner roadwheel to rotate at a slower speed than the outer roadwheel during cornering.

Note: *There is no recommendation from Volkswagen with regards to changing or checking the transmission oil level. The transmission should be checked for leaks at service time and the conscientious owner might consider changing the oil on a high mileage vehicle.*

2 Gear change housing and cable – removal, refitting and adjustment

Note: *The selector cable end fittings changed diameter from mid 2006. If replacing an early transmission with a later one, the selector levers (on the transmission) will need to be swapped over to the newer transmission.*

Removal

1 Remove the air filter housing as described in Chapter 4A Section 3 (petrol engines) or (diesel engines).

2 Remove the battery and battery tray as described Chapter 5A Section 3.

3 Prise out the clips securing the inner cable to the lever on the transmission, and the outer cable to the support bracket **(see illustrations)**. Withdraw the cable from the support bracket and discard the clips – new ones must be fitted.

4 Working inside the vehicle, remove the centre console, as described in Chapter 11 Section 26.

5 Remove the insulation around the base of the gear change housing to access the mounting nuts.

6 Raise the front of the vehicle and support it securely on axle stands (see *Jacking and vehicle support*).

7 Remove the centre tunnel front heat shield from the underside of the vehicle to gain access to the base of the selector lever housing and cables. It may be necessary to separate the exhaust downpipe from the intermediate pipe with reference Chapter 4C Section 9 (petrol engines) or Chapter 4D Section 9 (diesel engines).

8 Working back inside the vehicle, undo the retaining nuts, then remove the bracket and lower the gear change housing downwards. Withdraw it complete with selector cables out from under the vehicle. It may be useful having the aid of an assistant at this point, to be under the vehicle when lowering the gear change housing.

9 To remove the cables from the gear change housing, prise out the clips securing the inner cable to the lower part of the gear lever, and then withdraw the securing clips from the outer cable to the gear change housing. Withdraw the cables and discard the clips – new ones must be fitted.

Refitting

10 Refitting is the reversal of the removal procedure, noting the following points:
a) Ensure that the cables are correctly routed and secured, as noted on removal.
b) Take care not to bend or kink the cables.
c) Carry out the cable adjustment procedure described below before reconnecting the cable at the transmission end.
d) When refitting the cables, use new clips.

2.3a Release the gearchange inner cable retaining clips …

2.3b …and slide out the outer cable securing clips

2.12 Push the collar down and lock in position

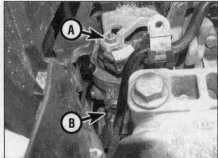

2.13 Press down on (A), then push in locking pin (B)

2.14a Remove the trim, before unclipping the gaiter

2.14b Locking the gear lever in position using a drill bit

2.15 Release the two locking collars back into position

Adjustment

11 Remove the engine top cover (where fitted), air cleaner housing, and air ducting for access to the top of the transmission. If required remove the battery and battery tray, as described in Chapter 5A Section 3.

12 With the gearchange set in the neutral position, push the two locking collars (one on each cable) forwards to compress the springs, turn them clockwise (looking from the driver's seat) to lock into position **(see illustration)**.

13 Press down on the selector shaft on the top of the transmission, and push the locking pin into the transmission while turning it

clockwise, until it engages and the selector shaft cannot move **(see illustration)**.

14 Working inside the vehicle, unclip the gear lever gaiter from the centre console -as described in Chapter 11 Section 26. Still in the neutral position, move the gear lever as far to the left as possible and insert the locking pin (or drill bit) through the hole in the base of the gear lever and into the hole in the housing **(see illustrations)**.

15 Working back in the engine bay, turn the two locking collars on the cables anti-clockwise so that the springs will release them back into position and lock the cables **(see illustration)**.

16 With the cable adjustment set, the locking pin in the transmission housing, can now be turned anti-clockwise to its original position pointing upwards.

17 Inside the vehicle, remove the locking pin from the gear lever, and then check the operation of the selector mechanism. When the gear lever is at rest in neutral, it should be central, ready to select 3rd or 4th. The gear lever gaiter can now be refitted to the centre console.

18 Refit the air ducting, air filter housing, battery tray, battery and engine cover (where fitted).

3 Transmission – removal and refitting

Note: *A safe method of supporting the engine while the transmission is removed will be required. The recommended method of removing the transmission is to support the both the engine and transmission with a support bar that fits across the engine bay. Support bars are readily available in the aftermarket, however the special extensions and fittings specified by Volkswagon are not available **(see illustration)**. Where the correct support bar is not available the best alternative for the home mechanic is to support the engine with an engine crane (installed from the side) and then support the transmission with a trolley jack or ideally a transmission jack **(see illustration)**. Whichever method is used the aid of an assistant is essential.*

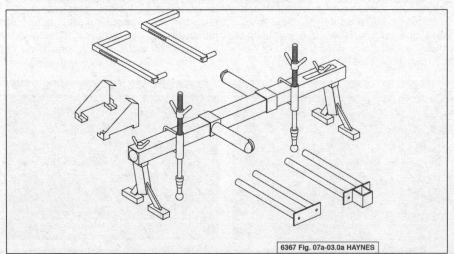

6367 Fig. 07a-03.0a HAYNES

3.0a Engine support bar and extensions

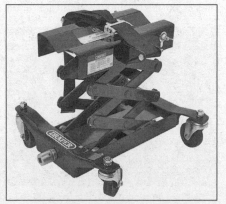

3.0b Transmission support jack

3.5 Mounting bracket securing bolts

3.6 Remove the support bracket

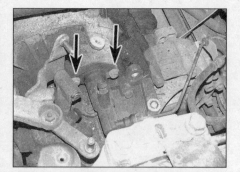

3.7a Unbolt and then...

3.7b ...remove the slave cylinder

3.8 Prise up the clip a little and pull the hose from the connection – 02Q transmissions

3.13a Reversing light switch on the OAF type transmission...

Removal

1 Select a solid, level surface to park the vehicle upon. Give yourself enough space to move around it easily. Apply the handbrake and chock the rear wheels.

2 Raise the front of the vehicle and support it securely on axle stands (see *Jacking and vehicle support*). Remove the engine/transmission undertray sections. Position a suitable container beneath the transmission, then unscrew the drain plug and drain the transmission oil.

3 Remove the engine top cover (where fitted). Remove the air filter housing as described in Chapter 4A Section 3 (petrol engines) or (diesel engines) On turbocharged models remove the charge air pipes from the turbocharger/intercooler.

4 Remove the battery and battery tray with reference to Chapter 5A Section 3. Where necessary, unclip the EVAP canister (petrol models) or the fuel filter (diesel models). This will avoid any damage to the filter or canister as the engine is moved. There is no need to disconnect the fuel lines.

5 Release the retaining clips and disconnect the gear selector cables from the gear selector levers, as described in Section 2. Unbolt and remove the cable mounting bracket from the top of the transmission **(see illustration)**.

6 Undo the retaining bolts, and then remove the support bracket **(see illustration)**, from the top of the transmission.

7 On models with the clutch slave cylinder on the top of the transmission, undo the retaining bolts, remove the cylinder and secure it out of harms way **(see illustrations)**.

8 On 02Q transmissions, seal the slave cylinder flexible hose using a hose clamp, then prise out the clip and pull the fluid pipe from the bleeder connection on the slave cylinder supply pipe **(see illustration)**.

9 Unbolt the earth cable from the engine/transmission or subframe.

10 With reference to Chapter 5A Section 8, remove the starter motor.

11 Remove the upper bolts securing the transmission to the engine.

12 Remove the left-hand wheel arch liner.

13 Disconnect the wiring from the reversing light switch **(see illustrations)**, and where fitted, the neutral position switch on vehicles with a start/stop system.

14 Where fitted unbolt the metal shield from around the right-hand driveshaft. On models with gas discharge headlights (Xenon) disconnect the ride height sensor and secure it to the side.

15 Slacken the clamp securing the exhaust intermediate pipe to the rear section and undo the front mounting bracket bolts **(see illustrations)**. This will allow the engine to be moved forwards and backwards during the transmission removal and alignment

3.13b ...and on the O2S transmission

3.15a Slacken the exhaust clamp...

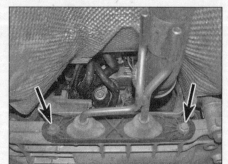

3.15b ...and undo the exhaust mounting bracket bolts

3.16 Support the driveshafts

3.18 Unbolt the engine rear mounting torque arm

3.19 Remove the plate (where fitted)

procedures. On turbocharged models the exhaust system must be disconnected completely at the turbocharger as described in Chapter 4C Section 9 (petrol engines) or Chapter 4D Section 9 (diesel engines).

16 Unscrew and remove the bolts securing the driveshafts to the transmission output flanges. On models with an intermediate shaft, remove the shaft as described in Chapter 8 Section 4. Tie the right-hand driveshaft to one side, and then tie the left-hand driveshaft to the suspension strut, so that the shaft is as high as possible **(see illustration)**. Alternatively, completely remove the driveshafts as described in Chapter 8 Section 2.

17 Remove the radiator cooling fans (as described in Chapter 3 Section 5). Cover the rear of the radiator/cooling fans to prevent it from being damaged as the engine/transmission is moved forward. On models with an AC pressure switch at the front of the engine, remove the

switch – there is a schraeder valve below the switch, so no refrigerant should escape.

18 Unbolt the engine rear mounting torque arm from the bottom of the transmission **(see illustration)**.

19 Where applicable, unbolt the flywheel cover plate from the transmission bellhousing **(see illustration)**.

20 Using a suitable hoist (see the note at the beginning of this Section) support the weight of the engine and then unscrew the bolts securing the transmission mounting to the body. Also, unbolt the mounting bracket from the transmission **(see illustrations)**. A block of wood should be fitted between the back of the engine and the subframe to push the engine forward as the assembly is lowered.

21 Lower the engine/transmission assembly slightly and, using a trolley jack/transmission jack, support the transmission. Position the jack so that it can be withdrawn from

the left-hand side of the car. As the engine/transmission is moved, make sure any wiring or hoses are not damaged.

22 Unscrew and remove the remaining lower transmission to engine mounting bolts, including the bolt located on the left-hand rear of the engine **(see illustration)**.

23 Carefully pull the transmission directly away from the engine, taking care not to allow its weight to rest on the clutch friction disc hub. A second person is helpful to pull the left-hand end of the engine as far forwards as possible.

24 On models with driveshafts that have a bolted flange, it will be necessary to manouvre the right-hand driveshaft flange, from around the flywheel on removal. If required, hold the drive flange in position and undo the centre securing bolt, then the drive flange can be removed **(see illustrations)**, make removal of the transmission easier.

3.20a Transmission end mounting bracket

3.20b Remove the bracket from the transmission

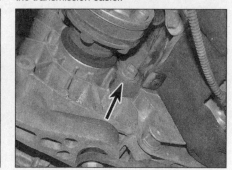

3.22 Undo the rear mounting bolt

3.24a Hold the drive flange in position ...

3.24b ...whilst removing the centre bolt ...

3.24c ...then slide the drive flange from the transmission

⚠ *Warning: Support the transmission to ensure that it remains steady on the jack head. Keep the transmission level until the input shaft is fully withdrawn from the clutch friction disc.*

25 When the transmission is clear of the locating dowels and clutch components, lower the transmission to the ground and withdraw from under the car.

Refitting

26 Refitting the transmission is essentially a reversal of the removal procedure, but note the following points:

a) *Apply a smear of high melting-point grease to the clutch friction disc hub splines; take care to avoid contaminating the friction surfaces.*

b) *In order to align the transmission with the flywheel, gently pull the engine forward as the transmission is manoeuvred into place.*

c) *Tighten the transmission to engine bolts to the specified torque.*

d) *Refer to the relevant part of Chapter 2 and tighten the engine mounting bolts to the correct torque.*

e) *Refer to Chapter 8 Section 2, and then tighten the driveshaft bolts to the specified torque.*

f) *On completion, refer to Section 2 and check the gearchange linkage/cable adjustment.*

g) *Refill the transmission with the correct grade and quantity of oil. Refer to 'Lubricants and fluids' and Section 4.*

| 4 | Transmission oil – removal refilling and level checking |

Oil level checking

02Q and 0AF transmissions

1 The vehicle must be on level ground to check the transmission oil level. This can be accomplished by using wheel ramps to raise the front of the vehicle and then raising the rear of the vehicle using a jack and axle stands (see *Jacking and vehicle support*). The alternative method is to remove the air filter housing and check the level from above. Access is restricted, but possible.

2 Where the vehicle has been jacked up to access the filler plug, remove the engine undershield.

3 The oil level should reach the lower edge of the filler/level hole. A certain amount of oil will have gathered behind the filler/level plug, and will trickle out when it is removed **(see illustrations)**. This does not necessarily indicate that the level is correct. To ensure that a true level is established, wait until the initial trickle has stopped, then add oil as necessary until a trickle of new oil can be seen emerging. The level will be correct when the flow ceases; use only good-quality oil of the specified type.

4 If the transmission has been overfilled so that oil flows out when the filler/level plug is

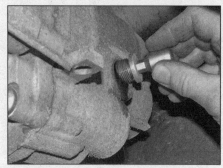

4.3a Removing the oil level/filler plug on 02Q transmissions

removed, check that the car is completely level (front-to-rear and side-to-side), and allow the surplus to drain off into a suitable container.

5 When the oil level is correct, refit the filler/ level plug and tighten it to the specified torque. Note that two versions of the oil filler plug are fitted and they have different torque settings.

6 Wipe off any spilt oil then refit the engine undertray (if removed), tighten the retaining screws securely. Lower the car to the ground. If access has been gained from above, refit the air filter housing.

0A4 and 02S transmissions

7 On these transmissions, it is impossible to check the level of the oil through the 'filler' plug, due to the engine/transmission installation angle; the level of the fluid is above the lower edge of the filler hole. The only method of ensuring the correct fluid level is to completely drain and refill the transmission – as described below.

Draining and refilling

02Q and 0AF transmissions

8 Raise the front of the vehicle and support it securely on axle stands (see *Jacking and vehicle support*). Remove the engine/ transmission undershield. Position a suitable container beneath the transmission, then unscrew the drain plug and drain the transmission oil **(see illustration)**. On 0AF transmissions the drain plug is below the filler/ level plug on the differential housing.

9 Remove the filler/level plug from the transmission **(see illustrations 4.3a and 4.3b)**.

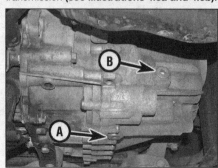

4.8 The oil drain plug (A) and filler level plug (B) on 02Q transmissions

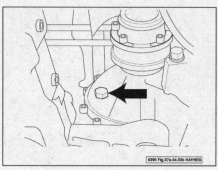

4.3b The plug location on 0AF transmissions

10 Refit the drain plug and tighten it to the specified torque. Note that two versions of the oil filler plug are fitted and they have different torque settings.

11 Fill the transmission with oil until it is just below the level of the filler plug. Allow time for the oil to drain down into the transmission and then re-check the level.

12 Fit the engine undershield and lower the vehicle to the ground (if only the front was raised). Remove the air filter housing and (with the vehicle on level ground) check the oil level from above. Add oil as required and then refit the filler/level plug. Again note that two versions of the filler/level plug are fitted.

13 Refit the air filter housing.

0A4 and 02S transmissions

14 The vehicle must be on level ground to check/fill the transmission. This can be accomplished by using wheel ramps to raise the front of the vehicle and then raising the rear of the vehicle using a jack and axle stands (see *Jacking and vehicle support*).

15 Undo the fasteners and remove the engine undertray.

16 Remove the air filter housing as described in Chapter 4A Section 3 (petrol engines) or (diesel engines).

17 In order to drain the transmission, the pivot pin must be removed from the underside of the transmission casing. However, to prevent the position of the selector forks being altered, press the selector shaft down, then turn the angled locking rod upwards, and lock the shaft in position **(see illustration)**.

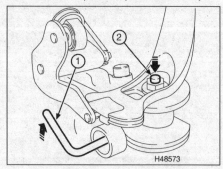

4.17 Press down the selector shaft (2), and rotate the locking rod (1) upwards

18 Place a container under the transmission casing.

19 Unscrew the drain plug from the base of the differential housing and allow the oil to drain **(see illustration)**.

20 Then undo the retaining bolt, pull out the pivot pin nearest the passenger's roadwheel and allow the oil to drain **(see illustration 4.19)**. The pivot pin O-ring seal must be renewed.

21 When the oil has finished draining, clean the surrounding area, refit the drain plug and tighten it to the specified torque.

22 Rotate the selector-shaft locking rod to its original position.

23 Unscrew the reversing light switch from the top of the transmission casing.

24 Using a 600 mm length of 10 mm (external) diameter hose, and funnel, add 1.7 litres of new oil to the transmission.

25 Refit the reversing light switch, and tighten it to the specified torque.

5 Transmission overhaul – general information

1 The overhaul of a manual transmission is a complex (and often expensive) task for the DIY home mechanic to undertake, which requires access to specialist equipment. It involves dismantling and reassembly of many small components, measuring clearances precisely and if necessary, adjusting them by selecting shims and spacers. Internal transmission

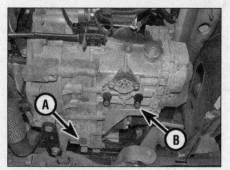

4.19 Oil drain plug (A) and pivot pin (B)

components are also often difficult to obtain and in many instances, extremely expensive. Because of this, if the transmission develops a fault or becomes noisy, the best course of action is to have the unit overhauled by a specialist repairer or to obtain an exchange reconditioned unit.

2 Nevertheless, it is not impossible for the more experienced mechanic to overhaul the transmission if the special tools are available and the job is carried out in a deliberate step-by-step manner, to ensure nothing is overlooked.

3 The tools necessary for an overhaul include internal and external circlip pliers, bearing pullers, a slide hammer, a set of pin punches, a dial test indicator and possibly a hydraulic press. In addition, a large, sturdy workbench and a vice will be required.

4 During dismantling of the transmission, make careful notes of how each component is fitted to make reassembly easier and accurate.

5 Before dismantling the transmission, it will help if you have some idea of where the problem lies. Certain problems can be closely related to specific areas in the transmission, which can make component examination and renewal easier. Refer to the Fault finding Section in this manual for more information.

6 Reversing light switch – testing, removal and refitting

Testing

1 Ensure that the ignition switch is turned to the OFF position.

2 Unplug the wiring harness from the reversing light switch at the connector. The switch is located on the front or top of the casing **(see illustrations)**

3 Connect the probes of a continuity tester, or multimeter set to the resistance measurement function, across the terminals of the reversing light switch.

4 The switch contacts are normally open, so with any gear other than reverse selected, the tester/meter should indicate an open circuit or infinite resistance. When reverse gear is selected, the switch contacts should close, causing the tester/meter to indicate continuity or zero resistance.

5 If the switch does not operate correctly, it should be renewed.

Removal

6 Ensure that the ignition switch is turned to the OFF position.

7 Unplug the wiring harness from the reversing light switch at the connector.

8 Unscrew the switch from the transmission casing, and recover the sealing ring.

Refitting

9 Refitting is a reversal of removal.

6.2a The reverse light switch on 0A4 transmissions...

6.2b ...and on 0AF transmissions

Chapter 7 Part B
DSG semi-automatic transmission

Contents

Degrees of difficulty

Easy, suitable for novice with little experience	**Fairly easy,** suitable for beginner with some experience	**Fairly difficult,** suitable for competent DIY mechanic	**Difficult,** suitable for experienced DIY mechanic	**Very difficult,** suitable for expert DIY or professional

Specifications

General

Description	DSG (Direct Shift Gearbox) semi-automatic 6- or 7-forward speeds and 1 reverse. Integral transaxle differential/final drive transmission with multi-plate clutch or dual-clutch.

Transmission type:
1.6 litre diesel (6-speed)	02E
2.0 litre diesel (7-speed)	0AM
2.0 litre petrol (6-speed)	02E

Transmission capacity:
Type 02E – 6-speed transmission	5.2 litres
Type 0AM – 7 speed transmission	1.7 litres

Clutch type:
6-speed (02E) transmission	Multi-plate (wet clutch), 4 outer plates, 4 inner plates & 1 drive plate.
7-speed (0AM) transmission...............................	Two (Dual-clutch) dry friction discs, with spring loaded centre hub.
Clutch operation...	Mechatronic unit (mechanical and electronic).

Torque wrench settings

	Nm	lbf ft
Driveshaft flange bolt.....................................	30	22
Transmission-to-engine bolts:		
M10 bolts..	40	30
M12 bolts..	80	59
Transmission mounting bracket-to-casing bolts*:		
Stage 1..	40	30
Stage 2..	Angle-tighten a further 90°	
Transmission mounting-to-bracket bolts*:		
Stage 1..	60	44
Stage 2..	Angle-tighten a further 90°	
Drain plug (0AM transmission).............................	30	22
Drain/level plug (02E transmission)	45	31
Oil level tube (02E transmission)	3	2

*Do not reuse

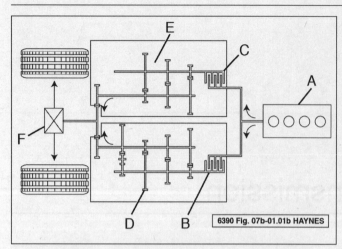

1.1a Schematic view of the six speed transmission (02E)

A Engine
B Clutch 1
C Clutch 2

D Sub transmission
1 (gears 1, 3, 5
and reverse

E Sub transmission 2
(gears 2, 4 and 6)
F Final drive

1.1b Schematic view of the seven speed transmission (0AM)

A Engine
B Dual clutch
C Final drive

D Sub transmission
1 (gears 2, 4, 6
and reverse)

E Sub transmission
2 (gears 1, 3, 5
and 7)

1 General Information

1 The VW semi-automatic Direct Shift Gearbox (DSG) has six or seven-forward speeds (and one reverse). In contrast to traditional automatic transmissions where a fluid flywheel (torque converter) transmits the power from the engine to the gearbox, the DSG has a multi-plate clutch in the six-speed transmission and a twin-clutch in the seven-speed transmission **(see illustrations)**. The main advantages of the DSG transmission, is near-instant gear changes, with seamless, highly efficient drive, resulting in less exhaust emissions and improved fuel consumption.

2 On the front of the transmission housing is a mechatronic unit, this is made up of mechanical and electronic components. The electronic part of the unit uses information from sensors (e.g. engine speed, road speed, driving mode etc.) to determine the optimum gear and shift commands, the mechanical part then selects the correct gear required. On 6-speed (02E) transmissions, the unit is inside the transmission housing with a pressed steel cover, bolted to the front of the transmission casing. On 7-speed (0AM) transmissions, it is a sealed unit, and is bolted to the front of the transmission **(see illustration)**

3 On 6-speed (02E) transmissions, the clutch is of a multi-plate type, which runs in the transmission oil. It is made up of four outer plates, which have teeth on the outer edge to locate in the clutch housing, four inner plates with teeth on the inner edge that locate on the clutch inner hub, and one drive plate that is located on the outside of the clutch plate assembly. The four larger outer plates (C1) operate 1st, 3rd, 5th and reverse gears, and then the four smaller inner plates (C2) operate

2nd, 4th and 6th gears. With this system whilst 'C1' has a gear engaged, then 'C2' will pre-select the next gear ready to change gear. The assembly requires special care, as all the components in the assembly are balanced together, during manufacture. The clutch assembly is sealed inside the bell housing by a clutch end cover, which also forms a seal, to prevent loss of oil. If the end cover is removed for any reason, it will need to be renewed.

4 On 7-speed (0AM) transmissions, the clutch is of a dual-clutch dry plate type, which has two friction discs and a spring loaded centre hub **(see illustration)**. The clutch friction disc nearest the engine is operated by the larger outer release lever (K1) and operates 1st, 3rd, 5th & 7th gears. The clutch friction disc nearest the transmission is operated by the smaller inner release lever (K2) and operates 2nd, 4th, 6th & Reverse gears. With this system whilst 'K1' has a gear engaged, then 'K2' will pre-select the next gear ready to change gear. The clutch assembly can only be purchased as a complete unit, and if renewed, new release levers will also need to be renewed.

5 A fault diagnosis system is integrated into the control unit, but analysis can only be undertaken with specialised equipment. It

is important that any transmission fault be identified and rectified at the earliest possible opportunity. A VAG/Seat dealer or suitable equipped specialist can 'interrogate' the ECM fault memory for stored fault codes, enabling him to pinpoint the fault quickly. Once the fault has been corrected and any fault codes have been cleared, normal transmission operation is restored.

6 Because of the need for special test equipment, the complexity of some of the parts, and the need for scrupulous cleanliness when these transmissions, the work which the owner can do is limited. Most major repairs and overhaul operations should be left to a VAG/Seat dealer or specialist, who will be equipped with the necessary equipment for fault diagnosis and repair. The information in this Chapter is therefore limited to a description of the removal and refitting of the transmission as a complete unit. The removal, refitting and adjustment of the selector cable is also described.

7 In the event of a transmission problem occurring, consult a VAG/Seat dealer or transmission specialist before removing the transmission from the vehicle, since the majority of fault diagnosis is best carried out with the transmission still in the vehicle.

1.2 Mechatronic unit – 0AM transmissions

1.4 Clutch assembly – 0AM transmissions

2 Transmission – removal and refitting

Warning: *The mechatronic unit on the front of the 7-speed (0AM) transmission is a sealed unit. If any oil is lost from this mechatronic unit, it will need to be renewed, as it cannot be refilled. Remove the breather cap from the top of the unit and cover with blanking plug, to prevent any loss of oil(see illustration).*

Removal

1 The transmission is removed downwards from the engine compartment. First, select a solid, level surface to park the vehicle upon. Give yourself enough space to move around it easily. Select P, apply the handbrake, and chock the rear wheels.

2 Loosen the front wheel bolts, and the driveshaft hub bolts. Do not undo the hub bolt more than 90° at this point; otherwise the wheel bearing could be damaged, whilst the weight is still on the wheels.

3 Raise the front of the vehicle and rest it securely on axle stands (see *Jacking and vehicle support*). Remove the front wheels. Allow a suitable working clearance underneath for the eventual withdrawal of the transmission. Undo the fasteners and remove the engine/transmission undershield.

4 Remove the air filter housing as described in Chapter 4A Section 3 (petrol engine) or (diesel engine) and then remove the battery, complete with the shrouds and support tray (Chapter 5A Section 3.

5 Where fitted remove the breather hose from the top of the transmission.

6 Undo the retaining clips and remove the turbocharger intake hose and charge air pipe to the intercooler as described in Chapter 4C Section 6 (petrol engines) or Chapter 4D Section 5 (diesel engines)

7 On some 6 speed models a protective cover is fitted to the base of the transmission. Remove this where fitted.

8 Remove the starter motor as described in Chapter 5A Section 8 and on 7 speed models remove the bolt now accessible through the starter motor housing.

2.0 Breather cap on mechatronic unit – 0AM transmissions

9 Disconnect the selector cable from the selector shaft lever on the top of the transmission, as described in Section 4. Position the cable to one side, noting that the retaining clip/circlip must be renewed.

10 On 6-speed (02E) transmissions there is a fluid cooler located on top of the transmission. Clamp off the cooler hoses with brake hose type clamps. Release the retaining clips and detach the hoses from the cooler **(see illustration)**. Take great care not to spill any coolant into the transmission unit, as this will cause damage.

11 Disconnect the wiring plug connectors from the front of the transmission **(see illustrations)**. Detach the wiring loom and remove the mounting brackets from the transmission.

a) *6-speed 02E transmissions – rotate the collar anti-clockwise to disconnect.*

2.11a Wiring plug and earth cable – 02E transmissions

2.11c ...on the front of the transmission – 0AM transmissions

2.10 Clamp the transmission cooler hoses (arrowed) – 02E transmissions

b) *7-speed 0AM transmissions – slide the locking lever upwards to disconnect.*

12 Disconnect the earth cable from either the transmission or the vehicle body **(see illustration)**.

13 Remove the upper engine-to-transmission mounting bolts.

14 Remove the radiator cooling fan assembly as described in Chapter 3 Section 5.

15 Undo the bolts and slide rearwards the exhaust pipe connecting piece between the front and rear sections of the exhaust system. Also undo the retaining bolts and remove the exhaust front mounting bracket **(see illustration)**. **Note:** *On some models the exhaust must be detached at the turbocharger and removed.*

16 Where fitted, disconnect the wiring plug from the engine oil level/temperature sensor

2.11b Wiring plug connections ...

2.12 Undo the earth cable securing nut

2.15 Undo the exhaust mounting retaining bolts

2.16 Disconnect the oil level/temperature sensor wiring connector

2.17 Fasten the driveshaft to one side

on the sump **(see illustration)**. This will prevent the sensor from damage when the engine moves forward.

17 With reference to Chapter 8 Section 2, unscrew and remove the bolts securing the driveshafts to the transmission output flanges. Tie the right-hand driveshaft to one side, and then tie the left-hand driveshaft to the suspension strut, so that the shaft is as high as possible **(see illustration)**. Alternatively, completely remove the driveshaft.

18 Support the engine with a hoist or support bar located on the front wing inner channels. Depending on the engine, temporarily remove components as necessary to attach the hoist.

19 Position a trolley jack underneath the transmission, and raise it to just take the weight of the unit.

20 Undo and remove the bolts securing the left-hand gearbox mounting. Place a wooden wedge between the engine block and the subframe. The block should be large enough to push the engine forward, but not so large that the Alternator and/or the compressor are in contact with the bodywork or any other components. By controlling both the engine hoist/support bar and the trolley jack, lower the transmission slightly. To make removal easier undo the retaining bolts and remove the mounting bracket from the top of the transmission casing.

21 On 6-speed (02E) transmissions, undo the retaining bolt and remove the small cover plate located above the right-hand driveshaft flange.

22 Lower the engine/transmission until there is sufficient clearance between the upper edge of the transmission and the left-hand chassis member.

23 Unscrew and remove the lower bolts securing the transmission to the engine, noting the bolt locations, as they are of different sizes and lengths.

24 On models with driveshafts that have a bolted flange, hold the flange in position and undo the flange securing bolt, then the flange can be removed **(see illustrations)**.

25 Check that all the fixings and attachments are clear of the transmission. Enlist the aid of an assistant to help in guiding and supporting the transmission during its removal.

26 The transmission is located on engine alignment dowels, and if stuck on them, it may be necessary to carefully tap and prise the transmission free of the dowels to allow separation. Once the transmission is disconnected from the location dowels, swivel the unit out and lower it out of the vehicle.

⚠ *Warning: Support the transmission to ensure that it remains steady on the jack head.*

27 When the transmission is clear of the locating dowels and clutch components, lower the transmission to the ground and withdraw from under the car. Make sure that the transmission does not fall and lose any transmission oil. Also make sure that the clutch assembly comes away with the transmission, and stays inside the bell-housing.

Refitting

28 Refitting the transmission is essentially a reversal of the removal procedure, but note the following points:

a) *On 7-speed (0AM) transmissions, renew the needle bearing in the end of the crankshaft.*

b) *When reconnecting the transmission to the engine, ensure that the location dowels are in position, and that the transmission is correctly aligned with them before pushing it fully into engagement with the engine.*

c) *Tighten all retaining bolts to their specified torque wrench settings.*

d) *Be sure to guide the selector cable into the support bracket as the transmission is refitted – renew the retaining clips.*

e) *Adjust the selector cable, as described in Section 4.*

f) *Refer to Chapter 8 Section 2, and then tighten the driveshaft bolts to the specified torque.*

g) *On completion, check the coolant level.*

h) *If a new transmission unit has been fitted, it will be necessary to have the transmission ECM 'matched' to the engine management ECM electronically, to ensure correct operation – seek the advice of your VAG/Seat dealer or suitably equipped specialist.*

i) *Refill the transmission with the correct grade and quantity of oil, as described in Section 6.*

3 Transmission overhaul – general information

1 In the event of a fault occurring, it will be necessary to establish whether the fault is electrical, mechanical or hydraulic in nature, before repair work can be contemplated. Diagnosis requires detailed knowledge of the transmission's operation and construction, as well as access to specialised test equipment, and so is deemed to be beyond the scope of this manual. It is therefore essential that problems with the automatic transmission be referred to a VAG/Seat dealer or specialist for assessment.

2 Although it is possible to remove the clutch assembly from out of the bell housing, it is not possible to fit a new assembly without

2.24a Hold the drive flange in position ...

2.24b ...whilst removing the centre bolt ...

2.24c ...then slide the drive flange from the transmission

DSG semi-automatic transmission 7B•5

4.4 Disconnect the wiring connector

4.6 Selector cable clip and outer cable circlip

4.7 Prise the inner cable from the ball joint

the use of special tools. When a new clutch is purchased, there are a number of shims and retaining clips of different thickness in the kit, therefore without the correct tools, it is not possible to correctly install the assembly back in place.

3 Note that a faulty transmission should not be removed before the vehicle has been assessed by a dealer or specialist, as fault diagnosis is best carried out with the transmission still in the vehicle.

---

4 Selector lever housing and cable – removal, refitting and adjustment

---

Note: *The selector lever housing and cable should not be separated, remove the housing and cable as a complete unit.*

Removal

1 Move the selector lever to the P position, and remove the battery and battery tray as described in Chapter 5A Section 3.

2 Remove the air filter housing as described in Chapter 4A Section 3 (petrol engine) or (diesel engine).

3 Working inside the vehicle, remove the

centre console and gear selector knob, as described in Chapter 11 Section 26.

4 Disconnect the wiring connector from the front of the gear selector housing **(see illustration)**.

6-speed (02E) transmission

5 Working inside the engine compartment, slacken the adjustment screw on the cable end fitting at the transmission end **(see illustration 4.16)**.

6 Prise out the clips securing the cable to the lever on the transmission, and the outer cable to the support bracket **(see illustration)**. Withdraw the cable from the support bracket and discard the clips – new ones must be fitted.

7-speed (0AM) transmission

7 Working inside the engine compartment, use a pair of long-nose pliers to release the cable end fitting from the ball head on the selector lever **(see illustration)**.

8 Prise out the clip securing the outer cable and withdraw the cable from the support bracket on the transmission **(see illustration)**. Discard the retaining clip – as a new one must be fitted.

All transmissions

9 Raise the front of the vehicle and support

it securely on axle stands (see *Jacking and vehicle support*).

10 Remove the centre tunnel front heat shield from the underside of the vehicle to gain access to the base of the selector lever housing and cable. It may be necessary to separate the exhaust downpipe from the intermediate pipe.

11 Working inside the vehicle, undo the retaining nuts **(see illustration)**, then remove the bracket and lower the gear selector housing downwards. Withdraw it complete with selector cable out from under the vehicle. It may be useful having the aid of an assistant at this point, to be under the vehicle when lowering the selector housing.

Refitting

12 Refitting is the reversal of the removal procedure, noting the following points:

a) *DO NOT grease the cable end fittings.*

b) *Ensure that the cable is correctly routed and secured, as noted on removal.*

c) *Take care not to bend or kink the cable.*

d) *Carry out the cable adjustment procedure described below before reconnecting the cable at the transmission end.*

e) *When refitting the outer cable to the support bracket, use new clips.*

4.8 Release the outer cable securing clip

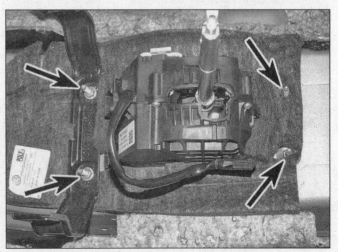

4.11 Gearchange housing mounting nuts

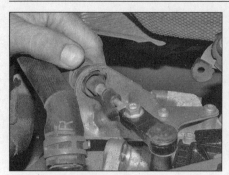

4.16a Secure the outer cable in the bracket ...

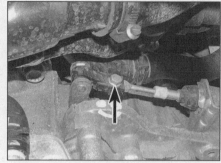

4.16b ...and slacken the cable adjustment screw

4.18 Push the lever in the direction show

Adjustment

13 Inside the car, move the selector lever to the P position.

14 If not already done, disconnect the cable from the selector lever on the transmission.

15 Move the selector lever inside the vehicle from 'P' to 'S' and back, repeatedly, to check that everything moves easily. Seat recommend that you do not grease the cable.

16 Reconnect the cable to the lever at the transmission, and then slacken the cable adjusting bolt **(see illustrations)**.

17 Check that both the selector lever inside

5.2 Prise up the selector lever gaiter surround trim

the car and the lever on the transmission are in their P positions. Gently rock the levers backwards and forwards to make sure the cable is settled. Do not move either lever out of the P position.

18 The transmission lever is in the P position when:

f) *6-speed 02E transmissions – it's pushed all the way back towards the bulkhead*

g) *7-speed 0AM transmissions – it's pushed back towards the selector cable mounting bracket* **(see illustration)**.

19 When in position, tighten the cable adjusting bolt.

20 Verify the operation of the selector lever by shifting through all gear positions and checking that every gear can be selected smoothly and without delay.

5 Emergency release of selector lever

1 If the vehicles battery is disconnected or discharged, it is possible to release the selector lever from its locked position. First, ensure the handbrake is fully applied.

2 Carefully prise up the selector lever gaiter

surround trim from the console and move it to one side **(see illustration)**.

3 Press the yellow plastic wedge downwards **(see illustration)**. It should now be possible to move the selector lever to the desired position.

6 Transmission oil renewal

7-speed dual clutch 0AM transmission

1 Take the vehicle on a short journey to warm the transmission oil, and then park the car on a level surface. For improved access to the drain plug, apply the handbrake, then jack up the front of the vehicle and support it on axle stands (see *Jacking and vehicle support*), but note that the rear of the vehicle should also be raised to ensure all oil is drained.

2 Undo the retaining screws and remove the engine undertray. Wipe clean the area around the transmission drain plug, which is situated on the lower rear of the transmission **(see illustration)**.

3 Place a container under the transmission casing, then unscrew the drain plug from the

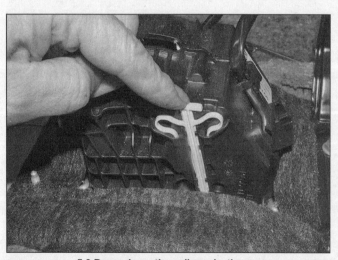

5.3 Press down the yellow plastic peg

6.2 Transmission oil drain plug – 0AM transmissions

6.7 Breather cap on selector cover – 0AM transmissions

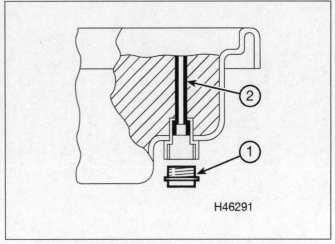

H46291

6.13 Transmission fluid level check – 02E transmissions

1 Level plug   2 Level tube

base of the differential housing, and allow the oil to drain.

4 Remove the battery and battery tray as described in Chapter 5A Section 3.

5 Remove the air cleaner housing as described in Chapter 4A Section 3 (petrol engine) or (diesel engine).

6 When the oil has finished draining, clean the surrounding area, refit the drain plug and tighten it to the specified torque.

7 Unclip the breather cap from the selector cover plate on the top of the transmission casing **(see illustration)**.

8 Using a length of hose, and funnel, add 1.7 litres of new oil to the transmission. Only use the VAG/Seat recommended oil, for 7-speed dual clutch 0AM transmission.

9 Refit the breather cap, making sure that it is secure. Renew if damaged.

10 The remainder of refitting is a reversal of removal.

6-speed dual clutch 02E transmission

Note: *An accurate fluid level check can only be made with the transmission fluid at a temperature of between 35°C and 45°C, and if it is not possible to ascertain this temperature, it is strongly recommended that the check be made by a VAG/Seat dealer who will have the instrumentation to check the temperature and to check the transmission electronics for fault codes. Over-filling or under-filling adversely affects the function of the transmission.*

11 Take the vehicle on a short journey to warm the transmission oil. And then park the vehicle on level ground and engage P with the selector lever. Raise the front and rear of the vehicle and support it on axle stands (see *Jacking and vehicle support*), ensuring the vehicle is kept level. Undo the retaining screws and remove the engine undertray to gain access to the base of the transmission unit.

12 Start the engine and run it at idle speed until the transmission fluid temperature reaches 35°C.

13 Unscrew the fluid level plug from the bottom of the transmission sump **(see illustration)**.

14 If fluid continually drips from the level tube as the fluid temperature increases, the fluid level is correct and does not need to be topped-up. Note that there will be some fluid already present in the level tube, and it will be necessary to observe when this amount has drained before making the level check. Make sure that the check is made before the fluid temperature reaches 45°C.

15 If no fluid drips from the level tube, even when the fluid temperature has reached 45°C, it will be necessary to add fluid. Seat technicians use an adapter which screws into the bottom of the transmission sump, however, a tube inserted up through the drain plug (into the space above the fluid); will allow fluid to be added. Ideally, the fluid should be allowed to cool before adding the fluid.

16 Check the condition of the seal on the level plug and renew it if necessary by cutting off the old seal and fitting a new one. Refit the plug and tighten to the specified torque.

17 Refit the engine undertray, tighten the retaining screws securely, and lower the vehicle to the ground.

18 Frequent need for topping-up indicates that there is a leak, which should be corrected as soon as possible.

Chapter 8
Driveshafts

Contents

Degrees of difficulty

Easy, suitable for novice with little experience		**Fairly easy,** suitable for beginner with some experience		**Fairly difficult,** suitable for competent DIY mechanic		**Difficult,** suitable for experienced DIY mechanic		**Very difficult,** suitable for expert DIY or professional	

Specifications

General

Driveshaft type	Steel shafts with outer constant velocity joints and inner tripod or constant velocity joints (according to type).

Type code differences:

VL 90 or VL 100	CV joints each end, inner joint diameter 90 mm or 100 mm bolted to transmission drive flanges on each side
VL 107 (bolt-on)	CV joints each end, inner joint diameter 107 mm bolted to transmission drive flanges on each side
VL 107 (slip-on).	CV joints each end, inner joint diameter 107 mm splined onto the transmission on each side
AAR 3300i/AAR 2600i.	CV outer joint, tripod inner joint body bolted to transmission drive flanges on each side

Lubrication

Overhaul and repair	Use only special grease supplied in sachets with gaiter/overhaul kits
Joint grease type	Refer to Seat dealer

Joint grease quantity:

Outer joint.	80 g
Inner joint	130 g

Torque wrench settings

	Nm	lbf ft
Driveshaft-to-transmission flange bolts:		
Stage 1	10	7
Stage 2:		
M8 x 48 bolts* – (VL90 & VL100 type joint).	40	30
M10 x 52 bolts* – (VL107 type joint)	70	52
M10 x 23 bolts – (AAR 3300i type joint)	70	52
Hub bolt: *		
Hex head bolt		
Stage 1	200	148
Stage 2 . . . . Angle-tighten a further 180°		
12 point bolt with NO ribbing under head		
Stage 1	200	148
Stage 2 . . . . Angle-tighten a further 180°		
12 point bolt with ribbing under head		
Stage 1	70	52
Stage 2 . . . . Angle-tighten a further 90°		
Intermediate driveshaft bearing bracket	20	15
Lower arm-to-balljoint nuts: *		
Cast steel lower arm.	60	44
Forged aluminium lower arm	100	74
Lower rear mounting-to-transmission bolts: *		
Stage 1	50	37
Stage 2 . . . . Angle-tighten a further 90°		
Wheel bolts.	120	89

*Use new bolts/nuts

1 General Information

1 Drive is transmitted from the differential to the front wheels by means of two steel driveshafts of either solid or hollow construction (depending on model, and which side of the vehicle). Both driveshafts are splined at their outer ends, to accept the wheel hubs, and are secured to the hub by a large bolt. The inner end of each driveshaft is either bolted to a transmission drive flange or splined directly onto the differential splined shaft.

2 The outer ends of each driveshaft are fitted with ball-bearing type constant velocity (CV) joints, to ensure the smooth and efficient transmission of drive at all the angles possible, as the roadwheels move up-and-down with the suspension, and as they turn from side to side under steering.

3 The inner ends of each driveshaft (except for the AAR 3300i and AAR 2600i type joints) are fitted with ball and cage type constant velocity (CV) joint. The AAR 3300i and the

AAR 2600i driveshafts have a triple roller type (tripod) joint fitted to the inner end of the driveshaft.

4 Hytrel (thermoplastic elastomer) gaiters are fitted over the CV joints with steel clips. These gaiters combine the flexibility of rubber with the strength and durability of thermoplastics. The gaiters contain the grease that lubricates the joints, and also protect the joints from the entry of dirt and debris.

2 Driveshafts – removal and refitting

Note: *A new hub bolt will be required on refitting. There are three different types of bolts fitted – two types of 12 point spline headed bolt and a standard hex bolt. One of the 12 point spline bolts has ribs **(see illustration)** under the face of the bolt head and one is smooth. The torque setting is different for each type of bolt (see Section), so it is important to make sure you check which is fitted.*

Removal

1 Remove the wheel trim/centre cap (as applicable) then apply the handbrake, and partially unscrew the relevant hub bolt with the vehicle resting on its wheels, by a maximum of 90° – note that the bolt is very tight, and a suitable extension bar will probably be required to aid unscrewing. Also slacken the road wheel securing bolts by half a turn. **Note:** *Do not loosen the bolt more than 90° with the vehicle standing on the ground, as the wheel bearings may be damaged.*

2 Apply the handbrake, then jack up the front of the vehicle and support it on axle stands (see). Remove the appropriate front roadwheel.

3 Remove the retaining screws and/or clips, and remove the undershields from beneath the engine/transmission unit to gain access to the driveshafts. Where necessary, also unbolt the heat shield from the transmission housing to improve access to the driveshaft inner joint **(see illustration)**.

4 Unscrew and remove the hub bolt **(see illustration)**. **Note:** *Discard the bolt and obtain*

2.0a Flange of bolt is ribbed

2.3 Remove the heat shield

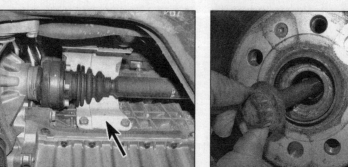

2.4 Remove the driveshaft/hub bolt

2.5 Undo the lower ball joint retaining nuts

2.6 Slide the driveshaft out from the hub assembly

2.7 Unbolt the lower rear mounting bracket

a new one, there are three types of bolt fitted, see note at the beginning of this Section.

5 Unscrew the three nuts securing the front suspension lower arm balljoint to the lower arm **(see illustration)**. Discard the nuts, as new ones must be used on refitting.

6 Lever the lower arm downwards to release it from the balljoint studs, then pull the hub carrier outwards, and at the same time withdraw the driveshaft outer constant velocity joint from the hub **(see illustration)**. If the joint splines are a tight fit in the hub, tap the joint out of the hub using a soft-faced mallet and drift. If this fails to free the driveshaft from the hub, the joint will have to be pressed out using a suitable tool bolted to the hub.

7 On some models, in order to gain the necessary clearance required to withdraw the left-hand driveshaft, it may be necessary to unbolt the rear engine transmission mounting from the subframe **(see illustration)**, and move the engine slightly.

8 Proceed as follows according to driveshaft type.

Caution: Support the driveshaft by suspending it with wire or string – do not allow it to hang under its own weight, or the joint may be damaged.

Inner joint with drive flange – VL90, VL100 and VL107

9 Mark the inner joint in relation to the drive flange for refitting. Using a multi-splined tool, unscrew and remove the six bolts securing

the inner driveshaft joint to the transmission flange and, recover the retaining plates from underneath the bolts **(see illustrations)**.

Inner joint splined to differential drive shaft – VL107

10 Position a container beneath the transmission to catch spilt oil, and then pull out the driveshaft. The internal driveshaft circlip may be tight in the transmission side gear, in which case careful use of a lever against the transmission casing will be required. Lever against a block of wood to prevent damage to the casing, and take care not to damage the oil seal as the driveshaft is being removed. **Note:** *Pull only on the inner joint body, not the driveshaft itself; otherwise the gaiter may be damaged.*

Inner joint triple roller (tripod) located in joint body – AAR 3300i and AAR 2600i

11 Mark the inner joint body and driveshaft in relation to each other, as the tripod joint will need to be refitted in the joint body in the same position. Loosen the larger retaining clip, ease off the rubber gaiter, and pull the triple roller out of the joint body.

All types

12 Manoeuvre the driveshaft out from underneath the vehicle and (where fitted) recover the gasket from the end of the inner constant velocity joint. **Note:** *Discard the gasket and obtain a new one.*

Caution: Do not allow the vehicle to rest on its wheels with one or both driveshaft(s) removed, as damage to the wheel bearings may result.

13 If moving the vehicle is unavoidable, temporarily insert the outer end of the driveshaft(s) in the hub(s), and tighten the driveshaft retaining bolt(s); in this case, the inner end(s) of the driveshaft(s) must be supported, for example by suspending with string from the vehicle underbody.

Refitting

14 Where applicable, check the condition of the circlip on the inner end of the driveshaft, and if necessary, renew it.

15 As applicable, clean the splines on each end of the driveshaft and in the hub and apply a little oil, and where applicable wipe clean the oil seal in the transmission casing. Check the oil seal and if necessary renew it as described in Chapter or. Smear a little oil on the lips of the oil seal before fitting the driveshaft.

Inner joint with bolted drive flange – VL90, VL100 and VL107

16 Ensure that the transmission flange and inner joint mating surfaces are clean and dry. Where necessary, fit a new gasket to the joint by peeling off its backing foil and sticking it in position **(see illustration)**.

17 Manoeuvre the driveshaft into position, aligning the previously made marks **(see**

2.9a Make alignment marks...

2.9b ...and remove the drive flange bolts and plates

2.16 Locate a new gasket on the inner joint

2.17 Align the markings made on removal when refitting

illustration), and then align the inner joint holes with those on the transmission flange. Refit the new retaining bolts and locking plates, and then tighten the retaining bolts to the specified torque.

Inner joint splined to differential drive shaft – VL107

18 Locate the inner end of the driveshaft into the transmission – turn the driveshaft as necessary to engage the splines. Press in the driveshaft until the internal circlip engages the groove. Check that the circlip is engaged by attempting to pull out the driveshaft with only moderate force.

Inner joint triple roller (tripod) located in joint body – AAR 3300i and AAR 2600i

19 Fill the inner joint with the specified quantity of grease, and then locate the driveshaft tripod into the joint body, aligning the previously made marks. Ease the gaiter onto the joint body, and secure the gaiter with a new retaining clip.

All types

20 With the lower arm levered downwards, engage the outer joint with the hub. Fit the new hub bolt and use it to draw the joint fully into position.
21 Align the balljoint studs with the holes in the lower arm, then release the arm and fit the three new nuts. Tighten the nuts to the specified torque.
22 Where applicable, refit the lower rear mounting-to-transmission bolts, and then tighten the new bolts to the specified torque.
23 Tighten the driveshaft bolt to the Stage 1 torque setting. **Note:** *The bolt must be tightened with the wheel clear of the ground.*

There are three different types of bolts fitted, one is ribbed **(see illustration 2.0)** *under the face of the bolt head and one is smooth. See note at the beginning of this Section.*
24 Refit the roadwheel and lower the vehicle to the ground, then angle-tighten the driveshaft bolt through the Stage 2 angle (see Section).
25 Once the driveshaft bolt is correctly tightened, tighten the wheel bolts to the specified torque and refit the wheel trim/centre cap.

3 Driveshaft gaiters – renewal

1 Remove the driveshaft from the car, as described in Section 2. Continue as described under the relevant sub-heading. Driveshafts with a tripod type inner joint can be identified by the shape of the inner CV joint; the driveshaft retaining bolt holes are in extensions from the joint, giving it a six-pointed star-shaped exterior, in contrast to the smooth, circular shape of the ball-and-cage joint **(see illustrations)**.

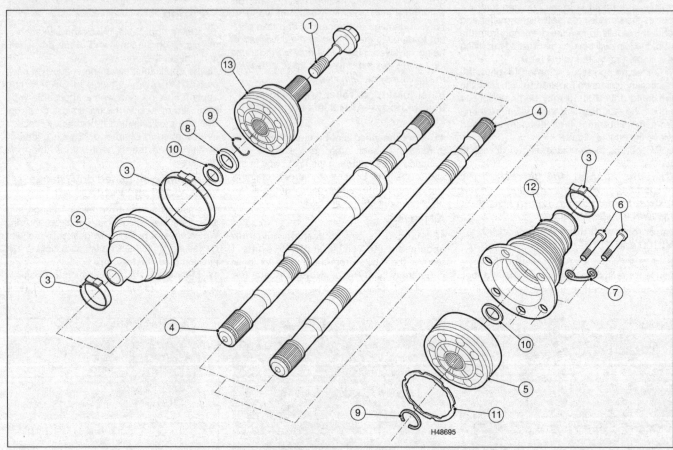

3.1a Driveshaft components – models with VL90 or VL100 CV joint

1 Hub bolt	4 Driveshaft	7 Bolt retaining plate
2 Outer joint gaiter	5 Inner joint	8 Thrust washer
3 Gaiter securing clip	6 Flange bolts	9 Circlip

10 Washer	12 Inner joint gaiter
11 Gasket	13 Outer joint

H48695

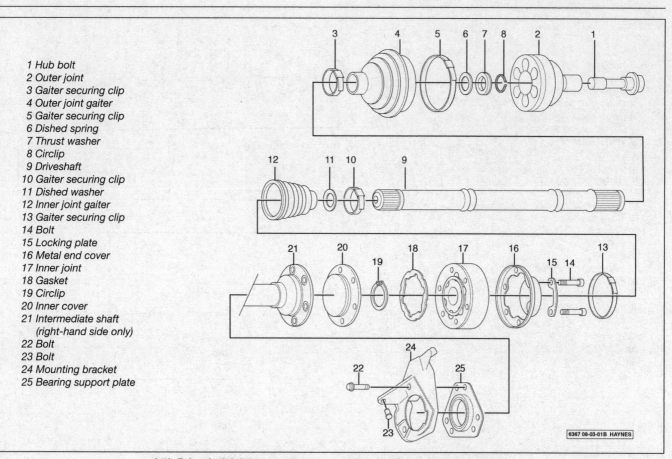

1 Hub bolt
2 Outer joint
3 Gaiter securing clip
4 Outer joint gaiter
5 Gaiter securing clip
6 Dished spring
7 Thrust washer
8 Circlip
9 Driveshaft
10 Gaiter securing clip
11 Dished washer
12 Inner joint gaiter
13 Gaiter securing clip
14 Bolt
15 Locking plate
16 Metal end cover
17 Inner joint
18 Gasket
19 Circlip
20 Inner cover
21 Intermediate shaft
   (right-hand side only)
22 Bolt
23 Bolt
24 Mounting bracket
25 Bearing support plate

6367 08-03-01B HAYNES

3.1b Driveshaft joint components – models with VL107 CV joint (bolt-on)

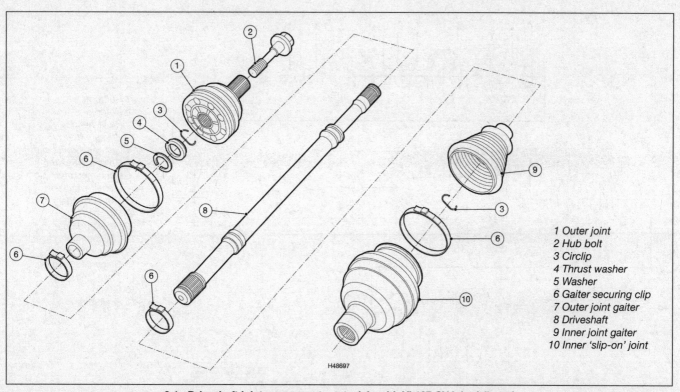

1 Outer joint
2 Hub bolt
3 Circlip
4 Thrust washer
5 Washer
6 Gaiter securing clip
7 Outer joint gaiter
8 Driveshaft
9 Inner joint gaiter
10 Inner 'slip-on' joint

H48697

3.1c Driveshaft joint components – models with VL107 CV joint (slip-on)

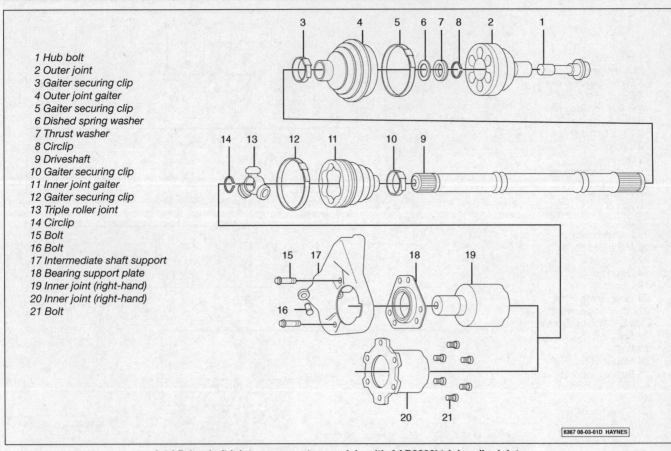

1 Hub bolt
2 Outer joint
3 Gaiter securing clip
4 Outer joint gaiter
5 Gaiter securing clip
6 Dished spring washer
7 Thrust washer
8 Circlip
9 Driveshaft
10 Gaiter securing clip
11 Inner joint gaiter
12 Gaiter securing clip
13 Triple roller joint
14 Circlip
15 Bolt
16 Bolt
17 Intermediate shaft support
18 Bearing support plate
19 Inner joint (right-hand)
20 Inner joint (right-hand)
21 Bolt

6367 08-03-01D HAYNES

3.1d Driveshaft joint components – models with AAR3300i triple roller joints

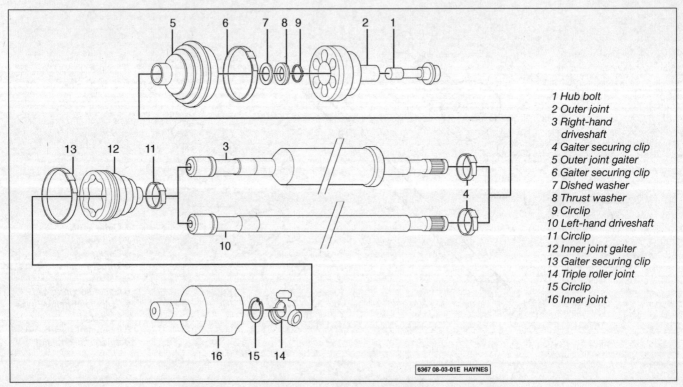

1 Hub bolt
2 Outer joint
3 Right-hand
   driveshaft
4 Gaiter securing clip
5 Outer joint gaiter
6 Gaiter securing clip
7 Dished washer
8 Thrust washer
9 Circlip
10 Left-hand driveshaft
11 Circlip
12 Inner joint gaiter
13 Gaiter securing clip
14 Triple roller joint
15 Circlip
16 Inner joint

6367 08-03-01E HAYNES

3.1e Driveshaft components – models with AAR2600i triple roller joints

3.2 Release the outer joint gaiter clips...

3.3 ...and slide the gaiter away from the joint

3.4 Use a mallet to drive the outer joint from the driveshaft

3.5 Removing the circlip, thrustwasher and dished washer

3.6 Removing the outer gaiter

3.11 Temporarily tape over the splines to protect the new gaiter

Outer CV joint gaiter – All types of driveshafts

2 Secure the driveshaft in a vice equipped with soft jaws, and release the two outer joint gaiter retaining clips **(see illustration)**. If necessary, the retaining clips can be cut to release them.

3 Slide the rubber gaiter down the shaft to expose the constant velocity joint, and scoop out excess grease **(see illustration)**.

4 Using a soft-faced mallet, tap the joint off the end of the driveshaft **(see illustration)**.

5 Remove the circlip from the driveshaft groove, and slide off the thrustwasher and dished washer, noting which way around it is fitted **(see illustration)**.

6 Slide the rubber gaiter off the driveshaft and discard it **(see illustration)**.

7 Thoroughly clean the constant velocity joint(s) using paraffin, or a suitable solvent, and dry thoroughly. Carry out a visual inspection as follows.

8 Move the inner splined driving member from side-to-side to expose each ball in turn at the top of its track. Examine the balls for cracks, flat spots or signs of surface pitting.

9 Inspect the ball tracks on the inner and outer members. If the tracks have widened, the balls will no longer be a tight fit. At the same time, check the ball cage windows for wear or cracking between the windows.

10 If on inspection any of the constant velocity joint components are found to be worn or damaged, it will be necessary to renew the complete joint assembly. If the joint is in satisfactory condition, obtain a new gaiter and retaining clips, a constant velocity joint

circlip and the correct type of grease. Grease is often supplied with the joint repair kit – if not, use good-quality molybdenum disulphide grease.

11 Tape over the splines on the end of the driveshaft, to protect the new gaiter as it is slid into place **(see illustration)**.

12 Slide the new gaiter onto the end of the driveshaft, then remove the protective tape from the driveshaft splines.

13 Slide on the dished washer, making sure its convex side is innermost, followed by the thrustwasher.

14 Pack the joint with half the quantity of the specified type of grease. Work the grease well into the bearing tracks whilst twisting the joint, and fill the rubber gaiter with the remaining half **(see illustrations)**.

15 Fit a new circlip to the driveshaft, then tap

3.14a Pack half of the grease in the joint...

3.14b ...and the remaining half in the gaiter

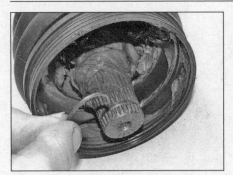

3.15a Fit a new circlip...

3.15b ...then refit the outer joint

3.16 Seat the gaiter on the outer joint and driveshaft, then lift its inner lip to equalise the air pressure

the joint onto the driveshaft until the circlip engages in its groove **(see illustrations)**. Make sure that the joint is securely retained by the circlip.

16 Ease the gaiter over the joint, and ensure that the gaiter lips are correctly located on both the driveshaft and constant velocity joint. Lift the outer sealing lip of the gaiter to

equalise air pressure within the gaiter **(see illustration)**.

17 Fit the large metal retaining clip to the gaiter. Pull the clip as tight as possible, and locate the hooks on the clip in their slots. Remove any slack in the gaiter retaining clip by carefully compressing the raised section of the clip. In the absence of the special tool, a pair of side-cutters may be used, taking care not to cut the clip **(see illustrations)**. Secure the small retaining clip using the same procedure.

18 Check the constant velocity joint moves freely in all directions, then refit the driveshaft to the vehicle, as described in Section 2.

Inner gaiter with bolted drive flange – VL90, VL100 and VL107

19 Secure the driveshaft in a vice equipped with soft jaws, then release the gaiter small securing clip, securing the gaiter to the driveshaft **(see illustration)**.

20 Using a hammer and a small drift, carefully drive the gaiter metal ring from the joint outer member **(see illustration)**.

21 Slide the gaiter down the driveshaft to expose the constant velocity joint, and scoop out excess grease.

22 Remove the circlip from the end of the driveshaft using circlip pliers **(see illustration)**.

23 Press or drive the driveshaft from the joint, taking great care not to damage the joint. Recover the dished washer fitted between the constant velocity joint and the gaiter **(see illustrations)**.

3.17a Fit the large metal retaining clip...

3.17b ...and use a suitable tool to tighten it

3.19 Release the gaiter small securing clip...

3.20 ...drive the metal ring from the joint outer member

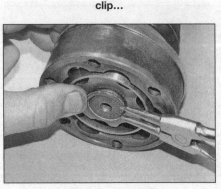

3.22 Remove the circlip...

3.23a ...followed by the joint...

3.23b ...dished washer...

3.24 ...and gaiter

3.25a Tilt the splined hub and cage to remove the ball-bearings...

3.25b ...then separate the hub from the cage

3.25c Inner CV joint gaiter repair kit

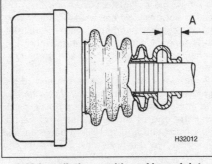

3.29 Installation position of inner joint gaiter on left-hand driveshaft

A = 17.0 mm

3.30a Pack the inner joint with half of the grease...

24 Slide the gaiter from the end of the driveshaft **(see illustration)**.

25 Proceed as described previously in paragraphs 7 to 12 **(see illustrations)**.

26 Slide the dished washer onto the driveshaft, making sure its convex side is innermost.

27 Fit the joint to the end of the driveshaft, noting that the chamfered edge of the internal splines on the joint should face towards the driveshaft. Drive or press the joint into position until it contacts the shoulder on the driveshaft.

28 Fit a new circlip to retain the joint on the end of the driveshaft.

29 If the left-hand driveshaft is being worked on, mark the final installation position of the gaiter outboard end on the driveshaft using tape or paint – do not scratch the surface of the driveshaft **(see illustration)**.

30 Pack the joint with the half the recommended quantity of grease (see Specifications), and then pack the gaiter with the remaining half **(see illustrations)**.

31 Slide the gaiter up the driveshaft, and press or drive the gaiter metal ring onto the joint outer member. To ensure the bolt holes are correctly positioned, temporarily fit a couple of flange bolts **(see illustrations)**.

32 If the left-hand driveshaft is being worked on, slide the outboard end of the gaiter into position using the mark made previously (see paragraph 29), then secure the outer gaiter securing clip in position as described in paragraph 17.

33 If the right-hand driveshaft is being worked on, slide the outboard end of the gaiter into position on the driveshaft, then

3.30b ...then pack the gaiter with the remaining half

secure the outer gaiter securing clip in position as described in paragraph 17 **(see illustration)**.

34 Check the driveshaft joint moves freely in

3.31a Temporarily fit one of the flange bolts to ensure the bolt holes are correctly aligned...

3.31b ...then drive the metal ring onto the joint outer member

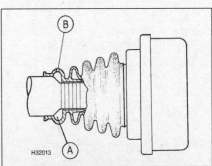

3.33 Installation position of inner joint gaiter on right-hand driveshaft

A Vent chamber in gaiter   B Vent hole

all directions, then refit the driveshaft to the vehicle, as described in Section 2.

Inner gaiter with splined joint to transmission – VL107

35 Secure the driveshaft in a vice equipped with soft jaws, and release the two inner joint gaiter retaining clips. If necessary, the retaining clips can be cut to release them.

36 Slide the rubber gaiter down the shaft to expose the constant velocity joint, and scoop out excess grease.

37 Using a soft-faced mallet, tap the joint off the end of the driveshaft.

38 Remove the circlip from the end of the driveshaft groove, and then slide the rubber gaiter off the driveshaft and discard it.

39 Proceed as described previously in paragraphs 7 to 12.

40 Pack the joint with half the quantity of the specified type of grease. Work the grease well into the bearing tracks whilst twisting the joint, and fill the rubber gaiter with the remaining half.

41 Fit a new circlip to the driveshaft, then tap the joint onto the driveshaft until the circlip engages in its groove. Make sure that the joint is securely retained by the circlip.

42 Ease the gaiter over the joint, and ensure that the gaiter lips are correctly located on both the driveshaft and constant velocity joint. Lift the outer sealing lip of the gaiter to equalise air pressure within the gaiter.

43 Fit the large metal retaining clip to the gaiter, as described in paragraph 17. Secure the small retaining clip using the same procedure.

44 Check the constant velocity joint moves freely in all directions, then refit the driveshaft to the vehicle, as described in Section 2.

Inner gaiter with triple roller (tripod) joint – AAR 3300i and AAR 2600i

45 Secure the driveshaft in a vice equipped with soft jaws, and clean out the old grease from around the triple roller joint and inside the gaiter.

46 Remove the circlip from the end of the driveshaft.

47 Press or drive the driveshaft from the tripod, taking great care not to damage the surfaces of the roller locating arms.

48 Slide the outer member and the rubber gaiter from the end of the driveshaft.

49 Thoroughly clean the joint components using paraffin, or a suitable solvent, and dry thoroughly. Carry out a visual inspection as follows.

50 Inspect the tripod rollers and the joint outer member for signs of wear, pitting or scuffing on their mating surfaces. Check that the joint rollers rotate smoothly, with no traces of roughness.

51 If the rollers or outer member shown signs of wear or damage, it will be necessary to renew the complete driveshaft, since the joint is not available separately. If the joint is in satisfactory condition, obtain a repair kit, consisting of a new gaiter, retaining clips, circlip, and the correct type and quantity of grease.

52 Tape over the splines on the end of the driveshaft, to protect the new gaiter as it is slid into place, and then slide the new gaiter over the end of the driveshaft and secure the smaller end in place with a new retaining clip. Remove the protective tape from the driveshaft splines.

53 Press or drive the tripod onto the end of the driveshaft until it contacts the stop, ensuring that the marks made on the end of the driveshaft and the tripod before dismantling are aligned. Note that the chamfered edge of the internal splines on the tripod should face towards the driveshaft.

54 Fit the new circlip to retain the tripod on the end of the driveshaft.

55 Check the triple roller joint moves freely, then refit the driveshaft to the vehicle, as described in Section 2.

4 Intermediate driveshaft – removal and refiiting

1 Some models have an intermediate driveshaft fitted to the right-hand side of the transmission. The shaft is supported by a bracket bolted to the rear of the engine block.

The bracket has a bearing held in place by a support bracket.

Removal

2 Remove the right-hand driveshaft as described in Section 2.

3 Remove the bolts from the bearing bracket and pull the shaft from the transmission.

4 Using a small pick or similar remove the oil seal from the transmission. It must be replaced.

5 If required the bearing can now be pressed of the shaft using a puller or hydraulic press.

Refitting

6 Fit a new oil seal to the transmission and then refit the shaft. Tighten the bearing support bracket to the specified torque and then refit the driveshaft.

5 Driveshaft overhaul – general information

1 If any of the checks described in this Chapter reveal wear in any driveshaft joint, first remove the roadwheel trim or centre cap (as applicable) and check that the hub bolt is tight. If the bolt is loose, obtain a new one, and tighten it to the specified torque (see Section 2). If the bolt is tight, refit the centre cap/trim, and repeat the check on the other hub bolt.

2 Road test the vehicle, and listen for a metallic clicking from the front of the vehicle as the vehicle is driven slowly in a circle on full-lock. If a clicking noise is heard, this indicates wear in the outer constant velocity joint; this means that the joint must be renewed.

3 If vibration consistent with roadspeed is felt through the car when accelerating, there is a possibility of wear in the inner constant velocity joints.

4 To check the joints for wear, remove the driveshafts, then dismantle them as described in Section 3. If any wear or free play is found, the affected joint must be renewed. Refer to a Seat dealer or aftermarket supplier for information on the availability of driveshaft components.

Chapter 9
Braking system

Contents

Degrees of difficulty

Easy, suitable for novice with little experience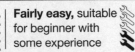
Fairly easy, suitable for beginner with some experience
Fairly difficult, suitable for competent DIY mechanic
Difficult, suitable for experienced DIY mechanic
Very difficult, suitable for expert DIY or professional

Specifications

Front brakes
Caliper type .... FN3, FS3, FNR-G or P4-38/42
Disc diameter:
  FS3 .... 280 mm
  FN3:
    15" wheels .... 288 mm
    16" wheels .... 312 mm
  FNR-G/P4-38/42 .... 345 mm
Disc thickness.
  New:
    FS3 .... 22.0 mm
    FN3 .... 25.0 mm
    FNR-G/P4-38/42. .... 30.0 mm
  Minimum permissible thickness:
    FS3 .... 19.0 mm
    FN3 .... 22.0 mm
    FNR-G/P4-38/42. .... 27 mm
Maximum disc run-out...................................... 0.1 mm
Brake pad lining thickness (all models):
  New .... 14.0 mm
  Minimum.. 2.0 mm

Rear disc brakes

Caliper type:

C I 38	1KQ/1KD models
C II 38 HR	1KZ/1KY models
C II 41 HR (TRW)	1KF/1KJ models
ZOH 38 BIR III (Bosch)	1KS models

Disc diameter:

C I 38	255 mm
C II 38 HR	286 mm
C II 41 HR (TRW)	260 mm (1KF) or 286 mm (1KJ)
ZOH 38 BIR III (Bosch)	272 mm

Disc thickness:

New:

C I 38/ZOH 38 BIR III	10.0 mm
C II 38 HR/C II 41 HR	12.0 mm

Minimum thickness:

C I 38/ZOH 38 BIR III	08.0 mm
C II 38 HR/C II 41 HR	10.0 mm
Maximum disc run-out	0.1 mm

Brake pad lining thickness:

All versions	12 mm
Minimum	2.0 mm

Torque wrench settings

	Nm	lbf ft
ABS control unit retaining bolts	8	6
ABS control unit mounting bracket nuts	20	15
ABS wheel sensor retaining bolts	8	6
Brake pedal pivot shaft nut*	25	18
Brake light switch	5	4
Brake pedal mounting bracket*	25	18
Brake disc shield bolts	12	9
Front brake caliper:		
Guide pins	30	22
Mounting bracket bolts (FN3 calipers)	190	141
Mounting bracket bolts (FNR G Cupra calipers)	200	148
Hydraulic brake line to caliper banjo bolt	35	26
Hydraulic brake line union nuts	14	10
Master cylinder mounting nuts*	25	18
Rear brake caliper:		
Guide pin bolts*	35	26
Mounting bracket bolts (multi-point socket head): *		
Stage 1	90	66
Stage 2	Angle-tighten a further 90°	
Roadwheel bolts	120	89
Servo unit mechanical vacuum pump	10	8
Servo unit mounting bolts*	25	18

*Use new fasteners

---

1 General information and precautions

General information

1 The braking system is of servo-assisted, diagonal dual-circuit hydraulic type. The arrangement of the hydraulic system is such that each circuit operates one front and one rear brake from a tandem master cylinder. Under normal circumstances, both circuits operate in unison, but, if there is hydraulic failure in one circuit, full braking force will still be available at two wheels. On petrol engines, vacuum for the servo unit is supplied from the inlet manifold, however, on diesel engines a combined fuel lift pump and vacuum pump is driven off the end of the camshaft.

2 All models covered by this manual are equipped with disc brakes at the front and rear. ABS is fitted as standard to all models (refer to Section 19 for further information on ABS operation).

3 The front disc brakes are actuated by single-piston sliding type calipers, which ensure that equal pressure is applied to each disc pad.

4 The rear brakes are also actuated by single-piston sliding calipers, which incorporate independent mechanical handbrake mechanisms as well.

Precautions

5 • When servicing any part of the system, work carefully and methodically; also observe scrupulous cleanliness when overhauling any part of the hydraulic system. Always renew components in axle sets (where applicable) if in doubt about their condition, and use only genuine VAG/Seat parts, or at least those of known good quality. Note the warnings given in Safety First 0, ! and at relevant points in this Chapter concerning the dangers of asbestos dust and hydraulic fluid.

2 Hydraulic system – bleeding

Warning: Hydraulic fluid is poisonous; wash off immediately and thoroughly in the case of skin contact, and seek immediate

medical advice if any fluid is swallowed or gets into the eyes. Certain types of hydraulic fluid are flammable, and may ignite when allowed into contact with hot components; when servicing any hydraulic system, it is safest to assume that the fluid is flammable, and to take precautions against the risk of fire as though it is petrol that is being handled. Hydraulic fluid is also an effective paint stripper, and will attack plastics; if any is spilt, it should be washed off immediately, using copious quantities of fresh water. Finally, it is hygroscopic (it absorbs moisture from the air) – old fluid may be contaminated and unfit for further use. When topping-up or renewing the fluid, always use the recommended type, and ensure that it comes from a freshly opened sealed container.

Note: *Seat specify that at least 0.25 litre of brake fluid should be expelled from each caliper.*

General

1 The correct operation of any hydraulic system is only possible after removing all air from the components and circuit; this is achieved by bleeding the system. Since the clutch hydraulic system also uses fluid from the brake system reservoir, it should also be bled at the same time by referring to Chapter 6.

2 During the bleeding procedure, add only clean, unused hydraulic fluid of the recommended type; never re-use fluid that has already been bled from the system. Ensure that sufficient fluid is available before starting work.

3 If there is any possibility of incorrect fluid being already in the system, the brake components and circuit must be flushed completely with uncontaminated, correct fluid, and new seals should be fitted to the various components.

4 If hydraulic fluid has been lost from the system, or air has entered because of a leak, ensure that the fault is cured before continuing further.

5 Park the vehicle on level ground, then chock the wheels and release the handbrake.

6 Check that all pipes and hoses are secure, unions tight and bleed screws closed. Clean any dirt from around the bleed screws.

7 Unscrew the master cylinder reservoir cap, and top the reservoir up to the MAX level line; refit the cap loosely, and remember to maintain the fluid level at least above the MIN level line throughout the procedure, or there is a risk of further air entering the system.

8 There is a number of one-man, do-it-yourself brake bleeding kits currently available from motor accessory shops. It is recommended that one of these kits is used whenever possible, as they greatly simplify the bleeding operation, and reduce the risk of expelled air and fluid being drawn back into the system. If such a kit is not available, the basic (two-man) method must be used, which is described in detail below.

9 If a kit is to be used, prepare the vehicle as described previously, and follow the kit manufacturer's instructions, as the procedure may vary slightly according to the type being used; generally, they are as outlined below in the relevant sub-section.

10 Whichever method is used, the same sequence must be followed (paragraph 12) to ensure the removal of all air from the system.

Bleeding sequence

11 If the system has been only partially disconnected, and suitable precautions were taken to minimise fluid loss, it should be necessary only to bleed that part of the system.

12 If the complete system is to be bled, then it should be done working in the following sequence:
a) Left-hand front brake.
b) Right-hand front brake.
c) Left-hand rear brake.
d) Right-hand rear brake.

13 If the system is completely empty of fluid, repeat the sequence up to five times, or until the brake pedal feels firm and has an acceptable amount of travel.

Bleeding

Basic (two-man) method

14 Collect together a clean glass jar of reasonable size, a suitable length of plastic or rubber tubing which is a tight fit over the bleed screw, and a ring spanner to fit the screw. The help of an assistant will also be required.

15 Remove the dust cap from the first screw in the sequence **(see illustration)**. Fit the spanner and tube to the screw, place the other end of the tube in the jar, and pour in sufficient fluid to cover the end of the tube.

16 Ensure that the master cylinder reservoir fluid level is maintained at least above the MIN level line throughout the procedure.

17 Have the assistant fully depress the brake pedal several times to build-up pressure, and then maintain it on the final downstroke.

18 While pedal pressure is maintained, unscrew the bleed screw (approximately one turn) and allow the compressed fluid and air to flow into the jar. The assistant should maintain pedal pressure, following it down to the floor if necessary, and should not release

it until instructed to do so. When the flow stops, tighten the bleed screw again, have the assistant release the pedal slowly, and recheck the reservoir fluid level.

19 Repeat the steps given in paragraphs 16 and 17 until the fluid emerging from the bleed screw is free from air bubbles. If the master cylinder has been drained and refilled, and air is being bled from the first screw in the sequence, allow approximately five seconds between cycles for the master cylinder passages to refill.

20 When no more air bubbles appear, tighten the bleed screw securely, remove the tube and spanner, and refit the dust cap. Do not overtighten the bleed screw.

21 Repeat the procedure on the remaining screws in the sequence, until all air is removed from the system and the brake pedal feels firm again.

Using a one-way valve kit

22 As their name implies, these kits consist of a length of tubing with a one-way valve fitted, to prevent expelled air and fluid being drawn back into the system; some kits include a translucent container, which can be positioned so that the air bubbles can be more easily seen flowing from the end of the tube.

23 The kit is connected to the bleed screw, which is then opened. The user returns to the driver's seat, depresses the brake pedal with a smooth, steady stroke, and slowly releases it; this is repeated until the expelled fluid is clear of air bubbles **(see illustration)**.

24 Note that these kits simplify work so much that it is easy to forget the master cylinder reservoir fluid level; ensure that this is maintained at least above the MIN level line at all times.

Using a pressure-bleeding kit

25 These kits are usually operated by the reservoir of pressurised air contained in the spare tyre. However, note that it will be probably necessary to reduce the pressure to less than 1.0 bar (14.5 psi); refer to the instructions supplied with the kit.

26 By connecting a pressurised, fluid-filled container to the master cylinder reservoir, bleeding can be carried out simply by opening each screw in turn (in the specified sequence),

2.15 Remove the dust cap from the first bleed screw in the sequence

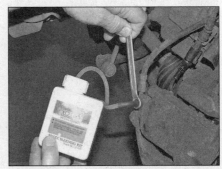

2.23 Bleeding a brake using a one-way valve kit

3.6a Brake hose retaining clip

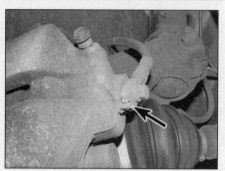

3.6b Locating peg for brake hose end fitting

hydraulic system as described in Section 2. Wash off any spilt fluid, and check carefully for fluid leaks.

and allowing the fluid to flow out until no more air bubbles can be seen in the expelled fluid.

27 This method has the advantage that the large reservoir of fluid provides an additional safeguard against air being drawn into the system during bleeding.

28 Pressure-bleeding is particularly effective when bleeding 'difficult' systems, or when bleeding the complete system at the time of routine fluid renewal.

All methods

29 When bleeding is complete, and firm pedal feel is restored, wash off any spilt fluid, tighten the bleed screws securely, and refit their dust caps.

30 Check the hydraulic fluid level in the master cylinder reservoir, and top-up if necessary (see *Weekly checks* Section 5).

31 Discard any hydraulic fluid that has been bled from the system; it will not be fit for re-use.

32 Check the feel of the brake pedal. If it feels at all spongy, air must still be present in the system, and further bleeding is required. Failure to bleed satisfactorily after a reasonable repetition of the bleeding procedure may be due to worn master cylinder seals.

3 Hydraulic pipes and hoses – renewal

Note: *Refer to the note in Section 2 concerning the dangers of hydraulic fluid.*

1 If any pipe or hose is to be renewed, minimise fluid loss by first removing the master cylinder reservoir cap, then tightening it down onto a piece of polythene to obtain an airtight seal. Alternatively, flexible hoses can be sealed, if required, using a proprietary brake hose clamp; metal brake pipe unions can be plugged (if care is taken not to allow dirt into the system) or capped immediately they are disconnected. Place a wad of rag under any union that is to be disconnected, to catch any spilt fluid.

2 If a flexible hose is to be disconnected, where applicable unscrew the brake pipe union nut before removing the spring clip which secures the hose to its mounting bracket.

3 To unscrew the union nuts, it is preferable to obtain a brake pipe spanner of the correct size; these are available from most large motor accessory shops. Failing this, a close-fitting open-ended spanner will be required, though if the nuts are tight or corroded, their flats may be rounded-off if the spanner slips. In such a case, a self-locking wrench is often the only way to unscrew a stubborn union, but it follows that the pipe and the damaged nuts must be renewed on reassembly. Always clean a union and surrounding area before disconnecting it. If disconnecting a component with more than one union, make a careful note of the connections before disturbing any of them.

4 If a brake pipe is to be renewed, it can be obtained, cut to length and with the union nuts and end flares in place, from Seat dealers. All that is then necessary is to bend it to shape, following the line of the original, before fitting it to the car. Alternatively, most motor accessory shops can make up brake pipes from kits, but this requires very careful measurement of the original, to ensure that the new pipe is of the correct length. The safest answer is usually to take the original to the shop as a pattern.

5 On refitting, do not overtighten the union nuts. It is not necessary to exercise brute force to obtain a sound joint.

6 Ensure that the pipes and hoses are correctly routed, with no kinks, and that they are secured in the clips or brackets provided **(see illustrations)**. After fitting, remove the polythene from the reservoir, and bleed the

4 Front brake pads – renewal

⚠️ *Warning: Renew both sets of front brake pads at the same time – never renew the pads on only one wheel, as uneven braking may result. Note that the dust created by wear of the pads may contain asbestos, which is a health hazard. Never blow it out with compressed air, and don't inhale any of it. An approved filtering mask should be worn when working on the brakes. DO NOT use petrol or petroleum-based solvents to clean brake parts; use a proprietary brake cleaner only.*

1 Several types of front caliper are fitted to the Leon range. There are minor differences, but removal and refitting of the brake pads is essentially the same for all types of caliper. Note however that FS3 type brakes have and inner and outer brake pad with spring clips attached. On these brakes the pads are removed with the caliper **(see illustrations)**.

2 Apply the handbrake, then slacken the front roadwheel nuts. Jack up the front of the vehicle and support it on axle stands. Remove both front roadwheels.

3 Follow the accompanying photos **(illustrations 4.3a to 4.3x)** for the actual pad renewal procedure. Be sure to stay in order and read the caption under each illustration, and note the following points:

a) *New pads may have an adhesive foil on the backplates. Remove this foil prior to installation.*

b) *Thoroughly clean the caliper guide surfaces, and apply a little brake assembly (polycarbamide) grease. DO NOT use a copper based grease.*

c) *When pushing the caliper piston back to accommodate new pads, keep a close eye on the fluid lever in the reservoir.*

d) *If there is any doubt as to the condition of the brake pads, always replace them.*

4.1a Removing the outer pad from FS3 type front brakes

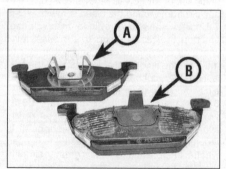

4.1b Note that the brake pads have different spring clips and must be fitted correctly. A = Inner (piston side) pad and B = outer pad

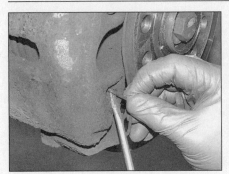

4.3a Use a screwdriver to carefully prise off the caliper retaining spring

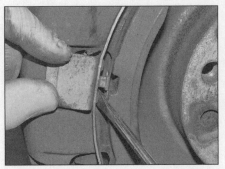

4.3b On FNR-G type brakes, unhook the spring from the pad, noting how it locks into the pad

4.3c Where fitted, disconnect the wiring plug from the brake pad warning light

4.3d Prise out the rubber caps…

4.3e …and use a hex key to undo the caliper guide bolts

4.3f Lift the caliper and inner pad from the disc

4.3g Pull the inner brake pad from the caliper piston…

4.3h …and lift the outer pad from the caliper bracket

4.3i Push the piston back into the caliper using a piston retraction tool or G-clamp…

4.3j …however, best practise is to clamp the hose, release the nipple and push the fluid into a brake bleeding bottle. Remember to tighten up the bleed nipple

4.3k Clean the pad mounting surfaces with a wire brush and brake cleaner

4.3l Apply a little brake lubricant ('cera-tec') grease to the pad mounting surfaces. DO NOT use a copper based grease

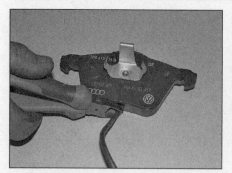

4.3m Where not required cut off the brake pad wear warning light cable

4.3n Fit the outer pad...

4.3o ...and then fit the inner pad to the caliper piston

4.3p On FNR-G type brakes fit both pads to the caliper

4.3q Where fitted peel off the protective cover from the adhesive surface of the outer pad

4.3r Slide the caliper with the inner pad fitted over the disc and outer pad

4.3s Refit the caliper guide bolts and tighten them to the specified torque...

4.3t ...and then refit the rubber caps

4.3u Refit the caliper retaining spring...

4.3v ...ensuring that it is located correctly on the caliper mounting bracket

4.3w On FNR-G type brakes hold the bottom of the spring in place with self locking pliers...

4.3x ...and then lift the top of the spring into place. Ensure the locking tab drops into the brake pad as the spring engages

4 Depress the brake pedal repeatedly, until the pads are pressed into firm contact with the brake disc, and normal (non-assisted) pedal pressure is restored.

5 Repeat the above procedure on the remaining front brake caliper.

6 If the bleed nipple was opened to push the pistons into the caliper check around the nipple for fluid leaks.

7 Refit the roadwheels, then lower the vehicle to the ground and tighten the roadwheel nuts to the specified torque.

8 Check the hydraulic fluid level as described in *Weekly checks* Section 5.

Caution: New pads will not give full braking efficiency until they have bedded-in. Be prepared for this, and avoid hard braking as far as possible for the first hundred miles or so after pad renewal.

5 Front brake caliper – removal, overhaul and refitting

Note: Before starting work, refer to the note at the beginning of Section 2 concerning the dangers of hydraulic fluid, and to the warning at the beginning of Section concerning the dangers of asbestos dust.

Removal

1 Apply the handbrake, then jack up the front of the vehicle and support it on axle stands (see *Jacking and vehicle support*). Remove the appropriate roadwheel.

2 Minimise fluid loss by first removing the master cylinder reservoir cap, and then tightening it down onto a piece of polythene, to obtain an airtight seal. Alternatively, use a brake hose clamp, a G-clamp or a similar tool to clamp the flexible hose.

3 Clean the area around the union, and then loosen the brake hose union nut.

4 Remove the brake pads as described in Section 4.

5 Unscrew the caliper from the end of the brake hose and remove it from the vehicle.

Overhaul

6 With the caliper on the bench, wipe away all traces of dust and dirt, but avoid inhaling the dust, as it is injurious to health.

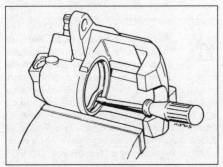

5.8 Use a small screwdriver to extract the caliper piston hydraulic seal

7 Withdraw the partially ejected piston from the caliper body, and remove the dust seal.

8 Using a small screwdriver, extract the piston hydraulic seal, taking great care not to damage the caliper bore **(see illustration)**.

9 Thoroughly clean all components, using only methylated spirit, isopropyl alcohol or clean hydraulic fluid as a cleaning medium. Never use mineral-based solvents such as petrol or paraffin, as they will attack the hydraulic system rubber components. Dry the components immediately, using compressed air or a clean, lint-free cloth. Use compressed air to blow clear the fluid passages.

10 Check all components, and renew any that are worn or damaged. Check particularly the cylinder bore and piston; these should be renewed if they are scratched, worn or corroded in any way (note that this means the renewal of the complete caliper body assembly). Similarly check the condition of the spacers/guide pins and their bushes/bores (as applicable); both spacers/pins should be undamaged and (when cleaned) a reasonably tight sliding fit in their bores. If there is any doubt about the condition of any component, renew it.

11 If the assembly is fit for further use, obtain the appropriate repair kit; the components are available from Seat dealers in various combinations.

12 Renew all rubber seals, dust covers and caps disturbed on dismantling as a matter of course; these should never be re-used.

13 On reassembly, ensure that all components are clean and dry.

14 Thinly coat the piston and piston seal with brake fitting paste (VAG part no G 052 150 A2). This should be included in the caliper overhaul/repair kit.

15 Fit the new piston (fluid) seal, using only your fingers (no tools) to manipulate it into the cylinder bore groove. Fit the new dust seal to the piston, and refit the piston to the cylinder bore using a twisting motion; ensure that the piston enters squarely into the bore. Press the piston fully into the bore, then press the dust seal into the caliper body.

Refitting

16 Screw the caliper fully onto the flexible hose union.

6.4 Using a DTI gauge to measure disc run-out

17 Refit the brake pads as described in Section 4.

18 Securely tighten the brake pipe union nut.

19 Remove the brake hose clamp or polythene, as applicable, and bleed the hydraulic system as described in Section 2. Note that, providing the precautions described were taken to minimise brake fluid loss, it should only be necessary to bleed the relevant front brake.

20 Refit the roadwheel, then lower the vehicle to the ground and tighten the roadwheel bolts to the specified torque.

6 Brake disc – inspection, removal and refitting

Note: Before starting work, refer to the note at the beginning of Section concerning the dangers of asbestos dust.

Note: If either disc requires renewal, BOTH should be renewed at the same time, to ensure even and consistent braking. New brake pads should also be fitted.

Front brake disc

Inspection

1 Apply the handbrake, then jack up the front of the car and support it on axle stands (see *Jacking and vehicle support*). Remove the appropriate front roadwheel.

2 Slowly rotate the brake disc so that the full area of both sides can be checked; remove the brake pads if better access is required to the inboard surface. Light scoring is normal in the area swept by the brake pads, but if heavy scoring or cracks are found, the disc must be renewed.

3 It is normal to find a lip of rust and brake dust around the perimeter of the disc; this can be scraped off if required. If, however, a lip has formed due to excessive wear of the brake pad swept area, then the disc thickness must be measured using a micrometer. Take measurements at several places around the disc, at the inside and outside of the pad swept area; if the disc has worn at any point to the specified minimum thickness or less, the disc must be renewed.

4 If the disc is thought to be warped, it can be checked for run-out. Either use a dial gauge mounted on any convenient fixed point, while the disc is slowly rotated, or use feeler blades to measure (at several points all around the disc) the clearance between the disc and a fixed point, such as the caliper mounting bracket. If the measurements obtained are at the specified maximum or beyond, the disc is excessively warped, and must be renewed; however, it is worth checking first that the hub bearings are in good condition. If the run-out is excessive, the disc must be renewed **(see illustration)**.

5 Check the disc for cracks, especially around the wheel bolt holes, and any other wear or damage, and renew if necessary.

6.7a Remove the bolts and...

6.7b ...lift off the caliper mounting bracket

6.8a Undo the retaining screw...

Removal

6 Remove the brake pads as described in Section 4. Using a piece of wire or string, tie the caliper to the front suspension coil spring, to avoid placing any strain on the brake hose.

7 Unscrew the two bolts securing the brake caliper mounting bracket to the hub (see illustrations).

8 Use chalk or paint to mark the relationship of the disc to the hub, then remove the screw securing the brake disc to the hub, and remove the disc (see illustrations). If it is tight, apply penetrating fluid, and tap its rear face gently with a hide or plastic mallet. The use of excessive force could cause the disc to be damaged.

Refitting

9 Refitting is the reverse of the removal procedure, noting the following points:
a) Ensure that the mating surfaces of the disc and hub are clean and flat.
b) Align (if applicable) the marks made on removal, and securely tighten the disc retaining screw.
c) If a new disc has been fitted, use a suitable solvent to wipe any preservative coating from the disc, before refitting the caliper.
d) Slide the caliper into position over the disc, making sure the pads pass either side of the disc. Tighten the caliper bracket mounting bolts to the specified torque.
e) Fit the pads as described in Section 4.
f) Refit the roadwheel, then lower the vehicle to the ground and tighten the roadwheel bolts to the specified torque. On

completion, repeatedly depress the brake pedal until normal (non-assisted) pedal pressure returns.

Rear brake disc
Inspection

10 Firmly chock the front wheels, then jack up the rear of the car and support it on axle stands. Remove the appropriate rear road-wheel.

11 Inspect the disc as described in paragraphs 1 to 5.

Removal

12 Unscrew the two bolts securing the brake caliper mounting bracket in position, then slide the caliper assembly off the disc. Using a piece of wire or string, tie the caliper to the rear suspension coil spring, to avoid placing any strain on the hydraulic brake hose.

13 Use chalk or paint to mark the relationship of the disc to the hub, then remove the screw securing the brake disc to the hub, and remove the disc (see illustration). If it is tight, apply penetrating fluid, and tap its rear face gently with a hide or plastic mallet. The use of excessive force could cause the disc to be damaged.

Refitting

14 Refitting is a reversal of the removal procedure, noting the following points:
a) Ensure that the mating surfaces of the disc and hub are clean and flat.
b) Align (if applicable) the marks made on removal, and securely tighten the disc retaining screw.
c) If a new disc has been fitted, use a suitable

solvent to wipe any preservative coating from the disc, before refitting the caliper.
d) Slide the caliper into position over the disc, making sure the pads pass either side of the disc. Tighten the caliper bracket mounting bolts to the specified torque. If new discs have been fitted and there is insufficient clearance between the pads to accommodate the new, thicker disc, it may be necessary to push the piston back into the caliper body as described in Section 4.
e) Refit the roadwheel, then lower the vehicle to the ground and tighten the roadwheel bolts to the specified torque. On completion, repeatedly depress the brake pedal until normal (non-assisted) pedal pressure returns.

7 Front brake disc shield – removal and refitting

Removal

1 Remove the brake disc as described in Section 6.

2 Unscrew the securing bolts, and remove the brake disc shield.

Refitting

3 Refitting is a reversal of removal. Tighten the shield retaining bolts to the specified torque. Refit the brake disc with reference to Section 6.

8 Rear brake pads – renewal

 Warning: Renew both sets of rear brake pads at the same time – never renew the pads on only one wheel, as uneven braking may result. Note that the dust created by wear of the pads may contain asbestos, which is a health hazard. Never blow it out with compressed air, and don't inhale any of it. An approved filtering mask should be worn when working on the brakes. DO NOT use petrol or petroleum-based solvents to clean brake parts; use brake cleaner or methylated spirit only.

6.13 Remove the rear brake disc

6.8b ...and remove the front brake disc

8.1 Type C1-38 brakes have the anti-rattle spring on the brake pad

1 There are several types of rear brake caliper fitted. All calipers (except the C1-38 type) have anti-rattle shims. C1-38 type brakes have brake pads that have springs fitted to the pad **(see illustration)**. Removal and refitting of the brake pads is essentially the same, regardless of the type of caliper fitted.

2 Chock the front wheels, slacken the rear road wheel nuts, then jack up the rear of the vehicle and support it on axle stands (see *Jacking and vehicle support*). Remove the rear wheels.

3 With the handbrake lever fully released, follow the accompanying photos **(see illustrations 8.3a to 8.3q)** for the actual pad renewal procedure. Be sure to stay in order and read the caption under each illustration, and note the following points:

a) *If re-installing the original pads, ensure they are fitted to their original position.*
b) *Thoroughly clean the caliper guide surfaces and guide bolts. Note that the shorter guide bolt is fitted to the top position.*
c) *If new pads are to be fitted, use a piston retraction tool to push the piston back and twist it clockwise at the same time – keep an eye on the fluid level in the reservoir whilst retracting the piston. Note however that best practise is to clamp the brake flexible hose, open the bleed nipple and push the fluid out, into a brake bleeding bottle.*

Caution: Pushing back the piston causes a reverse-flow of brake fluid, which has been known to 'flip' the master cylinder rubber seals, resulting in a total loss of braking. To avoid this, clamp the caliper flexible hose and open the bleed screw – as the piston is pushed back, the fluid can be directed into a suitable container using a hose attached to the bleed screw. Close the screw just before the piston is pushed fully back, to ensure no air enters the system.

4 Depress the brake pedal repeatedly, until the pads are pressed into firm contact with the brake disc, and normal (non-assisted) pedal pressure is restored.

5 Repeat the above procedure on the remaining brake caliper.

6 If necessary, adjust the handbrake as described in Section 14.

7 Refit the roadwheels, then lower the vehicle to the ground and tighten the roadwheel nuts to the specified torque.

8 Check the hydraulic fluid level as described in *Weekly checks* Section 5.

Caution: New pads will not give full braking efficiency until they have bedded-in. Be prepared for this, and avoid hard braking as far as possible for the first hundred miles or so after pad renewal.

9   Rear brake caliper – removal, overhaul and refitting

Note: *Before starting work, refer to the note at the beginning of Section 2 concerning the dangers of hydraulic fluid, and to the warning at the beginning of Section concerning the dangers of asbestos dust.*

Removal

1 Chock the front wheels, then jack up the rear of the vehicle and support on axle stands (see *Jacking and vehicle support*). Remove the relevant rear wheel.

2 Minimise fluid loss by first removing the master cylinder reservoir cap, and then tightening it down onto a piece of polythene, to obtain an airtight seal. Alternatively, use a brake hose clamp, a G-clamp or a similar tool to clamp the flexible hose.

3 Clean the area around the union on the

8.3a Counterhold the guides...

8.3b ...and remove the bolts

8.3c Lift off the caliper

8.3d Remove the brake pads...

8.3e ...and then (where fitted) recover the shims

8.3f Retract the piston using a piston wind back tool

8.3g Select the correct end fitting for the brake tool

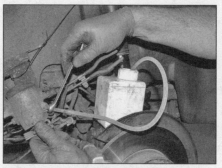

8.3h Best practise is always to clamp the hose, open the nipple and push the fluid into a bottle

8.3i Use a wire brush and brake cleaner to thoroughly clean the caliper and the brake pad mounting points

8.3j Check that the guide pins slide freely – clean and lubricate as required

8.3k Fit the new shims (on models that use shims) and then…

8.3l …apply a little brake grease ('cera-tec' or similar) to the mounting points. DO NOT use a copper based grease.

8.3m Fit the new inner brake pad…

8.3n …followed by the outer brake pad

8.3o Refit the caliper and then…

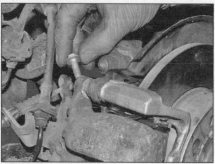

8.3p …fit the new caliper guide bolts

8.3q Tighten the bolts to the specified torque

caliper, and then loosen the brake hose union nut.

4 Lift the caliper from the brake pads as described in Section 8.

5 Unscrew the caliper from the end of the flexible hose and remove it from the vehicle.

Overhaul

Note: *It is not possible to overhaul the brake caliper handbrake mechanism. If the mechanism is faulty, or fluid is leaking from the handbrake lever seal the caliper assembly must be renewed.*

6 With the caliper on the bench, wipe away all traces of dust and dirt, but avoid inhaling the dust, as it is injurious to health.

7 Using a small screwdriver, carefully prise out the dust seal from the caliper, taking care not to damage the piston.

8 Remove the piston from the caliper bore by rotating it in an anti-clockwise direction. This can be achieved using a suitable pair of circlip pliers engaged in the caliper piston slots. Once the piston turns freely but does not come out any further, the piston can be withdrawn by hand.

9 Using a small screwdriver, extract the piston hydraulic seal(s), taking care not to damage the caliper bore.

10 Withdraw the guide pins from the caliper, and remove the guide sleeve gaiters.

11 Thoroughly clean all components, using a proprietary brake cleaner. Dry the components immediately, using compressed air or a clean, lint-free cloth. Use compressed air to blow clear the fluid passages.

12 Check all components, and renew any that are worn or damaged. Check particularly the cylinder bore and piston; these should be renewed (note that this means the renewal of the complete caliper body assembly) if they are scratched, worn or corroded in any way. Similarly check the condition of the spacers/guide pins and their bushes/bores (as applicable); both spacers/pins should be undamaged and (when cleaned) a reasonably tight sliding fit in their bores. If there is any doubt about the condition of any component, renew it.

13 If the assembly is fit for further uses obtain the appropriate repair kit; the components are available from Seat dealers in various combinations.

14 Renew all rubber seals, dust covers and caps disturbed on dismantling as a matter of course; these should never be re-used.

15 On reassembly, ensure that all components are clean and dry.

16 Smear a thin coat of brake fitting paste (VAG part no G 052 150 A2) on the piston, seal and caliper bore. This should be included in the overhaul/repair kit. Fit the new piston (fluid) seal, using only the fingers (no tools) to manipulate into the cylinder bore groove.

17 Fit the new dust seal to the piston groove, then refit the piston assembly. Turn the piston in a clockwise direction, using the method employed on dismantling, until it is fully retracted into the caliper bore.

18 Press the dust seal into position in the caliper housing.

19 Apply the grease supplied in the repair kit to the guide pins. Fit the new gaiters to the guide pins and fit the pins to the caliper ensuring that the gaiters are correctly located in the grooves on both the pins and caliper.

20 Prior to refitting, fill the caliper with fresh hydraulic fluid by slackening the bleed screw and pumping the fluid through the caliper until bubble-free fluid is expelled from the union hole.

Refitting

21 Screw the caliper fully onto the flexible hose union.

22 Refit the caliper over the brake pads as described in paragraphs 10 to 12 of Section 8.

23 Securely tighten the brake pipe union nut.

24 Remove the brake hose clamp or remove the polythene from the fluid reservoir, as applicable, and bleed the hydraulic system as described in Section 2. Note that, providing the precautions described were taken to minimise brake fluid loss, it should only be necessary to bleed the relevant rear brake.

25 Connect the handbrake cable to the caliper, and adjust the handbrake as described in Section 14.

26 Refit the roadwheel, then lower the vehicle to the ground and tighten the roadwheel bolts to the specified torque. On completion, check the hydraulic fluid level as described in *Weekly checks* Section 5.

10 Brake pedal – removal and refitting

Removal

1 Disconnect the battery negative lead as described in Chapter 5A Section 3.

2 Where fitted, remove the driver's side lower trim panel.

3 Undo the retaining screw and remove the footwell air vent from above the pedal assembly **(see illustration)**.

4 On early models disconnect and then remove the brake light switch as described in Section 18.

5 It is now necessary to release the brake pedal from the ball on the vacuum servo pushrod. To do this, a VAG/Seat special tool is available, but a suitable alternative can be improvised. Note that the plastic lugs in the pedal are very stiff, and it will not be possible to release them by hand. Depress and hold down the pedal, then, using the tool, release the securing lugs, and pull the pedal from the servo pushrod **(see illustrations)**.

10.3 Remove the footwell air ducting

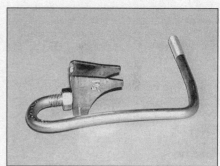
10.5a Improvised special tool constructed from a modified exhaust clamp, used to release the brake pedal from the servo pushrod

10.5b Using the tool to release the brake pedal from the servo pushrod

10.5c Rear view of the brake pedal showing plastic lugs securing pedal to servo pushrod

10.6 Remove the nut

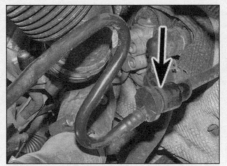

12.1 Typical Non-return valve in the vacuum pipe

12.3 Disconnect the hose from the vacuum pump

6 Release the retaining clip, then undo and remove the pivot shaft nut **(see illustration).**
7 Slide the pivot shaft bolt out, until the pedal is free, and then remove the pedal.
8 Carefully clean all components, and renew any that are worn or damaged.

Refitting

9 Prior to refitting, apply a smear of multi-purpose grease to the pivot shaft and pedal bearing surfaces.
10 Pull the servo unit pushrod down, and at the same time manoeuvre the pedal into position, ensuring that the pivot bush is correctly located.
11 Hold the servo unit pushrod, and push the pedal back onto the pushrod ball. Make sure the pedal is securely fastened to the pushrod.
12 Insert the pedal pivot bolt and tighten the retaining nut to the specified torque.
13 The remainder of the refitting procedure is the opposite of the removal; refer to the relevant Chapters where applicable.

11 Servo unit –
testing, removal and refitting

Testing

1 To test the operation of the servo unit, depress the footbrake several times to exhaust the vacuum, then start the engine whilst keeping the pedal firmly depressed. As the engine starts, there should be a noticeable 'give' in the brake pedal as the vacuum builds-up. Allow the engine to run for at least two minutes, and then switch it off. If the brake pedal is now depressed, it should feel normal, but further applications should result in the pedal feeling firmer, with the pedal stroke decreasing with each application.
2 If the servo does not operate as described, first inspect the servo unit non-return valve as described in Section 12. Check the operation of the vacuum pump as described in Section 21.
3 If the servo unit still fails to operate satisfactorily, the fault lies within the unit itself. Repairs to the unit are not possible – if faulty, the servo unit must be renewed.

Removal

4 Remove the master cylinder as described in Section 13.
5 Where applicable remove the heat shield from the servo, then carefully ease the vacuum hose out from the sealing grommet in the front of the servo. Where applicable, also disconnect the wiring from the servo vacuum sensor, then extract the retaining circlip with a screwdriver, and withdraw the sensor from the servo.
6 With reference to Chapter 11 Section 27, where fitted, remove the driver's side trim panel below the dash.
7 Where fitted, unscrew the two retaining screws and remove the connecting plate between the clutch and brake pedals (manual transmission models only). Also, where fitted, remove the air duct and cover for access to the servo mounting nuts.
8 It is now necessary to release the brake pedal from the ball on the vacuum servo pushrod. To do this, a VAG/Seat special tool is available, but a suitable alternative can be improvised (see Section 10). Note that the plastic lugs in the pedal are very stiff, and it will not be possible to release them by hand. Using the tool, release the securing lugs, and pull the pedal from the servo pushrod.
9 To improve access, remove the plenum chamber front panel as described in Chapter 12 Section 19.
10 To gain enough room to remove the brake servo the engine must be moved forward.
11 Jack up and support the front of the vehicle (see *Jacking and vehicle support*), remove the engine undershield and then remove the engine pendulum mounting.
12 Install an engine crane and support the engine. Remove the right-hand engine mounting and then lower the engine slightly (110mm maximum).
13 Working in the footwell, undo the nuts securing the servo unit to the bulkhead, then return to the engine compartment and manoeuvre the servo unit out of position, and recover the gasket where fitted.

Refitting

14 Check the servo unit vacuum hose sealing grommet for signs of damage or deterioration, and renew if necessary.
15 Where applicable, fit a new gasket to the

rear of the servo unit, and then reposition the unit in the engine compartment.
16 From inside the vehicle, ensure that the servo unit pushrod is correctly engaged with the brake pedal, and push the pedal onto the pushrod ball. Check the pushrod ball is securely engaged, then refit the servo unit mounting nuts and tighten them to the specified torque.
17 As applicable, refit the connecting plate, air duct and cover.
18 Refit the facia trim panels.
19 Carefully ease the vacuum hose back into position in the servo, taking great care not to displace the sealing grommet. Refit the heat shield to the servo and, where applicable, refit the vacuum sensor and wiring.
20 Refit the master cylinder as described in Section 13 of this Chapter.
21 On completion, start the engine and check for air leaks at the vacuum hose-to-servo unit connection.

12 Servo non-return valve –
testing, removal and refitting

1 The non-return valve is located in the vacuum hose leading from the vacuum pump to the brake servo **(see illustration).**

Removal

2 Ease the vacuum hose out of the servo unit, taking care not to displace the grommet.
3 Note the routing of the hose, then slacken the retaining clip(s) and disconnect the opposite end of the hose assembly from the manifold (petrol engines) or vacuum pump (diesel engines) **(see illustration)** and remove it from the car.

Testing

4 Examine the check valve and vacuum hose for signs of damage, and renew if necessary.
5 The valve may be tested by blowing through it in both directions; air should flow through the valve in one direction only; when blown through from the servo unit end of the valve. Renew the valve if this is not the case.
6 Examine the servo unit rubber sealing grommet for signs of damage or deterioration, and renew as necessary.

Refitting

7 Ensure that the sealing grommet is correctly fitted to the servo unit.

8 Ease the hose union into position in the servo, taking great care not to displace or damage the grommet.

9 Ensure that the hose is correctly routed, and connect it to the inlet manifold/vacuum pump, ensuring the hose is secured in the retaining clips.

10 On completion, start the engine and check the valve-to-servo unit connection for signs of air leaks.

13 Master cylinder – removal, overhaul and refitting

Note: *Before starting work, refer to the warning at the beginning of Section 2 concerning the dangers of hydraulic fluid. A new master cylinder O-ring will be required on refitting.*

Removal

1 Disconnect the battery negative lead as described in Chapter 5A Section 3.

2 Remove the engine top cover and air inlet trunking. On models fitted with a diesel particulate filter (DPF) remove the heat shield.

3 Remove the timing belt upper cover by unclipping it.

4 On models fitted with a DPF disconnect the wiring plug and remove the pressures sensor complete with the mounting bracket.

5 Where fitted remove the heat shield from the front of the plenum chamber.

6 Remove the master cylinder reservoir cap, disconnecting the wiring plug from the brake fluid level warning switch **(see illustration)**, and then syphon the hydraulic fluid from the reservoir. **Note:** *Do not syphon the fluid by mouth, as it is poisonous; use a syringe or an old antifreeze tester. Also, disconnect the wiring connector from the brake light switch, at the lower part of the master cylinder.*

7 On manual transmission models, disconnect and plug the clutch master cylinder supply hose from the side of the brake fluid reservoir **(see illustration)**.

8 Wipe clean the area around the brake pipe unions on the side of the master cylinder, and place absorbent rags beneath the pipe unions to catch any leaking fluid. Make a note of the correct fitted positions of the unions, then unscrew the union nuts and carefully withdraw the pipes. Plug or tape over the pipe ends and master cylinder orifices, to minimise the loss of brake fluid, and to prevent the entry of dirt into the system. Wash off any spilt fluid immediately with cold water.

9 On models fitted with a master cylinder mounted brake warning light switch, disconnect the wiring plug from the switch. Unbolt and then remove the switch.

10 Unscrew and remove the two nuts and washers securing the master cylinder to the vacuum servo unit, remove the heat shield (where fitted), then withdraw the unit from the engine compartment **(see illustration)**. Remove the O-ring from the rear of the master cylinder, and check it for damage, renew if required.

Overhaul

11 If the master cylinder is faulty, it must be renewed. Repair kits are not available.

12 The only items that can be renewed are the mounting seals for the fluid reservoir; if these show signs of deterioration, prise them out with a screwdriver. Lubricate the new seals with clean brake fluid, and press them into the master cylinder ports.

Refitting

13 Remove all traces of dirt from the master cylinder and servo unit mating surfaces, and fit a new O-ring to the groove on the master cylinder body.

14 Fit the master cylinder to the servo unit, ensuring that the servo unit pushrod enters the master cylinder bore centrally. Refit the

13.6 Filler cap incorporating brake fluid level warning switch

heat shield (where applicable) and the master cylinder mounting nuts and then tighten them to the specified torque.

15 Wipe clean the brake pipe unions, then refit them to the master cylinder ports and tighten them securely.

16 Refit the hydraulic fluid reservoir; making sure it is entered correctly in the rubber grommets.

17 On manual transmission models, reconnect the clutch master cylinder supply hose to the reservoir.

18 Refill the master cylinder reservoir with new fluid, and bleed the complete hydraulic system as described in Section 2.

19 Reconnect the wiring to the brake level sender unit and brake light switch as applicable.

20 Refit the engine cover and air trunking where necessary, and then reconnect the battery negative lead.

14 Handbrake – adjustment

1 To check the handbrake adjustment, first apply the footbrake firmly several times to establish correct pad-to-disc clearance, then

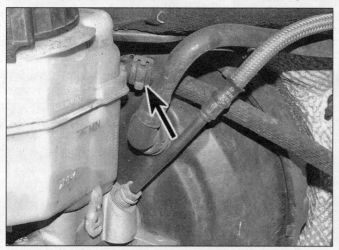

13.7 Clutch fluid supply hose and union

13.10 Master cylinder securing nuts

14.3a Removing the compartment to...

14.3b ...access the adjuster nut

14.5 Slacken the adjuster nut

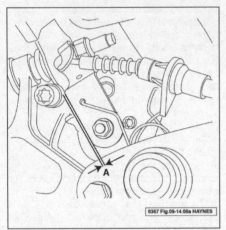

14.6a Turn the adjuster nut until a gap (A) of between 1.0 and 1.5 mm can be seen (C type calipers)

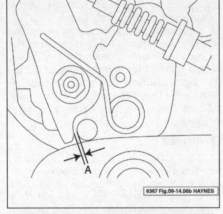

14.6b Turn the adjuster nut until a gap (A) of between 1.0 and 1.5 mm can be seen (Bosch caliper)

apply and release the handbrake several times.

2 Applying normal moderate pressure, pull the handbrake lever to the fully applied position, counting the number of clicks from the handbrake ratchet mechanism. If adjustment is correct, there should be approximately 4 to 7 clicks before the handbrake is fully applied. If this is not the case, adjust as follows.

3 Remove the rear compartment from the centre console as described in Chapter 11 Section 26 to gain access to the adjuster **(see illustrations)**.

4 If adjustment is required slacken the rear wheel nuts. Chock the front wheels, then jack up the rear of the vehicle and support it on axle stands (see *Jacking and vehicle support*). Remove both rear wheels.

5 With the handbrake fully released, slacken the handbrake adjuster nut until both the rear caliper handbrake levers are back against their stops **(see illustration)**.

6 From this point, tighten the adjusting nuts until both handbrake levers just move off the caliper stops. Ensure that the gap between each caliper handbrake lever and its stop is

between 1.0 and 1.5 mm, and ensure both the right and left-hand gaps are equal **(see illustrations)**. Check that both wheels/discs rotate freely, and then check the adjustment by applying the handbrake fully and counting the clicks from the handbrake ratchet (see paragraph 2). If necessary, re-adjust.

7 Once adjustment is correct refit the rear wheels and lower the vehicle to the ground. Refit the cup holder.

15 Handbrake lever – removal and refitting

Removal

1 Remove the centre console as described in Chapter 11 Section 26.

2 If desired, remove the handbrake lever cover sleeve by depressing the locating tag with a screwdriver, then sliding the sleeve from the lever.

3 Disconnect the wiring plug from the handbrake 'on' warning light switch **(see illustration)**.

4 Slacken the handbrake cable adjuster nut sufficiently to allow the ends of the cables to be disengaged from the equaliser plate and then remove the air distribution ducting **(see illustrations)**.

5 Unclip the wiring loom for the rear 12v power outlet. Unscrew the retaining nuts, and withdraw the lever and bracket assembly from the floor.

15.3 Handbrake warning light switch

15.4a Handbrake cables and equaliser plate

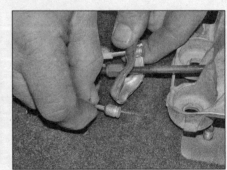

15.4b Disengage the inner cables

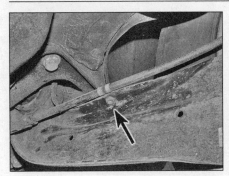

16.4a Unbolt the bracket (early models)

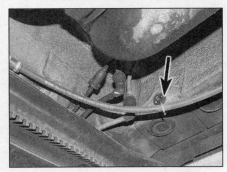

16.4b Unclip the cable (all models)

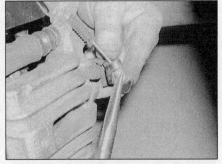

16.5 Release the cable inner from its lever and withdraw the cable from the caliper

Refitting

6 Refitting is a reversal of removal, bearing in mind the following points.

a) *Prior to refitting the centre console, adjust the handbrake as described in Section 14.*

b) *Check the operation of the handbrake 'on' warning switch prior to refitting the centre console.*

16 Handbrake cables – removal and refitting

Removal

Note: *There are minor differences in the removal and refitting of the cables between earlier and later models. However removal and refitting is essentially the same for all models.*

1 Remove the compartment form the rear of the centre console (as described in Chapter 11 Section 26) to gain access to the handbrake lever. The handbrake cable consists of two sections, a right and a left-hand section, which are linked to the lever by an equaliser plate. Each section can be removed individually.

2 Chock the front wheels, then jack up the rear of the car and support it on axle stands.

3 Slacken the handbrake cable adjuster nut sufficiently to allow the ends of the cables to be disengaged from the equaliser plate, see Section 15.

4 Working back along the length of the cable,

note its correct routing, and free it from all the relevant guides and retaining clips **(see illustrations)**.

5 Disengage the inner cable from the caliper handbrake lever, then remove the outer cable retaining clip and detach the cable from the caliper **(see illustration)**. Withdraw the cable from underneath the vehicle.

Refitting

6 Refitting is a reversal of removal, bearing in mind the following points.

a) *When locating the handbrake cable sheath in the guide on the rear trailing arm, the cable clamping ring must lie in the middle of the clip.*

b) *Before refitting the centre console, adjust the handbrake as described in Section 14.*

17 Handbrake warning light switch – removal and refitting

Removal

1 Remove the centre console, with reference to Chapter 11 Section 26 if necessary.

2 Disconnect the wiring plug from the switch.

3 Release the securing lugs and (where fitted) the cable tie. Remove the switch from the handbrake lever assembly **(see illustration 15.3)**.

Refitting

4 Refitting is a reversal of removal.

18 Brake light switch – removal and refitting

Note: *Early models (up to 11/05 approximately) have the switch fitted to the brake pedal. Later models have the switch fitted to the master cylinder.*

Brake pedal mounted switch

Removal

Note: *The switch has been updated. Always replace the old style switch with the new version. The new version can be identified by the brightly coloured tip of the plunger.*

1 Where fitted, remove the drivers side lower trim panel as described in Chapter 11 Section 27.

2 If required, remove the screw and pull out the air distribution duct.

3 Disconnect the wiring plug **(see illustration)** depress the brake pedal and then rotate the switch 45 degrees anti-clockwise to remove it.

Refitting

4 Refitting is a reversal of removal, but do not refit the old style switch – always fit the new style switch regardless of the condition of the old switch.

Master cylinder mounted switch

Removal

5 The brake light switch is located on the lower part of the master cylinder **(see illustration)**. Remove the engine top cover, and where applicable, remove the air inlet trunking to improve access. On models fitted with a particulate filter remove the heat shield and differential pressure sensor (complete with the sensor mounting bracket).

6 Disconnect the wiring from the switch.

7 Unscrew the mounting bolt, then pull the switch from the bottom of the master cylinder, and remove it from the locking lug at the top.

Refitting

8 Refitting is a reversal of removal, but tighten the mounting bolt to the specified torque.

18.3 Disconnect the wiring plug

18.5 Brake light switch

19 Anti-lock braking system (ABS) – general information and precautions

1 The anti-lock braking system (ABS) fitted as standard to all models, prevents wheel lock-up under heavy braking, and not only optimises stopping distances, but also improves steering control. By electronically monitoring the speed of each roadwheel in relation to the other wheels, the system can detect when a wheel is about to lock-up, before control is actually lost. The brake fluid pressure applied to that wheel's brake caliper is then decreased and restored ('modulated') several times a second until control is regained. The system components are: four wheel-speed sensors, a hydraulic unit with integral Electronic Control Unit (ECU), brake lines and a dashboard-mounted warning light. The four wheel-speed sensors are active sensors integrated into the wheel bearings. Unlike the older generation 'passive' wheel speed sensors, 'active' sensors can determine if the vehicle has stopped and if the sensor is faulty. Active sensors produce a square wave form which is then transmitted to the ECU, and used to calculate the rotational speed of each wheel. The ECU has a self-diagnostic facility, to inhibit the operation of the ABS if a fault is detected, lighting the dashboard-mounted warning light. The braking system will then revert to conventional, non-ABS operation. If the nature of the fault is not immediately obvious upon inspection, the vehicle must be taken to a Seat dealer (or suitably equipped garage), who will have the diagnostic equipment required to interrogate the ABS ECU electronically.

2 There are two ABS systems fitted to the models covered in this Manual.

3 The Mark 70 version includes a traction control system (TCS), which uses the basic ABS system, with an additional pump and valves fitted to the hydraulic actuator. If wheelspin is detected at a speed below 30 mph, one of the valves opens, to allow the pump to pressurise the relevant brake, until the spinning wheel slows to a rotational speed corresponding to the speed of the vehicle.

This has the effect of transferring torque to the wheel with most traction. At the same time, the throttle plate is closed slightly, to reduce the torque from the engine.

4 The Mark 60 version includes electronic differential locking (EDL) and an electronic stability programme (ESP). The EDL system applies the brake of the spinning wheel in order to transfer torque to the wheel with the better grip. On models with ESP, the system recognises critical driving conditions and stabilises the vehicle by individual wheel braking and by intervention in the engine control, which occurs independently of the position of the brake and accelerator pedals.

5 The operation of the ABS system is entirely dependent on electrical signals. To prevent the system responding to any inaccurate signals, a built-in safety circuit monitors all signals received by the ECU. If an inaccurate signal or low battery voltage is detected, the ABS system is automatically shut down, and the warning light on the instrument panel is illuminated, to inform the driver that the ABS system is not operational. Normal braking will still be available, however.

6 If a fault does develop in the ABS system, the car must be taken to a Seat dealer or suitably equipped garage for fault diagnosis and repair.

20 Anti-lock braking system (ABS) components – removal and refitting

Hydraulic unit and electronic control

Removal

1 The hydraulic unit (valve block) and electronic control (modulator) are combined in a single assembly. The unit is mounted on the left-hand side of the engine bay behind the battery.

2 Disconnect and then remove the battery as described in Chapter 5A Section 3.

3 Unclip and remove the battery shrouds and then remove the battery tray (see illustration).

4 A brake pedal depressor must now be installed. Jack up and support the front

left-hand wheel and the rear left-hand wheel. Remove both wheels and connect a hose and collection bottle to both front and rear bleed nipples. Open the bleed nipples and depress the brake pedal a minimum of 60 mm. Hold the brake pedal in this position using the brake pedal depressing tool. Do not remove the brake pedal depressor. Close the bleed nipples.

5 Identify and mark the position of the brake lines on the ABS hydraulic unit. Place shop towels below the unit and then disconnect the wiring plug from the ABS modulator (see illustration). Seal the plug in a clean plastic bag if there is any chance it will be contaminated by brake fluid.

6 Remove the brake lines from the unit. Seal all the brake lines and the hydraulic unit openings with suitable plugs. Unbolt the complete assembly from the mounting bracket and remove it from the vehicle.

7 On the bench unbolt the electronic control modulator from the valve block. Recover the seals. Seal the modulator in a plastic bag and keep the hydraulic unit (valve block) upright at all times.

Refitting

8 Refitting is a reversal of removal noting the following points:
a) The electronic control unit (modulator) can only be removed and refitted a maximum of two times.
b) If a new control unit is fitted it must be programmed to the vehicle using suitable diagnostic equipment.
c) Bleed the brakes, using pressure bleeding equipment if possible.

Front wheel sensor

Removal

9 Chock the rear wheels, then firmly apply the handbrake, jack up the front of the car and support on axle stands (see Jacking and vehicle support). Remove the appropriate front roadwheel.

10 Disconnect the electrical connector from the sensor by carefully lifting up the retaining tag, and pulling the connector from the sensor (see illustration).

11 Slacken and remove the hexagon socket-head bolt securing the sensor to the

20.3 Remove the battery tray

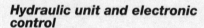

20.5 Release the wiring plug

20.10 Disconnect the wiring connector...

20.11 ...and unbolt the front wheel sensor

21.9 Disconnect the small vacuum pipe (common rail engine)

21.10 Remove the vacuum hose and rubber seal (common rail)

hub carrier, and remove the sensor from the car (see illustration).

Refitting

12 Ensure that the sensor and hub carrier sealing faces are clean.

13 Apply a thin coat of multi-purpose grease to the mounting hole inner surface, then fit the sensor to the hub carrier. Refit the retaining bolt and tighten it to the specified torque.

14 Ensure that the sensor wiring is correctly routed and retained by all the necessary clips, and reconnect the wiring connector.

15 Refit the roadwheel, then lower the car to the ground and tighten the roadwheel bolts to the specified torque.

Rear wheel sensor

Removal

16 Chock the front wheels, then jack up the rear of the car and support it on axle stands (see *Jacking and vehicle support*). Remove the appropriate roadwheel.

17 Remove the sensor as described in paragraphs 4 and 5 for the front wheel speed sensor.

Refitting

18 Refit the sensor as described above in paragraphs 6 to 9.

Reluctor rings

19 The reluctor rings are integral with the wheel bearings (front and rear) and can only be inspected after removal of the hub

assembly. If faulty, the bearings must be renewed as described in Chapter 10 Section 3 (front) and Chapter 10 Section 10 (rear).

21 Vacuum pump – testing, removal and refitting

Testing

1 The basic test should be to first exhaust the vacuum by pressing the brake pedal repeatedly – the pedal will feel hard with little travel once the vacuum is exhausted. Start the engine whilst keeping your foot on the brake. The pedal should move down slightly as the vacuum builds up. Further comprehensive testing is described below.

2 A comprehensive test of the braking system vacuum pump requires a vacuum gauge. First, remove the engine top cover to access the vacuum hose.

3 Disconnect the vacuum hose from the pump, and connect the gauge to the pump union using a suitable length of hose.

4 Start the engine and allow it to idle, and then measure the vacuum created by the pump. As a guide, after one minute, a minimum of approximately 500 mm Hg should be recorded. If the vacuum registered is significantly less than this, it is likely that the pump is faulty. However, seek the advice of a VW dealer before condemning the pump.

5 Reconnect the vacuum hose. Overhaul of

the vacuum pump is not possible, since no major components are available separately for it. If faulty, the complete pump assembly must be renewed.

Removal

Note: *There are several variants of the vacuum pumped fitted, depending on the engine type. All 'PD' engines have tandem pumps that act as lift pumps to supply the injectors with fuel. Removal and refitting of all variants is essentially the same.*

Note: *A new pump O-ring will be required on refitting.*

6 Remove the engine cover and then remove the air filter housing as described in.

7 On models fitted with a tandem pump, anticipate some fuel spillage as the pump is removed. Place rags or shop towels below the pump to absorb any spillage.

8 Disconnect the fuel lines and immediately seal them.

9 Disconnect the small vacuum pipe from the vacuum hose connector on the top of the vacuum pump (see illustration).

10 Pull the vacuum hose from the top of the pump, noting the rubber seal may stay on the vacuum pump outlet pipe (see illustration).

11 Unscrew the mounting bolts and withdraw the vacuum pump from the cylinder head (see illustrations). Recover the O-ring seal and discard, as a new one will be required for refitting.

Refitting

12 Fit the new O-ring seal to the vacuum pump, and apply a smear of oil to aid installation.

13 Manoeuvre the vacuum pump into position, making sure that the slot in the pump drive gear aligns with the slot on the pump driveshaft.

14 Refit the pump retaining bolts, and tighten to the specified torque.

15 Reconnect the vacuum hoses and secure with the retaining clip (where fitted). If the vacuum hose seal is still fitted to the outlet pipe on the vacuum pump, remove it and fit it inside the end of the vacuum hose before refitting.

16 Refit the air intake pipe and engine top cover.

21.11a Vacuum pump securing bolts (1.9 PD engine)

21.11b Vacuum pump securing bolts (1.6 CR engine)

Chapter 10
Suspension and steering systems

Contents

Degrees of difficulty

Easy, suitable for novice with little experience	**Fairly easy,** suitable for beginner with some experience	**Fairly difficult,** suitable for competent DIY mechanic	**Difficult,** suitable for experienced DIY mechanic	**Very difficult,** suitable for expert DIY or professional

Specifications

Front suspension
Type . . . .   Independent, with MacPherson struts incorporating coil springs, telescopic shock absorbers and anti-roll bar

Rear suspension
Type . . . .   Trailing arm with Multi-link transverse arms, separate gas-filled telescopic shock absorbers, coil springs and anti-roll bar

Steering
Type . . . .   Rack-and-pinion. Electro-mechanical power assistance standard

Wheel alignment and steering angles*
Front wheel:
Camber . . . .   - 0°37' ± 30'
Maximum difference between sides . . . .   0°30'
Castor . . . .   7° 40' ± 30'
Maximum difference between sides . . . .   0°30'
Toe setting . . . .   0°10' ± 10'
Toe-out on turns (20° left or right) . . . .   1° 38' ± 20'
Ride height . . . .   380 mm
Rear wheel:
Camber angle . . . .   -1°20' ± 30'
Maximum difference between sides . . . .   0°20'
Toe setting . . . .   0°+10' ± 10'
Maximum difference between sides . . . .   0°20'
Ride height . . . .   382 mm
*Figures are for standard suspension. Refer to Seat for alternative suspension and the latest figures.

Torque wrench settings

	Nm	lbf ft
Front suspension		
Anti-roll bar link nuts*	65	48
Anti-roll bar to subframe:		
Stage 1	20	15
Stage 2	Angle-tighten a further 90°	
Hub bolt*		
Heaxgon head bolt		
Stage 1	200	148
Stage 2	Angle-tighten a further 180°	
12 point bolt (with ribs under head)		
Stage 1	70	52
Stage 2	Angle-tighten a further 90°	
12 point bolt (with NO ribs under head)		
Stage 1	200	148
Stage 2	Angle-tighten a further 180°	
Hub to wheel bearing housing		
Stage 1	70	52
Stage 2	Angle-tighten a further 90°	
Lower arm:		
To bracket (front): *		
Stage 1	70	52
Stage 2	Angle-tighten a further 180°	
To lower balljoint nuts: *		
Cast steel lower arm	60	44
Forged aluminium/sheet steel lower arm	100	74
Mounting bracket to body: *		
Stage 1	70	52
Stage 2	Angle-tighten a further 90°	
Mounting bracket to bracket (subframe): *		
Stage 1	50	37
Stage 2	Angle-tighten a further 90°	
Rear engine mounting bolts: *		
To subframe:		
Stage 1	100	74
Stage 2	Angle-tighten a further 90°	
To transmission:		
Stage 1	50	37
Stage 2	Angle-tighten a further 90°	
Subframe-to-underbody/bracket bolts*:		
Stage 1	70	52
Stage 2	Angle-tighten a further 90°	
Splash plate to wheel bearing housing	10	7
Suspension strut:		
Bottom clamp*:		
Stage 1	70	52
Stage 2	Angle-tighten a further 90°	
Upper mounting*:		
Stage 1	15	11
Stage 2	Angle-tighten a further 90°	
Upper piston rod nut*	60	44
Vehicle level sender to subframe and lower arm	9	7
Always replace the nuts/bolts		
Rear suspension		
ABS speed sensor	8	6
Anti-roll bar:		
To subframe bolts*:		
Stage 1	20	15
Stage 2	Angle-tighten a further 90°	
Anti-roll bar link nut *	45	33
Hub to wheel bearing housing bolt: *		
Stage 1	200	148
Stage 2	Angle-tighten a further 180°	
Lower transverse link to wheel bearing housing bolt & nut: *		
Stage 1	70	52
Stage 2	Angle-tighten a further 180°	

Torque wrench settings (continued)

	Nm	lbf ft
Rear suspension (continued)		
Radius rods:		
To body:		
Stage 1	40	30
Stage 2	Angle-tighten a further 90°	
To subframe:		
Stage 1	90	66
Stage 2	Angle-tighten a further 45°	
Shock absorber:		
To wheel bearing housing.	180	134
To body*:		
Stage 1	50	37
Stage 2	Angle-tighten a further 45°	
Piston rod nut*	25	18
Splash shield to wheel bearing housing.	12	9
Stone deflector to transverse link.	8	6
Subframe to body bolts: *		
Stage 1	70	66
Stage 2	Angle-tighten a further 180°	
Track control rod to subframe bolt & nut: *		
Stage 1	70	66
Stage 2	Angle-tighten a further 180°	
Track control rod to wheel bearing housing bolt & nut: *		
Stage 1	130	96
Stage 2	Angle-tighten a further 180°	
Trailing arm bolts: *		
To wheel bearing housing		
Stage 1	90	66
Stage 2	Angle-tighten a further 45°	
To mounting bracket:		
Stage 1	90	66
Stage 2	Angle-tighten a further 90°	
Mounting bracket to underbody:		
Stage 1	50	37
Stage 2	Angle-tighten a further 45°	
Transverse links to subframe	95	70
Upper transverse link to wheel bearing housing bolt & nut: *		
Stage 1	130	96
Stage 2	Angle-tighten a further 180°	
Vehicle level sender	5	4
*Always replace the nuts/bolts		
Steering		
Crash bar (above brake/clutch pedal) bolt	20	15
Steering column:		
To mounting bracket bolts	20	15
Mounting bracket to body	20	15
Strut to mounting bracket/body.	20	15
Universal joint to steering gear bolt*	30	22
Steering gear:		
To subframe*:		
Stage 1	50	37
Stage 2	Angle-tighten a further 90°	
Shield (Torx bolts).	6	4
Steering wheel to column*:		
Stage 1	30	22
Stage 2	Angle-tighten a further 90°	
Track rod end to track rod	70	66
Track rod end to wheel bearing housing nut: *		
Stage 1	20	15
Stage 2	Angle-tighten a further 90°	
Track rod to steering gear rack	100	74
*Always replace the nuts/bolts		
Roadwheels		
Roadwheel bolts.	120	89

1 General Information

1 The independent front suspension is of the MacPherson strut type, incorporating coil springs and integral telescopic shock absorbers. The struts are located by transverse lower suspension arms, which use rubber inner mounting bushes, and incorporate a balljoint at the outer ends. The front wheel bearing housings, which carry the wheel bearings, brake calipers and the hub/disc assemblies, are attached to the MacPherson struts by clamp bolts, and connected to the lower arms through the balljoints. A front anti-roll bar is fitted to all models. The anti-roll bar is rubber-mounted, and is connected to both lower suspension arms by short links.

2 The rear suspension consists of a trailing arm, rubber-mounted at its front end to the underbody, a wheel bearing housing, lower main transverse link and coil spring, track control rod, upper transverse link, and separate shock absorber. A rear anti-roll bar is fitted to all models. The anti-roll bar is rubber-mounted on the rear subframe, and is connected to the wheel bearing housings on each side by a short connecting link.

3 The safety steering column incorporates an intermediate shaft at its lower end. The intermediate shaft is connected to both the steering column and steering gear by universal joints, although the shaft is supplied as part of the column assembly and cannot be separated. Both the inner steering column and intermediate shaft have splined sections that collapse during a major frontal impact. The outer column is also telescopic with two sections, to facilitate reach adjustment.

4 The steering gear is mounted onto the front subframe, and is connected by two track rods, with balljoints at their inner and outer ends, to the steering arms projecting rearwards from the wheel bearing housings. The track rod ends are threaded to the track rods in order to allow adjustment of the front wheel toe setting. The steering gear has electro-mechanical assistance, and incorporates an integral control unit. It is only functional when the engine is running. There are no hydraulic components, and steering assistance is automatically matched to the vehicle speed, steering wheel torque and steering wheel angle.

5 All models are fitted with an Anti-lock Brake System (ABS), and can also be fitted with a Traction Control System (TCS), an Electronic Differential Lock (EDL) system and an Electronic Stability Program (ESP). The ABS may also be referred to as including EBD (Electronic Brake Distribution), which means it adjusts the front and rear braking forces according to the weight being carried. The TCS may also be referred to as ASR (Anti Slip Regulation).

6 The TCS system prevents the front wheels from losing traction during acceleration by reducing the engine output. The system is switched on automatically when the engine is started, and it utilises the ABS system sensors to monitor the rotational speeds of the front wheels.

7 The ESP system extends the ABS, TCS and EDL functions to reduce wheel spin in difficult driving conditions. It does this by using highly sensitive sensors that monitor the speed of the vehicle, lateral movement of the vehicle, the brake pressure, and the steering angle of the front wheels. If, for example, the vehicle is tending to oversteer, the brake will be applied to the front outer wheel to correct the situation. If the vehicle is tending to understeer, the brake will be applied to the rear inside wheel. The steering angle of the front wheels is monitored by an angle sensor on the top of the steering column.

8 The TCS/ESP systems should always be switched on, except when driving with snow chains, driving in snow or driving on loose surfaces, when some wheel spin may be advantageous. The ESP switch is located in the centre of the facia.

9 Some models are also fitted with an Electronic Differential Lock (EDL), which reduces unequal traction from the front wheels. If one front wheel spins 100 rpm or more faster than the other, the faster wheel is slowed down by applying the brake to that wheel. The system is not the same as the traditional differential lock, where the actual differential gears are locked. Because the system applies a front brake, in the event of a brake disc overheating the system will shut down until the disc has cooled. No warning light is displayed if the system shuts down. As is the case with the TCS system, the EDL system uses the ABS sensors to monitor front wheel speeds.

Note: *As a general rule all suspension component fixings, should only be fully tightened when the unladen vehicle is on it's wheels, with the tyres at the correct pressure. Always check the ride height before and after completion of any work* **(see illustration).** *The ride height can also be compared to the ride height given in the specifications. It is possible in most cases to raise the suspension (with a trolley jack) to the correct height to fully tighten the suspension bolts, however extreme caution is required as achieving the correct height may start to lift the vehicle off the axle stands.*

2 Front wheel bearing housing (swivel hub) – removal and refitting

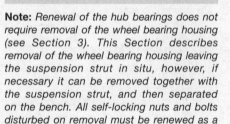

Note: *Renewal of the hub bearings does not require removal of the wheel bearing housing (see Section 3). This Section describes removal of the wheel bearing housing leaving the suspension strut in situ, however, if necessary it can be removed together with the suspension strut, and then separated on the bench. All self-locking nuts and bolts disturbed on removal must be renewed as a matter of course.*

Removal

1 Remove the wheel trim/hub cap (as applicable) and loosen the driveshaft retaining bolt (hub bolt) by 90° with the vehicle resting on its wheels. Also loosen the wheel bolts. **Note:** *Do not loosen the hub bolt more than 90° at this stage, or the wheel bearing may be damaged.*

2 Apply the handbrake, then jack up the front of the vehicle and support it on axle stands (see *Jacking and vehicle support*). Remove the front roadwheel.

3 Unscrew and remove the driveshaft retaining bolt. Check that the driveshaft moves freely in the hub before proceeding.

4 Remove the ABS wheel sensor as described in Chapter 9 Section 20. Also, unbolt the brake hose/wiring bracket from the strut **(see illustration).**

5 Remove the brake disc as described in Chapter 9 Section 6. This procedure includes removing the brake caliper; however do not disconnect the hydraulic brake hose from the caliper. Using a piece of wire or string, tie the caliper to the front suspension coil spring,

1.9 Check the ride height by measuring from the centre of the hub nut vertically to the wheel arch

2.4 Removing the brake hose/wiring bracket

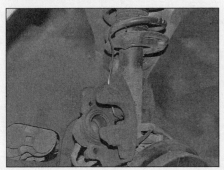

2.5 Suspend the caliper from the front suspension spring

2.8 Lower ball joint securing nuts

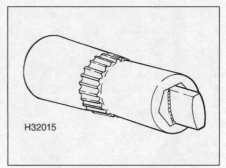

2.10a Tool used by Seat technicians to open up the split wheel bearing housing

2.10b A home made version of the factory tool can also be used

2.10c Using a cold chisel to open up the wheel bearing housing and release the suspension strut

2.10d Withdrawing the wheel bearing housing from the bottom of the suspension strut

to avoid placing any strain on the hydraulic brake hose **(see illustration)**.

6 Unbolt the splash plate from the wheel bearing housing.

7 Loosen the nut securing the steering track rod balljoint to the wheel bearing housing. To do this, fit a ring spanner to the nut, and then hold the balljoint pin stationary using an Allen key. With the nut removed, it may be possible to release the balljoint from the wheel bearing housing by turning the balljoint pin with an Allen key. If not, leave the nut on by a few turns to protect the threads, then use a universal balljoint separator to release the balljoint. Remove the nut completely once the taper has been released.

8 Unscrew the front suspension lower balljoint-to-lower arm retaining nuts **(see illustration)**, then lever the lower arm down to release the balljoint studs from the arm. Now use a soft-faced mallet to tap the driveshaft from the hub splines while pulling out the bottom end of the wheel bearing housing. If the driveshaft is tight on the splines, it may be necessary to use a puller bolted to the hub to remove it. Tie the driveshaft to one side.

9 Note which way round it is fitted, then unscrew the nut and remove the clamp bolt securing the wheel bearing housing to the bottom of the strut.

10 The wheel bearing housing must now be released from the strut. To do this, Seat technicians insert a special tool into the split wheel bearing housing, and turn it through 90° to open up the clamp. A similar tool can

be made out of an old screwdriver or bolt. Alternatively a suitable cold chisel can be driven into the split as a wedge. Slightly press inwards the top of the wheel bearing housing, and then push it downwards from the bottom of the strut **(see illustrations)**.

Refitting

11 Ensure that the driveshaft outer joint and hub splines are clean and dry, and then lubricate the splines with fresh engine oil. Also lubricate the threads and contact surface of the hub nut/bolt with oil.

12 Lift the wheel bearing assembly into position, and engage the hub with the splines on the outer end of the driveshaft. Fit the new hub bolt, tightening it by hand only at this stage.

13 Engage the wheel bearing housing with the bottom of the suspension strut; making sure that the hole in the side plate aligns with the holes in the split housing. Remove the tool used to open the split.

14 Insert the strut-to-wheel bearing housing clamp bolt from the front, and fit the new retaining nut. Tighten the nut to the specified torque.

15 Refit the lower arm balljoint to the lower arm, and tighten the nuts to the specified torque.

16 Refit the track rod balljoint to the wheel bearing housing, then fit a new retaining nut and tighten it to the specified torque. If necessary, hold the balljoint pin with an Allen key while tightening the nut.

17 Refit the splash plate and tighten the bolts.
18 Refit the brake disc and caliper with reference to Chapter 9 Section 6.
19 Refit the ABS wheel sensor as described in Chapter 9 Section 20.
20 Ensure that the outer joint is drawn fully into the hub, and then refit the roadwheel.
21 Have an assistant depress the brake pedal, and then tighten the driveshaft retaining bolt in the stages given in the Specifications. It is recommended that an angle gauge be used to ensure the correct tightening angle. **Note:** *The car must not be standing on its wheels when tightening the bolt, or the wheel bearing may be damaged.*
22 Lower the vehicle to the ground, tighten the roadwheel bolts, and refit the wheel trim/hub cap.

3 Front hub bearings – renewal

Note: *The bearing is a sealed, pre-adjusted and pre-lubricated, double-row roller type, and requires no maintenance. It is bolted to the wheel bearing housing.*

1 Remove the wheel trim/hub cap (as applicable) and loosen the driveshaft retaining bolt (hub bolt) with the vehicle resting on its wheels. **Note:** *Do not loosen the hub bolt more than 90° at this stage, or the wheel bearing may be damaged. Also loosen the wheel bolts.*

3.7a Removing the front hub/wheel bearing from the wheel bearing housing

3.7b Front hub/wheel bearing. Note that some models use a 3 bolt fixing

2 Apply the handbrake, then jack up the front of the vehicle and support it on axle stands (see *Jacking and vehicle support*). Remove the front roadwheel.

3 Unscrew and remove the driveshaft retaining bolt.

4 Remove the ABS wheel sensor as described in Chapter 9 Section 20.

5 Remove the brake disc as described in Chapter 9 Section 6. This procedure includes removing the brake caliper; however do not disconnect the hydraulic brake hose from the caliper. Using a piece of wire or string, tie the caliper to the front suspension coil spring, to avoid placing any strain on the hydraulic brake hose.

6 Press the driveshaft outer stub towards the transmission as far as possible, then unscrew and remove the wheel bearing retaining bolts from the rear of the housing.

7 Remove the hub/wheel bearing complete with hub from the outside of the housing while sliding it from the driveshaft splines **(see illustrations)**.

8 Fit the new wheel bearing to the housing and engage the hub splines with the driveshaft outer stub.

9 Refit the brake disc and caliper with reference to Chapter 9 Section 6.

10 Refit the ABS wheel sensor with reference to Chapter 9 Section 20.

11 Pull the driveshaft outer stub fully into the hub and fit a new hub bolt, hand-tight at this stage.

12 Ensure that the outer joint is drawn fully into the hub, and then refit the roadwheel.

13 Tighten the driveshaft retaining bolt in the stages given in the Specifications. It is recommended that an angle gauge be used to ensure the correct tightening angle. **Note:** *The*

car must not be standing on its wheels when tightening the bolt, or the wheel bearing will be damaged.

14 Lower the vehicle to the ground and tighten the roadwheel bolts.

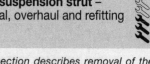

4 Front suspension strut – removal, overhaul and refitting

Note: *This section describes removal of the suspension strut leaving the wheel bearing housing in situ, however, if necessary it can be removed together with the wheel bearing housing, then separated on the bench. All self-locking nuts and bolts disturbed on removal must be renewed as a matter of course.*

Removal

1 Remove the wheel trim/hub cap (as applicable) and loosen the driveshaft retaining bolt (hub bolt) with the vehicle resting on its wheels. **Note:** *Do not loosen the hub bolt more than 90° at this stage, or the wheel bearing may be damaged. Also loosen the wheel bolts.*

2 Apply the handbrake, then jack up the front of the vehicle and support it on axle stands (see *Jacking and vehicle support*). Remove the appropriate roadwheel.

3 Unscrew the nut and disconnect the anti-roll bar link from the strut **(see illustration)**.

4 Release the ABS sensor wiring from the strut.

5 Unscrew and remove the driveshaft retaining bolt **(see illustration)**.

6 Unscrew the front suspension lower balljoint-to-lower arm retaining nuts, then lever the lower arm down to release the balljoint studs. Now use a soft-faced mallet to tap the driveshaft from the hub splines while pulling out the bottom end of the wheel bearing housing **(see illustrations)**. If the driveshaft is tight on the splines, it may be necessary to use a puller bolted to the hub to remove it. Tie the driveshaft to one side, then refit the lower balljoint to the lower arm and secure with the nuts, hand-tightened.

7 Note which way round it is fitted, then unscrew the nut and remove the clamp bolt

4.3 Disconnect the anti-roll bar drop link from the strut

4.5 Unscrew and remove the driveshaft retaining bolt

4.6a Unscrew the securing nuts...

4.6b ...detach the lower arm from the balljoint studs...

4.6c ...and withdraw the driveshaft from the hub splines

4.7 Remove the clamp bolt...

4.8 ...then use a suitable tool to expand the split, so that the strut can be pulled from the wheel bearinghousing

4.11a Unscrew the upper mounting bolts...

securing the wheel bearing housing to the bottom of the strut **(see illustration)**.

8 The wheel bearing housing must now be released from the strut. To do this, Seat technicians insert a special tool into the split wheel bearing housing, and turn it through 90° to open up the clamp. A similar tool such as an Allen key can be used, or alternatively a suitable cold chisel can be driven into the split as a wedge. Slightly press inwards the top of the wheel bearing housing, and then push it downwards from the bottom of the strut **(see illustration)**. Support the wheel bearing housing to one side without straining the hydraulic brake hose.

9 Remove the wiper arms, windscreen cowl and the plenum chamber front panel as described in Chapter 12 Section 19. Access can be improved if the wiper motor is also removed.

10 To ensure correct refitting, mark the strut upper mounting in relation to the body. If the reason for removing the strut is overhaul, loosen the upper mounting centre nut one turn, while holding the piston rod with an Allen key.

11 Support the strut, then unscrew the upper mounting bolts and lower the strut from under the wheel arch **(see illustrations)**.

4.11b ...and lower the strut from under the wheel arch

Overhaul

⚠ *Warning: Before attempting to dismantle the suspension strut, a suitable tool to hold the coil spring in compression must be obtained. Adjustable coil spring compressors are readily available, and are recommended for this operation. Any attempt to dismantle the strut without such a tool is likely to result in damage or personal injury.*

12 With the strut removed from the car, clean

4.13 Compress the coil spring with the compressor tool

away all external dirt. If necessary, mount it upright in a vice during the dismantling procedure.

13 Fit the spring compressor, and compress the coil spring until all tension is relieved from the upper spring seat **(see illustration)**.

14 Unscrew and remove the upper centre retaining nut, whilst retaining the strut piston with a suitable Allen key, then remove the mounting, thrust bearing, and coil spring **(see illustrations)**.

15 Remove the protective sleeve and upper

4.14a Unscrew the upper centre nut...

4.14b ...then remove the mounting, thrust bearing and coil spring

4.15a Remove the protective sleeve...

4.15b ...and upper bearing race...

4.15c ...then remove the bump stop from the upper mounting

bearing race, and then remove the bump stop from the upper mounting **(see illustrations)**.

16 With the strut assembly now completely dismantled, examine all the components for wear, damage or deformation, and check the bearing for smoothness of operation. Renew any of the components as necessary.

17 Examine the strut for signs of fluid leakage. Check the strut piston for signs of pitting along its entire length, and check the strut body for signs of damage. While holding it in an upright position, test the operation of the strut by moving the piston through a full stroke, and then through short strokes of 50 to 100 mm. In both cases, the resistance felt should be smooth and continuous. If the resistance is jerky, or uneven, or if there is any visible sign of wear or damage to the strut, renewal is necessary.

18 If any doubt exists about the condition of the coil spring, carefully remove the spring compressors, and check the spring for distortion and signs of cracking. Renew the spring if it is damaged or distorted, or if there is any doubt as to its condition.

19 Inspect all other components for signs of damage or deterioration, and renew as necessary.

20 Assemble the bump stop to the upper mounting, and then refit the upper bearing race and protective sleeve to the mounting. The larger diameter of the bump stop must be against the upper mounting.

21 Fit the coil spring (together with the compressor tool) onto the strut, making sure its lower (larger diameter) end is correctly located against the spring seat stop.

22 Refit the thrust bearing and upper mounting, then screw on a new retaining nut. Tighten the nut to the specified torque while holding the piston rod with an Allen key.

Refitting

23 Manoeuvre the strut into position under the wheel arch, and locate in the suspension strut turret in the previously noted position. If a new strut is being fitted, locate it so that one of the two arrows marked on the upper mounting point's forwards, and the bolt holes are aligned. Insert the bolts and tighten to the specified torque.

24 Refit the plenum chamber cover and wiper arms.

25 Engage the wheel bearing housing with the bottom of the suspension strut; making sure that the hole in the side plate aligns with the holes in the split housing. Raise the housing, while pressing it inwards to assist entry. Use a trolley jack if necessary. When fully entered, remove the tool used to open the split.

26 Insert the new strut-to-wheel bearing housing bolt from the front, and fit the new retaining nut. Tighten the nut to the specified torque.

27 Unscrew the nuts from the lower balljoint and lever down the lower arm to release it from the wheel bearing housing. Insert the outer end of the driveshaft through the hub and engage it with the splines. Fit the new hub bolt, tightening it by hand only at this stage.

28 Refit the lower arm balljoint to the lower arm, and tighten the nuts to the specified torque. Where applicable, refit the headlight range control sensor arm to the lower arm and tighten the nut.

29 Refit the ABS wheel sensor as described in Chapter 9 Section 20.

30 Ensure that the outer joint is drawn fully into the hub, and then refit the roadwheel.

31 Have an assistant depress the brake pedal, and then tighten the driveshaft retaining bolt in the stages given in the Specifications. It is recommended that an angle gauge be used to ensure the correct tightening angle. **Note:** *The car must not be standing on its*

5.4 Remove the outer mounting bolt

wheels when tightening the bolt, or the wheel bearing may be damaged.

32 Lower the vehicle to the ground and tighten the roadwheel bolts.

5 Front suspension lower arm – removal, overhaul and refitting

Note: *Before dismantling any suspension component, check the vehicle ride height by measuring the distance from the centre of the front hub to the wheel arch. Note down this figure.*
Note: *The lower arm is available in either cast steel, pressed sheet steel or aluminum. When renewing the arm, make sure the correct type is fitted according to model. VAG subframe locating pins (T10096) or similar are required for the work in this Section, to ensure correct front wheel alignment. All self-locking nuts and bolts disturbed on removal must be renewed as a matter of course.*

Removal

1 Apply the handbrake, then jack up the front of the vehicle and support it on axle stands (see *Jacking and vehicle support*). Remove the appropriate front roadwheel and the engine compartment undertray.

2 Unscrew the front suspension lower balljoint-to-lower arm retaining nuts, then lever down the lower arm to release the arm from the balljoint studs **(see illustration 4.6a and 4.6b)**.

3 At this stage, Seat technicians use locating pins T10096 in place of the rear outer subframe/rear bush mounting bolt to ensure correct front wheel alignment. If these pins are not available, only remove and refit one lower arm at a time, and mark the position of the subframe/bush accurately with dabs of paint.

4 Unscrew and remove the rear outer mounting bolt **(see illustration)** and, where available, substitute it with a locating pin tightened to 20 Nm (15 lbf ft).

5 Unscrew and remove the front mounting bolt, then support the lower arm and unscrew the two rear inner mounting bolts **(see**

5.5a Undo the inner front mounting bolt ...

5.5b ...and the inner rear mounting bolts

5.7a A suitable setup for bush removal

5.7b The new bush must enter the arm off centre

5.7c Two sockets are used to allow clearance for the lip of the bush

5.7d Adjust the position of the bush so that it protrudes equally on both sides

illustrations). Remove the lower arm from beneath the car.

Overhaul

6 Thoroughly clean the lower arm, then check carefully for cracks or any other signs of wear or damage, paying particular attention to the rubber mounting bushes. If the rear bush requires renewal it is supplied complete with the housing and can be pushed easily off the lower arm.

7 When fitting a new front mounting bush, it must be initially tilted with one lip in the bore. As the bush is inserted, it will straighten up. Make sure the bush is centred in its bore **(see illustrations)**.

Refitting

8 Locate the lower arm on the subframe and insert the front mounting bolt loosely.

9 Insert the two rear outer mounting bolts loosely, then position the inner bolt hole in the exact position noted during removal. If a location pin was used on removal, the bracket will be correctly positioned on the pin. Where no pin was used align the arm with the previously made paint marks Tighten the two outer bolts to the specified torque, then remove the pin and refit the inner bolt, and tighten to the specified torque.

10 Lightly tighten the front mounting bolt.

11 Lever down the lower arm and locate the balljoint studs in their holes. Fit the new nuts and tighten to the specified torque.

12 Raise the arm using a suitable jack until the suspension is in the unladen position. Measure the distance from the hub nut centre to the wheel arch and adjust the distance to

that noted down at the beginning of the work **(see illustration)**.

Caution: Exercise extreme care as the vehicle may be lifted off the axle stand during this procedure. An alternative approach is to refit the wheel and tighten the arm bolts with the vehicle on the ground (or wheel ramps).

13 Tighten the front mounting bolts to the specified torque.

14 Refit the roadwheel and undertray, and lower the car to the ground.

---

6 Front suspension lower arm balljoint – removal, inspection and refitting

Note: *All self-locking nuts and bolts disturbed on removal must be renewed as a matter of course.*

Removal

Method 1

1 Remove the wheel bearing housing as described in Section 2.

2 Unscrew and remove the balljoint retaining nut **(see illustration)**, then release the balljoint from the wheel bearing housing using a universal balljoint separator. Withdraw the balljoint.

Method 2

3 Remove the wheel trim/hub cap (as applicable) and loosen the driveshaft retaining bolt (hub bolt) with the vehicle resting on its

5.12 Raise the arm until the distance from the hub centre to the wheel arch is the same as noted down

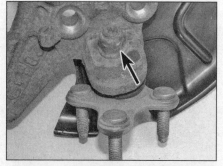

6.2 Front suspension lower arm balljoint retaining nut

wheels. **Note:** *Do not loosen the hub bolt more than 90° at this stage, or the wheel bearing may be damaged. Also loosen the wheel bolts.*

4 Apply the handbrake, then jack up the front of the vehicle and support it on axle stands (see *Jacking and vehicle support*). Remove the appropriate roadwheel.

5 Unscrew and remove the driveshaft retaining bolt.

6 Unscrew the front suspension lower balljoint-to-lower arm retaining nuts, then lever the lower arm down to release the balljoint studs. Now use a soft-faced mallet to tap the driveshaft from the hub splines while pulling out the bottom end of the wheel bearing housing. If the driveshaft is tight on the splines, it may be necessary to use a puller bolted to the hub to remove it. It is not necessary to remove the driveshaft completely from the hub. Retain the wheel bearing housing away from the lower arm by inserting a block of wood between the strut and the inner body panel.

7 Unscrew and remove the balljoint retaining nut, then release the balljoint from the wheel bearing housing using a universal balljoint separator. Withdraw the balljoint.

Inspection

8 With the balljoint removed, check that it moves freely, without any sign of roughness. Check also that the balljoint rubber gaiter shows no sign of deterioration, and is free from cracks and splits. Renew as necessary.

Refitting

Method 1

9 Fit the balljoint to the wheel bearing housing and fit the new retaining nut. Tighten the nut to the specified torque setting, noting that the balljoint shank can be retained with an Allen key if necessary to prevent it from rotating.

10 Refit the wheel bearing housing with reference to Section 2.

Method 2

11 Fit the balljoint to the wheel bearing housing and fit the new retaining nut. Tighten the nut to the specified torque setting, noting that the balljoint shank can be retained with

an Allen key if necessary to prevent it from rotating.

12 Remove the wooden block and move the strut inwards, then refit the balljoint to the lower arm using new nuts, and tighten them to the specified torque.

13 Refit the driveshaft retaining bolt and tighten it sufficiently to draw the driveshaft fully into the hub, then refit the roadwheel.

14 Have an assistant depress the brake pedal, and then tighten the driveshaft retaining bolt in the stages given in the Specifications. It is recommended that an angle gauge be used to ensure the correct tightening angle. **Note:** *The car must not be standing on its wheels when tightening the bolt, or the wheel bearing may be damaged.*

15 Lower the vehicle to the ground and tighten the roadwheel bolts.

---

7 Front anti-roll bar – removal and refitting

Note: *As the subframe must be lowered during this procedure, VAG/Seat subframe locating pins (T10096) or similar are required ensuring correct front wheel alignment. All self-locking nuts and bolts disturbed on removal must be renewed as a matter of course.*

Removal

1 Apply the handbrake, then jack up the front of the vehicle and support it on axle stands (see *Jacking and vehicle support*). Remove both front roadwheels and the engine compartment undertray.

2 Inside the car, undo the nuts and remove the trim beneath the foot pedals for access to the steering column universal joint. Unscrew the clamp bolt and pull the universal joint from the steering gear pinion.

3 The anti-roll bar may be removed with or without the side links. Unscrew the nuts securing the links to the struts or anti-roll bar as required **(see illustration)**.

4 Working on each side in turn, unscrew the front suspension lower balljoint-to-lower arm retaining nuts, then disconnect the track rod ends with reference to Section 24.

5 Unscrew the bolts securing the anti-roll bar clamps to the subframe **(see illustrations)**. Mark the anti-roll bar to indicate which way round it is fitted, and the position of the rubber mounting bushes; this will aid refitting.

6 Unscrew and remove the engine/ transmission rear mounting bolts from the transmission.

7 Support the subframe with a trolley jack and block of wood. If not using the special VAG/Seat locating pins T10096, accurately mark the position of the subframe to ensure correct wheel alignment.

8 Unscrew the mounting bolts and slightly lower the subframe, taking care not to damage the electrical wiring. Where available, fit the locating pins to facilitate refitting.

9 Lift the anti-roll bar forwards over the bracket, and lower it to the floor.

10 Carefully examine the anti-roll bar components for signs of wear, damage or deterioration, paying particular attention to the rubber mounting bushes. Renew worn components as necessary.

Refitting

11 Refitting is a reversal of removal but tighten all nuts and bolts to the specified torque where given. When refitting the subframe, align it with the marks made on removal, or use the special location pins before tightening the mounting bolts. To assist entry of the steering gear gaiter through the bulkhead, apply a soapy solution to it. Have the front wheel alignment checked at the earliest opportunity.

---

8 Front anti-roll bar drop link – removal and refitting

Note: *All self-locking nuts and bolts disturbed on removal must be renewed as a matter of course.*

Removal

1 Apply the handbrake, then jack up the front of the vehicle and support it on axle stands (see *Jacking and vehicle support*). Remove the relevant front roadwheel.

7.3 Disconnect the drop link from the anti-roll bar

7.5a Undo the two mounting bolts...

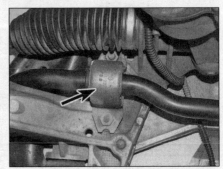

7.5b ...to release the anti-roll bar clamp on the subframe

8.2a Counterhold the shank, unbolt...

8.2b ...and remove the link from the strut

8.2c Repeat the procedure at the anti-roll bar

2 Unscrew and remove the nuts securing the link to the strut and anti-roll bar **(see illustrations)**.

3 Inspect the link rubbers for signs of damage or deterioration. If evident, renew the link complete.

Refitting

4 Refitting is a reversal of removal, but tighten the nuts to the specified torque.

9 Rear wheel bearing housing – removal and refitting

Removal

Note: *Before dismantling any suspension component, check the vehicle ride height by measuring the distance from the centre of the front hub to the wheel arch. Note down this figure.*

1 Chock the front wheels, then jack up the rear of the vehicle and support it on axle stands (see *Jacking and vehicle support*).

2 Remove the rear coil spring as described in Section 14.

3 Remove the rear hub as described in Section 10.

4 Unbolt the splash plate from the rear wheel bearing housing.

5 Disconnect the wiring, then unscrew the mounting bolt and remove the ABS sensor from the rear wheel bearing housing.

6 Unscrew the bolt securing the rear shock absorber to the rear wheel bearing housing.

7 Unscrew the bolts securing the upper transverse link and lower transverse links to the rear wheel bearing housing.

8 Unscrew the bolt securing the rear track control rod to the rear wheel bearing housing.

9 Support the rear wheel bearing housing, and then unscrew the mounting bolts from the trailing arm. Also, undo the nut and disconnect the anti-roll bar link from the trailing arm.

10 Withdraw the rear wheel bearing housing from the car.

Refitting

11 Attach the rear wheel bearing housing to the rear track control rod, and the upper and lower transverse links, and hand-tighten the bolts.

12 Fit the wheel bearing housing to the trailing arm and insert the rear, upper bolt loosely.

13 Insert the two remaining bolts securing the trailing arm to the rear wheel bearing housing, and tighten them to the specified torque.

14 Refit the splash plate and tighten the bolts to the specified torque.

15 Refit the rear hub with reference to Section 10.

Caution: Exercise extreme care as the vehicle may be lifted off the axle stand during this procedure. An alternative approach is to refit all the components and tighten the suspension bolts with the vehicle on the ground (or wheel ramps).

16 Place a trolley jack under the hub.

Position the centre of the rear hub to the ride-height distance noted at the beginning of the procedure, using the trolley jack to adjust the position **(see illustration)**.

17 Tighten the following bolts to their specified torque, in the order given:

a) *Track control rod.*
b) *Lower transverse link.*
c) *Upper transverse link. Position the washer so that it clears the splash plate.*

18 Refit the shock absorber lower mounting bolt and tighten to the specified torque.

19 Remove the trolley jack, then refit the rear coil spring with reference to Section 14.

20 Refit the ABS sensor and tighten the mounting bolt. Reconnect the wiring.

21 Refit the roadwheel, then lower the vehicle to the ground, tighten the roadwheel bolts, and refit the wheel trim/hub cap. Have the rear wheel alignment checked and if necessary adjusted by a Seat dealer or suitably equipped garage.

10 Rear hub/wheel bearings – checking and renewal

Note: *The rear wheel bearings cannot be renewed independently of the rear hub, because the outer races are formed in the hub itself. If excessive wear is evident, the rear hub must be renewed complete. The rear hub bolt must always be renewed after removal.*

Removal

1 Chock the front roadwheels, then jack up the rear of the vehicle and support on axle stands (see *Jacking and vehicle support*). Release the handbrake and remove the relevant rear roadwheel.

2 Remove the rear brake caliper and mounting bracket with reference to Chapter 9 Section 9. Do not disconnect the hydraulic brake pipe. Move the caliper just clear of the brake disc, without bending the hydraulic pipe excessively, and support it with welding rod or on an axle stand.

3 Undo the crosshead screw then withdraw the brake disc from the hub.

4 Remove the dust cap from the centre of the hub using a screwdriver or cold chisel **(see illustration)**.

9.16 Adjust the wheel arch to hub distance as required before tightening the fixings

10.4 Removing the dust cap

5 Unscrew and remove the hub bolt, using a multi-spline tool. Note that it is tightened to a high torque and a socket extension bar may be required to loosen it. The bolt must be renewed whenever removed.

6 Using a suitable puller, pull the hub and bearings from the stub axle. The bearing inner race will remain on the stub axle, and a puller will be required to remove it; use a sharp cold chisel to move the race away from the stub axle base so that the puller legs can fully engage the race.

7 Examine the hub and bearings for wear, pitting and damage. If any damage is evident, renew the hub complete.

Refitting

8 Wipe clean the stub axle, then check that the bearing races are adequately lubricated with suitable grease. Check that the inner bearing race is located correctly in the hub. Also make sure that the ABS rotor is pressed firmly onto the inner end of the hub.

9 Locate the hub as far as possible on the stub axle.

10 Screw on the new bolt and tighten it to the specified torque.

11 Check the dust cap for damage and renew it if necessary. Use a hammer to carefully tap the cap into the hub. **Note:** *A badly fitting dust cap will allow moisture to enter the bearing, reducing its service life.*

12 Refit the brake disc and tighten the crosshead screw.

13 Refit the rear brake mounting bracket and caliper with reference to Chapter 9 Section 9.

14 Refit the roadwheel and lower the vehicle to the ground.

11 Rear track control rod – removal and refitting

Removal

1 Chock the front roadwheels, then jack up the rear of the vehicle and support on axle stands (see *Jacking and vehicle support*). Remove the relevant rear roadwheel.

2 Remove the rear coil spring as described in Section 14.

3 Undo the retaining nut and disconnect the link rod from the end of the anti-roll bar. Undo the anti-roll bar mounting bracket bolts and move the anti-roll bar to access the inner mounting bolt for the track control arm. See Section 16 for further information.

4 Note the fitted position of the rear track control rod, with either the 'closed' side facing forwards or the 'open' side facing downwards. Also, note which way round the mounting bolts are fitted.

5 Unscrew and remove the mounting bolts and nuts **(see illustration)** and withdraw the track control rod from under the vehicle.

Refitting

6 Refitting is a reversal of removal, but tighten the bolts to the specified torque and position the rod and bolts as previously noted. Have the rear wheel alignment checked and if necessary adjusted by a Seat dealer or suitably equipped garage.

12 Rear transverse links – removal and refitting

Upper link

Removal

1 Chock the front roadwheels, then jack up the rear of the vehicle and support on axle stands (see *Jacking and vehicle support*). Remove the roadwheel.

2 Remove the rear coil spring as described in Section 14.

3 Release the ABS speed sensor wiring from the upper link, and then unscrew the bolt securing the link to the wheel bearing housing.

4 At the inner end of the upper link, mark the position of the eccentric bolt and subframe in relation to each other. This alignment determines the camber setting of the rear wheels.

5 Note which way round the eccentric bolt is fitted, then unscrew and remove it and withdraw the upper link.

Refitting

6 Refitting is a reversal of removal, but delay fully-tightening the mounting bolts until the rear suspension is in the loaded position – with the weight of the vehicle on the wheels. Make sure the eccentric bolt is correctly aligned as previously noted, and also position the 'star' washer to provide a clearance between one of its points and the splash plate. Have the rear wheel alignment checked and if necessary adjusted by a Seat dealer or suitably equipped garage.

Lower transverse link

Removal

7 Chock the front roadwheels, then jack up the rear of the vehicle and support on axle stands (see *Jacking and vehicle support*). Remove the roadwheel.

8 Remove the rear coil spring as described in Section 14.

9 Unscrew the bolt securing the lower transverse link to the wheel bearing housing.

10 On models with headlight range control, unscrew the nut and disconnect the sensor arm from the link.

11 At the inner end of the lower link, mark the position of the eccentric bolt and subframe in relation to each other **(see illustration)**. This alignment determines the camber setting of the rear wheels.

12 Unscrew and remove the inner bolt and withdraw the lower transverse link from under the car.

Refitting

13 Refitting is a reversal of removal, but delay fully-tightening the mounting bolts until the rear suspension is set to the correct ride-height given in Section 9. Make sure the eccentric bolt is correctly aligned as previously noted, and also position the 'star' washer to provide a clearance between one of

11.5 Track control inner mounting

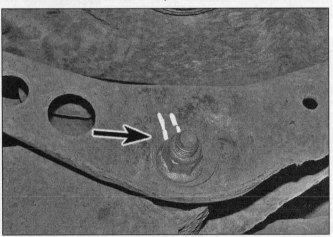

12.11 Eccentric bolt marked up for adjustment of the rear suspension

13.6 Rear trailing arm mounting bracket

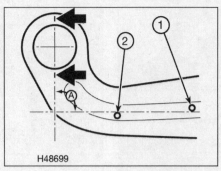

13.9a Make a vertical line along holes (1) and (2) on the arm as shown...

Angle (A) = 90°

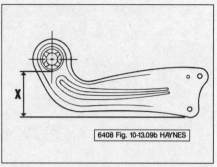

13.9b On some versions the bush must be aligned by measuring the distance as shown

Distance (X) = 114 mm

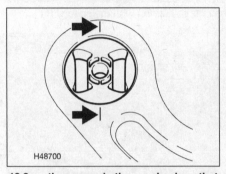

13.9c ...then press in the new bush so that the centre line is as shown

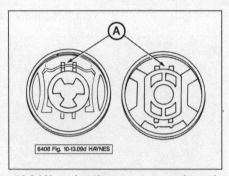

13.9d Note that there are two versions of the bush. In both cases the centre of the protrusions (A) is the alignment point

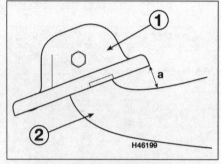

13.10 Mounting bracket assembly to trailing arm

1 Mounting bracket   2 Trailing arm
a = 36.0 mm

its points and the splash plate. Have the rear wheel alignment checked and if necessary adjusted by a Seat dealer or wheel alignment specialist.

13 Rear trailing arm and bracket – removal, overhaul and refitting

Removal

1 Chock the front roadwheels, then jack up the rear of the vehicle and support on axle stands (see *Jacking and vehicle support*). Remove the roadwheel.
2 Remove the rear coil spring as described in Section 14.
3 Unscrew the bolt securing the handbrake cable support to the trailing arm.
4 Unscrew the nut and detach the anti-roll bar link.
5 Unscrew the bolts securing the trailing arm to the rear wheel bearing housing.
6 Mark the position of the trailing arm front mounting bracket in relation to the underbody **(see illustration)**.
7 Support the front mounting bracket on a trolley jack, then unscrew the bolts, lower the assembly and withdraw the rear trailing arm and bracket from under the vehicle.

Overhaul

8 Thoroughly clean the trailing arm and bracket, then unscrew the front pivot bolt and separate the arm from the bracket. Check carefully for cracks or any other signs of wear or damage, paying particular attention to the rubber mounting bush.
9 If the bush requires renewal, take the arm to a Seat dealer or suitably equipped garage. Alternatively, a hydraulic press and suitable spacers may be used to press the bush out of the arm, and to install the new one. Dip the bush in a mild solution of washing-up liquid and water. When fitting the new bush to the front of the trailing arm, it is important to position it correctly. Make a vertical line on the arm as shown in the accompanying illustration, and then press in the new bush so that the line is between the two projections shown **(see illustrations)**.
10 With the new bush in position, locate the front of the arm in the bracket, and insert the bolt. Position the arm in relation to the bracket as shown **(see illustration)** then tighten the bolt/nut to the specified torque. There are two types of bush.

Refitting

11 Fit the trailing arm to the wheel bearing housing and insert the bolts loosely.
12 Fit the anti-roll bar link and screw on the nut loosely.

13 Raise the front mounting bracket and locate it on the underbody in its previously noted position. Insert the new bolts and tighten to the specified torque.
14 Lower the jack then tighten the arm-to-housing bolts to the specified torque.
15 Tighten the anti-roll bar link nut.
16 Refit the handbrake cable support and tighten the bolt.
17 Refit the rear coil spring with reference to Section 14.
18 Refit the roadwheel and lower the vehicle to the ground. Have the rear wheel alignment checked and if necessary adjusted by a Seat dealer or suitably equipped garage.

14 Rear coil spring – removal and refitting

 Warning: Adjustable coil spring compressors are readily available, and are recommended for this operation. Any attempt to remove the coil spring without such a tool is likely to result in damage or personal injury.
Note: *To ensure even rear suspension, both rear coil springs should be renewed at the same time.*

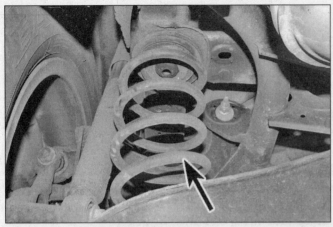

14.2a Rear coil spring

14.2b Compress the rear coil spring with the special tool...

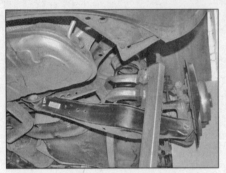

14.2c ...then remove it from the trailing arm and underbody

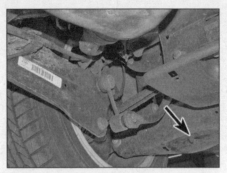

14.4a Rear coil spring seat located in the lower transverse link...

14.4b ...and underbody

Removal

1 Chock the front roadwheels, then jack up the rear of the vehicle and support on axle stands (see *Jacking and vehicle support*). Remove the relevant rear roadwheel.

2 Fit the tool to the coil spring and compress it until it can be removed from the trailing arm and underbody **(see illustrations)**. With the coil spring on the bench, carefully release the tension of the tool and remove it.

3 With the coil spring removed, recover the upper and lower spring seats and check them for damage. Obtain new ones if necessary, but note that they are different, the lower one having a location pin that enters a hole in the lower transverse link. Also clean thoroughly the spring locations on the underbody and trailing arm.

Refitting

4 Refitting is a reversal of removal, but make sure that the lug on the lower spring seat engages in the hole in the lower transverse link and the lower end of the coil spring abuts the stop on the seat. Also make sure the upper seat is correctly engaged with the lug on the underbody **(see illustrations)**. **Note:** *To ensure even rear suspension, both rear coil springs should be renewed at the same time.*

15 Rear shock absorber – removal and refitting

Note: *All self-locking nuts and bolts disturbed on removal must be renewed as a matter of course.*
Note: *To ensure even rear suspension, both rear shock absorbers should be renewed at the same time.*

Removal

1 Before removing the shock absorber, an idea of how effective it is can be gained by depressing the rear corner of the car. If the shock absorber is in good condition, the body should rise then settle in its normal position. If

the body oscillates more than this, the shock absorber is defective.

2 Chock the front roadwheels, then jack up the rear of the vehicle and support on axle stands (see *Jacking and vehicle support*). Remove the relevant rear roadwheel.

3 Position a trolley jack beneath the trailing arm, and raise the arm so that the shock absorber is slightly compressed. Note on some models, it may be necessary to remove the plastic stone protection guard first.

4 Unscrew the lower mounting bolt, then unscrew the upper mounting bolts and withdraw the shock absorber **(see illustrations)**.

5 With the shock absorber on the bench, remove the cap, then unscrew the nut from the top of the piston rod and remove the upper

15.4a Unscrew the rear shock absorber lower mounting bolt...

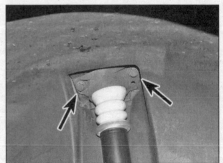

15.4b ...and upper mounting bolts

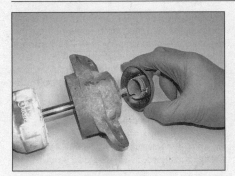

15.5a Remove the cap...

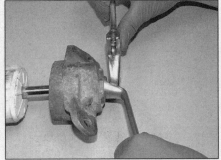

15.5b ...unscrew the nut while holding the piston...

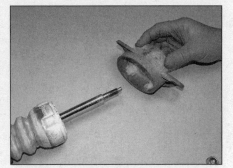

15.5c ...remove the upper mounting bracket...

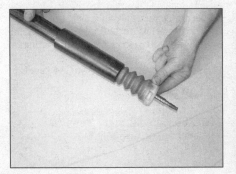

15.5d ...and bump stop

15.5e Recover the collar (where fitted) from the protective sleeve if the bump stop is to be replaced

3 To make access easier, undo the retaining nuts and securing clips and remove the plastic shield from below the anti-roll bar – where fitted.
4 Mark the anti-roll bar to indicate which way round it is fitted, and the position of the rubber mounting bushes; this will aid refitting.
5 Unscrew the bolts securing the anti-roll bar clamps to the rear subframe, and recover the clamps **(see illustration)**.

Refitting

6 Refitting is a reversal of removal but tighten all nuts and bolts to the specified torque.

---

17 Rear anti-roll bar drop link – removal and refitting

Removal

1 Chock the front roadwheels, then jack up the rear of the vehicle and support on axle stands (see *Jacking and vehicle support*). Remove the relevant rear roadwheel.
2 Note that upper part of the connecting link to the trailing arm has a balljoint, and the lower part of the connecting link to the end of the anti-roll bar, is fitted with a rubber bush **(see illustration)**.
3 Unscrew the nut securing the ball joint to the trailing arm, then undo the bolt and nut from the end of the anti-roll bar to withdraw the link from under the vehicle.
4 Inspect the link rubber/balljoint for signs of damage or deterioration. If evident, renew the link complete.

Refitting

5 Refitting is a reversal of removal, but renew the retaining nuts and tighten to the specified torque.

---

18 Vehicle level sender – removal and refitting

Removal

1 The front sender for the headlight range control system is located on the left-hand side

---

mounting bracket, followed by the bump stop, support ring (where fitted), protective tube, and protective cap **(see illustrations)**. **Note:** *Two types of bump stop are supplied; a short version with a support ring, and a long version without a support ring.*
6 If necessary, the action of the shock absorber can be checked by mounting it upright in a vice. Fully depress the rod, and then pull it up fully. The piston rod must move smoothly over its complete length.

Refitting

7 Locate the components removed from the top of the shock absorber in their correct order, and screw on a new nut. Tighten the nut and fit the cap.
8 Locate the shock absorber in the rear wheel arch, then insert the upper mounting bolts and tighten to the specified torque.

9 Extend the shock absorber if necessary, and insert the lower mounting bolt loosely.
10 Fully tighten the shock absorber lower mounting bolt.
11 Refit the roadwheel and lower the vehicle to the ground.

---

16 Rear anti-roll bar – removal and refitting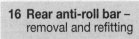

Removal

1 Chock the front roadwheels, then jack up the rear of the vehicle and support on axle stands (see *Jacking and vehicle support*). Remove both rear roadwheels.
2 Working on each side in turn, unscrew the nut and detach the connecting links from the anti-roll bar, see Section 17.

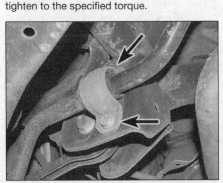

16.5 Rear anti-roll bar mounting clamp bolts

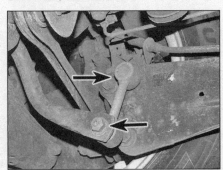

17.2 Rear anti-roll bar drop link joints

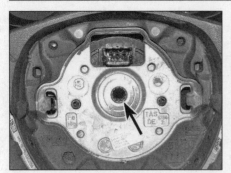

19.5a Using a splined tool …

19.5b …unscrew and remove the bolt

19.6 The steering wheel should be marked for its central position

of the front subframe, and incorporates an arm and link attached to the left-hand front lower suspension arm. The rear sender is bolted to the left-hand side of the rear subframe, and an arm and link is attached to a bracket on the lower transverse link.

2 To remove the front sender, apply the handbrake then jack up the front of the vehicle and support it on axle stands (see *Jacking and vehicle support*). Remove the front roadwheel, and then note the position of the sender on the lower arm. Unscrew the nut and disconnect the link and bracket from the lower arm. Disconnect the wiring then unscrew the bolt and remove the sender from the front subframe.

3 To remove the rear sender, chock the front roadwheels and then jack up the rear of the vehicle and support on axle stands (see *Jacking and vehicle support*). Disconnect the wiring from the sender. Unscrew the bolts securing the link and bracket to the lower

transverse link, then unscrew the bolts and remove the sender from the rear subframe.

Refitting

4 Refitting is a reversal of removal, but tighten the mounting bolts to the specified torque. If necessary, have the sender outputs checked by a Seat dealer. This work requires the use of special equipment not available to the home mechanic.

19 Steering wheel – removal and refitting

⚠️ **Warning: During the airbag removal and refitting procedures, avoid sitting in the front seats.**

Removal

1 Set the front wheels in the straight-ahead position, and release the steering lock by inserting the ignition key.

2 Disconnect the battery negative (earth) lead as described in Chapter 5A Section 3.

3 Adjust the steering column to its highest position, then extend it into the passenger compartment as far as possible, and lock it in this position.

Caution: To prevent any discharge of static electricity into the airbag circuit, temporarily touch the vehicle bodywork before disconnecting the wiring.

4 Remove the driver's airbag as described in Chapter 12 Section 27.

⚠️ **Warning: Position the airbag in a safe and secure place, away from the work area.**

5 Using a multi-spline socket, unscrew and remove the retaining bolt, while holding the steering wheel stationary **(see illustrations)**. The bolt can be reused up to five times – mark the bolt with a centre punch every time it is reused. Discard the retaining bolt after removing more than five times.

6 Check if the steering wheel is marked in relation to the column. If not, use a dab of paint to mark them, and then ease the steering wheel from the column splines by firmly rocking it side-to-side **(see illustration)**.

Refitting

7 Locate the steering wheel on the column splines making sure that the previously made marks are correctly aligned.

8 Refit the retaining bolt and tighten to the specified torque while holding the steering wheel stationary.

9 Refit the driver's airbag with reference to Chapter 12 Section 27.

10 Reconnect the battery negative (earth) lead.

20 Steering column – removal, inspection and refitting

Removal

1 Disconnect the battery negative lead as described in Chapter 5A Section 3.

2 Remove the steering wheel as described in Section 19.

3 Return the steering to the straight-ahead position. Adjust the steering column to its lowest position and extend it into the passenger compartment as far as possible, and then lock it in this position.

4 Remove the upper shroud and lower shrouds from the steering column as described in Chapter 11 Section 25.

5 Where fitted, remove the lower drivers side panel from the facia.

6 Undo the retaining screw and remove the footwell heater ducting from above the pedals **(see illustration)**.

7 Remove the switch assembly from the top of the steering column as described in Chapter 12 Section 5.

8 Disconnect the wiring connector from the

20.6 Remove the footwell air ducting

20.8a Disconnect the wiring plug…

20.8b …and release the wiring loom securing clips

20.9a Undo the earth wire securing nut …

20.9b …release the retaining clips …

20.9c …and remove the wiring loom bracket

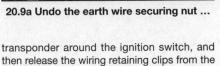

transponder around the ignition switch, and then release the wiring retaining clips from the steering column (see illustrations).

9 Follow the wiring loom down the length of the steering column, undo the earth cable retaining nut, release the retaining clips and remove the wiring loom from under the steering column, noting its fitted position (see illustrations).

10 Working in the footwell, release the plastic fixings and remove the cover from the base of the column (see illustration). On later models with manual transmission remove the crash bar (next to the clutch pedal).

11 Undo the clamp bolt from the steering column lower universal joint. Unscrew the clamp bolt and pull the universal joint from the steering gear pinion. Note that the pinion shaft has a cut-out to enable fitting of the clamp bolt, and the splined pinion shaft incorporates a flat making it impossible to assemble the joint to the shaft in the wrong position (see illustration). Discard the clamp bolt; a new one should be used on refitting.

12 Note that the inner and outer columns, and the intermediate shaft, are telescopic, to facilitate the reach adjustment. It is important to keep the splined sections of the inner steering column engaged with each other while the steering column is removed. If they become detached due to the outer column sections being separated, especially on a vehicle, which has completed a high mileage, it is possible that rattling noises may occur.

13 Support the steering column, then unscrew and remove the mounting bolts.

20.10 Remove the cover

Withdraw the steering column from inside the car (see illustrations).

Caution: Do not carry the steering column by suspending it from the universal joint or intermediate shaft, as this will damage the universal joint and steering column bushes. Also, do not bend the joints by more than 90°.

14 If necessary, remove the ignition switch/ steering column lock with reference to Section 21.

Inspection

15 The steering column is designed to collapse in the event of a front-end crash, to prevent the steering wheel injuring the driver. Before refitting the steering column, examine the column and mountings for signs of damage and deformation.

16 Check the inner column sections for

20.11 The column pinch bolt

signs of free play in the column bushes. If any damage or wear is found on the steering column bushes, the column must be renewed as an assembly.

17 The intermediate shaft is permanently attached to the inner column and cannot be renewed separately. Inspect the universal joints for excessive wear. If evident, the complete steering column must be renewed.

Refitting

18 Where removed, refit the ignition switch/ steering column lock/switch carrier with reference to Section 21.

19 Refit the steering column to the bulkhead bracket, making sure it located on the locating lug. Insert the mounting bolts, and tighten to the specified torque, starting with the upper bolts first and then the lower bolts.

20 Attach the universal joint on the steering gear pinion splines, making sure the road wheels and steering wheel are in the correct position. Insert the new clamp bolt, and tighten to the specified torque.

21 Reconnect the earth cable and tighten the retaining nut, then refit the wiring harness under the steering column.

22 Reconnect the wiring to the ignition switch transponder and secure the wiring in its retaining clips.

23 Refit the footrest trim and footwell vent under the steering column.

24 Refit the light switch assembly with reference to Chapter 12 Section 5.

25 Where removed refit the lower facia panel, as described in Chapter 11 Section 25.

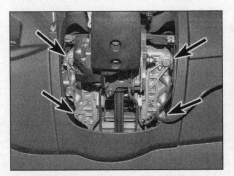

20.13a Undo the mounting bolts…

20.13b …and withdraw the steering column

26 Refit the lower and upper shrouds, as described in Chapter 11 Section 25.
27 Refit the steering wheel with reference to Section 19.
28 Reconnect the battery negative lead.

---

21 Ignition switch/lock cylinder – removal and refitting

Ignition switch

Removal

1 Disconnect the battery negative lead as described in Chapter 5A Section 3.
2 Remove the steering wheel as described in Section 19.
3 Unclip and remove the upper shroud from the steering column and then remove the lower shroud from the steering column, as described in Chapter 11 Section 25.
4 Insert the key and turn to the 'on' position. Release the retaining clip and disconnect the wiring plug from the rear of the ignition switch **(see illustration)**.
5 Pull back the label, then using a small screwdriver prise the plastic cover from the rear of the switch **(see illustration)**.
6 Using two small screwdrivers in the slots in the outer housing release the two retaining clips, and then withdraw the ignition electrical switch **(see illustrations)**.

Refitting

7 Refit the switch to the steering lock housing

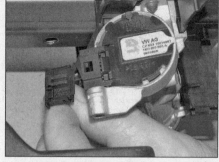
21.4 Disconnect the wiring plug

and press in until the two retaining clips engage. Making sure the slot in the rear of the switch is aligned correctly.
8 Refit the plastic cover to the rear of the switch and stick down the label.
9 Reconnect the wiring plug to the ignition switch.
10 Refit the upper and lower shrouds, and tighten the screws.
11 Refit the steering wheel with reference to Section 19.
12 Reconnect the battery negative lead.

Lock cylinder

Removal

13 Carry out the procedures as described in paragraphs 1 and 3.
14 Disconnect the wiring connector from the transponder around the ignition switch **(see illustration 20.8a)**.

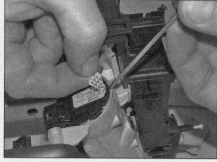
21.5 Peel back the label

15 Insert the ignition key and turn the lock cylinder to the drive/on position (which is 90° from off position).
16 Insert a piece of wire 1.2 mm in diameter in the drilling next to the ignition key, slide it in to release the locking lever, then withdraw the lock cylinder from the housing **(see illustrations)**. To make the piece of wire locate in the locking lever easier, file an angle on the end of the wire.

Refitting

17 Refit the lock cylinder with the ignition key in the Drive position, then remove the wire.
18 Reconnect the wiring connector to the transponder.
19 Refit the upper and lower shrouds, and tighten the screws.
20 Refit the steering wheel with reference to Section 19.
21 Reconnect the battery negative lead.

21.6a Insert two thin screwdrivers in the recesses...

21.6b ...to release the two securing clips...

21.6c ...and withdraw the electrical switch

21.16a Hole provided in the ignition switch

21.16b insert a thin rod through the hole ...

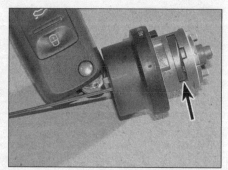
21.16c ...to release the locking lever

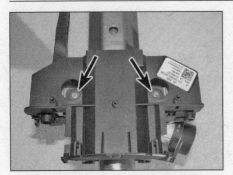

21.22 Steering lock housing shear bolts

Ignition switch/lock cylinder housing

Removal

22 The ignition switch/lock housing is integral with the switch assembly carrier, which is secured to the steering column with shear-head bolts **(see illustration)**.

23 To remove the housing first remove the steering column as described in Section 20.

24 The switch/lock carrier is secured by shear-head bolts, and the heads are broken off in the tightening procedure. To remove the old bolts, either drill them out, or use a sharp cold chisel to cut off their heads or turn them anti-clockwise. Withdraw the carrier from the steering column.

25 When refitting locate the switch/lock carrier on the steering column, and insert the new shear-head bolts. Tighten the bolts until their heads break off.

26 The remaining procedure is a reversal of removal.

---

22 Steering gear assembly – removal, overhaul and refitting

Note: *As the subframe must be lowered during this procedure, VAG/Seat subframe locating pins (T10096) or similar are required to ensure correct front wheel alignment. New subframe mounting bolts, track rod balljoint nuts, steering gear retaining bolts, and an intermediate shaft universal joint clamp bolt will be required on refitting.*

Note: *Several versions of the steering rack are fitted to the Leon range. Where an earlier version is fitted replacement steering racks may be the later version. All the versions are compatible with all vehicles, but a modified wiring loom will be required where a later generation steering rack is fitted to an earlier vehicle.*

Removal

1 Disconnect the battery negative lead (refer to Chapter 5A Section 3).

2 Apply the handbrake, then jack up the front of the vehicle and support it on axle stands positioned on the underbody, leaving the subframe free (see *Jacking and vehicle support*). Position the steering straight-ahead, and then remove both front roadwheels. Also remove the engine compartment undertray.

3 Inside the vehicle, undo the screws and remove the plastic cover for access to the universal joint connecting the steering inner column to the steering gear pinion (see Section 20). Unscrew and remove the clamp bolt, and pull the universal joint from the pinion splines. **Note:** *The steering gear pinion incorporates a cut-out for the clamp bolt, and therefore the joint can only be fitted in one position. Discard the clamp bolt; a new one should be used on refitting.*

4 Remove the battery and battery tray with reference to Chapter 5A Section 3.

5 Unclip the plastic cover from the wiring loom, which runs along the left-hand chassis leg under the battery tray. Trace the wiring loom from the steering rack, undo the retaining nuts and disconnect the wiring connections so that the wiring loom can be removed with the steering rack assembly **(see illustrations)**.

6 Disconnect the track rod ends from the steering arms on the front wheel bearing housings, as described in Section 24.

7 Working on each side in turn, disconnect the anti-roll bar links to the struts, as described in Section 8.

8 Working on each side in turn, unscrew the front suspension lower balljoint-to-lower arm retaining nuts, then lever the lower arms down to release the balljoint studs, as described in Section 6.

9 Unscrew and remove the engine/transmission rear mounting bolts from the transmission **(see illustration)**.

10 Unbolt the exhaust system mounting

22.5a Unclip the wiring cover …

22.5b …disconnect the wiring from the fusebox…

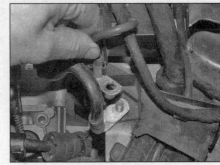

22.5c …disconnect the earth cable …

22.5d …disconnect the wiring plug connector …

22.5e …and lift the wiring loom out of position

22.9 Undo the lower rear mounting bracket bolts

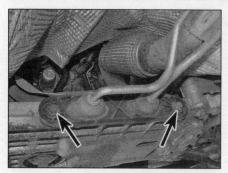

22.10 Undo the exhaust mounting bracket bolts

22.11a Mark the position of the front...

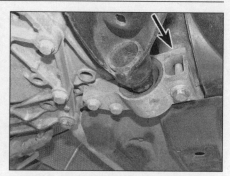

22.11b ...and rear subframe mounting positions

from the subframe **(see illustration)**. Where applicable unbolt the exhaust system heat shield from the subframe.

11 Support the subframe with a trolley jack and block of wood. If not using the special VAG/Seat locating pins T10096, accurately mark the position of the subframe to ensure correct wheel alignment **(see illustrations)**.

12 Unscrew the mounting bolts and slightly lower the subframe **(see illustration)**, taking care not to damage the electrical wiring, as it is being lowered. Where available, fit the VAG/Seat locating pins to facilitate refitting.

13 Lower the subframe together with the steering gear to the floor, then undo the retaining screws and remove the heat shield from over the steering gear.

14 Disconnect all wiring from the steering gear assembly **(see illustrations)**. Release the wiring loom from the clips along the steering rack and subframe, noting its fitted position.

Caution: Do not touch the wiring terminals on the electronic control unit, as a static electricity discharge may damage the internal components.

15 Unbolt the steering gear from the subframe **(see illustration)**, and carefully place it on the floor taking care not to damage the electronic control unit.

Overhaul

16 Examine the steering gear assembly for signs of wear or damage, and check that the rack moves freely throughout the full length of its travel, with no signs of roughness or excessive free play between the steering gear pinion and rack.

17 It is not possible to overhaul the steering gear assembly housing components, and if it is faulty, the assembly must be renewed. The only components that can be renewed individually are the steering gear gaiters, the

track rod end balljoints and the track rods, as described later in this Chapter.

Refitting

18 Refitting is a reversal of removal, noting the following points:

a) *Align the subframe with the marks made on removal, or use the special VAG/Seat location pins before tightening the mounting bolts.*

b) *Make sure that all mounting bolts are tightened to the specified torque.*

c) *Make sure wiring connections are secure and that the wiring loom is routed correctly.*

d) *Have the front wheel alignment checked at the earliest opportunity.*

e) *On models with ESP, the steering angle sensor basic settings must be set by a Seat dealer (or suitably equipped garage) using specialist diagnostic equipment.*

f) *If a new steering gear has been fitted, it must also be adapted to the vehicle by a Seat dealer (or suitably equipped garage) using suitable diagnostic equipment.*

23 Steering gear gaiters and track rods – renewal

Steering gear rubber gaiters

1 Remove the track rod end balljoint as described in Section 24. Also, unscrew the locking nut from the track rod arm, after noting its position.

22.12 Subframe front and rear mounting bolts (one side shown)

22.14a Disconnect the wiring connector on the rear ...

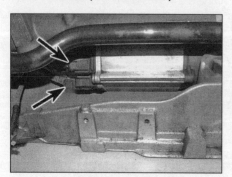

22.14b ...and the front of the steering electric motor ...

22.14c ...then undo the wiring loom securing bolt

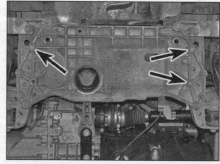

22.15 Steering rack mounting bolts

2 Wipe clean the rubber gaiter to prevent entry of dirt or moisture. Note the fitted position of the gaiter on the track rod, then release the retaining clips and slide the gaiter off the steering gear housing and track rod (**see illustration**).

3 Wipe clean the track rod and the steering gear housing, and then apply a film of suitable grease to the surface of the rack. To do this, turn the steering wheel as necessary to fully extend the rack from the housing, then reposition it in its central position.

4 Carefully slide the new gaiter onto the track rod, and locate it on the steering gear housing. Position the gaiter as previously noted on removal, making sure that it is not twisted, then lift the outer sealing lip of the gaiter to equalise air pressure within the gaiter.

5 Secure the gaiter in position with new retaining clips. Where crimped-type clips are used, pull the clip as tight as possible, and locate the hooks in their slots. Remove any slack in the clip by carefully compressing the raised section. In the absence of the special crimping tool, a pair of side-cutters may be used, taking care not to cut the clip.

6 Screw on the locking nut, then refit the track rod end balljoint as described in Section 24.

Track rods

7 Remove the relevant steering gear rubber gaiter as described earlier. If there is insufficient working room with the steering gear mounted in the car, remove it as described in Section 22 and hold it in a vice while renewing the track rod.

8 Hold the steering rack stationary with one spanner on the flats provided, then loosen the balljoint nut with another spanner. Fully unscrew the nut and remove the track rod from the rack.

9 Locate the new track rod on the end of the steering rack and screw on the nut. Hold the rack stationary with one spanner and tighten the balljoint nut to the specified torque. A crow's foot adapter may be required since the track rod prevents access with a socket, and care must be taken to apply the correct torque in this situation.

10 Refit the steering gear or rubber gaiter with reference to the earlier paragraphs or Section 22. On completion check and, if necessary, adjust the front wheel alignment as described in Section 25.

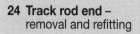

24 Track rod end – removed and refitting

Note: *A new balljoint retaining nut will be required on refitting.*

Removal

1 Apply the handbrake, then jack up the front of the vehicle and support it on axle stands (see *Jacking and vehicle support*). Remove the relevant roadwheel.

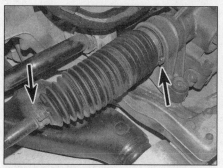

23.2 Steering rack gaiter securing clips

24.3 Slacken the lock nut

24.4a Using an Allen key to hold the ball joint shank while loosening the nut

24.4b Release the track rod ball joint

2 If the track rod end is to be re-used, mark its position in relation to the track rod to facilitate refitting.

3 Unscrew the track rod end locknut by a quarter of a turn (**see illustration**). Do not move the locknut from this position, as it will serve as a handy reference mark on refitting.

4 Loosen and remove the nut securing the track rod end balljoint to the wheel bearing housing, and release the balljoint tapered shank using a universal balljoint separator. Note that the balljoint shank has a hexagon hole – hold the shank with an Allen key while loosening the nut (**see illustrations**).

5 Counting the exact number of turns necessary to do so, unscrew the track rod end from the track rod (**see illustration**).

6 Carefully clean the balljoint and the threads. Renew the balljoint if its movement is sloppy or too stiff, if excessively worn, or if damaged in any way; carefully check the stud taper and

threads. If the balljoint gaiter is damaged, the complete balljoint assembly must be renewed; it is not possible to obtain the gaiter separately.

Refitting

7 Screw the track rod end onto the track rod by the number of turns noted on removal. This should bring the track rod end to within a quarter of a turn of the locknut, with the alignment marks that were made on removal (if applicable) lined up. Tighten the locknut.

8 Refit the balljoint shank to the steering arm on the wheel bearing housing, then fit a new retaining nut and tighten it to the specified torque. Hold the shank with an Allen key if necessary.

9 Refit the roadwheel, then lower the car to the ground and tighten the roadwheel bolts to the specified torque.

10 Check and, if necessary, adjust the front wheel toe setting as described in Section 25.

24.4c Use a ball joint separator to release the track rod ball joint if it is a tight fit

24.5 Unscrewing the track rod end from the track rod

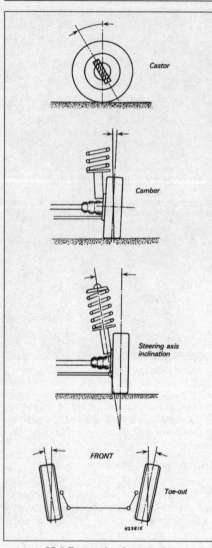

25.1 Front wheel geometry

25 Wheel alignment and steering angles – general information

Definitions

1 A car's steering and suspension geometry is defined in three basic settings – all angles are expressed in degrees; the steering axis is defined as an imaginary line drawn through the axis of the suspension strut, extended where necessary to contact the ground **(see illustration)**.

2 Camber is the angle between each roadwheel and a vertical line drawn through its centre and tyre contact patch, when viewed from the front or rear of the car. Positive camber is when the roadwheels are tilted outwards from the vertical at the top; negative camber is when they are tilted inwards.

3 The front Camber angle is only adjustable by loosening the front suspension subframe mounting bolts and moving it slightly to one side. The camber angle can be checked using a camber checking gauge. This will also alter the Castor angle.

4 The rear Camber angle is only adjustable by loosening the upper transverse inner mounting nut, then turning the eccentric bolt (this can be turned left or right to a maximum range of 90°). The camber angle can be checked using a camber checking gauge. This will also alter the Castor angle.

5 Castor is the angle between the steering axis and a vertical line drawn through each roadwheel's centre and tyre contact patch, when viewed from the side of the car. Positive castor is when the steering axis is tilted so that it contacts the ground ahead of the vertical; negative castor is when it contacts the ground behind the vertical. Slight castor angle adjustment is possible by loosening the

front suspension subframe bolts and moving it slightly to one side. This also alters the Camber angle.

6 Castor is not easily adjustable, and is given for reference only; while it can be checked using a castor checking gauge, if the figure obtained is significantly different from that specified, the car must be taken for careful checking by a professional, as the fault can only be caused by wear or damage to the body or suspension components.

7 Toe is the difference, viewed from above, between lines drawn through the roadwheel centres and the car's centre-line. Toe-in is when the roadwheels point inwards, towards each other at the front, while toe-out is when they splay outwards from each other at the front.

8 The front wheel toe setting is adjusted by screwing the track rod(s) in/out of the outer balljoint(s) to alter the effective length of the track rod assembly.

9 Rear wheel toe setting is adjusted by loosening the lower transverse inner mounting nut, and then turning the eccentric bolt (this can be turned left or right to a maximum range of 90°).

10 If the figures obtained are significantly different from that specified, the car must be taken for careful checking by a professional, as the fault can only be caused by wear or damage to the body or suspension components.

Checking and adjustment

11 Due to the special measuring equipment necessary to check the wheel alignment, and the skill required to use it properly, the checking and adjustment of these settings is best left to a Seat dealer or similar expert. Note that most tyre-fitting centres now possess sophisticated checking equipment.

Chapter 11
Bodywork and fittings

Contents

Degrees of difficulty

Easy, suitable for novice with little experience	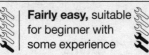	**Fairly easy,** suitable for beginner with some experience	**Fairly difficult,** suitable for competent DIY mechanic	**Difficult,** suitable for experienced DIY mechanic	**Very difficult,** suitable for expert DIY or professional

Specifications

Torque wrench settings	Nm	lbf ft
Bonnet hinge nuts	22	16
Door hinges:		
Stage 1*	20	15
Stage 2	Angle-tighten a further 90°	
Upper hinge nut	14	10
Door lock	20	15
Crossmember bolts/nuts	20	15
Front window glass clamp bolts.	8	6
Front seat belt adjuster bolt	20	15
Front seat belt lower mounting bolt (microencapsulated)*.	40	30
Front seat belt stalk to seat frame bolt (microencapsulated)*	20	15
Front seat mounting retaining bolts	40	30
Inertia reel mounting bolts	40	30
Seat belt and stalk anchorage bolts.	40	30
Tailgate hinge bolts.	10	7

*Renew the bolts

1 General Information

1 The body shell is made of pressed-steel sections, with many of the structural panels being made from high tensile steel. Most components are welded together, either by traditional spot welding or where increased rigidity is required by continuous laser welding. Structural adhesives are used to bond some body panels.

2 Extensive use is made of plastic materials, mainly in the interior, but also in exterior components. The front and rear bumpers, and front grille, are injection-moulded from a synthetic material that is very strong and yet light. Plastic components such as wheel arch liners are fitted to the underside of the vehicle, to improve the body's resistance to corrosion.

2 Maintenance – bodywork and underframe

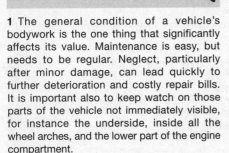

1 The general condition of a vehicle's bodywork is the one thing that significantly affects its value. Maintenance is easy, but needs to be regular. Neglect, particularly after minor damage, can lead quickly to further deterioration and costly repair bills. It is important also to keep watch on those parts of the vehicle not immediately visible, for instance the underside, inside all the wheel arches, and the lower part of the engine compartment.

2 The basic maintenance routine for the bodywork is washing – preferably with a lot of water, from a hose. This will remove all the loose solids, which may have stuck to the vehicle. It is important to flush these off in such a way as to prevent grit from scratching the finish. The wheel arches and underframe need washing in the same way, to remove any accumulated mud, which will retain moisture and tend to encourage rust. Paradoxically enough, the best time to clean the underframe and wheel arches is in wet weather, when the mud is thoroughly wet and soft. In very wet weather, the underframe is usually cleaned of large accumulations automatically, and this is a good time for inspection.

3 Periodically, except on vehicles with a wax-based underbody protective coating, it is a good idea to have the whole of the underframe of the vehicle steam-cleaned, engine compartment included, so that a thorough inspection can be carried out to see what minor repairs and renovations are necessary. Steam cleaning is available at many garages, and is necessary for the removal of the accumulation of oily grime, which sometimes is allowed to become thick in certain areas. If steam-cleaning facilities are not available, there are some excellent grease solvents available which can be brush-applied; the dirt can then be simply hosed off. Note that these methods should not be used on vehicles with wax-based underbody protective coating, or the coating will be removed. Such vehicles should be inspected annually, preferably just prior to Winter, when the underbody should be washed down, and any damage to the wax coating repaired. Ideally, a completely fresh coat should be applied. It would also be worth considering the use of such wax-based protection for injection into door panels, sills, box sections, etc, as an additional safeguard against rust damage, where such protection is not provided by the vehicle manufacturer.

4 After washing paintwork, wipe off with a chamois leather to give an unspotted clear finish. A coat of clear protective wax polish will give added protection against chemical pollutants in the air. If the paintwork sheen has dulled or oxidised, use a cleaner/polisher combination to restore the brilliance of the shine. This requires a little effort, but such dulling is usually caused because regular washing has been neglected. Care needs to be taken with metallic paintwork, as special non-abrasive cleaner/polisher is required to avoid damage to the finish. Always check that the door and ventilator opening drain holes and pipes are completely clear, so that water can be drained out. Brightwork should be treated in the same way as paintwork. Windscreens and windows can be kept clear of the smeary film that often appears, by the use of proprietary glass cleaner. Never use any form of wax or other body or chromium polish on glass.

3 Maintenance – upholstery and carpets

1 Mats and carpets should be brushed or vacuum-cleaned regularly, to keep them free of grit. If they are badly stained, remove them from the vehicle for scrubbing or sponging, and make quite sure they are dry before refitting. Seats and interior trim panels can be kept clean by wiping with a damp cloth. If they do become stained (which can be more apparent on light-coloured upholstery), use a little liquid detergent and a soft nail brush to scour the grime out of the grain of the material. Do not forget to keep the headlining clean in the same way as the upholstery. When using liquid cleaners inside the vehicle, do not over-wet the surfaces being cleaned. Excessive damp could get into the seams and padded interior, causing stains, offensive odours or even rot.

2 If the inside of the vehicle gets wet accidentally, it is worthwhile taking some trouble to dry it out properly, particularly where carpets are involved. Do not leave oil or electric heaters inside the vehicle for this purpose.

4 Minor body damage – repair

Scratches

1 If the scratch is very superficial, and does not penetrate to the metal of the bodywork, repair is very simple. Lightly rub the area of the scratch with a paintwork renovator, or a very fine cutting paste, to remove loose paint from the scratch, and to clear the surrounding bodywork of wax polish. Rinse the area with clean water.

2 Apply touch-up paint to the scratch using a fine paintbrush; continue to apply fine layers of paint until the surface of the paint in the scratch is level with the surrounding paintwork. Allow the new paint at least two weeks to harden, and then blend it into the surrounding paintwork by rubbing the scratch area with a paintwork renovator or a very fine cutting paste. Finally, apply wax polish.

3 Where the scratch has penetrated right through to the metal of the bodywork, causing the metal to rust, a different repair technique is required. Remove any loose rust from the bottom of the scratch with a penknife, and then apply rust-inhibiting paint to prevent the formation of rust in the future. Using a rubber or nylon applicator, fill the scratch with bodystopper paste. If required, this paste can be mixed with cellulose thinners to provide a very thin paste that is ideal for filling narrow scratches. Before the stopper-paste in the scratch hardens, wrap a piece of smooth cotton rag around the top of a finger. Dip the finger in cellulose thinners, and quickly sweep it across the surface of the stopper-paste in the scratch; this will ensure that the surface of the stopper-paste is slightly hollowed. The scratch can now be painted over as described earlier in this Section.

Dents

4 When deep denting of the vehicle's bodywork has taken place, the first task is to pull the dent out, until the affected bodywork almost attains its original shape. There is little point in trying to restore the original shape completely, as the metal in the damaged area will have stretched on impact, and cannot be reshaped fully to its original contour. It is better to bring the level of the dent up to a point, which is about 3 mm below the level of the surrounding bodywork. In cases where the dent is very shallow anyway, it is not worth trying to pull it out at all. If the underside of the dent is accessible, it can be hammered out gently from behind, using a mallet with a wooden or plastic head. Whilst doing this, hold a suitable block of wood firmly against the outside of the panel, to absorb the impact from the hammer blows and thus prevent a large area of the bodywork from being 'belled-out'.

5 Should the dent be in a section of the

bodywork that has a double skin, or some other factor making it inaccessible from behind, a different technique is called for. Drill several small holes through the metal inside the area – particularly in the deeper section. Then screw long self-tapping screws into the holes, just sufficiently for them to gain a good purchase in the metal. Now the dent can be pulled out by pulling on the protruding heads of the screws with a pair of pliers.

6 The next stage of the repair is the removal of the paint from the damaged area, and from an inch or so of the surrounding 'sound' bodywork. This is accomplished most easily by using a wire brush or abrasive pad on a power drill, although it can be done just as effectively by hand, using sheets of abrasive paper. To complete the preparation for filling, score the surface of the bare metal with a screwdriver or the tang of a file, or alternatively, drill small holes in the affected area. This will provide a really good 'key' for the filler paste.

7 To complete the repair, see the Section on filling and respraying.

Rust holes or gashes

8 Remove all paint from the affected area, and from an inch or so of the surrounding 'sound' bodywork, using an abrasive pad or a wire brush on a power drill. If these are not available, a few sheets of abrasive paper will do the job most effectively. With the paint removed, you will be able to judge the severity of the corrosion, and therefore decide whether to renew the whole panel (if this is possible) or to repair the affected area. New body panels are not as expensive as most people think, and it is often quicker and more satisfactory to fit a new panel than to attempt to repair large areas of corrosion.

9 Remove all fittings from the affected area, except those, which will act as a guide to the original shape of the damaged bodywork (e.g. headlight shells etc). Then, using tin snips or a hacksaw blade, remove all loose metal and any other metal badly affected by corrosion. Hammer the edges of the hole inwards, in order to create a slight depression for the filler paste.

10 Wire-brush the affected area to remove the powdery rust from the surface of the remaining metal. Paint the affected area with rust-inhibiting paint, if the back of the rusted area is accessible, treat this also.

11 Before filling can take place, it will be necessary to block the hole in some way. This can be achieved by the use of aluminium or plastic mesh, or aluminium tape.

12 Aluminium or plastic mesh, or glassfibre matting, is probably the best material to use for a large hole. Cut a piece to the approximate size and shape of the hole to be filled, then position it in the hole so that its edges are below the level of the surrounding bodywork. It can be retained in position by several blobs of filler paste around its periphery.

13 Aluminium tape should be used for small or very narrow holes. Pull a piece off the roll, trim it to the approximate size and shape required, then pull off the backing paper (if used) and stick the tape over the hole; it can be overlapped if the thickness of one piece is insufficient. Burnish down the edges of the tape with the handle of a screwdriver or similar, to ensure that the tape is securely attached to the metal underneath.

Filling and respraying

14 Before using this Section, see the Sections on dent, deep scratch, rust holes and gash repairs.

15 Many types of bodyfiller are available, but generally speaking, those proprietary kits, which contain a tin of filler paste and a tube of resin hardener, are best for this type of repair. A wide, flexible plastic or nylon applicator will be found invaluable for imparting a smooth and well-contoured finish to the surface of the filler.

16 Mix up a little filler on a clean piece of card or board – measure the hardener carefully (follow the maker's instructions on the pack), otherwise the filler will set too rapidly or too slowly. Using the applicator, apply the filler paste to the prepared area; draw the applicator across the surface of the filler to achieve the correct contour and to level the surface. As soon as a contour that approximates to the correct one is achieved, stop working the paste – if you carry on too long, the paste will become sticky and begin to 'pick-up' on the applicator. Continue to add thin layers of filler paste at 20-minute intervals, until the level of the filler is just proud of the surrounding bodywork.

17 Once the filler has hardened, the excess can be removed using a metal plane or file. From then on, progressively finer grades of abrasive paper should be used, starting with a 40-grade production paper, and finishing with a 400-grade wet-and-dry paper. Always wrap the abrasive paper around a flat rubber, cork, or wooden block – otherwise the surface of the filler will not be completely flat. During the smoothing of the filler surface, the wet-and-dry paper should be periodically rinsed in water. This will ensure that a very smooth finish is imparted to the filler at the final stage.

18 At this stage, the dent should be surrounded by a ring of bare metal, which in turn should be encircled by the finely 'feathered' edge of the good paintwork. Rinse the repair area with clean water, until all of the dust produced by the rubbing-down operation has gone.

19 Spray the whole area with a light coat of primer – this will show up any imperfections in the surface of the filler. Repair these imperfections with fresh filler paste or bodystopper, and once more smooth the surface with abrasive paper. Repeat this spray-and-repair procedure until you are satisfied that the surface of the filler, and the feathered edge of the paintwork, are perfect.

Clean the repair area with clean water, and allow to dry fully.

20 The repair area is now ready for final spraying. Paint spraying must be carried out in a warm, dry, windless and dust-free atmosphere. This condition can be created artificially if you have access to a large indoor working area, but if you are forced to work in the open, you will have to pick your day very carefully. If you are working indoors, dousing the floor in the work area with water will help to settle the dust, which would otherwise be in the atmosphere. If the repair area is confined to one body panel, mask off the surrounding panels; this will help to minimise the effects of a slight mis-match in paint colours. Bodywork fittings (e.g. chrome strips, door handles etc) will also need to be masked off. Use genuine masking tape, and several thicknesses of newspaper, for the masking operations.

21 Before commencing to spray, agitate the aerosol can thoroughly, and then spray a test area (an old tin, or similar) until the technique is mastered. Cover the repair area with a thick coat of primer; the thickness should be built up using several thin layers of paint, rather than one thick one. Using 400-grade wet-and-dry paper, rub down the surface of the primer until it is really smooth. While doing this, the work area should be thoroughly doused with water, and the wet-and-dry paper periodically rinsed in water. Allow to dry before spraying on more paint.

22 Spray on the top coat, again building up the thickness by using several thin layers of paint. Start spraying at one edge of the repair area, and then, using a side-to-side motion, work until the whole repair area and about 2 inches of the surrounding original paintwork is covered. Remove all masking material 10 to 15 minutes after spraying on the final coat of paint.

23 Allow the new paint at least two weeks to harden, then, using a paintwork renovator, or a very fine cutting paste, blend the edges of the paint into the existing paintwork. Finally, apply wax polish.

Plastic components

24 With the use of more and more plastic body components by the vehicle manufacturers (e.g. bumpers. spoilers, and in some cases major body panels), rectification of more serious damage to such items has become a matter of either entrusting repair work to a specialist in this field, or renewing complete components. Repair of such damage by the DIY owner is not really feasible, owing to the cost of the equipment and materials required for effecting such repairs. The basic technique involves making a groove along the line of the crack in the plastic, using a rotary burr in a power drill. The damaged part is then welded back together, using a hot-air gun to heat up and fuse a plastic filler rod into the groove. Any excess plastic is then removed, and the area rubbed down to a smooth finish. It is important that a filler rod of the correct plastic is used, as body components can be made of a variety of different types (e.g. polycarbonate, ABS, polypropylene).

25 Damage of a less serious nature (abrasions, minor cracks etc) can be repaired by the DIY owner using a two-part epoxy filler repair material. Once mixed in equal proportions, this is used in similar fashion to the bodywork filler used on metal panels. The filler is usually cured in twenty to thirty minutes, ready for sanding and painting.

26 If the owner is renewing a complete component himself, or if he has repaired it with epoxy filler, he will be left with the problem of finding a suitable paint for finishing which is compatible with the type of plastic used. At one time, the use of a universal paint was not possible, owing to the complex range of plastics encountered in body component applications. Standard paints, generally

6.2 Remove the hidden fixing

speaking, will not bond to plastic or rubber satisfactorily. However, it is now possible to obtain a plastic body parts finishing kit, which consists of a pre-primer treatment, a primer and coloured top coat. Full instructions are normally supplied with a kit, but basically, the method of use is to first apply the pre-primer to the component concerned, and allow it to dry for up to 30 minutes. Then the primer is applied, and left to dry for about an hour before finally applying the special-coloured top coat. The result is a correctly coloured component, where the paint will flex with the plastic or rubber, a property that standard paint does not normally possess.

5 Major body damage – repair

1 Where serious damage has occurred, or large areas need renewal due to neglect, it means that complete new panels will need welding-in, and this is best left to professionals. If the damage is due to impact, it will also be necessary to check completely the alignment of the body shell, and this can only be carried out accurately by a VW dealer using special jigs. If the body is left misaligned, it is primarily dangerous, as the car will not handle properly, and secondly, uneven stresses will be imposed on the steering, suspension and possibly transmission, causing abnormal wear, or complete failure, particularly to such items as the tyres.

6 Front bumper cover – removal and refitting

Note: *Depending on the year of production and the trim level slight changes to the removal and refitting procedures may be necessary.*

Removal

1 Apply the handbrake, then jack up the front of the vehicle and support it on axle stands (see *Jacking and vehicle support* Section 5). For ease of access, remove both front road wheels.

2 Remove the fixings from the front edges of both wing liners **(see illustrations)**.

3 At the base of the bumper cover remove the lower fixings and then remove the outer upper fixings. Some early models (up to 09/2006) may have additional fixings at each corner of the lower grille **(see illustrations)**.

4 Release the trailing edge of the bumper cover from the front wing – at both sides **(see illustration)**.

5 Unclip the trailing edges of the bumper from the front wing mounting bracket and then pull the bumper forward. Remove the remaining central upper fixings and then with the aid of an assistant pull the bumper cover forward. Where fitted disconnect the headlight washers and front foglight wiring plugs as the cover is removed **(see illustrations)**.

6 If required the upper and lower grilles can

6.3a Remove the lower fixings...

6.3b ...the upper fixings...

6.3c ...and where fitted the lower grille fixings

6.4 Release the bumper cover from the front wings

6.5a Disconnect the wiring plugs...

6.5b ...and remove the bumper cover

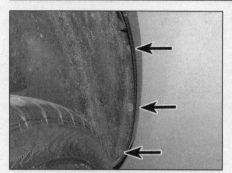

7.3a Remove the liner fixings

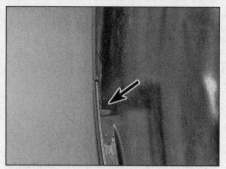

7.3b Remove the partially hidden upper fixing

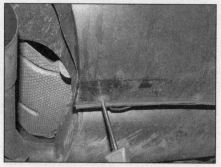

7.3c Remove the lower fixings

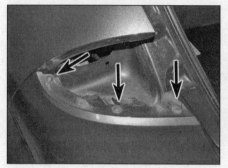

7.4 Remove the upper retaining screws

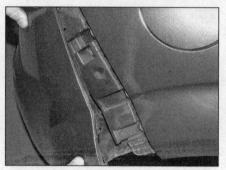

7.5a Release the bumper cover from the rear wing

7.5b Disconnect the number plate wiring plug

now be removed from the cover. Note that removal of the upper grille is possible with the bumper cover still on the car. The lower grille can also be removed, but if the bumper cover is still on the vehicle the upper grille must be removed first to gain access to the lower grille retaining clips.

Refitting

7 Refitting is a reverse of the removal procedure, ensuring that the bumper ends engage correctly with the locating clips on the edge of the front wing panels.

7 Rear bumper cover – removal and refitting

Note: *Depending on the model, it is possible that slight changes to the removal and refitting procedures may be necessary.*

Removal

1 To improve access, chock the front wheels, and then jack up the rear of the vehicle and support it on axle stands (see *Jacking and vehicle support* Section 5).
2 Remove the rear light clusters as described in Chapter 12 Section 8.
3 Remove the fixings securing the wheel arch liners to the bumper ends and then remove the lower fixings **(see illustrations)**.

4 Remove the fixings from the upper edge of the cover **(see illustration)**.
5 With the help of an assistant, release the bumper cover from the rear wing panels at both sides. The cover is a tight fit and should be carefully worked free from the wing mounted support bracket. Withdraw the bumper from the rear of the vehicle and disconnect the wiring plug for the number plate lamp **(see illustrations)**. Where fitted disconnect the wiring from the parking sensors.

Refitting

6 Refitting is a reverse of the removal procedure, ensuring that the bumper ends engage correctly with the rear wing panels, as the bumper is refitted.

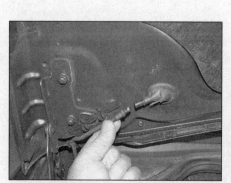

8.2a Disconnect and unclip the washer hose

8 Bonnet – removal, refitting and adjustment

Removal

1 Open the bonnet and using a pencil or felt tip pen, mark the outline of each bonnet hinge relative to the bonnet, to use as a guide on refitting.
2 Unclip the washer hose and on models fitted with heated washer jets disconnect the wiring plug. Remove the bonnet soundproofing by releasing the clips **(see illustrations)**.
3 Work around the bonnet and release the

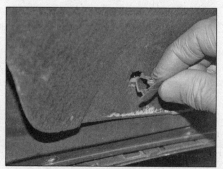

8.2b Remove the clips from the soundproofing

8.3a Squeeze the clips with pliers to release them

8.3b Lift off the panel...

8.3c ...disconnect the hose and remove the panel

8.4 Remove the upper bolts

8.5 Mark the position of the hinge

wear and free play at the pivots, and if necessary renew. Each hinge is secured to the body by bolts (see illustration). Mark the position of the hinge on the body then undo the retaining bolts and remove the hinge. On refitting, align the new hinge with the marks and tighten the retaining bolts.

Refitting and adjustment

6 With the aid of an assistant, offer up the bonnet and loosely fit the retaining bolts. Align the hinges with the marks made on removal, and then tighten the retaining bolts/nuts securely.

7 Refit the washer hose and wiring in the reverse order of removal.

8 Close the bonnet, and check for alignment with the adjacent panels. If necessary, unscrew the hinge bolts and re-align the bonnet. Once the bonnet is correctly aligned, tighten the hinge bolts. Check that the bonnet fastens and releases satisfactorily.

upper moulding by releasing the clips and screws (see illustrations).

4 Undo the under bonnet retaining nuts and then remove the upper retaining bolts (see

illustration). With the aid of an assistant carefully lift the bonnet clear. Store the bonnet out of the way in a safe place.

5 Inspect the bonnet hinges for signs of

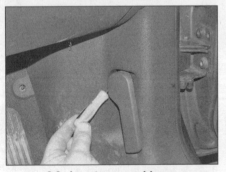

9.3a Insert a screwdriver...

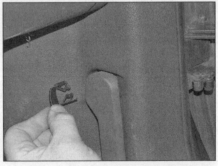

9.3b ...to release the securing clip

9.4a Undo the plastic retaining nut...

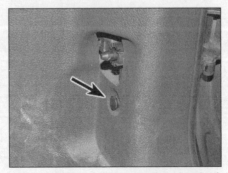

9.4b ...and remove the screw/clip ('scrivet)

9 Bonnet release cable – removal and refitting

Removal

1 The bonnet release cable is in two sections, with a coupling located over the right-hand headlight. To remove the shorter front section cable, remove the bonnet lock as described in Section 10.

2 To remove the rear section, first release the coupling from over the right-hand headlight, as described in Section 10, paragraph 4

3 Working inside the vehicle, locate the release lever and pull it out slightly, then insert a small screwdriver into the gap between the release lever and its securing clip. Let the lever return to its original position, then release the clip with a screwdriver (note that the clip may fall behind the trim) (see illustrations).

4 Remove the plastic screw and then remove the 'scrivet' (see illustrations). Prise up the sill trim and remove it.

9.5 Unscrew the release handle mounting bracket screws

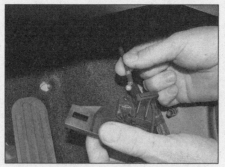

9.6 Unclip the release cable from the lever

9.11 Slide the securing clip into the handle before refitting

5 Undo the two retaining screws and withdraw the bonnet release lever from the bottom of the A-pillar **(see illustration)**.

6 Release the outer cable from the mounting bracket, and then detach the inner cable from the lever **(see illustration)**.

7 Working along the length of the cable and noting its correct routing, free it from any retaining clips/ties and release the cable sealing grommet from the bulkhead.

8 Tie a length of string to the end of the cable inside the vehicle, and then withdraw the cable through into the engine compartment.

9 Once the cable is free, untie the string and leave it in position in the vehicle; the string can then be used to draw the new cable back into position.

Refitting

10 Tie the inner end of the string to the end of the cable, and then use the string to draw the bonnet release cable back from the engine compartment. Once the cable is through, untie the string.

11 Refitting is a reversal of removal. **Note:** *Before refitting the bonnet release lever, fit the securing clip back into the lever first* **(see illustration)** *and then push the lever back into place.*

12 Ensure the rubber grommet in the bulkhead is fitted correctly, and the cable is correctly routed and secured to all the relevant retaining clips.

13 Before closing the bonnet, check the operation of the release lever and cable.

10 Bonnet lock – removal and refitting

Removal

1 Open the bonnet and disconnect the wiring for the contact micro switch at the connector located by the right-hand headlight **(see illustration)**.

2 Disconnect the bonnet cable at the join adapter located by the right-hand headlight. Unclip the cover and release the two parts of the cable **(see illustrations)**. This will allow the cable to be disconnected from the bonnet lock.

3 Note the location of the mounting bolts. Unscrew the bolts and remove them **(see illustration)**.

10.1 Disconnecting the bonnet lock wiring connector

4 Prise the lock from the crossmember, and withdraw the cable and wiring from the front panel, release them from any retaining clips. If required lift the cover and unhook the release cable end fitting from the lock lever.

Refitting

5 Refitting is a reversal of removal. Check that the bonnet fastens and releases satisfactorily. If adjustment is necessary, loosen the bonnet lock retaining bolts, and adjust the position of the lock to suit. Finally, tighten the bolts.

11 Door – removal, refitting and adjustment

Note: *The hinge bolts must always be renewed if loosened.*

10.2a Unclip the cable from the slam panel

10.2b Open the holder and...

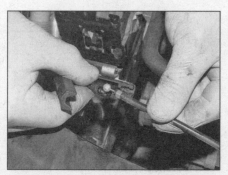

10.2c ...disconnect the two parts of the cable

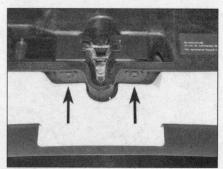

10.3 Remove the bolts

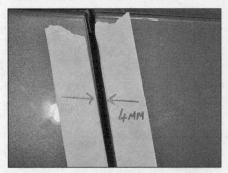

11.1 Make alignment marks and measure the door gaps

11.3 Support the door with a jack

11.4 Upper door hinge nut

11.5 Remove the lower hinge bolts

11.12 Door lock striker mounted on the C-pillar

Removal

1 Before removing the door, measure the door gaps and mark them as required **(see illustration)**. Make alignment marks on both hinges.

2 Open the door then disconnect the wiring at the A- or B-pillar as described in Section 12.

3 Support the door on a jack – protect the door by covering the head of the jack with card or a clean rag **(see illustration)**.

4 With an assistant supporting the door, remove the nut from the upper hinge bolt **(see illustration)**.

5 Slacken the lower hinge bolts and then with an assistant supporting the door fully remove the bolts. Lift the door of the upper hinge bolt and with the aid of an assistant remove it **(see illustration)**.

6 Examine the hinges for signs of wear or damage. If renewal is necessary, the upper hinge can only be removed (or adjusted) after the facia panel is removed Section 27. The lower hinge can be removed after the kick panel is removed Section 25. Before removing the hinges, accurately mark their position to ensure correct refitting.

Refitting

7 Where renewed, fit the hinges and tighten the bolts to the specified torque.

8 With the aid of an assistant, offer up the door to the vehicle and locate it on the guide bolt. Fit the new hinge bolts and tighten to the specified torque.

9 Reconnect the wiring plug and secure with the locking lever, then refit the rubber bellows.

Adjustment

10 Close the door and check the door alignment with the surrounding body panels. There must be an even gap all around, and the door must be level with the surrounding body panels.

11 Adjustment can only be made at the A-pillar (or B-pillar). There is no adjustment possible at the door itself.

12 Check that the striker enters the door lock centrally as the door is closed, and if necessary adjust the position of the striker by loosening its mounting bolts **(see illustration)**.

12 Door inner trim panel – removal and refitting

Note: *The door trim panels are removed with the window regulator and the power window motor. On the front doors the door lock is also removed at the same time.*

Removal

1 Switch off the ignition, before disconnecting any wiring connectors.

Front door

2 Remove the upper trim panel and then position the window glass, so that the glass retaining clamp bolts are visible. Slacken the clamp bolts, but do not remove them. Release the window glass and tape it up in the fully closed position **(see illustrations)**.

3 At the A-pillar, unclip the gaiter from the wiring plug, release the connector from

12.2a Prise up and remove the trim

12.2b Slacken the clamp bolts

12.2c Tape up the door glass in the closed position

12.3a Peel back the gaiter…

12.3b …and release the wiring plug from the A-pillar

12.3c Release the locking clip, separate the plug…

the pillar, pull up the locking tabs and separate the two halves of the connector. Push the wiring plug into the door frame (see illustrations).

4 Remove the door handle and release the lock fixings as as described in Section 13.

5 Unclip the door grab handle cover using a trim tool and then remove the speaker grille (see illustrations).

6 Work round the panel and remove the fixings from inside of the pull handle, the speaker and the upper and lower regulator. Note that the window regulator lower bolts do not require complete removal as the regulator slots into the door frame.

7 The panel is hooked in place and can be removed by lifting the panel up and moving it toward the front of the vehicle. At the same time the window regulator must be worked free from the inside of the door frame. Fully release the panel so that the lock can be manoeuvred out from the door. Whilst supporting the panel, unclip the wiring loom from the retaining clip (see illustrations).

Rear doors

8 On vehicles fitted with manual rear windows release the winder handle by sliding the locking clip away from the handle. Release the handle from the shaft.

9 Lever out the pull handle trim and then remove the now exposed screw (see illustrations).

10 Using a trim tool, prise free the panel

12.3d …and then push the plug into the door frame

12.5a Remove the cover…

12.5b …and speaker grille

12.7a Unclip the wiring loom. Note the hooks on the door panel

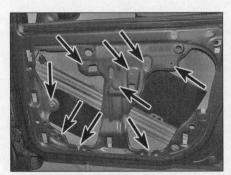

12.7b The door with the panel removed showing the mounting points.

12.9a Lever out the trim piece…

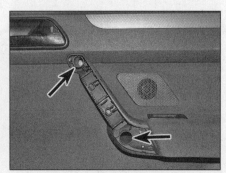

12.9b …and remove the screw

12.10a Remove the panel

12.10b Unhook the release cable...

12.10c ...and disconnect the wiring plug

12.11 Slacken the clamp bolts

12.12a Remove the grille and...

insert. Disconnect the door handle release cable as the panel is removed and (where fitted) disconnect the tweeter wiring plug (see illustrations).

11 Position the window glass so that the glass clamps are accessible (see illustration). Slacken the clamps and pull the window glass up. Tape the glass up in the fully closed position.

12 Unclip the speaker grille and then remove the panel mounting screws (see illustrations).

13 At the B-pillar disconnect the wiring plug, using the same method as described in this section for the front door trim panel.

14 Remove the screws from the panel (see illustration). Note that the lower regulator screw slots into the door frame and does not need to be fully removed

15 Lift the panel up from the door frame and at the same time work the regulator free from the inside of the panel (see illustration).

16 Support the panel, disconnect the lock wiring plug and unclip the wiring loom from the door frame as required (see illustrations).

Refitting

17 The Refitting of the trim panel is then a reversal of removal. Check the operation of the door electrical equipment.

18 On models with manual windows refit the winder handle so that it is approximately 30 degrees below horizontal with the window fully closed.

12.12b ...then remove the screws

12.14 Remove the fixings

12.15 Releasing the panel and regulator from the lower slotted mounting

12.16a Unplug the lock wiring plug

12.16b Unclip the wiring loom

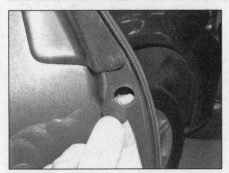

13.1 Remove the cover

13.2a Slacken the retaining screw…

13.2b …and pull out the lock cylinder

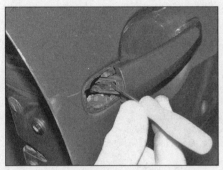

13.3a Unhook the release cable…

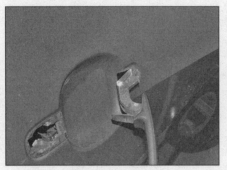

13.3b …and then remove the handle

13.4 Remove the bolts

13 Door handle and lock components – removal and refitting

Note: *The exterior handle and cylinder lock can be removed from the outside of the door. At the front the door lock is removed with the door trim panel, at the rear the lock is removed after the trim panel has been removed.*

Removal

Front door lock cylinder and handle

1 Open the door, then remove the blanking plug from the rear edge of the door to access the retaining screw (see illustration).

2 Pull out the door release handle slightly and slacken the retaining screw whilst pulling on the lock cylinder (see illustrations).
Note: *Do not remove the screw completely. Only slacken the bolt sufficiently to release the lock cylinder. If the bolt is slackened too far (or removed) the locking plate may well become detached from the support bracket.*

3 Pull out the handle and using a screwdriver unhook the release cable from the handle. Unhook and then slide out the handle (see illustrations). Recover the gasket.

Front door lock and support bracket

4 Remove the door handle as described above and then remove the lock mounting bolts (see illustration). Note that the lock will

not fall into the door frame as it is fixed to to the door trim panel.
5 The door inner trim must now be removed as described in Section 12.
6 Protect the front of the door panel with an old blanket or similar placed on the bench and then unclip the lock from the panel (see illustrations).
7 The door handle support bracket can now be removed from the door skin after the single bolt is removed. Recover the gaskets as the bracket is removed. Also where required the inside release handle can be removed from the door panel.

Rear door handle

8 Remove the door trim panel as described in Section 12.

13.6a Disconnect the wiring plug…

13.6b …and then unclip the lock from the panel

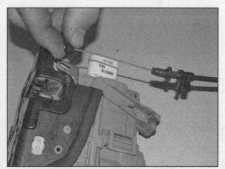

13.6c If required the bowden cable can now be removed from the lock

13.9a Open the locking clip and...

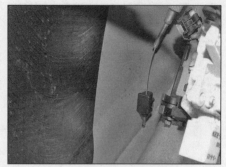

13.9b ...release the cable

13.10a Pull out the rubber guide

13.10b Remove the screws

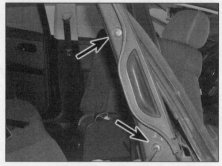

13.11 Remove the screws

13.12a Remove the hidden fixing and then...

9 Working inside the door frame, unclip the release cable from the door lock **(see illustrations).**

10 Pull up the rear edge of the window guide channel rubber (there is no need to remove it completely) and remove the now exposed fixings **(see illustrations).**

11 Remove the blanking plugs and then remove the fixings from the outer edge of the door **(see illustration).**

12 Working through the door frame remove the lower fixing and then pull up the handle to remove it from the door **(see illustrations).**

Rear door lock

13 Remove the door trim panel as described in Section 12.

14 Unclip the release cable from the exterior handle as described above.

15 Remove the mounting bolts **(see illustrations)** and recover the lock from the door frame.

16 The release cable can now be removed from the lock where required.

Refitting

17 Refitting is a reversal of removal, bearing in mind that on the front door the door release cable for the outer handle must have little or no slack in the cable when fitted. On the rear door the bowden cable for the outer handle must also be clipped in place with no slack

in the cable. Always check the operation of all locks and handles before closing the door.

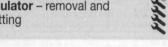

14 Door window glass and regulator – removal and refitting

Removal

Front door window glass

1 Remove the door handle cover and the upper trim panel as described in Section 12.

2 Move the window glass up or down as

13.12b ...remove the handle

13.15a Remove the bolts and...

13.15b ...remove the lock

14.2 Slacken the glass clamp bolts

14.4 Lift out the glass

14.6 Slacken the bolts

required to access the glass clamps **(see illustration)**.

3 Lower the glass to the mid position – the glass must remain in the clamps.

4 Lift the rear of the window glass so that it is tilted. Lift it upwards (whilst rotating it towards

the front of the door) and remove it from the outside of the door **(see illustration)**.

Rear door window glass

5 Remove the door upper trim panel as described in Section 12.

6 If required, lower the window glass until the glass clamp bolts are visible **(see illustration)**. Support the glass, loosen the bolts and then fully lower the glass carrier.

7 Protect the bodywork if necessary and then prise up and remove the outer weather seal **(see illustration)**.

8 Pull up the glass and rotate it to the rear of the vehicle. Slide the glass forward, clear of the front guide channel. Lift and tilt the glass forward, up and out of the door frame **(see illustrations)**.

Window motor and regulator

Note: *The window motor and regulator are removed with door trim panel.*

9 Remove the door inner trim panel as described in Section 12 and tape the window up in the fully closed position **(see illustration)**.

10 Place the door panel on a protected surface and unclip the regulator from the door panel. Unbolt the motor cover and then unbolt the motor (or winder mechanism) from the door panel and remove it **(see illustrations)**.

Refitting

11 Refitting is a reversal of removal.

14.7 Remove the weatherseal

14.8a Rotate the glass

14.8b Lift the window glass from the rear door

14.9 Tape the window glass in position to the frame

15 Tailgate and support struts
– removal and refitting

Removal

Tailgate

1 Open the tailgate and remove the fixings **(see illustration)**.

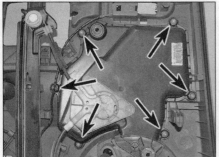

14.10a Remove the cover bolts

14.10b Remove the bolts and unplug the motor (rear door shown)

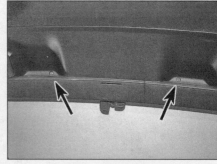

15.1 Remove the fixings

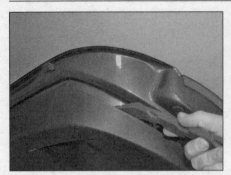

15.2a Unclip the panel…

15.2b …and remove it from the tailgate

15.2c Note the emergency release lever….

15.2d …accessible via the slot in the panel

15.3 Disconnect the hose

15.4a Disconnect the multiple wiring plugs

2 Carefully lever between the panel and the tailgate with a plastic trim tool. Work around the outside of the panel, and when all the clips are released, remove the panel **(see illustrations)**.
3 Disconnect the screen washer hose **(see illustration)**.
4 Free the wiring grommets from the tailgate/bodyshell and then disconnect the wiring plugs **(see illustrations)**. Pull out the previously disconnected screen washer hose.
5 Using a suitable marker pen, draw around the outline of each hinge marking its correct position on the tailgate **(see illustration)**.
6 With the help of an assistant to support the tailgate, remove the support struts as described below.
Caution: The tailgate is heavy. At least one assistant will be required to remove it.

7 Unscrew and remove the bolts securing the hinges to the tailgate. Where necessary, recover the gaskets that are fitted between the hinge and vehicle body.
8 Inspect the hinges for signs of wear or damage and renew if necessary. The hinges are secured to the vehicle by nuts or bolts (depending on model), which can be accessed once the headlining rear cover strip has been removed.

Support struts

⚠ *Warning: The support struts are filled with a gas under pressure. They must not be opened and they and must be disposed of safely.*

9 With the help of an assistant, support the tailgate in the open position.
10 Using a small flat-bladed screwdriver lift

the locking clip, and pull the gas support strut off its balljoint mounting on the tailgate **(see illustration)**. Repeat the procedure on the lower strut mounting and remove the strut from the vehicle body. **Note:** *If the gas strut is to be re-used, the locking clip must not be taken all the way out, or the clip will be damaged.*

Refitting

Tailgate

11 Refitting is the reverse of removal, aligning the hinges with the marks made before removal. Tighten retaining bolts to the specified torque.
12 On completion, close the tailgate and check its alignment with the surrounding panels. If necessary slight adjustment can be

15.4b Release the loom at the conduit and pull it from the tailgate.

15.5 Mark the position of the hinges

15.10 Lift locking clip upwards, then pull strut off upper balljoint

16.2 Disconnect the wiring from the tailgate lock

16.3 Undo the lock retaining screws

16.5 Remove the tailgate handle/release unit nuts...

made by unscrewing the retaining bolts and repositioning the tailgate on its hinges. If the tailgate buffers are in need of adjustment, continue as follows.

Support struts

13 Refitting is a reverse of the removal procedure, ensuring that the strut is securely retained by its retaining clips.

16 Tailgate lock components – removal and refitting

Removal

Note: *If the lock is inoperative, the tailgate can be opened manually (see illustration 15.2d).*

Tailgate lock

1 Open up the tailgate and remove the trim panel as described in Section 15.
2 Disconnect the wiring from the lock **(see illustration)**.
3 Undo the retaining bolts and remove the lock from the lower edge of the tailgate **(see illustration)**.

Tailgate handle/release unit

4 Open up the tailgate and remove the trim panel as described in Section 15.
5 Disconnect the wiring connector and then undo the nuts holding the unit to the tailgate **(see illustration)**.
6 The handle/release unit can now be withdrawn from the tailgate **(see illustration)**.

16.6 ...and withdraw the release handle from the tailgate

Refitting

7 Refitting is a reversal of removal, however, before refitting the trim panel, check the operation of the lock components.

17 Central locking components – description, removal and refitting

Description

1 The central locking system consists of the following main components. Note that the central locking and anti-theft alarm systems share some components (see Chapter 12 Section 25):
a) Convenience system onboard supply control unit located behind the driver's side facia panel.
b) Door control units integrated in the window regulator motors.
c) Electric door lock actuators integrated in the door locks.

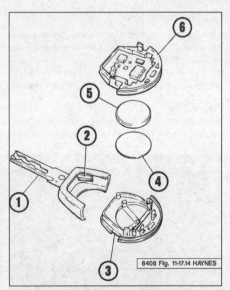

17.14 Fixed blade key

1 Key blade	4 Cushion
2 Transponder housing	5 Battery
3 Battery housing	6 Transmitter housing

d) Fuel tank filler cap flap actuator located behind the luggage compartment trim.
e) Tailgate lock actuator located in the tailgate, together with the release button.
f) Anti-theft alarm horn located in the plenum chamber.
g) Bonnet contact switch located on the bonnet lock.
h) Remote control transmitter on the ignition key fob.

Removal

2 Before working on any electrical circuit, disconnect the battery negative lead (refer to Chapter 5A Section 3).

Convenience system control unit

3 Remove the glovebox as described in Section 27.
4 Disconnect the wiring plug, unclip the module and remove it.

Door control unit

5 Remove the window regulator motor as described in Section 14.

Electric door lock actuator

6 Remove the door lock (see Section 13).

Fuel tank filler cap flap actuator

7 The actuator actuator can be accessed through the flap in the right-hand side panel.
8 Reach through and disconnect the wiring plug from the actuator.
9 Slacken the mounting bolts, and then slide the actuator upwards and release it from the inner body panel. There is no need to remove the bolts completely as the actuator is mounted on keyhole slots.
10 Withdraw the actuator upward, complete with operating rod.

Tailgate/boot lid lock activator

11 Remove the tailgate handle/release unit as described in Section 16.

Bonnet contact switch

12 Remove the bonnet lock as described in Section 10. If necessary, the operating cable may remain attached to the lock.
13 On the lock, release the tab and push the switch from the slotted holes.

Remote control transmitter battery

Note: *Depending on the trim level and year of construction, the vehicle may be supplied with either a rigid key fob or a folding key fob.*

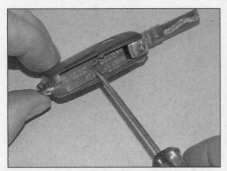

17.16a Using a small screwdriver, release the clip...

17.16b ...and open the battery cover

17.17a Remove the battery...

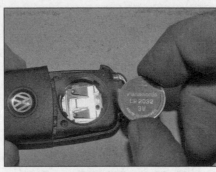

17.17b ...noting which way it is fitted

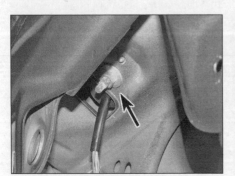

18.2 Tape the glass in the closed position

Standard type key

14 Using a screwdriver inserted in the slot, separate the transmitter unit cover from the key **(see illustration)**.

15 Pry open the two halves of the battery/transmitter housing and then remove the battery, noting the orientation of the positive/negative poles.

Folding blade key

16 Using a screwdriver inserted in the slot, separate the transmitter unit cover from the key **(see illustrations)**.

17 Carefully prise out the battery, noting which way round it is fitted **(see illustrations)**.

Refitting

18 Refitting is a reversal of removal, except where a new fuel tank actuator is required

on 07/2012 on vehicles. On these vehicles the mounting keyhole slots must be modified to accept the new actuator. This requires the cutting of two additional vertical slots at the end of the keyholes. These should be cut so that the distance from the centre line of the keyhole slot to the base of the new vertical slot is 6 mm. Corrosion protection and remedial paintwork will be required after the slots are cut. On completion check the operation of the central locking system.

18 Mirrors and associated components – removal and refitting

Removal

Exterior mirror

1 Remove the door upper trim panel as described in Section 12.

2 Slacken the window glass clamps (Section 14) and then pull up the glass and tape it securely in the closed position **(see illustration)**.

3 Turn on the ignition and lower the glass clamps to the centre of the door panel.

4 Remove the main door trim panel as described in Section 12.

5 Disconnect the wiring plug and then unbolt the mirror from the door **(see illustrations)**.

Mirror glass

Note: *The mirror glass is clipped into place. Removal of the glass without the VAG group special tools may result in breakage of the glass. T20043 will be required for early models (up to 2009) and 3370 will be required for later models. Note however that standard bodywork trim tools should also be suitable. Wear protective gloves and glasses to prevent personal injury.*

Early models (up to 2009)

6 Press in the outer edge of the mirror glass so that the inner edge is furthest from the housing. If required, protect the edge of the housing with masking tape, then insert the special tool and lever the mirror glass from its mounting clips. Take great care when removing the glass; do not use excessive force, as the glass is easily broken **(see illustration)**.

18.5a Undo the bolt and...

18.5b ...remove the mirror

18.6 Remove the exterior mirror glass...

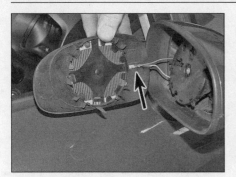

18.7 ...and disconnect the wiring plugs

18.13a Unclip and then...

18.13b ...remove the cover

7 Disconnect the wiring connectors from the mirror heating element **(see illustration)**.

Later models (2009 on)

8 Push the lower edge of the glass into the housing.
9 Insert the special tool (or a hooked plastic trim tool) over the top of the glass.
10 Push in the base of the mirror glass whilst pulling the top out with the the hooked tool.
11 Disconnect the wiring connectors from the mirror heating element.

Mirror housing

Vehicles built up to 2009

12 Remove the mirror glass as described in this Section.
13 Using a combination of a small screwdriver (to release the inner tabs) and a trim tool (to lever off the cover) remove the cover **(see illustrations)**.

Vehicles built from 2009

14 Remove the mirror glass as described in this Section.
15 Using a small screwdriver release the retaining tabs and then rotate the the complete mirror away from the door. Remove the housing.
16 Work around the housing and free off the locking tabs with a screwdriver. Unclip and then remove the housing.

Mirror switch

17 Refer to Chapter 12 Section 5.

Refitting

18 Refitting is the reverse of the relevant

removal procedure. When refitting the mirror glass, press firmly at the centre taking care not to use excessive force, as the glass is easily broken.
Caution: Wear protective gloves and glasses to prevent personal injury.

Interior mirror

Removal

19 If rear view mirror is fitted with a rain sensor ensure the ignition is off (remove the key).
20 Using a plastic trim tool unclip the cover **(see illustration)**.
21 Remove the complete interior light panel as described in Chapter 12 Section 7. Note that removal is possible with the light in position, but removing it gives easy access to the locking spring assembly.
22 Using a flat blade screwdriver release the tension on the spring and at the same time pull down the mirror **(see illustration)**.
23 On models fitted with an auto dimming mirror, disconnect the wiring plug as the mirror is removed.

Refitting

24 The retaining spring must be de-tensioned using a pair of waterpump pliers (or similar). This allows the spring pressure to be relieved, making installation easier **(see illustration)**.
25 Reconnect the wiring plug (where fitted) and then insert the mirror into the base plate housing. Push the opposite side to the spring tab into the housing first and then (using a

screwdriver or similar) rotate and push the opposite side into position. Ensure that the spring tail engages into the slot in the mirror base.
26 Refit the cover and then refit the light unit.

19 Windscreen and rear window glass – general information

1 These areas of glass are bonded in position with a special adhesive. Renewal of such fixed glass is a difficult, messy and time-consuming task, which is beyond the scope of the home mechanic. It is difficult, unless one has plenty of practice, to obtain a secure, waterproof fit. Furthermore, the task carries a high risk of breakage; this applies especially to the laminated glass windscreen. In view of this, owners are strongly advised to have this sort of work carried out by one of the many specialist windscreen fitters.

20 Sunroof – general information

1 Due to the complexity of the sunroof mechanism, considerable expertise is needed to repair, renew or adjust the sunroof components successfully. Removal of the roof first requires the headlining to be removed, which is a complex and tedious operation, and not a task to be undertaken lightly.

18.20 Unclip the cover

18.22 Release the mirror

18.24 Compress the spring to reset it

Therefore, any problems with the sunroof should be referred to a Seat dealer. On models with an electric sunroof, if the sunroof motor fails to operate, first check the relevant fuse. If the fault cannot be traced and rectified, the sunroof can be opened and closed manually using an Allen key to turn the motor spindle (a suitable key is supplied with the vehicle, and should be clipped onto the inside of the sunroof motor trim). To gain access to the motor, unclip the rear of the trim cover to open. Unclip the Allen key, and then insert it fully into the motor opening (against spring pressure). Rotate the key to move the sunroof to the required position.

21 Body exterior fittings – removal and refitting

Wheel arch liners and body under-panels

1 The various plastic covers fitted to the underside of the vehicle are secured in position by a mixture of screws, nuts and retaining clips and removal will be fairly obvious on inspection. Work methodically around the panel removing its retaining screws and releasing its retaining clips until the panel is free and can be removed from the underside of the vehicle. Most clips used on the vehicle are simply prised out of position.

Remove the wheels to ease the removal of the wheel arch liners **(see illustration)**.
2 On refitting, renew any retaining clips that may have been broken on removal, and ensure that the panel is securely retained by all the relevant clips and screws.

Body trim strips and badges

3 The various body trim strips and badges are held in position with a special adhesive tape and locating lugs. Removal requires the trim/badge to be heated, to soften the adhesive, and then carefully lifted away from the surface. Due to the high risk of damage to the vehicle's paintwork during this operation, it is recommended that this task should be entrusted to a VW dealer.

22 Seats – removal and refitting

Note: *Refer to the warnings in Chapter 12 Section 26, on airbags.*

Removal

Front seats

Note: *The amount of wiring connectors under the seat will vary depending on the vehicle specification.*

1 Disconnect the battery negative lead (refer to Section 5A Section 3).

2 Slide the seat forwards as far as possible and unscrew the rear mounting bolts **(see illustration)**.

⚠️ *Warning: As a precaution against unintentional electrostatic discharge into the airbag, briefly touch part of the vehicle body before disconnecting the wiring.*

3 Tilt the seat back slightly and unbolt the protective cover to access the wiring connectors **(see illustrations)** Disconnect the wiring connectors. Seat technicians fit an adapter to the airbag wiring connector as a safety precaution; however, wrap the wiring connectors with insulation tape or similar, to prevent any contact of the wiring.
4 Slide the seat rearwards as far as possible and unscrew the front mounting bolts **(see illustration)**.
5 Remove the headrests from the seat backs; this will give more room when removing the seat from the vehicle.
6 When removing the seat from the vehicle, do not lift the seat by the seat belt stalk or by the seat adjustment levers. If necessary, have an assistant help to remove the seat, as it is heavy, and surrounding trim panels may be otherwise damaged. Make sure that the seat runners do not damage the paintwork as the seat is withdrawn.
7 Remove the mounting bolts and tuck the loom under the carpet to protect it from damage.

21.1 Removing a rear wheel arch liner

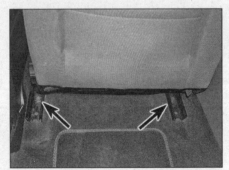
22.2 Seat rear mounting bolts

22.3a Remove the cover...

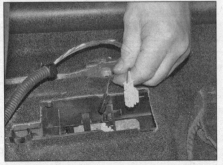
22.3b ...and disconnect the wiring plugs

22.3c Unclip the loom from the bracket

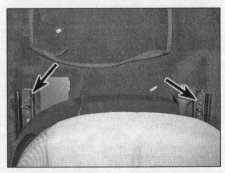
22.4 Seat front mounting bolts

Rear seats

8 Open the tailgate and remove the parcel shelf.

9 Remove the cover, unbolt the clamp and remove it. Lift up the single seat and recover the spacer **(see illustrations)**.

10 Lift up and unhook the single seat and then remove it. Lift up the double seat and then unbolt the centre seat belt mounting **(see illustrations)**. Remove the seat

11 The seat squab can now be removed. Note that the squab can also be removed with the seat back in place. At the rear of the seat squab unclip the covers and then pull up the squab to remove it **(see illustrations)**.

Refitting

12 Refitting is a reversal of removal, but tighten the mounting bolts to the specified torque where given.

23 Seat belt tensioning mechanisms – general information

1 All models covered in this manual are fitted with a front seat belt tensioner system incorporated in each of the inertia reels. Rear seat belt inertia reels with the tensioner system are only fitted to some models, other models having standard inertia reels.

2 The system is designed to instantaneously take up any slack in the seat belt in the case of a sudden frontal impact, therefore reducing the possibility of injury to the front seat occupants. The seat belt tensioner is triggered by a frontal impact above a predetermined force. Lesser impacts, including impacts from behind, will not trigger the system.

3 When the system is triggered, the explosive gas in the tensioner mechanism retracts and locks the seat belt. This prevents the seat belt moving and keeps the occupant firmly in position in the seat. Once the tensioner has been triggered, the seat belt will be permanently locked and the assembly must be renewed.

4 There is a risk of personal injury if the system is triggered inadvertently when working on the vehicle, and it is therefore strongly recommended that any work involving the seat belt inertia reels be entrusted to a Seat dealer. Note the following warnings before contemplating any work on the front seat belts.

 Warning: Do not expose the tensioner mechanism to temperatures in excess of 100°C.

 Warning: Before removing any components, disconnect the battery negative lead as described in Section.

 Warning: If the tensioner mechanism is dropped, it must be renewed, even it has suffered no apparent damage.

 Warning: Do not allow any solvents to come into contact with the tensioner mechanism.

 Warning: Do not attempt to open the tensioner mechanism as it contains explosive gas.

 Warning: Tensioners must be discharged before they are disposed of, but this task should be entrusted to a Seat dealer.

 Warning: Some of the bolts securing the seat belts are microencapsulated and need to be heated with a hot air blower before they are slackened. Microencapsulated means that the threads of the bolts are coated with a locking compound from new. New bolts will be required for refitting.

24 Seat belt components – removal and refitting

 Warning: Refer to Section 23 before proceeding.

Front seat belt removal

1 Disconnect the battery negative lead (as described in Section 5A Section 3).

2 Remove the upper and lower trim panels from the B-pillar with reference to Section 25.

3 Pull back the carpet, then unscrew and remove the seat belt lower anchor mounting

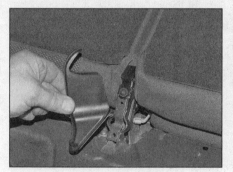

22.9a Unclip the bracket cover

22.9b Recover the plastic spacer

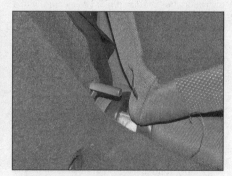

22.10a Release the seat back from the pivot

22.10b Unbolt the centre inertia reel

22.11a Remove the covers

22.11b Pull up the front of the squab to release it

24.3a Unbolt and then...

24.3b ...remove the bolt and bracket, noting the orientation of the bracket

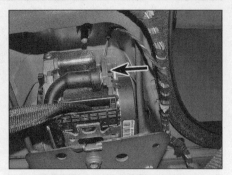

24.4 Release the locking peg, to disconnect wiring connector

24.5 Front seat belt inertia reel mounting bolt

24.6 Front seat belt guide

bolt and remove the belt from the floor **(see illustrations)**. Note that the bolt may be microencapsulated and must be heated with a hot air blower before being loosened. Protect the seat belt with a dampened cloth while heating the bolt. Discard the bolt and obtain a new one. Also, note that the threads of the corresponding nut must be cleaned with a tap before fitting the new bolt.

⚠️ *Warning: As a precaution against unintentional electrostatic discharge, briefly touch part of the vehicle body before disconnecting the wiring.*

4 Release the locking clip and disconnect the wiring connector from the inertia reel **(see illustration)**.

5 Unscrew the mounting bolt and remove the inertia reel from the bottom of the B-pillar **(see illustration)**.

6 Undo the screws and remove the belt guide from the B-pillar **(see illustration)**.

7 Unscrew and remove the bolt securing the seat belt upper anchor to the height adjuster on the B-pillar **(see illustration)**. The seat belt assembly can then be removed from the vehicle.

8 To remove the belt height adjustment, remove the securing bolt from the top of the adjuster and lift upwards to release from the pillar.

Front seat belt stalk removal

9 Remove the front seat assembly as

described in Section 22. On a clean surface, turn the seat over.

10 Slide the seat runners forward to access the seat belt stalk retaining bolt, under the seat frame. Unscrew and remove the bolt securing the stalk to the seat, and remove the stalk **(see illustration)**. Note that the bolt is microencapsulated and must always be replaced. Also, note that the threads of the corresponding nut must be cleaned with a tap before fitting the new bolt.

Rear seat belt removal

11 Remove the load area side trim panels as described in Section 25.

12 Remove the bolt **(see illustration)**. Note that the bolt may be microencapsulated. Discard the bolt and obtain a new one. Also, note where a microencapsulated bolt is used, the threads of the corresponding nut must be cleaned with a tap before fitting the new bolt.

13 At the base of the C-pillar, unbolt and remove the lower mounting and then remove the belt and reel from the vehicle.

Rear centre belt removal

14 The reel is fitted into the centre rear seat back. Remove the seat as described in Section 22.

15 Place the seat on a protective surface and

24.7 Front seat belt upper anchor

24.10 Turn over the seat to remove the stalk bolt

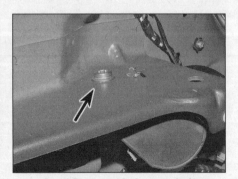

24.12 Rear seat belt upper bolt

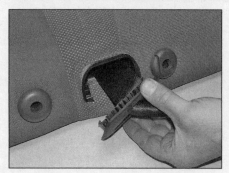

24.15a Unclip the seat belt guide

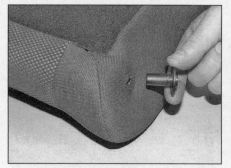

24.15b Remove the pivots (at both sides)

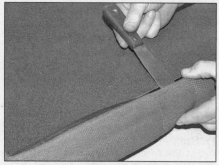

24.15c Unclip and...

remove the rear cover to access the mounting bolt **(see illustrations)**.

16 Partially remove the seat back cushion and then unbolt and remove the seat belt reel **(see illustration)**.

Rear seat belt stalks

17 Remove or tip forward the rear seat back. Unbolt and remove the stalk **(see illustration)** (and the centre belt lower anchor, if working on the centre seat)

Refitting

18 Refitting is a reversal of the removal procedure, ensuring that all the seat belt units are located correctly and mounting bolts are securely tightened to their specified torque. Check all the trim panels are securely retained by all the relevant retaining clips. When refitting the B-pillar upper trim panels, ensure that the height adjustment peg engages correctly with the trim panel.

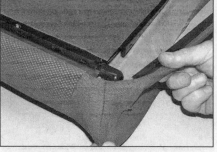

24.15d ...work free the seat fabric

24.16 Unbolt the reel

25 Interior trim –
removal and refitting

⚠️ **Warning: Refer to the warnings in Chapter 12 Section 26, on airbags.**

Interior trim panels

1 The interior trim panels are secured using either screws or various types of trim fasteners, usually studs or clips.
2 Check that there are no other panels

overlapping the one to be removed; usually there is a sequence that has to be followed, and this will only become obvious on close inspection.
3 Remove all obvious fasteners, such as screws. If the panel will not come free, it will be held by hidden clips or fasteners. These are usually situated around the edge of the panel and can be prised up to release them; note, however, that they can break quite easily so new ones should be available. The best way of releasing such clips, without the correct type of tool, is to use a large flat-bladed screwdriver. If a considerable amount of trim requires removal a wise investment will be a comprehensive set of plastic trim tools **(see illustration)**. Note in many cases that the adjacent sealing strip must be prised back to release a panel.
4 When removing a panel, never use excessive force or the panel may be damaged; always check carefully that all fasteners or other

relevant components have been removed or released before attempting to withdraw a panel.
5 On models with side airbags, disconnect the battery negative lead (refer to Chapter 5A Section 3).
6 Refitting is the reverse of the removal procedure; secure the fasteners by pressing them firmly into place and ensure that all disturbed components are correctly secured to prevent rattles.

A-pillar

Upper section

7 Starting from the top of the panel and using a plastic trim tool release the panel **(see illustrations)**.
8 Recover the retaining clips, inspect them and replace as required.
9 Refitting is a reversal of removal.

24.17 Unbolt the stalk

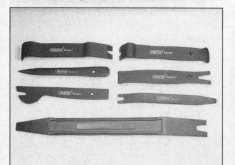

25.3 A useful selection of trim removal tools will avoid damage to the paintwork or trim panel

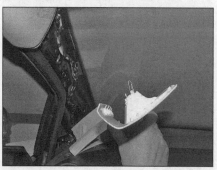

25.7 Remove the A-pillar trim

Lower section

10 The left hand lower panel can be removed by unclipping it after partially removing the front edge of the sill panel. If working on the right-hand panel, the bonnet release lever must be removed first, as described in Section 9.

B-pillar

Note: *The B-pillar trim is in two sections. The lower section must be removed first.*

11 Remove the door sill trims as described in this Section. At the base of the B-pillar remove the fixings and then unclip the upper **(see illustrations)**. On models fitted with an alarm disconnect the wiring plug from the interior sensor switch as the panel is removed.
12 Remove the fixings from the upper section of the panel **(see illustrations)**.
13 Feed the seatbelt buckle through the panel to remove it completely.

25.11a Remove the lower fixings...

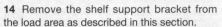

25.11b ... and unclip the panel

C-pillar

Note: *The upper section of the panel is also the load area/rear glass trim panel. This panel must be removed first.*

25.12a Remove the upper panel fixings and then...

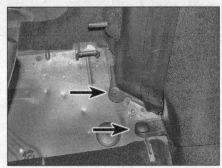

25.12b ...unclip the panel – note the locking tabs that locate the panel behind the headlining

14 Remove the shelf support bracket from the load area as described in this section.
15 Fold down the rear seats and then remove the 'Airbag' blanking plug from the upper section. Remove the fixing **(see illustration)** and the rotate the base of the panel to remove it. Care should be taken as the headlining is easily damaged as the panel is removed.
16 Remove the lower nuts and the upper fixing screws from the lower section of the panel. Use a trim tool and unclip the panel **(see illustrations)**.
17 Refitting is a reversal of removal.

Load area panels

Shelf support panel

18 Open the tailgate and remove the parcel shelf.
19 Remove the fixings and lift out the panel **(see illustrations)**. Where fitted disconnect

25.15 Remove the upper fixing

25.16a Remove the lower fixings...

25.16b ...and the upper fixings (one shown)

25.16c Remove the panel

25.19a Remove the fixings

25.19b Rotate the panel to unhook and remove it

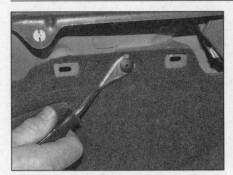

25.23a Remove the upper trim clip

25.23b Remove the panel

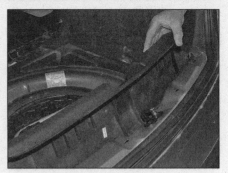

25.26 Remove the panel

the wiring plug from the load area lamp as the panel is removed.

20 Refitting is a reversal of removal.

Lower side panel

21 Remove the shelf support panel as described above and then remove the C-pillar upper panel, again as described above.

22 Remove the load area tailgate slam panel as described below.

23 Remove the luggage tie down brackets and then remove the fixing screws. Remove the panel (see illustrations).

24 Refitting is a reversal of removal.

Tailgate slam panel

25 Remove the load area floor covering and the spare wheel.

26 Prise out the trim clips from the base of the panel. Unclip and remove the panel (see illustration).

27 Refitting is a reversal of removal.

Sill panel

Note: The sill panel runs the full length of the vehicle. It must be removed before the B-pillar lower panel can be removed.

28 Access can be improved if the rear seat squab is first removed as as described in Section 22.

29 Remove the lower C-pillar panel first (as described in this Section). Note that because of the way the C-pillar panel interlocks with the sill panel it must be removed first (see illustrations).

30 Refitting is a reversal of removal.

Steering column shrouds

31 Remove the steering wheel as described in Chapter 10 Section 19. Note that it is also possible to remove the shrouds with the steering wheel in position, but the column switches (apart from the key reader) can not be removed with the steering wheel in position.

32 Fully extend the column and then unclip the bellows section from below the instrument panel (see illustration).

33 Remove the upper fixing screws and release the upper panel (see illustration).

34 Remove the single fixing from the lower section and then manoeuvre the lower section

25.29a Prise up the panel. Note the locking hole for the lower C-pillar panel

25.29b Work the panel free along the door steps

25.29c Slide out the seatbelt webbing and remove the panel

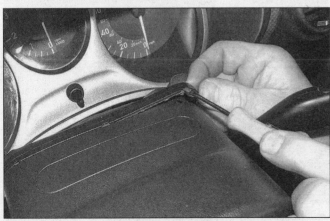

25.32 Unclip the bellows

25.33 Remove the upper shroud

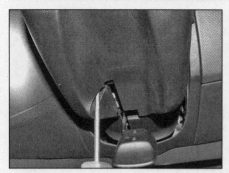

25.34a Remove the screw and then...

25.34b ...work the shroud past the lever to remove it

25.37 Remove the screws

past the column adjustment lever and remove it **(see illustrations)**.

35 Refitting is a reversal of removal.

Glovebox

36 Switch off the ignition.

37 Open the glovebox door and remove the upper fixings **(see illustration)**.

38 Where fitted remove the lower cover from below the glovebox and then remove the lower screws.

39 Partially release the glovebox and disconnect the wiring plugs from the PAD (Passenger Airbag Deactivation) switch and the glovebox lamp. Remove the glovebox.

40 Refitting is the reverse of removal.

Carpets

41 The passenger compartment floor carpet is in one piece and is secured at its edges by screws or clips, usually the same fasteners used to secure the various adjoining trim panels.

42 Carpet removal and refitting is reasonably straightforward but very time-consuming because all adjoining trim panels must be removed first, as must components such as the seats, the centre console and seat belt lower anchorages.

Headlining

43 The headlining is clipped to the roof and can be withdrawn only once all fittings such as the grab handles, sun visors, interior lights, sunroof (if fitted), and related upper trim panels have been removed.

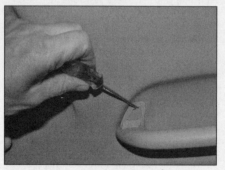

25.44a Carefully unclip the covers...

25.44b ...and undo the retaining screws

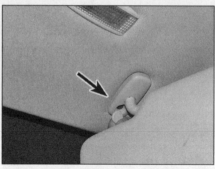

25.44c Unclip the cover to access the screw for the sunvisor

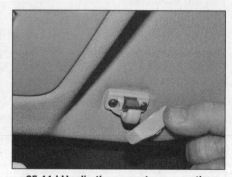

25.44d Unclip the cover to access the screw for the sunvisor locating mount

44 To remove the sun visors and grab handles, the plastic covers have to be unclipped first, to gain access to the securing screws **(see illustrations)**

45 Note that headlining removal requires considerable skill and experience if it is to be carried out without damage and is therefore best entrusted to an expert.

26 Centre console – removal and refitting

Removal

Note: *It is possible to remove the centre console with both front seats in position, however removal of the console is slightly easier if they are both removed first (as described in Section 22).*

1 Switch off the ignition. If the front seats are not to be removed slide them back as far as possible.

2 Remove the fixings from the driver's foot rest. Remove the footrest and then unclip the both front side panels from the console **(see illustrations)**.

26.2a Remove the footrest...

26.2b ...and then remove both front side panels

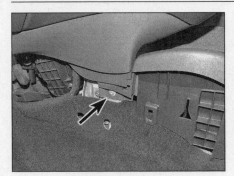

26.3 Remove the screws

26.4a Remove the compartment...

26.4b ...and disconnect the wiring plug

3 At the front (on both sides) remove the screws **(see illustration)**.

4 Prise up and remove the front storage compartment, disconnect the wiring plug and remove the compartment **(see illustrations)**.

5 Prise up and remove the main switch panel. Disconnect the wiring plugs as the switch is removed **(see illustrations)**.

6 Prise up and remove the handbrake lower cover.

7 Remove the embellisher from the gear lever **(see illustration)** gaiter and then unclip the gaiter from the console. If the gaiter will not release, leave it in position until the console can be partially turned over and then release (by cutting off) the gear knob retaining clamp. The gaiter can then be removed with the knob and the gaiter removed from the console on the bench.

8 At the rear of the console prise free the storage compartment. Where fitted disconnect the wiring plug as the compartment is removed.

9 Remove the rear mounting bolts and then lift up the rear of the console. Remove the console by pulling it up and backwards, whilst carefully maneuvering it over the gear lever **(see illustrations)**.

10 With the console on the bench the gaiter can be easily removed as all the retaining clips are now accessible **(see illustration)**.

26.5a Remove the switch panel

26.5b Disconnect the wiring plugs

26.7 Remove the embellisher

26.9a Remove the screws

26.9b Where the gaiter could not be released, turn over the console and cut off the gear knob clamp

26.9c Removing the console with the gaiter and gear knob

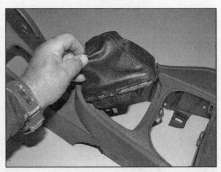

26.10 Remove the gaiter

26.11a Remove the rear screws...

26.11b ...and the centre screws

26.11c Lift out the subframe

11 If full access to the gear lever or handbrake are required, the centre console subframe can now be removed **(see illustrations)**.

Refitting

12 Refitting is a reversal of removal, except where the gear knob has been removed with the gaiter. If this method was used to remove the console a new clamp will be required.

27 Facia panel assembly – removal and refitting

Note: Refer to the warnings in Chapter 12 Section 26 for airbags.
Note: To avoid any damage to the facia panel, removal of the steering column (Chapter 10 Section 20) is highly recommended.

Removal

1 Disconnect the battery negative lead (as described in Chapter 5A Section 3).
2 Remove the steering column shrouds as described in Section 25.
3 Remove the steering column control unit and switch assembly, as described in Chapter 12 Section 5.
4 Prise out the trim panels from each end of the facia and remove the lower mounting screw from the fusebox **(see illustration)**.
5 Remove the A-pillar upper trim panels, as described in Section 25.
6 Remove the glovebox as described in Section 25 and then disconnect the passenger airbag. Remove the airbag to crossmember mounting bolts.
7 Remove the centre console as described in Section 26.
8 Remove the Audio or SatNav unit, as

described in Chapter 12 Section 22 and then remove the instrument panel (Chapter 12 Section 11)
9 Working from the bottom of the panel and up the sides remove the facia centre panel/ heater control panel using a plastic trim tool **(see illustrations)**.
10 On climatronic models remove the sunlight sensor from the top of the facia **(see illustrations)**. Disconnect the wiring plug as the sensor is removed and then tie a length of cord to the wiring plug as an aid for reassembly.
11 Remove the lighting switch as described in Chapter 12 Section 5.
12 Where fitted remove the lower cover from beneath the steering wheel.
13 On models fitted with climate control, disconnect the wiring plugs from the left and right air vent temperature sensors. These

27.4 Unclip the cover from the side of the facia

27.9a Remove the outer screws...

27.9b ...the upper screws...

27.9c ...and work the panel free

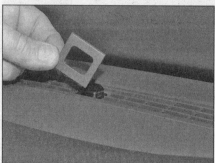

27.10a Remove the cover...

27.10b..and unclip the sensor

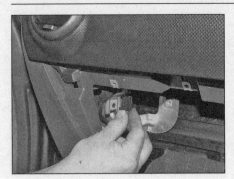

27.13a Disconnect the temperature sensor (left-hand)...

27.13b ...and then the right-hand

27.14a Remove the cover...

are accessed through the glovebox and the lighting switch **(see illustrations)**.

14 Remove the cover from the diagnostic plug housing and then unclip the plug from the bracket **(see illustrations)**.

15 Working logically from one side to the other, locate the facia cover screws. Unscrew the retaining screws from the facia assembly **(see illustrations)**. As the facia is being removed, disconnect any loom retaining clips.

16 Make a final check that all the screws and wiring have been removed and unclipped and then with the aid of an assistant remove the fascia panel **(see illustrations)**.

Refitting

17 Refitting is a reversal of the removal procedure, noting the following points:

a) *Ensure the facia guides engage correctly with the clips on the bulkhead. As the facia is being fitted, check that all wiring is routed as noted during removal.*

b) *Insert all of the retaining screws hand-tight, then close both front doors and check that the facia is positioned centrally between the door trims. If it needs to be moved one way or another, place a wad of cloth between the facia and door trim, then close the door. This should move the facia as required. When central, fully tighten the securing screws.*

c) *On completion, reconnect the battery and check that all the electrical components and switches function correctly.*

27.14b ...and unclip the plug

27.15b Undo the lower facia centre retaining screws...

27.15d Remove the screw from the centre of the binnacle

27.15a The slightly hidden screws in the end panels

27.15c ...and the upper ones

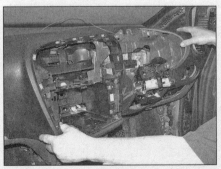

27.16a Remove the fascia (note the cord attached to the sunlight sensor as a refitting aid)

27.16b With the facia removed the crossmember is now accessible

28.3 Remove the bolt

28.6 Remove both central support struts

28.8 Unbolt the relay housing

28 Crossmember – removal and refitting

Note: *The most likely reason to remove the crossmember is to gain access to the heater housing and in particular the air conditioning components. If this is the case then the AC system must be degassed by a garage equipped with an AC service station. Many garages and specialists will offer a mobile service at a reasonable cost.*

Removal

1 Remove the facia panel as described in Section 27.

2 Unbolt the steering column from the crossmember (as described in Chapter 10 Section 20). There is no need to remove it completely, but if it is left attached to the steering rack and laid on the floor then care must be taken not to damage the column switches etc.

3 Remove the wiper arms and windscreen cowl panels as described in Chapter 12 Section 19. Remove the insulation and then remove the plenum chamber/scuttle bulkhead panel. Remove the single bolt from the bulkhead **(see illustration)**.

4 Working inside remove the central air distribution ducts from the top of the heater housing.

5 Unbolt the crash protector from the brake pedal.

6 Mark the position of the central support struts and then remove them **(see illustration)**.

7 Locate and remove the bolts securing the heater housing to the crossmember.

8 Unbolt the relay housing from the crossmember **(see illustration)**. On models fitted with Zenon headlights unbolt the control module form the crossmember.

9 Mark the position of the crossmember in relation to the A-pillars **(see illustration)** and then unbolt the crossmember. Note

28.9 Mark the position of the crossmember and then remove the bolts (early version shown)

that later models have a slightly different mounting, but removal is essentially the same. Check that all sections of the wiring loom are unclipped from the crossmember and then with the aid of an assistant remove the crossmember.

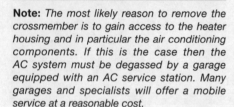

Chapter 12
Body electrical system

Contents

Degrees of difficulty

Easy, suitable for novice with little experience	**Fairly easy,** suitable for beginner with some experience	**Fairly difficult,** suitable for competent DIY mechanic	**Difficult,** suitable for experienced DIY mechanic	**Very difficult,** suitable for expert DIY or professional

Specifications

System type	12 volt negative earth
Fuses	See *Wiring diagrams*

Bulbs

	Wattage	Type
Brake light*	21W	P21W
Brake light – High level	LED	
Cornering light	35W	H8
Cornering light (models with bi-Zenon bulbs)	55W	H7
Door entry light (in door)	5W	Festoon
Footwell light	5W	W5W
Front direction indicators	21W	PY21W
Front direction indicator side repeater	5W	W5W
Front foglight (up to 2009)	55W	H11
Front foglight (2009 on)	55W	HB4
Front sidelight	5W	W5W
Glovebox light	3W	Festoon
Headlight:		
Halogen headlights		
Dipped beam	55W	H7
Main beam	55W	H1
Gas discharge headlights (Zenon)		
Dipped beam	35W	D1S
Main beam	55W	H7
Main/dipped (bi-Zenon models)	35W	D1S (H510)

Bulbs (continued)

	Wattage	Type
Interior light (reading)	5W	W5W
Interior light (courtesy)	10W	Festoon
Interior light (vanity mirror)	5W	Festoon
Load area light	10W	Festoon
Number plate light	5W	Festoon (C5W)
Rear direction indicators	21W	PY21W/P21
Rear foglight	21W	P21
Rear sidelight*	5W	R5W
Reversing light	21W	P21W

*LED type on some models

Torque wrench settings

	Nm	lbf ft
Crash sensor mounting bolt	9	7
Passenger airbag and bracket retaining screws	9	7
Wiper arm securing nut:		
Front	20	15
Rear	12	9

1 General information and precautions

⚠ **Warning: Before carrying out any work on the electrical system, read through the precautions given in Safety First! at the beginning of this manual, and in Chapter 5A Section 1.**

1 The electrical system is of 12-volt negative earth type. Power for the lights and all electrical accessories is supplied by a lead-acid type battery, which is charged by the alternator.

2 This Chapter covers repair and service procedures for the various electrical components not associated with the engine. Information on the battery, alternator and starter motor can be found in Chapter 5A.

3 It should be noted that prior to working on any component in the electrical system, the ignition and all electrical consumers must be switched off. Additionally, where stated, the battery negative lead must be disconnected, as described in Chapter 5A Section 1.

4 Some models are fitted with gas discharge headlight systems, which include automatic range control to reduce the possibility of dazzling oncoming drivers. Note the special precautions which apply to these systems as given in Section 6.

2 Electrical fault finding – general information

Note: *Refer to the precautions given in Safety First! and in Chapter 5A Section 1 before starting work. The following tests relate to testing of the main electrical circuits, and should not be used to test delicate electronic circuits (such as anti-lock braking systems), particularly where an electronic control module is used.*

General

1 A typical electrical circuit consists of an electrical component; any switches, relays, motors, fuses, fusible links or circuit breakers related to that component, and the wiring and connectors which link the component to both the battery and the chassis. To help to pin-point a problem in an electrical circuit, wiring diagrams are included at the end of this Chapter. **Note:** *Many of the circuits are controlled by computerised systems (for instance, the windscreen wipers will only operate with the bonnet closed), so before assuming there are faults, it is worthwhile checking if specific conditions apply.*

2 Before attempting to diagnose an electrical fault, first study the appropriate wiring diagram to obtain a complete understanding of the components included in the particular circuit concerned. The possible sources of a fault can be narrowed down by noting if other components related to the circuit are operating properly. If several components or circuits fail at one time, the problem is likely to be related to a shared fuse or earth connection.

3 Electrical problems usually stem from simple causes, such as loose or corroded connections, a faulty earth connection, a blown fuse, a melted fusible link, or a faulty relay (refer to Section 3 for details of testing relays). Visually inspect the condition of all fuses, wires and connections in a problem circuit before testing the components. Use the wiring diagrams to determine which terminal connections will need to be checked in order to pin-point the trouble spot.

4 The basic tools required for electrical fault finding include a circuit tester or voltmeter (a 12-volt bulb with a set of test leads can also be used for certain tests); a self-powered test light (sometimes known as a continuity tester); an ohmmeter (to measure resistance); a battery and set of test leads; and a jumper wire, preferably with a circuit breaker or fuse incorporated, which can be used to bypass

suspect wires or electrical components. Before attempting to locate a problem with test instruments, use the wiring diagram to determine where to make the connections.

5 To find the source of an intermittent wiring fault (usually due to a poor or dirty connection, or damaged wiring insulation), a wiggle test can be performed on the wiring. This involves wiggling the wiring by hand to see if the fault occurs as the wiring is moved. It should be possible to narrow down the source of the fault to a particular section of wiring. This method of testing can be used in conjunction with any of the tests described in the following sub-Sections.

6 Apart from problems due to poor connections, two basic types of fault can occur in an electrical circuit – open-circuit, or short-circuit.

7 Open-circuit faults are caused by a break somewhere in the circuit, which prevents current from flowing. An open-circuit fault will prevent a component from working, but will not cause the relevant circuit fuse to blow.

8 Short-circuit faults are caused by a short somewhere in the circuit, which allows the current flowing in the circuit to escape along an alternative route, usually to earth. Short-circuit faults are normally caused by a breakdown in wiring insulation, which allows a feed wire to touch either another wire, or an earthed component such as the bodyshell. A short-circuit fault will normally cause the relevant circuit fuse to blow.

Finding an open-circuit

9 To check for an open-circuit, connect one lead of a circuit tester or voltmeter to either the negative battery terminal or a known good earth.

10 Connect the other lead to a connector in the circuit being tested, preferably nearest to the battery or fuse.

11 Switch on the circuit, bearing in mind that some circuits are live only when the ignition switch is moved to a particular position.

12 If voltage is present (indicated either by

the tester bulb lighting or a voltmeter reading, as applicable), this means that the section of the circuit between the relevant connector and the battery is problem-free.

13 Continue to check the remainder of the circuit in the same fashion.

14 When a point is reached at which no voltage is present, the problem must lie between that point and the previous test point with voltage. Most problems can be traced to a broken, corroded or loose connection.

Finding a short-circuit

15 To check for a short-circuit; first disconnect the load(s) from the circuit (loads are the components which draw current from a circuit, such as bulbs, motors, heating elements, etc).

16 Remove the relevant fuse from the circuit, and connect a circuit tester or voltmeter to the fuse connections.

17 Switch on the circuit, bearing in mind that some circuits are live only when the ignition switch is moved to a particular position.

18 If voltage is present (indicated either by the tester bulb lighting or a voltmeter reading, as applicable), this means that there is a short circuit.

19 If no voltage is present, but the fuse still blows with the load(s) connected, this indicates an internal fault in the load(s).

Finding an earth fault

20 The battery negative terminal is connected to earth – the metal of the engine/transmission and the car body – and most systems are wired so that they only receive a positive feed, the current returning through the metal of the car body. This means that the component mounting and the body form part of that circuit. Loose or corroded mountings can therefore cause a range of electrical faults, ranging from total failure of a circuit, to a puzzling partial fault. In particular, lights may shine dimly (especially when another circuit sharing the same earth point is in operation), motors (e.g. wiper motors or the radiator cooling fan motor) may run slowly, and the operation of one circuit may have an apparently unrelated effect on another. Note that on many vehicles, earth straps are used between certain components, such as the engine/transmission and the body, usually where there is no metal-to-metal contact between components due to flexible rubber mountings, etc.

21 To check whether a component is properly earthed, disconnect the battery (refer to the warnings given in the Reference section at the rear of the manual) and connect one lead of an ohmmeter to a known good earth point. Connect the other lead to the wire or earth connection being tested. The resistance reading should be zero; if not, check the connection as follows.

22 If an earth connection is thought to be faulty, dismantle the connection and clean back to bare metal both the bodyshell and the wire terminal or the component earth connection mating surface. Be careful to

3.2 Remove the end panel to access the interior fusebox. Note the fuse puller

remove all traces of dirt and corrosion, and then use a knife to trim away any paint, so that a clean metal-to-metal joint is made. On reassembly, tighten the joint fasteners securely; if a wire terminal is being refitted, use serrated washers between the terminal and the bodyshell to ensure a clean and secure connection. When the connection is remade, prevent the onset of corrosion in the future by applying a coat of petroleum jelly or silicone-based grease or by spraying on (at regular intervals) a proprietary ignition sealer or a water dispersant lubricant.

3 Fuses and relays – general information

Fuses and fusible links

1 Fuses are designed to break a circuit when a predetermined current is reached, in order to protect the components and wiring, which could be damaged by excessive current flow. Any excessive current flow will be due to a fault in the circuit, usually a short-circuit (see Section 2).

2 The main fuses are located in the fusebox on the end of driver's side of the facia. Prise off the facia end cover to access the fuses **(see illustration)**.

3 To remove a fuse, first switch off the circuit concerned (or the ignition), and then pull the fuse out of its terminals **(see illustration)**.

4 The wire within the fuse should be visible; if the fuse has blown it will be broken or melted.

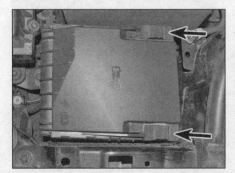

3.8a Slide the locking tabs in the direction of the arrows…

3.3 Removing a fuse from the fusebox

5 Always renew a fuse with one of the correct rating; never use a fuse with a different rating from that specified.

6 Refer to the wiring diagrams for details of the fuse ratings and the circuits protected. The fuse rating is stamped on the top of the fuse; the fuses are also colour-coded as follows.

Colour	Rating
Light brown	5A
Brown	7.5A
Red	10A
Blue	15A
Yellow	20A
White or clear	25A
Green	30A
Orange	40A

7 Never renew a fuse more than once without tracing the source of the trouble. If the new fuse blows immediately, find the cause before renewing it again; a short to earth as a result of faulty insulation is most likely. Where a fuse protects more than one circuit, try to isolate the fault by switching on each circuit in turn (where possible) until the fuse blows again. Always carry a supply of spare fuses of each relevant rating on the vehicle.

8 Additional fuses and relays are located in the fusebox located on the left-hand side of the engine compartment. Release the two securing clips and withdraw the cover to gain access to the fuses and relays **(see illustrations)**.

3.8b …to access the fuse and relay box in the engine compartment

3.9 Fusible link on the front of the fuse/relay box

3.12 Relays located behind the facia (shown with facia removed)

9 To renew a fusible link, first disconnect the battery negative terminal. Unscrew the retaining nuts then remove the blown link from the front of the engine compartment fusebox **(see illustration)**. Fit the new link to its terminals and reconnect the leads. Ensure the link and leads are correctly seated then refit the retaining nuts and tighten securely. Clip the cover back into position then reconnect the battery.

Relays

10 A relay is an electrically operated switch, which is used for the following reasons:
a) *A relay can switch a heavy current remotely from the circuit in which the current is flowing, allowing the use of lighter-gauge wiring and switch contacts.*
b) *A relay can receive more than one control input, unlike a mechanical switch.*
c) *A relay can have a timer function – for example, the intermittent wiper relay.*

11 Most of the relays are located on the relay plate behind the driver's side facia, however, additional relays are located in the engine compartment fusebox **(see illustration 3.8)**.
12 Access to the relays can be obtained after removing the cover from the diagnostic socket and by reaching up from beneath the facia **(see illustration)**. Identification details of the relays are given at the start of the wiring diagrams.
13 If a circuit or system controlled by a relay develops a fault, and the relay is suspect, operate the system. If the relay is functioning, it should be possible to hear it click as it is energised. If this is the case, the fault lies with the components or wiring of the system. If the relay is not being energised, then either the relay is not receiving a main supply or a switching voltage, or the relay itself is faulty. Testing is by the substitution of a known good unit, but be careful – while some relays are identical in appearance and in operation, others look similar but perform different functions.
14 To remove a relay, first ensure that the

relevant circuit is switched off. The relay can then simply be pulled out from the socket, and pushed back into position.
15 The direction indicator/hazard flasher relay is integral with the hazard warning switch. Refer to Section 5 for the switch removal procedure.

4 Electrical connectors

1 Most electrical connections on these vehicles are made with multiwire plastic connectors. The mating halves of many connectors are secured with locking clips molded into the plastic connector shells. The mating halves of some large connectors, such as some of those under the instrument panel, are held together by a bolt through the center of the connector.
2 To separate a connector with locking clips, use a small screwdriver to pry the clips apart carefully, then separate the connector halves. Pull only on the shell, never pull on the wiring harness, as you may damage the individual wires and terminals inside the connectors. Look at the connector closely before trying to separate the halves. Often the locking clips are engaged in a way that

4.5a Most electrical connectors have a single release tab that you depress to release the connector

is not immediately clear. Additionally, many connectors have more than one set of clips.
3 Each pair of connector terminals has a male half and a female half. When you look at the end view of a connector in a diagram, be sure to understand whether the view shows the harness side or the component side of the connector. Connector halves are mirror images of each other, and a terminal shown on the right side end-view of one half will be on the left side end-view of the other half.
4 It is often necessary to take circuit voltage measurements with a connector connected. Whenever possible, carefully insert a small straight pin (not your meter probe) into the rear of the connector shell to contact the terminal inside, then clip your meter lead to the pin. This kind of connection is called "backprobing." When inserting a test probe into a terminal, be careful not to distort the terminal opening. Doing so can lead to a poor connection and corrosion at that terminal later. Using the small straight pin instead of a meter probe results in less chance of deforming the terminal connector. "T" pins are a good choice as temporary meter connections. They allow for a larger surface area to attach the meter leads too.
5 Typical electrical connectors **(see illustrations)**:

4.5b Some electrical connectors have a retaining tab which must be pried up to free the connector

4.5c Some connectors have two release tabs that you must squeeze to release the connector

4.5d Some connectors use wire retainers that you squeeze to release the connector

4.5e Critical connectors often employ a sliding lock (1) that you must pull out before you can depress the release tab (2)

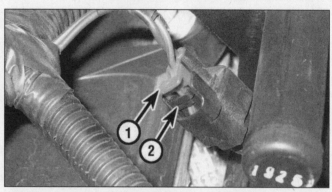

4.5f Here's another sliding-lock style connector, with the lock (1) and the release tab (2) on the side of the connector

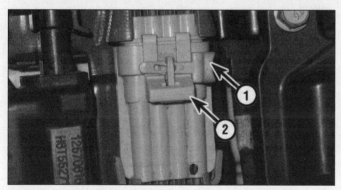

4.5g On some connectors the lock (1) must be pulled out to the side and removed before you can lift the release tab (2)

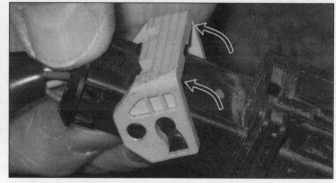

4.5h Some critical connectors, like the multi-pin connectors at the Electronic Control Module employ pivoting locks that must be flipped open

5 Switches – removal and refitting

Note: *Before working on any electrical systems switch off the ignition and all electrical consumers and remove the ignition key.*

Ignition switch

1 Refer to Chapter 10 Section 21.

Wiper and indicator/ cruise switches

2 Check that the front wheels are pointing straight-ahead and the steering wheel is in its centre position, then remove the steering wheel as described in Chapter 10 Section 19.

3 Remove the steering column shrouds, as described in Chapter 11 Section 25.

4 Disconnect the wiring connectors from the electronic control unit **(see illustration)**.

5 Remove the steering column electronics control unit, from below the steering column. To do this, undo the single retaining screw, then insert a 2.5 mm diameter rod (drill bit) through the hole provided (below the screw hole), and release the centre clip. Now use a screwdriver to release the rear clip. Pull down

5.4 Disconnect the wiring connector

5.5a Undo the retaining screw

5.5b Insert a 2.5mm drill bit to release the front securing clip

5.5c Release the rear clip and remove the control unit

5.5d The two securing clips

5.6a Release the two securing clips...

5.6b ...and remove the airbag clock spring/slip-ring unit

the control unit from the column switch carrier **(see illustrations)**.

6 The airbag clock spring/slip-ring must be held in its centre position while it is removed, to ensure correct refitting. Unclip the airbag clock spring/slip-ring from the combination switch carrier by lifting the retaining hooks **(see illustrations)**.

7 Remove the relevant switch by inserting a 1.0 mm feeler gauge through the slot provided to release the switch from the carrier. Release the securing clip at the rear of the switch and slide the switch forwards out from the housing **(see illustrations)**.

8 Refitting is a reversal of removal.

Models from 06/2010

Note: *On these models the lighting and wiper*

switch is one unit and cannot be renewed separately.

9 Disconnect the wiring connectors from the airbag clock spring/electronic control unit. Note this is one unit and cannot be separated.

10 On switch assembly manufactured by VALEO, undo the three retaining screws and withdraw the airbag clock spring/electronic control unit from over the steering column. Then undo the three securing screws and remove the switch assembly.

11 On switch assembly manufactured by KOSTAL, undo the retaining screw at the lower part of the assembly, then release the two upper securing clips, then withdraw the airbag clock spring/electronic control unit from over the steering column. Undo

the securing screw from the lower part of the assembly, and then remove the switch assembly.

12 Refitting is a reversal of removal. Note, as the switch assembly is being refitted, make sure the indicator switch is in the centre position; otherwise the reset lever if extended can break off as the switch is fitted.

Lighting switch

13 With the light switch in position O, press the switch centre inwards and turn it slightly to the right. Hold this position and pull the switch from the dash **(see illustration)**. Removing the facia end panel makes this task slightly easier.

5.7a Slide a feeler blade down to release the securing clip...

5.7b ...then unclip and remove the switch

5.13 Press the switch centre inwards and turn it slightly to the right to remove

5.14 Disconnect the wiring plug

5.21a Remove the storage compartment

5.21b Disconnect the wiring plug

14 As the switch is withdrawn from the dash, disconnect the wiring plug **(see illustration)**.
15 To refit the switch, first reconnect the wiring plug, then hold the switch and press the rotary part inwards and slightly to the right. Insert the switch into the dash, turn the rotary part to position O and release. Check the switch for correct operation.

Headlamp range control and instrument illumination switch

16 Remove the lighting switch as described in this section.
17 Unclip the switch from the main light control switch.
18 Refitting is a reversal of removal.

Air conditioning/rear window heating switches

19 The switches are integral with the heater control panel, and cannot be removed separately. Refer to Chapter 3 Section 9 for details of heater control panel removal and refitting.

Heater blower motor switch

20 The switch is integral with the heater control panel, and cannot be removed separately. Refer to Chapter 3 Section 9 for details of heater control panel removal and refitting.

Hazard warning

21 Remove the storage compartment from the front of the centre console. Disconnect the wiring plug as the compartment is removed. Unclip the switch from the panel **(see illustrations)**.
22 Refitting is a reversal of removal.

Centre console mounted switches

23 The number of switches mounted on the centre console varies according to the trim level and the year of manufacture. The following switches are mounted to the centre console switch panel:
● Mirror adjustment switch
● ESP (Electronic Stability Programme) switch
● Door lock open/close switch
● TPM (Tyre Pressure Monitoring) switch
● Parking aid switch
● Stop/Start switch
24 Prise free the switch panel. Disconnect the wiring plug(s) as the panel is removed **(see illustrations)**.
25 Release the retaining clips and withdraw the appropriate switch from the panel.
26 Refitting is a reversal of removal.

Passenger airbag disable/ deactivation switch (PAD)

Caution: Disconnect the battery and wait several minutes before disconnecting the

5.21c Remove the switch

switch. Rear the warnings given in this Section 26 when working on any part of the airbag system.
27 The switch is located in the glovebox. Remove the glovebox as described in Chapter 11 Section 25.
28 Using a trim tool placed at the rear of the switch, unclip the switch.
29 Refitting is a reversal of removal.

Glovebox light switch

30 Remove the glovebox, as described in Chapter 11 Section 25.
31 Release the retaining lug and slide the switch downwards to remove it from the rear of the glovebox.
32 Refitting is a reversal of removal.

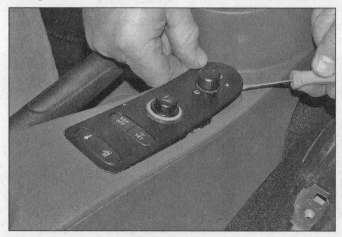

5.24a Prise up the panel...

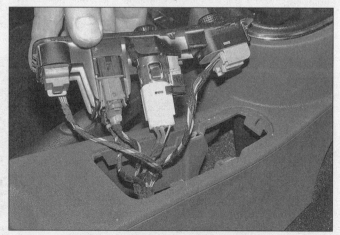

5.24b ...and disconnect the wiring plugs

5.33a Remove the cover from the handle

5.33b Remove the insert panel

5.34 Remove the switch (front shown, but the rear is similar)

Power window switch assembly

33 Using a trim tool, remove the pull handle cover and then remove the trim from the door panel **(see illustrations)**.

34 Locate the now exposed locking tabs, release them and fully prise out the panel **(see illustration)**.

35 Disconnect the wiring plug as the switch panel is removed **(see illustration)**.

36 Refitting is a reversal of removal.

Handbrake 'on' warning switch

37 Refer to Chapter 9 Section 17.

Brake light switch

38 Refer to Chapter 9 Section 18.

Reversing light switch

39 Refer to.

Courtesy light switches

40 The courtesy light switches are integrated into the door lock mechanisms and cannot be renewed independently. If the courtesy light switch is faulty, renew the door lock mechanism as described in Chapter 11 Section 13.

Luggage area light switch

41 The luggage compartment light switch is integrated into the tailgate lock mechanism, and cannot be renewed independently. If the luggage compartment light switch is faulty, renew the tailgate lock mechanism as described in Chapter 11 Section 16.

Interior monitoring deactivation switch

42 Carefully prise the switch from the B-pillar trim panel. Disconnect the wiring plug.

43 Refitting is a reversal of removal.

Rain and light sensor

44 The windscreen wipers/lights (depending on model) are automatically activated, when the sensor located in the interior mirror base detects droplets of water/darkness.

45 Unclip the outer cover and then slide up the inner cover.

46 Release the sensor and disconnect the wiring plug.

47 Refitting is a reversal of removal, but note that the windscreen must be thoroughly clean (or cleaned) before refitting the sensor.

---

6 Bulbs (exterior lights) – renewal

General

1 Whenever a bulb is renewed, note the following points:

a) *Switch off the ignition and all electrical consumers before commencing work.*

b) *Remember that if the light has just been in use the bulb may be extremely hot.*

c) *Always check the bulb contacts and holder, ensuring that there is clean metal-to-metal contact. Clean off any corrosion or dirt before fitting a new bulb.*

d) *Wherever bayonet-type bulbs are fitted ensure that the spring-tensioned arms bear firmly against the bulb contacts.*

e) *Always ensure that the new bulb is of the correct rating and that it is thoroughly clean before fitting it.*

Halogen headlight bulbs

2 If working on the left-hand headlight remove the air filter housing as described in Chapter 4A Section 3 (petrol engines) or (diesel engines).

3 If working on the right-hand headlight on vehicles with a petrol engine move the charcoal canister to the side as described in Chapter 4C Section 2. If working on a vehicle with a diesel engine move the fuel filter housing to the side as described in. In both cases there is no need to disconnect the pipework.

Headlight main beam

Note: *Do not touch the glass envelope of the bulb if it is to be re-used.*

4 Remove the cover from the rear of the headlight **(see illustration)**.

5 Hold the bulb in position and disconnect the wiring plug **(see illustration)**. Release the

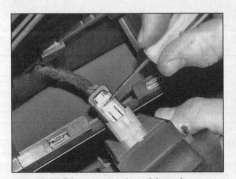

5.35 Disconnect the wiring plug

6.4 Remove the cover

6.5a Disconnect the wiring plug

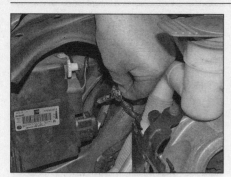
6.5b Remove the bulb

6.9a Release the spring and...

6.9b ...remove the cover

spring retaining clip and withdraw the bulb **(see illustration)**.

6 When handling the new bulb, use a tissue or clean cloth to avoid touching the glass with the fingers; moisture and grease from the skin can cause blackening and rapid failure of this type of bulb. If the glass is accidentally touched, wipe it clean using methylated spirit.

7 Install the new bulb; ensuring that it is located correctly in the headlight unit. Refit the spring clip and the wiring plug.

8 Refit the headlight cover, making sure that it is secure.

Headlight dip beam

Note: *Do not touch the glass envelope of the bulb if it is to be re-used.*

9 Working in the engine compartment, release the spring clip and remove the cover **(see illustrations)**.

10 Reach into the headlight and release the spring retaining clip and remove the bulb.

Disconnect the wiring plug from the bulb. Alternatively, disconnect the wiring plug and then release the spring clip before removing the bulb **(see illustrations)**.

11 When handling the new bulb, use a tissue or clean cloth to avoid touching the glass with the fingers; moisture and grease from the skin can cause blackening and rapid failure of this type of bulb. If the glass is accidentally touched, wipe it clean using methylated spirit.

12 Fit the wiring plug to the new bulb and then fit the bulb to the headlight, ensuring that the location lugs are aligned. Refit the bulb retaining spring clip.

13 Refit the headlight cover, making sure that it is secure.

Sidelight bulb

14 Remove the rear cover by pulling it free **(see illustration 6.4)**. The bulb is located below the main beam bulb.

15 Pull out the bulb, complete with the bulb holder.

16 The bulb is a push-fit in the holder and can be removed by depressing the locking tabs on the bulb holder. Access is limited and a pair of bent nose needle nose pliers will make removal considerably easier. Remove the bulb from the bulb holder **(see illustration)**.

17 Refitting is a reversal of removal, making sure that the headlight cover is securely refitted.

Front direction indicator

Note: *The direction indicator bulb is accessible without removing the air filter housing, EVAP canister (petrol models) or fuel filter (diesel models).*

18 Turn the bulbholder anti-clockwise and remove it from the headlight complete with the bulb **(see illustration)**.

19 Depress and then turn the bulb to release it from the bulb holder **(see illustration)**.

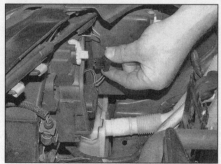
6.10a Disconnect the wiring plug

6.10b Release the spring clip

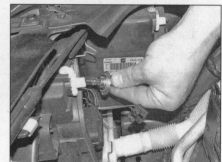
6.10c Remove the bulb

6.16 Remove the bulb

6.18 Remove the bulb and bulb holder

6.19 Remove the bulb

20 Fit the new bulb to the bulb holder, then fit the bulb holder to the headlight and turn clockwise to secure.

Gas discharge headlights

Warning: The headlight bulb contains gas at very high pressure, and it is recommended that gloves and eye protection be worn to prevent potential personal injury.

Note: *Do not touch the glass envelope of the bulb if it is to be re-used.*

Note: *Models fitted with gas discharge (Zenon) bulbs have a 'bi-zenon' type of bulb fitted. These bulbs provide (by means of a mechanical shutter) both main and dipped beams. Some models also have an additional standard halogen main beam bulb fitted. Daytime running lamps are also fitted to some versions.*

Dipped/main beam

Note: *Two versions of the bulb are fitted. Removal is similar regardless of the type fitted.*
21 Remove the headlight as described in Section 8.
22 Remove the screws from the rear of the headlight and then remove the cover.
23 Compress the retaining spring and push it upwards to release it. Disconnect the wiring plug and remove the bulb.
24 If working on a model fitted with a H510 type bulb, disconnect the wiring plug and then rotate the locking ring to release the bulb. Remove the bulb.
25 When handling the new bulb, use a tissue or clean cloth to avoid touching the glass with the fingers; moisture and grease from the skin can cause blackening and rapid failure of

this type of bulb. If the glass is accidentally touched, wipe it clean using methylated spirit.
26 Fit the new gas discharge bulb and wiring plug. Refit the spring clip or on H510 bulbs rotate the lock ring.
27 Refit the headlight cover and then refit the headlight.
Caution: After refitting a gas discharge headlamp, the basic setting of the Automatic Range Control system should be checked.

Main beam

28 If working on the left-hand headlight remove the air filter housing as described in Chapter 4A Section 3 (petrol engines) or (diesel engines).
29 If working on the right-hand headlight on vehicles with a petrol engine move the charcoal canister to the side as described in Chapter 4C Section 2. If working on a vehicle with a diesel engine move the fuel filter housing to the side as described in. In both cases there is no need to disconnect the pipework.
30 Pull off the rear cover.
31 Disconnect the wiring plug and then release the spring clip by pushing it upwards. Remove the bulb.

Daytime running bulb

32 Where fitted the bulb is located next to the front sidelight. Removal and refitting is the same procedure for the replacement of the front sidelight as described above for halogen headlights.

Sidelight bulb

33 The removal procedure is the same as described above for halogen headlights.

Front direction indicator

34 The removal procedure is the same as described above for halogen headlights.

Front foglight

35 If not already done so, turn off the ignition and remove the vehicle key. Remove the single screw and unclip the cover from the front of the foglight. Note that some models do not have a removable grille/cover. On these models if working on the right-hand foglight the wing liner must be removed to reach the bulb. If working on the left-hand foglight access is possible via the hole in the inner wing (beneath the headlight) after removing the air filter housing – where required.
36 Remove the fixings and pull the lamp forward **(see illustrations)**.
37 On later models (03/2009 on) turn the bulb holder anti-clockwise and remove it complete with the bulb from the rear of the foglight. **Note:** *The bulb is integral with the bulbholder. On early models (up to 03/2009) release the cover from the back of the foglight by unhooking the spring clip. Disconnect the wiring plug, release the bulb retaining spring clip and remove the bulb* **(see illustration)**.
38 Fit the new bulb using a reversal of the removal procedure.

Direction indicator side repeater

39 Protect the paintwork if necessary and then lever off the lamp using a trim tool **(see illustration)**.
40 Rotate the bulb holder, remove it and pull out the wedge type bulb **(see illustration)**.
41 Refitting is a reversal of removal.

6.36a Unclip the cover and...

6.36b ...remove the screws...

6.36c ...and pull out the complete lamp

6.37 Open the rear cover to access the bulb

6.39 Remove the lamp. Note the spring clip retainer

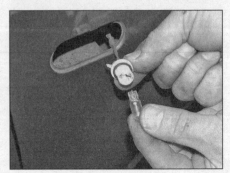

6.40 Remove the bulb

6.43a Remove the rear light bulbholder...

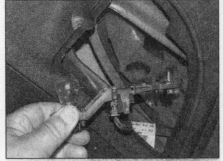

6.43b ...and remove the relevant bulb

6.46a Release the locking tabs...

6.46b ...and remove the bulb holder

Rear light cluster

Note: *Some models feature LED type lights in the rear lamps. If these fail the complete lamp must be replaced.*

Wing mounted lights

42 Remove the rear light cluster as described in Section 8.
43 Unclip the bulb holder from the cluster **(see illustrations)**.
44 Fit the new bulb using a reversal of the removal procedure.

Tailgate mounted lights

Note: *The tailgate lamp contains the rear side light and either the rear foglight or reversing light.*
45 Open the tailgate and remove the cover.
46 Release the retaining clips and remove the bulb holder from the rear of the light unit. **(see illustrations)**.
47 Fit the new bulb using a reversal of the removal procedure.

High-level brake light

Note: *The light is of LED design; therefore if faulty the complete unit must be renewed. Note that the LEDs are arranged in groups of four, and if just one group fails, the light still meets legal requirement. Failure of more than one group renders the light illegal.*
48 Remove the high-level light unit as described in Section 8.

Number plate light

49 Remove the screws and withdraw the light unit from the tailgate **(see illustration)**.
50 Remove the bulb holder and then remove the festoon type bulb **(see illustrations)**.
51 Fit the new bulb using a reversal of the removal procedure.

7 Bulbs (interior lights) – renewal

General

1 Whenever a bulb is renewed, note the following points:
a) *Switch off the ignition and all electrical consumers before commencing work.*

6.46c ...and remove the relevant bulb

b) *Remember that if the light has just been in use the bulb may be extremely hot.*
c) *Always check the bulb contacts and holder, ensuring that there is clean metal-to-metal contact between them. Clean off any corrosion or dirt before fitting a new bulb.*
d) *Wherever bayonet-type bulbs are fitted ensure that the live contact(s) bear firmly against the bulb contact.*
e) *Always ensure that the new bulb is of the correct rating and that it is completely clean before fitting it.*

Front courtesy/reading light

Note: *The design of the light varies according to the trim level and year of production. Removal for all variants is essentially the same procedure.*

6.49 Remove the screws

6.50a Remove the lamp and then...

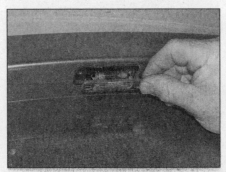

6.50b ...remove the bulb

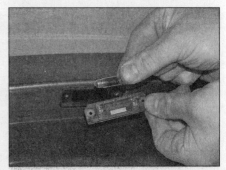

7.2 Remove the lens

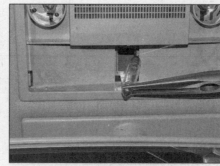

7.3 Remove the bulb

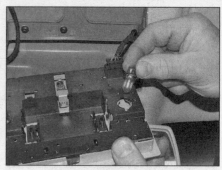

7.7 Remove the bulb holder

2 Using a small screwdriver, release the lens **(see illustration)**.
3 Unclip the festoon type bulb (courtesy light) **(see illustration)**.

7.8 Removing the bulb with a length of tubing

4 Fit the new bulb(s) using a reversal of the removal procedure.
5 On some variants, to access the reading light bulbs the complete assembly must be removed.
6 Release the complete lamp with a trim tool.
7 Rotate the bulb holder to access the bulb. Pull the wedge type bulb straight out from the bulb holder **(see illustration)**.
8 It is also possible to remove the bulb with a short length of flexible tube – screen washer hose for example **(see illustration)**.
9 Refitting is a reverse of the removal procedure.

Rear courtesy/reading lights

Note: *There are slight variation in the design of the rear light, depending on the year of production and if an anti-theft alarm system is fitted or not. Removal and refitting of the bulbs is essentially the same for all models.*

10 Using a screwdriver, carefully release the locking lugs and remove the lens from the light unit **(see illustration)**.
11 Pull the festoon or wedge-type bulb from the light **(see illustration)**.
12 If required remove the light unit by releaseing the two retaining clips and removing it from the headlining. Disconnect the wiring connector as it is removed **(see illustrations)**.
13 Fit the new bulb and light unit using a reversal of the removal procedure.

Vanity mirror lights

14 Carefully prise the light unit from its location in the headlining **(see illustrations)**. The festoon-type bulb is a push-fit in the spring contacts
15 Refitting is a reverse of the removal procedure.

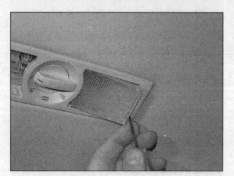

7.10 Unclip the light cover...

7.11 ...and pull out the festoon (or wedge-type) bulb

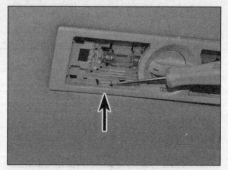

7.12a Release the locking tabs with a small screwdriver...

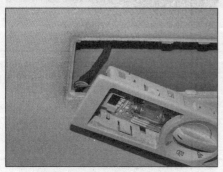

7.12b ...and remove the lamp

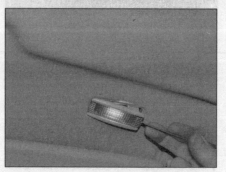

7.14a Prise out the vanity mirror light...

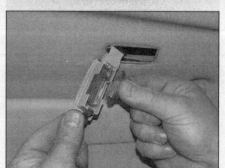

7.14b ...and remove the festoon-type bulb

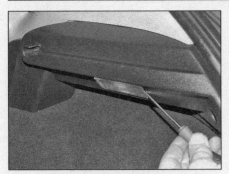

7.19 Remove the lamp with a small screwdriver

8.4a Remove the lower bolt...

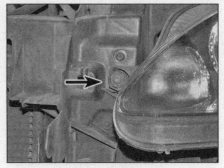

8.4b ...the inner...

Glovebox illumination light

16 Open the glovebox, then use a screwdriver to prise out the lens.

17 Disconnect the wiring plug and pull out the wedge-type bulb.

18 Fit the new bulb using a reversal of the removal procedure.

Luggage compartment light

19 Carefully prise the light unit from its location in the load area side panel **(see illustration)**. The festoon-type bulb is a push-fit in the spring contacts.

20 Fit the new bulb using a reversal of the removal procedure.

Heater/ventilation control panel illumination

21 The control panel is illuminated by LEDs built into the panel. Consequently, if a fault develops, renewal of the panel is necessary.

Switch illumination

22 The switch illumination bulbs are integral with the switches. If a bulb fails, the complete switch must be renewed.

> ### 8 Exterior light units –
> removal and refitting

1 Before removing any light units, switch off the ignition and all electrical consumers, and then remove the ignition key.

Headlight

2 Remove the front bumper cover, as described in Chapter 11 Section 6.

3 Mark the position of the headlight mountings to ensure correct alignment on refitting

4 Remove the upper, lower and inboard mounting bolts **(see illustrations)**.

5 Pull the headlight forward and disconnect the wiring plug **(see illustration)**.

6 Refitting is a reversal of removal, but on completion, check that the headlight is aligned flush with the surrounding bodywork. If not, turn the eccentric adjustment bushing on the inside of the headlight as required. Finally, have the headlight alignment checked at the earliest opportunity.

Caution: After refitting a gas discharge headlamp, the basic setting of the Automatic Range Control system should be checked. Because of the requirement for specialised equipment, this can only be carried out by a Seat dealer or suitably equipped specialist. The headlight bulb contains gas at very high pressure, and it is recommended that gloves and eye protection be worn to prevent potential personal injury.

Gas discharge bulb control unit/ starter

7 Remove the headlight as described above.

8 Undo the retaining screws, and remove the control unit from the base of the headlight.

8.4c ...and finally the upper

Note that the electrical connections are automatically separated when the unit is removed.

9 Refitting is a reversal of removal.

Direction indicator side repeater

10 The lamp is removed for bulb replacement as described in Section 6.

Front foglight

11 The foglight is removed for bulb replacement as described in Section 6.

Wing mounted lights

12 Open the tailgate, and remove the outer bolts **(see illustration)**.

13 Open the access panel and unscrew the thumb wheel from the rear of the lamp **(see illustration)**.

14 Pull the lamp forward, disconnect the

8.5 Disconnect the wiring plug

8.12 Remove the bolts

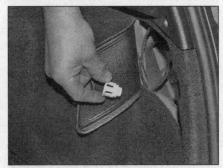

8.13 Unscrew the thumb wheel (shown removed)

8.14 Disconnect the wiring plug

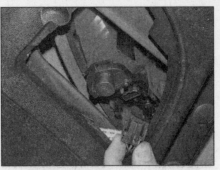

8.16 Disconnect the wiring plug

8.17a Remove the covers

8.17b Remove the outer fixings...

8.17c ...and the inner fixing

wiring plug and remove the lamp **(see illustration)**.

15 Refitting is a reversal of removal.

Tailgate mounted lights

16 Open the tailgate and then open the access panel. Disconnect the wiring plug **(see illustration)**.

17 Remove the blanking plugs and then unbolt the lamp. Remove the lamp **(see illustrations)**.

High level brake light

Note: *The light is of LED design; therefore if faulty the complete unit must be renewed. Note that the LEDs are arranged in groups of four, and if just one group fails, the light still meets legal requirement. Failure of more than one group renders the light illegal.*

18 Open the tailgate and remove the blanking plugs **(see illustration)**.

19 Using a flat bladed screwdriver through the access holes unclip the lamp **(see illustrations)**.

20 Fit the new light unit using a reversal of the removal procedure.

Rear number plate light

21 The procedure is described as part of the rear number plate light bulb renewal procedure in Section 6.

8.17d Remove the lamp

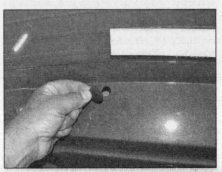

8.18 Remove the blanking plugs

| 9 | **Headlight beam adjustment components** – removal and refitting | |

Headlight adjustment switch

1 The switch is integral with the instrument illumination switch.

2 Removal and refitting of the switch assembly is covered in Section 5.

Headlight range adjustment motor

3 If working on the left-hand headlight remove the air filter housing as described in Chapter 4A Section 3 (petrol engines) or (diesel engines). If working on the right-hand headlight, remove the headlight as described in Section 8.

8.19a Unclip the lamp using the right-hand access point

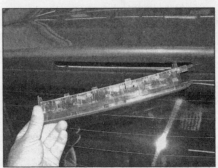

8.19b Lift out the brake light and disconnect the wiring plug

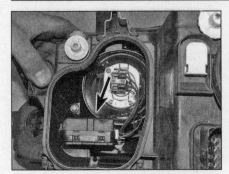
9.5 The range control motor

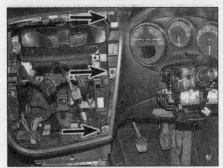
11.4 Remove the screws

11.5 Remove the panel

11.6a Remove the screws

11.6b Pull out the instrument panel and withdraw it from the facia (steering wheel removed for clarity)

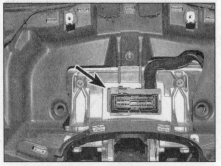
11.6c The captive wiring plug on early models

4 Remove the outer cover from the rear of the headlight.

5 Disconnect the wiring from the range control motor, then turn it clockwise and pull it back slightly, then press the ball-head downwards out of the cup, and remove the motor **(see illustration)**.

6 Refitting is a reversal of removal, making sure the ball joint mounting is correctly located in the headlight reflector.

Automatic range control ECU

Note: *Although it is possible to remove and refit the ECU, the new unit will need to be coded before it will function correctly. This task can only be carried out by a Seat dealer or suitably equipped specialist.*

7 The ECU is located behind the left-hand facia end panel. Prise the panel and remove the mounting bolts.

8 Release the locking lever and open the wiring connector so that it can be disconnected.

9 Refitting is a reversal of removal.

Vehicle level sender

10 Model fitted with gas discharge headlights (Zenon) have level sensors fitted to the front left-hand suspension arm and to the rear suspension.

11 Chock the rear road wheels, then jack up the front of the vehicle and support on axle stands (see *Jacking and vehicle support*). Remove the relevant left-hand front road.

12 Disconnect the wiring plug and remove the upper mounting bolt. Remove the lower bolt from the link arm bracket and withdraw the sensor.

13 The procedure for removing the rear sensor is identical.

14 Refitting is a reversal of removal, but note that the light leveling system will need calibrating using factory equipment.

10 Headlight beam alignment – general information

1 Accurate adjustment of the headlight beam is only possible using optical beam setting equipment and this work should be carried out by a Seat dealer or suitably-equipped workshop. All MOT stations will have this equipment.

2 For reference, the headlights can be manually adjusted using the adjuster assemblies fitted to the top of each light unit.

11 Instrument panel – removal and refitting

Removal

1 Switch off the ignition and all electrical consumers and remove the ignition key. Release the steering wheel adjustment handle, pull the wheel out as far as possible, and set it in the lowest position.

2 Remove the audio (or SatNav) unit as described in Section 22.

3 Remove the heating control panel as described in Chapter 3 Section 9.

4 Remove the now exposed fixings **(see illustration)**.

5 Protecting the facia as required and using a plastic trim tool work the panel trim free from the facia **(see illustration)**.

6 Release the upper and lower fixings and carefully pull the instrument panel from the facia. Note that the wiring plug is captive and releases as the panel is removed on early models (up to 02/09). On later models release the locking bar and disconnect the wiring plug as the panel is removed **(see illustrations)**.

Refitting

7 Refitting is a reversal of removal.

12 Instrument panel components – removal and refitting

1 It is not possible to dismantle the instrument panel. If any of the gauges are faulty, the complete instrument panel must be renewed.

15.3 Unclip the bulb holder

15.4a Work free the centre (earth) section...

15.4b ...and remove it

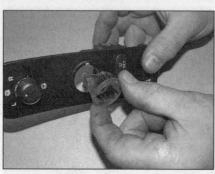

15.4c Remove the bezel

13 Service interval display – general information and resetting

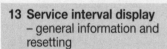

1 All models are equipped with a Service Interval Display (SID). This is displayed on the information panel in the centre of the instrument panel. After all necessary maintenance work has been completed the SID should be reset.

2 Vehicles may be set to fixed or variable service intervals depending on the production number. It is possible to alter the service intervals using diagnostic equipment.

3 The reset procedure is part of routine maintenance as described in Section (petrol engines) or Chapter 1B Section 8 (diesel engines)

14 Clock – removal and refitting

1 The clock is integral with the instrument panel, and cannot be removed separately. The instrument panel is a sealed unit, and if the clock, or any other components, are faulty, the complete instrument panel must be renewed. Refer to Section 11 to remove it.

15 12v Power outlets – removal and refitting

Removal

Note: *A special tool is available that will allow*

the retaining sleeve of the power outlet to be easily removed (T40148).

1 Disconnect the battery negative lead (refer to Chapter 5A Section 3).

2 Prise out the switch panel from the centre console. Disconnect the wiring plug.

3 Where fitted, remove the bulb holder **(see illustration)**.

4 Release the retaining clips and push the centre element of the lighter out of the illuminated bezel **(see illustrations)**. Note that considerable dexterity will be required to remove the central section of the outlet.

Refitting

5 Refitting is a reversal of removal.

16 Horn – removal and refitting

Removal

1 Switch off the ignition and all electrical consumers and remove the ignition key.

2 Jack up and support the front of the vehicle (see *Jacking and vehicle support*) Remove the engine undershield and then release the front edge of the wing liner (or remove it completely). Note on some models there are two horns, one at each side of the vehicle.

3 Reach up and disconnect the wiring plug and then unscrew the mounting bolt. Remove the horn together with the mounting bracket **(see illustrations)**.

4 If required, unscrew the nut on the top of the horn to remove the bracket.

Refitting

5 Refitting is a reversal of removal.

17 Speedometer sensor – general information

1 Unlike earlier models, no electronic speedometer sensor is fitted to the models covered in this manual. Vehicle speed is determined from the ABS wheel sensor signals, and processed by the engine management ECU.

18 Wiper arm – removal and refitting

Removal

1 Operate the wiper motor, then switch off so that the wiper arms return to the at-rest position. Alternatively, set the wiper blades to the service (vertical) position by operating the wipers within 10 seconds of switching off the ignition. **Note:** *The wiper motor will only operate with the bonnet closed. If required, open the bonnet then use a screwdriver to close the bonnet catch.*

16.3a The right-hand front horn and...

16.3b ...the left-hand (shown with bumper removed for clarity)

18.3a Remove the cover

18.3b Unscrew the spindle nut...

18.3c ...and remove the wiper arm

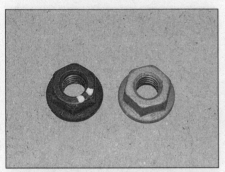

18.3d Note that the wiper arm nuts are coloured for left and right-hand threads

18.4a Prise up the cover and...

18.4b ...remove the nut...

2 Use a piece of masking tape to the glass along the edge of the wiper blade to use as an alignment aid on refitting.

Caution: The left-hand wiper arm has a left-hand thread. Turn the nut clockwise to remove it.

3 On the front wiper arms, prise off the wiper arm spindle nut cover, then slacken but do not completely remove the spindle nut. Lift the blade off the glass and carefully rock the wiper arm from side to side, until it releases from the spindle. If necessary, use a small puller to release the arm from the spindle. Remove the spindle nut and the wiper arm **(see illustrations). Note:** *If both windscreen wiper arms are to be removed at the same time mark them for identification; the arms are not interchangeable.*

4 On the rear wiper arms, unclip the spindle

nut cover, then slacken but do not completely remove the spindle nut. Lift the blade off the glass and carefully rock the wiper arm from side to side, until it releases from the spindle. Remove the spindle nut and wiper arm and on the rear wiper arms remove the washer jet **(see illustrations).**

5 If any of the arms are a tight fit on the spindle, the arm can be removed from the spindle using a small puller **(see illustration).**

Refitting

6 Ensure that the wiper arm and spindle splines are clean and dry, and then refit the arm to the spindle, aligning the wiper blade with the tape fitted on removal. Refit the spindle nut, tightening it securely, and clip the nut cover back in position.

Note: *Where the wipers were set to the service*

position in paragraph 1, they will only resume their normal position after actuating the wipers twice, or after starting a journey when the speed of the car is greater than 1 mph.

19 Windscreen wiper motor and linkage – removal and refitting

Removal

Note: *The wiper system uses two motors – there is no physical link between them. The motor control units are integrated into the wiper motors.*

1 Remove the wiper arms as described in Section 18.

2 Disconnect the battery negative lead (as described in Chapter 5A Section 3).

18.4c ...then remove the tailgate wiper arm

18.4d Remove the washer jet (rear screen only)

18.5 A small puller is needed to remove wiper arm

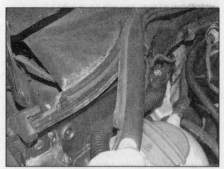

19.3a Remove the rubber seal...

19.3b ...work the cowl free...

19.3c ...and remove it

19.4a Remove the bolts...

19.4b ...and lift out the panel

19.5 Disconnecting the wiring plug

3 Pull off the rubber sealing strip from the top of the bulkhead and then remove the plenum chamber cover/windscreen cowl **(see illustrations)**. Note that the panel is a tight fit in the gutter at the base of the windscreen. Apply plenty of lubricant and slowly ease the panel free from the gutter.
Caution: Do not use a screwdriver to lever between the cowling and windscreen, as *this is likely to result in the windscreen cracking.*
4 Remove the bolts from each end of the plenum chamber front panel and then pull the panel up to remove it **(see illustrations)**.
5 Disconnect the wiring plug from the wiper motor **(see illustration)**.
6 Unscrew the mounting bolts and manoeuvre the windscreen wiper motor out from the plenum chamber **(see illustrations)**.
7 Where required, recover the washers and spacers from the motor mounting rubbers, noting their locations, then inspect the rubbers for signs of damage or deterioration, and renew if necessary.
8 If required the crank can be removed from the motor, but it is essential that the arm is replaced in exactly the same position. Make

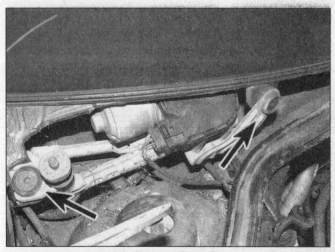

19.6a Remove the bolts...

19.6b ...and lift out the motor. Shown with the plenum chamber front panel partially removed

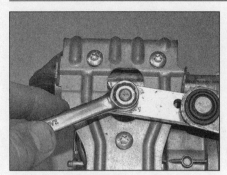

19.8a Undo the nut...

19.8b ...and prise off the linkage arm

19.8c Undo the 3 Torx bolts...

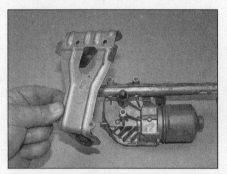

19.8d ...lift off the bracket...

19.8e ...and linkage arm

19.8f When refitting, set the crank arm 2.4 mm from the stop of the motor body

alignment marks between the crank arm and the motor body. If the motor is replaced then the crank arm should be set at 2.4mm from the stop on the motor body **(see illustrations)**.

Refitting

9 Refitting is a reversal of removal, bearing in mind the following points.

a) *If the motor has been separated from the linkage, ensure that the marks made on the motor spindle and linkage before removal are aligned, and ensure that the linkage is orientated as noted before removal.*

b) *Ensure that the washers and spacers are fitted to the motor mounting rubbers as noted before removal.*

c) *Make sure the locating peg aligns with the grommet in the bulkhead, when refitting.*

d) *Refit the wiper arms as described in Section 18.*

20 Rear wiper motor – removal and refitting

Removal

1 Remove the wiper arm as described in Section 18.

2 Recover the wiper motor shaft sealing ring.

3 Open the tailgate, then remove the trim panel as described in Chapter 11 Section 15.

4 Disconnect the washer fluid hose from the washer nozzle connector on the motor assembly **(see illustration)**.

5 Release the locking clip and unplug the wiring connector from the motor **(see illustration)**.

6 Unscrew the three nuts securing the motor, and then withdraw the assembly. Note the position of the rubber sealing ring, renew if necessary **(see illustration)**.

Refitting

7 Refitting is a reversal of removal, but ensure that the motor shaft rubber sealing ring/ grommet is correctly refitted to prevent water leaks, and refit the wiper arm with reference to Section 18.

20.4 Remove the screen washer hose from the tailgate wiper motor...

20.5 ...and then disconnect the wiring plug

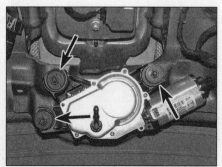

20.6 Tailgate wiper motor securing nuts

21.2a Undo the retaining bolt...

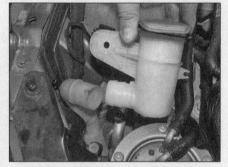

21.2b ...and remove the filler neck

21.4 Disconnect the wiring connector

21 Washer system components – removal and refitting

Washer fluid reservoir and pumps

Note: *Models fitted with Xenon headlights have an additional washer pump fitted above the windscreen washer pump. Where fitted this is removed after the windscreen washer pump has been removed.*

1 Switch off the ignition and all electrical consumers and remove the ignition key.

2 In the engine compartment, unscrew the mounting bolt and remove the top from the reservoir filler neck **(see illustrations)**.

3 Chock the rear wheels, then jack up the front of the vehicle and support on axle stands (see *Jacking and vehicle support*). Remove the right-hand wheel. Remove the wing liner and right-hand headlight as described in Section 8. Alternatively, remove the front bumper cover as described in Chapter 11 Section 6.

4 Disconnect the wiring connector from the top of the washer pump **(see illustration)**.

5 Note the position of the hoses, and then disconnect them from the washer pump **(see illustration)**. Position a suitable container beneath the reservoir to catch spilt fluid.

6 To remove the washer pump, pull it upwards from the reservoir and out from its rubber grommet **(see illustration)**.

7 Where fitted, disconnect the wiring plug from the fluid level sender unit.

8 Release the retaining clips and disconnect the wiring loom from the side of the washer reservoir.

9 Remove the fixings then remove the reservoir from the vehicle.

10 Refitting is a reversal of removal.

Windscreen washer jets

11 Open the bonnet, and pull the washer jet towards the front of the bonnet, and down to remove. Release the securing clip and disconnect the washer tube (and wiring where applicable), then remove the washer jet.

12 Refitting is a reversal of removal. Note that the aim of the jet can be adjusted using a screwdriver and turning the eccentric shaft at the base of the washer jet.

Tailgate washer jet

13 Switch off the ignition and all electrical consumers and remove the ignition key.

14 Unclip the cover from the wiper arm spindle for access to the washer jet **(see illustration)**, and pull the jet from the centre of the spindle.

15 On refitting, ensure that the jet is securely pushed into position. Check the operation of the jet. If necessary, adjust the nozzle, aiming the spray at a point slightly above the area of glass swept by the wiper blade.

Headlight pop-up washer jets

16 Switch off the ignition and all electrical consumers and remove the ignition key.

17 Carefully pull the washer jet out from the front bumper to its full extent, and hold it.

Carefully prise the end cap from the washer jet.

18 Still holding the washer jet, lift the securing clip slightly, and pull the jet from the lift cylinder.

19 Refitting is a reversal of removal. Operate the washers several times to bleed any trapped air.

Headlight pop-up washer jet lift cylinder

20 Remove the washer jet end cap as described in paragraphs 19 and 20.

21 Remove the front bumper as described in Chapter 11 Section 6.

22 Undo the two retaining screws, and withdraw the cylinder.

23 Clamp the hose, squeeze the retaining clip, and disconnect the hose.

24 Refitting is a reversal of removal. Operate the washers several times to bleed any trapped air.

22 Audio unit – removal and refitting

Note: *A variety of audio systems are fitted to the Leon range of vehicles. These range from a simple Radio through to an integrated navigation system (SatNav) and DVD player. Early models (up to 03/2009) require the use of special tools. On later models the once the trim panel has been removed standard tools are used. This Section only applies to standard-fit audio equipment.*

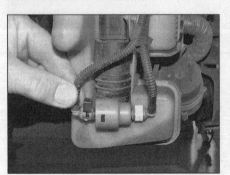

21.5 Disconnect the washer fluid hoses

21.6 Remove the pump

21.14 Remove the washer jet

22.4a Insert the keys into the slots at the base of the audio unit

22.4b Fit the correct tool or feeler gauges into the slots in the top of the unit

22.4c Pull out the unit. Note the location of the upper locking tab

Audio unit

Removal

1 The audio unit is equipped with an electronic anti-theft system linked to the instrument panel. If the voltage supply to the radio is temporarily disconnected, the radio will function again when the supply is reconnected, without entering the safety code number, provided the radio is located in the original vehicle. Should the radio operation be blocked, normal operation can be restored by entering the correct anti-theft code.

2 Where fitted, remove any CDs which may be in the unit. Switch off the ignition and all electrical consumers, and remove the ignition key.

Audio unit removal (early models)

3 If working on an early model obtain the special audio removal keys (VAG special tools T20196 and T20184). The tool T20184 is widely available in the aftermarket and T20196 can be improvised using two sets of feeler gauges.

4 On early models insert the special keys. The T20196 keys are fitted to the upper outer edge of the surround and the keys T20184 are fitted to the centre of the unit. Pull the lower set of keys backwards to release the audio unit **(see illustrations)**.

5 As the unit is removed, unlatch the wiring plug, disconnect the aerial plug and fully remove the audio unit **(see illustration)**.

Audio unit removal (from 03/2009)

6 Carefully unclip the trim panel from around the audio unit **(see illustrations)**.

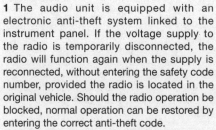

22.4d Detailed view of the lower release pins in position

7 Undo the mounting screws **(see illustration)** and pull out the unit until the wiring plugs can be disconnected. Release the locking clips and disconnect the wiring connectors from the rear of the unit.

8 Withdraw the radio/CD player unit from the facia.

Refitting

9 Refitting is a reversal of removal.

CD changer

Removal

10 The CD changer (where fitted) is located below the front passenger seat.

11 Switch off the ignition and all electrical consumers, and remove the ignition key.

12 Remove the front seat as described in Chapter 11 Section 22.

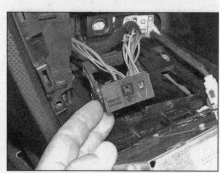

22.5 Disconnect the wiring plug(s)

13 Unbolt the protective cover, remove the main mounting bolts and release the CD changer. Disconnect the wiring plug as unit is removed.

Refitting

14 Refitting is a reversal of removal.

---

23 Loudspeakers –
removal and refitting

Treble/mid range speaker

1 Switch off the ignition and all electrical consumers, and remove the ignition key.

2 Remove the door trim panel as described in Chapter 11 Section 12.

3 Disconnect the wiring plug. The speaker is

22.6a Carefully release the surround...

22.6b ...and remove it

22.7 Undo the screws

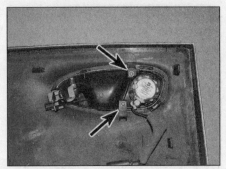

23.3 Remove the screws

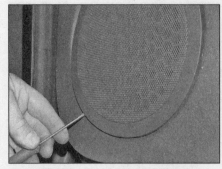

23.6 Unclip the speaker cover

23.7a Disconnect the wiring plug

23.7b Remove the screws and withdraw the speaker

held in place with screws – remove the screws and unclip the speaker **(see illustration)**. Remove the speaker from the housing.
4 Refitting is a reversal of removal.

Front and rear bass speakers

5 Switch off the ignition and all electrical consumers, and remove the ignition key.
6 Prise of the speaker grille with a trim tool or small screwdriver **(see illustration)**.
7 Disconnect the wiring plug from the loudspeaker, remove the screws and pull the speaker from the door frame/trim panel **(see illustrations)**.
8 Refitting is a reversal of removal.

24 Radio aerials, filters and modules – removal and refitting

Aerials

Roof mounted aerials

1 The aerial is mounted on the rear of the roof.
2 Partially lower the C-pillar panels as described in Chapter 11 Section 25.
3 Carefully lower the rear of the headlining, taking care not to damage it. Disconnect the wiring; note the fitted position of the wiring, as there may be more than one connector, depending on model.
4 Unscrew the securing nut and withdraw the aerial base from the roof. Hold the aerial base as the nut is being unscrewed to prevent the

base from rotating and scratching the roof panel. Recover the rubber spacer.
5 Refitting is a reversal of removal, but make sure that the two guide lugs on the rubber spacer are correctly located in the aerial base.

Tailgate glass mounted aerial

6 The aerial is integrated into the glass and only fitted to 03/2009 on models that also have higher specification audio units fitted.
7 If the aerial fail the tailgate glass must be replaced. This task is best left to a glazing specialist.

Amplifiers

8 The aerials that are integrated into the tailgate glass have an amplifier fitted.
9 Remove the tailgate trim to access the amplifier.
10 Disconnect the wiring plug, remove the bolt and remove the amplifier from the tailgate.
11 Refitting is a reversal of removal.

25 Anti-theft alarm system and engine immobiliser – general information

Note: *This information is applicable only to the anti-theft alarm system fitted by Seat as standard equipment.*

1 Models in the range are fitted with an anti-theft alarm system as standard equipment. The alarm has switches on all the doors (including the tailgate/boot lid), the bonnet and the ignition switch. If the tailgate,

bonnet or any of the doors are opened whilst the alarm is set, the alarm horn will sound and the hazard warning lights will flash. Some models are equipped with an internal monitoring system, which will activate the alarm system if any movement in the cabin is detected.
2 The alarm is set using the key in the driver's or passenger's front door lock or with the central locking remote control transmitter. The alarm system will then start to monitor its various switches approximately 30 seconds later.
3 All models are fitted with an immobiliser system, which is activated by the ignition switch. A transponder reading coil on the ignition switch reads a code contained within the ignition key. The system sends a signal to the engine management electronic control unit (ECU), which allows the engine to start if the code is correct. If an incorrect ignition key is used, the engine will not start.
4 If a fault is suspected with the alarm or immobiliser systems, the vehicle should be taken to a Seat dealer for examination. They will have access to a special diagnostic tester that will quickly trace any fault present in the system.

26 Airbag system – general information and precautions

 Warning: Before carrying out any operations on the airbag system, disconnect the battery negative terminal (Chapter 5A Section 3). When operations are complete, make sure no one is inside the vehicle when the battery is reconnected.

Warning: Note that the airbags must not be subjected to temperatures in excess of 90°C. When the airbag is removed, ensure that it is stored with the pad upwards to prevent possible inflation.

Warning: Do not allow any solvents or cleaning agents to contact the airbag assemblies. They must be cleaned using only a damp cloth.

Warning: The airbags and control unit are both sensitive to impact. If either is dropped or damaged they should be renewed.

1 A driver's airbag, passenger's airbag are fitted as standard to the Leon range. Some models will also have side, curtain and rear airbags. The airbag system consists of the airbag unit (complete with gas generator), which is fitted to the steering wheel (driver's side), facia (passenger's side), roof (where applicable), front seats (in the seatback), rear seats (unders the lower C-pillar panel), an impact sensor, the control unit and a warning light in the instrument panel.
2 The airbag system is triggered in the event

27.8a Insert a screwdriver and...

27.8b ...release the airbag. Note the locking spring

27.8c With the airbag removed, note the locking tabs on the steering wheel

of a heavy frontal or side impact above a predetermined force, depending on the point of impact. The airbag is inflated within milliseconds and forms a safety cushion between the driver and the steering wheel, the passenger and the facia, and in the case of side impact, between front seat occupants and the sides of the cabin. This prevents contact between the upper body and cabin interior, and therefore greatly reduces the risk of injury. The airbag then deflates almost immediately.

3 Every time the ignition is switched on, the airbag control unit performs a self-test. The self-test takes approximately 3 seconds and during this time the airbag warning light on the facia is illuminated. After the self-test has been completed the warning light should go out. If the warning light fails to come on, remains illuminated after the initial 3-second period or comes on at any time when the vehicle is being driven, there is a fault in the airbag system. The vehicle should then be taken to a Seat dealer or suitably equipped garage for examination at the earliest possible opportunity.

27 Airbag system components – removal and refitting

Note: *Refer to the warnings in Section 26 before carrying out the following operations.*
1 Disconnect the battery negative lead (refer to Chapter 5A Section 3) then continue as described under the relevant heading.
Caution: To prevent any discharge of static electricity into the airbag circuit, temporarily touch the vehicle bodywork before disconnecting the wiring from any airbag unit.

Driver's airbag

Note: *There are two types of airbag fitted and the removal procedure differs slightly. Before attempting to remove the airbag the type should be identified first. Pull out the column and using a small mirror examine the shape of the opening at the rear of the steering wheel. Type 1 steering wheels have a round opening, type 2 have a square opening.*

2 Set the front wheels to the straight-ahead position, and release the steering lock by inserting the ignition key.
3 Adjust the steering column to its highest position by releasing the adjustment handle, then extend the steering wheel as far as possible. Lock the column in this position.

Type 1

4 A special tool (T10027) is available to release the airbag, however removal is possible with a tapered punch – a long centre punch is idela. Car should be taken to avoid damage to the column cowl panels if a punch is used.
5 Insert the special tool or punch or approximately 8 mm into the hole in the upper rear of the steering wheel hub, then push the tool in and up to release the clip and free the airbag locking lug. Now turn the steering wheel through 180° and release the remaining airbag locking lug.
6 Carefully withdraw the airbag module from the steering wheel, release the locking clip and disconnect the wiring connector.
7 Turn the steering wheel to its central, straight-ahead position.

⚠️ *Warning: Position the airbag in a safe and secure place, away from the work area.*

Type 2

8 Insert a pin punch or screwdriver into the square hole at the rear of the steering wheel **(see illustrations)**. Move the punch upwards – away from the column – to release the spring

retainer. Rotate the steering wheel and release the opposite side.
9 Carefully withdraw the airbag module from the steering wheel, release the locking clip and disconnect the wiring connector **(see illustrations)**.
Warning: Position the airbag in a safe and secure place, away from the work area.
10 When refitting make sure the steering wheel is in the straight-ahead position, then locate the airbag module in position and reconnect the wiring. Carefully press in the module until both locking lugs are heard to engage. Reconnect the battery negative lead, ensuring that nobody is inside the vehicle as the lead is connected.

Passenger's airbag

11 Remove the passenger side glovebox with reference to Chapter 11 Section 27.
12 Release the locking clip and disconnect the wiring from the passenger's airbag.
13 Undo the mounting screws, then remove the airbag and side brackets from under the facia.

⚠️ *Warning: Position the airbag in a safe and secure place, away from the work area.*

14 Refitting is a reversal of removal, but tighten the mounting screws to the specified torque. Reconnect the battery negative lead, ensuring that nobody is inside the vehicle as the lead is connected.

Front seat side impact airbags

15 The side impact air bags are integral

27.9a Where fitted disconnect the wiring plugs from the steering wheels controls...

27.9b ...and then remove the main wiring plug

27.17 The control unit hidden beneath the heater housing

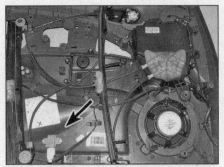

27.26a The front crash sensor...

27.26b ...and the C-pillar sensor

with the seats. As seat upholstery removal requires considerable skill and experience, if it is to be carried out without damage, it is best entrusted to an expert.

Roof curtain and rear side airbags

16 This work involves removing the headlining and major dismantling of interior trim panels, and is best entrusted to a Seat dealer.

Airbag control unit

17 The airbag control unit is located beneath the centre of the facia, under the heater housing **(see illustration)**.
18 Remove the trim from the passenger's side of the centre console with reference to Chapter 11 Section 26.
19 Remove the support bracket.
20 Release the locking lever and disconnect the wiring from the control unit.
21 Unscrew the nuts and remove the control unit from the vehicle.
22 Refitting is the reverse of removal making sure the wiring connector is securely reconnected. Reconnect the battery negative lead, ensuring that nobody is inside the vehicle as the lead is connected.

Airbag wiring contact unit

23 The airbag contact unit/clock spring is part of the steering column switch assembly, remove the contact unit as described in Section 5.

Passenger airbag on/ off switch

24 The switch is located inside the glovebox. Remove the switch with reference to Section 5.

Crash sensors

25 Crash sensors are fitted to the front door trim panels and at the base of both C-pillars. Remove the appropriate trim panel (as described in Chapter 11 Section 25 for the rear sensors or the door trim panel as described in Chapter 11 Section 12 for the front sensors).
26 Disconnect the wiring plug from the crash sensor and then remove the fixings **(see illustrations)**.

27 Refitting is a reversal of removal. Make sure that nobody is inside the vehicle when first switching on the ignition.

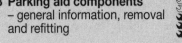

28 Parking aid components
– general information, removal and refitting

General information

1 The parking aid system is available as a standard fitment on higher specification models, and optional on other models. Four ultrasound sensors are located in the rear and (where fitted) the front bumper. These measure the distance to the closest object behind or in front of the car, and inform the driver using acoustic signals from a warning buzzer. The nearer the object, the more frequent the acoustic signals.
2 The system includes a control unit and self-diagnosis program, and therefore, in the event of a fault, the vehicle should be taken to a Seat dealer or suitably equipped garage.
3 Switch off the ignition and all electrical consumers and remove the ignition key, before removing any electrical components.

Control unit

4 The parking aid control unit for the sensors is located behind the left-hand load area side panel. Remove the panel as described in Chapter 11 Section 25 to access the control unit. The front sensor control unit is located above the convenience system onboard

supply control unit and relay carrier, behind the facia on the driver's side of the vehicle.
5 Depress the locking lugs and disconnect the wiring plugs from the control unit.
6 Unclip the unit from the mounting bracket above the relays.
7 Refitting is a reversal of removal.

Range/distance sensor

8 Remove the bumper cover as described in Chapter 11 Section 6 (front) or Chapter 11 Section 7 (rear).
9 Remove the sensor **(see illustration)**.
10 Disconnect the wiring plug **(see illustration)**.
11 Refitting is a reversal of removal. Press the sensor firmly into position until the retaining clips engage.

Warning buzzer

12 The warning buzzers are located behind the left-hand load area side panel for the rear sensors, and behind the front front relay plate for the front sensors.
13 To remove the rear warning buzzer, remove the left-hand inner trim, with reference to Chapter 11 Section 25. Disconnect the wiring, then release the two plastic rivets and remove the warning buzzer.
14 To remove the front warning buzzer, remove the right-hand facia end panel, as described in Chapter 11 Section 27. Unscrew and then lower the relay carrier plate. Disconnect the wiring, then release the two plastic rivets and remove the warning buzzer.
15 Refitting is a reversal of removal.

28.9 Unclip the sensor...

28.10 ...and disconnect the wiring plug

MAIN FUSE BOX IN ENGINE COMPARTMENT (UP TO OCTOBER 2008)

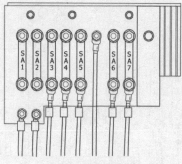

UP TO APRIL 2006

FUSE/RELAY	VALUE	DESCRIPTION	OEM NAME
SA1	200 A	Alternator 140 A, 150 A also used for alternator 90 A / 110 A	SA1
SA2	80 A	Electromechanical power steering motor	SA2
SA3	80 A	Radiator fan, 50 A also used	SA3
SA4	-	Not used	SA4
SA5	100 A	Auxiliary heater element	SA5
SA6	80 A	Fuses on fuse holder left dash panel (SC12-SC17, SC20, SC22-SC27, SC43, SC44, SC45)	SA6
SA7	-	Not used	SA7

FROM MAY 2006 TO OCTOBER 2006

FUSE/RELAY	VALUE	DESCRIPTION	OEM NAME
SA1	200 A	Alternator 140 A, 150 A also used for alternator 90 A / 110 A	SA1
SA2	80 A	Electromechanical power steering motor	SA2
SA3	50 A	Radiator fan	SA3
SA4	-	Driver door control unit, Front passenger door control unit, Rear left door control unit, Rear right door control unit, Convenience system central control unit	SA4
SA5	100 A	Fuses on fuse holder left dash panel (SC12-SC17, SC20, SC22-SC27, SC43, SC44, SC45)	SA5
SA6	100 A	Auxiliary heater element	SA6
SA7	-	Not used	SA7

FROM NOVEMBER 2006 TO OCTOBER 2008

FUSE/RELAY	VALUE	DESCRIPTION	OEM NAME
SA1	200 A	Alternator 140 A, 150 A also used for alternator 90 A / 110 A	SA1
SA2	80 A	Electromechanical power steering motor	SA2
SA3	50 A	Radiator fan	SA3
SA4	-	Driver door control unit, Front passenger door control unit, Rear left door control unit, Rear right door control unit, Convenience system central control unit	SA4
SA5	100 A	Fuses on fuse holder left dash panel (SC12-SC17, SC20, SC22-SC27, SC43, SC44, SC45)	SA5
SA6	80 A	High heat output relay	SA6
SA7	40 A	Low heat output relay	SA7

Fuses and relays

MAIN FUSE BOX IN ENGINE COMPARTMENT (FROM NOVEMBER 2008 TO SEPTEMBER 2012)

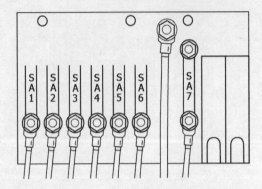

FROM NOVEMBER 2008 TO FEBRUARY 2009

FUSE/RELAY	VALUE	DESCRIPTION	OEM NAME
SA1	200 A	Alternator 140 A, 150 A also used for alternator 90 A / 110 A	SA1
SA2	80 A	Electromechanical power steering motor	SA2
SA3	50 A	Radiator fan	SA3
SA4	80 A	High heat output relay	SA4
SA5	80 A	Fuses on fuse holder left dash panel (SC12-SC17, SC20, SC22-SC27, SC43, SC44, SC45)	SA5
SA6	40 A	Low heat output relay	SA6
SA7	-	Driver door control unit, Front passenger door control unit, Rear left door control unit, Rear right door control unit, Convenience system central control unit	SA7

FROM MARCH 2009 TO SEPTEMBER 2012

FUSE/RELAY	VALUE	DESCRIPTION	OEM NAME
SA1	200 A	Alternator 140 A, 150 A also used for alternator 90 A / 110 A	SA1
SA2	80 A	Electromechanical power steering motor	SA2
SA3	50 A	Radiator fan	SA3
SA4	80 A	High heat output relay	SA4
SA5	80 A	Fuses on fuse holder left dash panel (SC12-SC17, SC20, SC22-SC27, SC43, SC44, SC45)	SA5
SA6	40 A	Low heat output relay	SA6
SA7	-	Driver door control unit, Front passenger door control unit, Rear left door control unit, Rear right door control unit, Onboard supply central control unit	SA7

FUSE AND RELAY BOX IN ENGINE COMPARTMENT

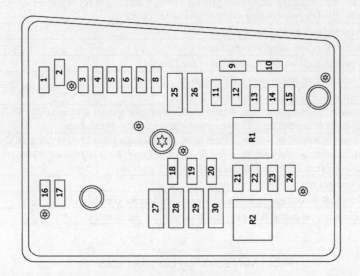

UP TO APRIL 2008

FUSE/RELAY	VALUE	DESCRIPTION	OEM NAME
1	25 A	Front passenger side wiper motor control unit	SB1
2	5 A	Steering column electronics control unit	SB2
3	5 A	Onboard power supply control unit	SB3
4	30 A	ABS control unit	SB4
5	15 A	Automatic transmission control unit	SB5
6	5 A	Control unit with display in dash panel insert	SB6
7	-	Not used	
8	15 A	Control unit with display for radio and navigation system or Radio	SB8
9	5 A	Mobile telephone control unit or Telephone transmitter and receiver unit	SB9
10	15 A	Terminal 30 voltage supply relay for Diesel, Engine control unit for petrol (5 A used)	SB10
11	-	Not used	
12	5 A	Data bus diagnostic interface	SB12
13	30 A	Engine control unit for Diesel	SB13
	25 A	Engine control unit for Petrol	
14	20 A	Ignition coils, Ignition transformer for Petrol	SB14
15	10 A	Lambda heater 1, Lambda heater 2, Lambda 3 heater for Petrol	SB15
	5 A	Fuel pump relay, Automatic glow period control unit for Diesel	
16	30 A	ABS control unit	SB16
17	15 A	Onboard power supply control unit	SB17
18	-	Not used	
19	30 A	Driver side wiper motor control unit	SB19
20	-	Not used	
21	15 A	Lambda 1, Lambda 2	SB21
22	5 A	Clutch position sensor, Cruise control system	SB22
23	15 A	Fuel pressure regulating valve for Petrol	SB23
	10 A	Exhaust gas recirculation valve, Charge pressure control solenoid valve, Exhaust gas recirculation cooler change-over valve for Diesel	

Fuses and relays (continued)

24	10 A	Radiator fan, Intake manifold flap motor for Diesel or Inlet camshaft control valve 1, Variable intake manifold change-over valve, Exhaust gas recirculation valve, Map-controlled engine cooling system thermostat for Petrol	SB24
25	40 A	Onboard power supply control unit	SB25
26	40 A	Onboard power supply control unit	SB26
27	50 A	Automatic glow period control unit for Diesel	SB27
	40 A	Secondary air pump motor for Petrol	
28	40 A	Terminal 15 voltage supply relay 2	SB28
29	30 A	Interior monitoring sensor, Alarm horn, Headlight washer system relay, Sliding sunroof adjustment control unit, Convenience system central control unit	SB29
	50 A	Interior monitoring sensor, Alarm horn, Headlight washer system relay, Sliding sunroof adjustment control unit, Rear left door control unit, Rear right door control unit, Convenience system central control unit	
30	40 A	X-contact relief relay	SB30
R1	-	Terminal 30 voltage supply relay or Engine current supply relay	A1
R2	-	Secondary air pump fuse or Additional coolant pump relay	A2

FROM MAY 2008 TO FEBRUARY 2009

FUSE/RELAY	VALUE	DESCRIPTION	OEM NAME
1	30 A	Front passenger side wiper motor control unit	SB1
2	-	Not used	-
3	5 A	Onboard power supply control unit	SB3
4	20 A	ABS control unit	SB4
5	15 A	Automatic transmission control unit	SB5
6	5 A	Control unit with display in dash panel insert, Steering column electronics control unit	SB6
7	40 A	Terminal 15 voltage supply relay 2	SB7
8	15 A	Control unit with display for radio and navigation system or Radio	SB8
9	5 A	Telephone transmitter and receiver unit, Navigation system interface	SB9
10	5 A	Terminal 30 voltage supply relay for Diesel, Engine control unit for petrol	SB10
11	-	Not used	
12	5 A	Data bus diagnostic interface	SB12
13	30 A	Engine control unit for Diesel	SB13
	15 A	Engine control unit for Petrol	
14	20 A	Ignition coils, Ignition transformer for Petrol	SB14
15	10 A	Lambda 1, Lambda 2, Lambda 3 for Petrol	SB15
	5 A	Fuel pump relay, Automatic glow period control unit for Diesel	
16	30 A	Onboard power supply control unit	SB16
17	15 A	Onboard power supply control unit	SB17
18	-	Not used	
19	30 A	Driver side wiper motor control unit	SB19
20	-	Not used	
21	10 A	Lambda probe heater for Diesel	SB21
	15 A	Lambda probe heater or Lambda probe after catalytic converter for Petrol	
22	5 A	Clutch position sender	SB22
23	15 A	Fuel pressure regulating valve for Petrol	SB23
	10 A	Exhaust gas recirculation valve, Charge pressure control solenoid valve, Exhaust gas recirculation cooler change-over valve for Diesel	

Fuses and relays (continued)

24	10 A	Radiator fan, Inlet camshaft control valve 1, Activated charcoal filter system solenoid valve 1, Variable intake manifold change-over valve for Petrol or Intake manifold flap motor for Diesel	SB24
25	40 A	ABS control unit	SB25
26	30 A	Onboard power supply control unit	SB26
27	40 A	Secondary air pump motor for Petrol	SB27
	50 A	Automatic glow period control unit for Diesel	
28	-	Not used	
29	50 A	Sliding sunroof adjustment control unit, Convenience system central control unit, Heated front seats control unit, Headlight washer system relay	SB29
30	50 A	X-contact relief relay	SB30
R1	-	Terminal 30 voltage supply relay or Engine current supply relay or Main relay	A1
R2	-	Secondary air pump fuse or Additional coolant pump relay	A2

FROM MARCH 2009 TO SEPTEMBER 2012

FUSE/RELAY	VALUE	DESCRIPTION	OEM NAME
1	30 A	Front passenger side wiper motor control unit	SB1
2	30 A	Automatic transmission control unit	SB2
3	5 A	Onboard supply control unit	SB3
4	20 A	ABS control unit	SB4
5	15 A	Automatic transmission control unit	SB5
6	5 A	Control unit in dash panel insert, Steering column electronics control unit	SB6
7	40 A	Terminal 15 voltage supply relay 2	SB7
8	15 A	Control unit with display for radio and navigation system or Radio	SB8
9	5 A	Telephone transmitter and receiver unit	SB9
10	5 A	Main relay for Petrol and Terminal 30 voltage supply relay for Diesel	SB10
11	-	Not used	
12	5 A	Data bus diagnostic interface	SB12
13	30 A	Engine control unit 15, A also used	SB13
14	20 A	Fuel pressure regulating valve, Fuel metering valve for Diesel or Ignition coils, Ignition transformer for Petrol	SB14
15	5 A	Automatic glow period control unit, Fuel pump relay or Electric fuel pump 2 relay for Diesel	SB15
	10 A	Lambda 1, Lambda 2, Lambda 3 for Petrol	
16	30 A	Onboard supply control unit	SB16
17	15 A	Dual tone horn relay	SB17
18	-	Not used	
19	30 A	Wiper motor control unit	SB19
20	-	Not used	
21	15 A	Lambda probe heater, Lambda probe 1 heater after catalytic converter for Petrol	SB21
	10 A	Lambda probe heater for Diesel	
22	5 A	Clutch position sender	SB22
23	15 A	Fuel pressure regulating valve for Petrol	SB23
	10 A	Exhaust gas recirculation valve, Charge pressure control solenoid valve, Exhaust gas recirculation cooler change-over valve for Diesel	

Fuses and relays (continued)

24	10 A	Radiator fan and Inlet camshaft control valve 1, Activated charcoal filter system solenoid valve 1, Variable intake manifold change-over valve for Petrol or Intake manifold flap motor for Diesel	SB24
25	40 A	ABS control unit	SB25
26	30 A	Onboard supply control unit	SB26
27	50 A	Secondary air pump motor for Petrol	SB27
	40 A	Automatic glow period control unit for Diesel	
28	-	Not used	
29	50 A	Sliding sunroof adjustment control unit, Convenience system central control unit, Heated front seats control unit, Headlight washer system relay	SB29
30	50 A	X-contact relief relay	SB30
R1	-	Terminal 30 voltage supply relay or Engine current supply relay or Main relay	A1
R2	-	Secondary air pump fuse or Additional coolant pump relay	A2

FUSE BOX IN PASSENGER COMPARTMENT

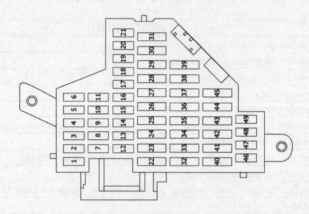

UP TO APRIL 2006

FUSE/RELAY	VALUE	DESCRIPTION	OEM NAME
1	10 A	Diagnosis connection and Engine control unit or Fuel pump control unit, Terminal 30 voltage supply relay, Current supply relay for engine control unit	SC1
2	5 A	ABS control unit or ABS control unit, Brake light switch	SC2
3	5 A	Airbag control unit, Front passenger side airbag deactivated warning lamp	SC3
4	5 A	Heater/heat output switch, Tyre pressure monitor button, TCS and ESP button, Reversing light switch, High-pressure sender, Climatronic control unit, Heated driver seat control unit, Heated front passenger seat control unit	SC4
5	5 A	Headlight range control regulator, Headlight range control, control unit, Left headlight range control motor, Right headlight range control motor	SC5
6	5 A	Control unit with display in dash panel insert, Power steering control unit, Data bus diagnostic interface and selector lever for automatic transmission, Tiptronic switch	SC6
7	5 A	Mobile telephone operating electronics control unit, Automatic anti-dazzle interior mirror, Heated windscreen	SC7
8	5 A	Trailer detector control unit	SC8

Fuses and relays (continued)

9	-	Not used	
10	-	Not used	
11	-	Not used	
12	10 A	Driver door control unit, Front passenger side door control unit	SC12
13	10 A	Diagnosis connection, Brake light switch, Lighting switch	SC13
14	5 A	Automatic gearbox control unit for Petrol	SC14
15	7.5 A	Onboard power supply control unit	SC15
16	10 A	Heater/heat output switch, Climatronic control unit, Air conditioning system control unit	SC16
17	5 A	Rain and light detector sensor	SC17
18	5 A	Parking aid control unit, Tiptronic switch, Selector lever for automatic transmission	SC18
19	-	Not used	
20	5 A	ABS control unit	SC20
21	10 A	Fuel pressure regulating valve or Not used	SC21
22	40 A	Fresh air blower	SC22
23	30 A	Driver door control unit, Front passenger side door control unit	SC23
24	25 A	Cigarette lighter	SC24
25	25 A	Onboard power supply control unit	SC25
26	30 A	12 V socket rear, 12 V socket 2 in luggage compartment	SC26
27	15 A	Fuel pump relay or Current supply relay for engine control unit or Fuel pump control unit	SC27
28	-	Not used	
29	10 A	Air mass meter, Intake manifold preheating heater element	SC29
30	20 A	Injectors for Petrol	SC30
31	20 A	Vacuum pump relay for Petrol with automatic gearbox	SC31
32	30 A	Rear left door control unit, Rear right door control unit	SC32
33	25 A	Sliding sunroof adjustment control unit	SC33
34	25 A	Convenience system central control unit	SC34
35	5 A	Interior monitoring sensor, Alarm horn, Convenience system central control unit	SC35
36	20 A	Headlight washer system relay	SC36
37	30 A	Heated driver seat control unit, Heated front passenger seat control unit	SC37
38	-	Not used	
39	20 A	Multifunction switch, Automatic gearbox control unit for Petrol	SC39
40	40 A	Heater/heat output switch	SC40
41	15 A	Rear window wiper motor	SC41
42	15 A	Onboard power supply control unit	SC42
43	15 A	Trailer detector control unit	SC43
44	20 A	Trailer detector control unit	SC44
45	15 A	Trailer detector control unit	SC45
46	5 A	Heater/heat output switch, Climatronic control unit, Air conditioning system control unit, Left washer jet heater element, Right washer jet heater element	SC46
47	10 A	Lambda probe heater	SC47
48	10 A	Lambda probe 1 heater after catalytic converter	SC48
49	7.5 A	Lighting switch	SC49

Fuses and relays (continued)

FROM MAY 2006 TO SEPTEMBER 2012

FUSE/RELAY	VALUE	DESCRIPTION	OEM NAME
1	10 A	Diagnosis connection, Headlight range control regulator, Headlight range control, control unit, Cornering light and headlight range control unit, Heated windscreen, Air mass meter, Intake manifold preheating heater element	SC1
2	10 A	Lighting switch, Brake light switch, Trailer detector control unit, Power steering control unit, Data bus diagnostic interface, Dash panel insert, Left headlight range control motor, Right headlight range control motor	SC2
3	5 A	Airbag control unit, Front passenger side airbag deactivated warning lamp	SC3
4	5 A	Heater/heat output switch, Tyre pressure monitor button, TCS and ESP button, Reversing light switch, High-pressure sender, Climatronic control unit, Air conditioning system control unit, Mobile telephone operating electronics control unit, Navigation system interface, Automatic anti-dazzle interior mirror, Left washer jet heater element, Right washer jet heater element	SC4
5	10 A	Power output module for right headlight	SC5
6	10 A	Power output module for left headlight	SC6
7	-	Not used	
8	-	Not used	
9	-	Not used	
10	-	Not used	
11	-	Not used	
12	15 A	Driver door control unit, Front passenger door control unit, Rear left door control unit, Rear right door control unit	SC12
13	10 A	Diagnosis connection, Lighting switch, Rain and light detector sensor	SC13
14	10 A	Heater/heat output switch, Climatronic control unit, Air conditioning system control unit, Parking aid control unit, Selector lever sensors control unit, Tiptronic switch	SC14
15	7.5 A	Onboard power supply control unit	SC15
16	-	Not used	
17	5 A	Interior monitoring sensor, Alarm horn	SC17
18	-	Not used	
19	-	Not used	
20	-	Not used	
21	-	Not used	
22	40 A	Fresh air blower for Climatronic	SC22
23	30 A	Driver door control unit, Front passenger side door control unit	SC23
24	-	Not used	
25	25 A	Onboard power supply control unit	SC25
26	30 A	Rear left door control unit, Rear right door control unit	SC26
27	15 A	Fuel pump relay for Diesel and Current supply relay, Fuel pump control unit for Petrol	SC27
28	30 A	Driver door control unit, Passenger door control unit, Rear left door control unit, Rear right door control unit, Convenience system central control unit	SC28
29	-	Not used	
30	20 A	Automatic transmission control unit, Multifunction switch for Petrol	SC30
31	20 A	Vacuum pump relay for Petrol	SC31
32	-	Not used	
33	25 A	Sliding sunroof adjustment control unit	SC33
34	25 A	Convenience system central control unit	SC34

Fuses and relays (continued)

35	-	Not used	
36	20 A	Headlight washer system relay	SC36
37	30 A	Heated front seats control unit	SC37
38	-	Not used	
39	-	Not used	
40	40 A	Heater/heat output switch, Air conditioning system control unit	SC40
41	20 A	Onboard power supply control unit, Rear window wiper motor	SC41
42	20 A	Cigarette lighter, 12 V socket rear, 12 V socket 2 in luggage compartment	SC42
43	15 A	Trailer detector control unit	SC43
44	20 A	Trailer detector control unit	SC44
45	15 A	Trailer detector control unit	SC45
46	-	Not used	
47	-	Not used	
48	-	Not used	
49	-	Not used	

RELAY BOX 1 IN PASSENGER COMPARTMENT

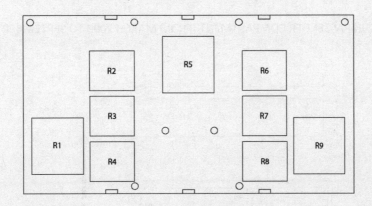

RELAY	DESCRIPTION	OEM NAME
R1	Terminal 15 voltage supply relay 2	B1
R2	Seat heater relay or Not used	B2
R3	Rear lid central locking relay	B3
R4	Terminal 30 voltage supply relay 2	B4
R5	Heated rear window relay	B5
R6	Dual tone horn relay	B6
R7	Double washer pump relay 1	B7
R8	Double washer pump relay 2	B8
R9	X-contact relief relay	B9

Fuses and relays (continued)

RELAY BOX 2 IN PASSENGER COMPARTMENT (UP TO FEBRUARY 2009)

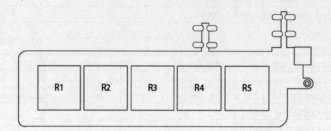

RELAY	DESCRIPTION	OEM NAME
R1	Headlight washer system relay	1
R2	Fuel pump relay or Circulation pump relay, Heated windscreen relay or Current supply relay for engine control unit	2
R3	Terminal 50 voltage supply relay	3
R4	Heated windscreen relay	4
R5	Terminal 50 voltage supply relay or Not used	5

RELAY BOX 2 IN PASSENGER COMPARTMENT (FROM MARCH 2009 TO SEPTEMBER 2012)

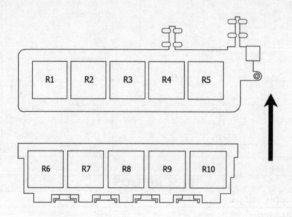

FROM MARCH 2009 TO OCTOBER 2009

RELAY	DESCRIPTION	OEM NAME
R1	Terminal 50 voltage supply relay	1
R2	Fuel pump relay or Electric fuel pump 2 relay or Current supply relay for engine control unit	2
R3	High heat output relay for Diesel	3
R4	Low heat output relay for Diesel	4
R5	Heated windscreen relay	5
R6	Not used	6
R7	X-contact relief relay	7
R8	Dual tone horn relay, Headlight washer system relay	8
R9	Heated rear window relay	9
R10	Terminal 15 voltage supply relay 2	10

Fuses and relays (continued)

FROM NOVEMBER 2009 TO SEPTEMBER 2012

RELAY	DESCRIPTION	OEM NAME
R1	Terminal 50 voltage supply relay	1
R2	Low heat output relay for Diesel	2
R3	High heat output relay for Diesel	3
R4	Heated rear window relay	4
R5	Heated windscreen relay	5
R6	Starter motor relay 2	6
R7	X-contact relief relay	7
R8	Dual tone horn relay, Headlight washer system relay	8
R9	Intake manifold preheating relay, Fuel pump relay, Electric fuel pump 2 relay or Current supply relay for engine control unit	9
R10	Terminal 15 voltage supply relay 2	10

Fuses and relays (continued)

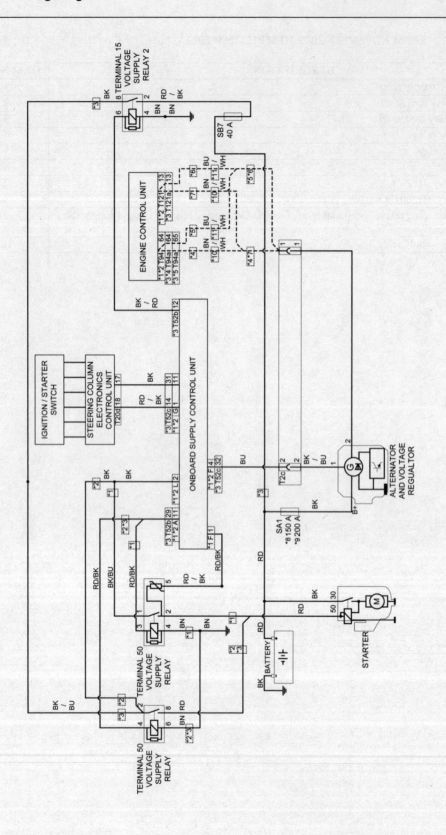

Starting and charging

*1 From April 2005 to April 2006
*2 From May 2006 to February 2009
*3 From March 2009 to September 2012
*4 For 2.0 L and 1.9 L Diesel
*5 For 1.6 L Diesel
*6 For 1.6 L Petrol
*7 For 2.0 L Petrol
*8 Models with 90 A / 110 A alternator
*9 Models with 140 A alternator
*10 Up to April 2008
*11 From May 2008

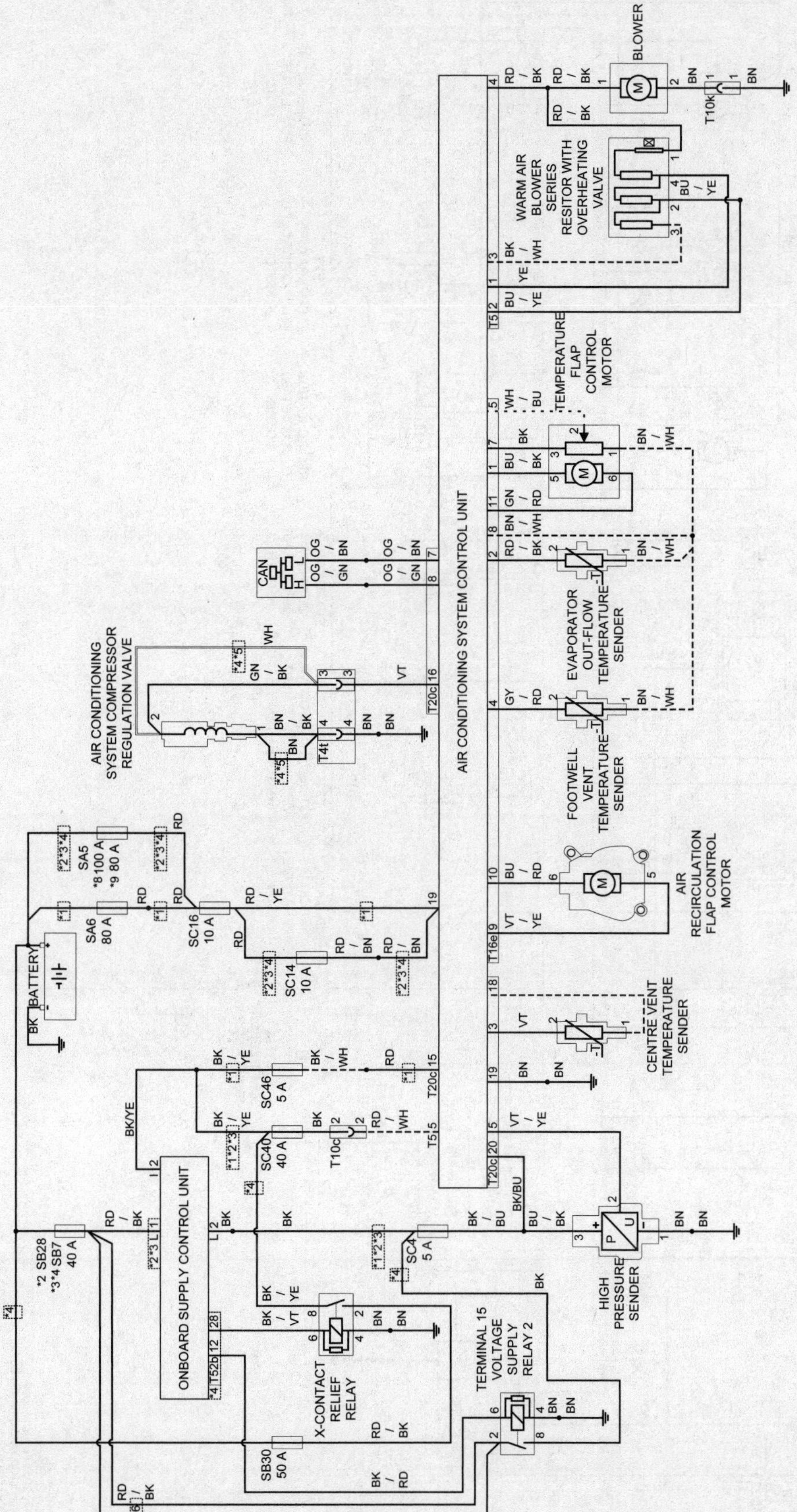

BLOWER

T10k

WARM AIR BLOWER SERIES RESISTOR WITH OVERHEATING VALVE

TEMPERATURE FLAP CONTROL MOTOR

AIR CONDITIONING SYSTEM CONTROL UNIT

AIR CONDITIONING SYSTEM COMPRESSOR REGULATION VALVE

CAN

T20c 16

T4f

EVAPORATOR OUT-FLOW TEMPERATURE SENDER

FOOTWELL VENT TEMPERATURE SENDER

AIR RECIRCULATION FLAP CONTROL MOTOR

CENTRE VENT TEMPERATURE SENDER

BATTERY

SA5
*8 100 A
*9 80 A

SA6
80 A

SC16
10 A

SC14
10 A

T16e

T20c 15

SC46
5 A

SC40
40 A

T10c

T5

T20c 20

SC4
5 A

HIGH PRESSURE SENDER

ONBOARD SUPPLY CONTROL UNIT

T52b

X-CONTACT RELIEF RELAY

TERMINAL 15 VOLTAGE SUPPLY RELAY 2

SB30
50 A

*2 SB28
*3 *4 SB7
40 A

Air conditioning, heating and cooling – Climatic

*1 Up to April 2006
*2 From May 2006 to April 2008
*3 From May 2008 February 2009
*4 From March 2009 to September 2012
*5 According to equipment
*6 Up to October 2009
*7 From November 2009
*8 Up to October 2008
*9 From November 2008

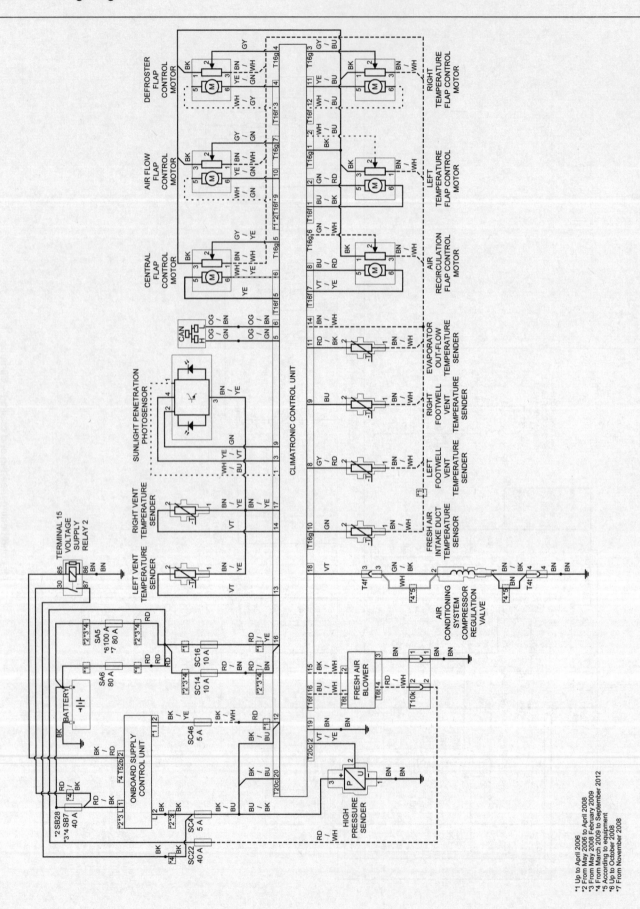

Air conditioning, heating and cooling – Climatronic

*1 Up to April 2006
*2 From May 2006 to April 2008
*3 From May 2008 February 2009
*4 From March 2009 to September 2012
*5 According to equipment
*6 Up to October 2008
*7 From November 2008

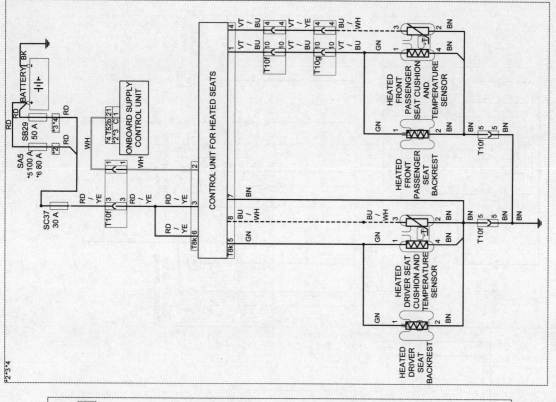

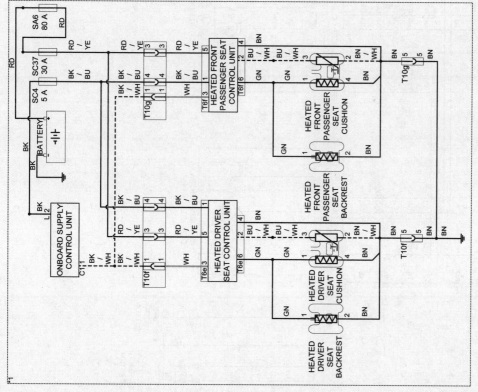

Air conditioning, heating and cooling – seat heating

*1 From April 2005 to October 2005
*2 From November 2005 to April 2006
*3 From May 2006 to February 2009
*4 From March 2009 to September 2012
*5 Up to October 2008
*6 From November 2008

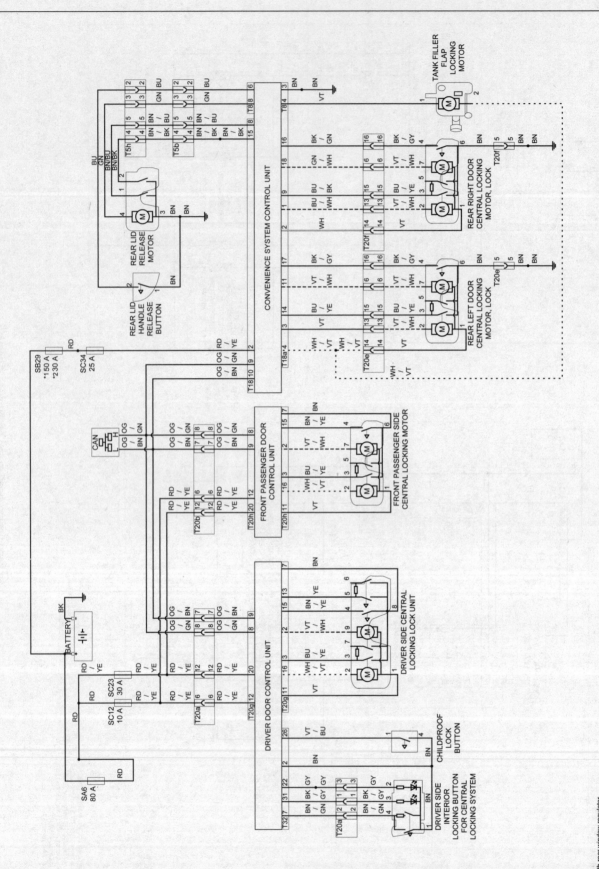

Power door locks – from April 2005 to April 2006

*1 With rear window regulator
*2 Without rear window regulator

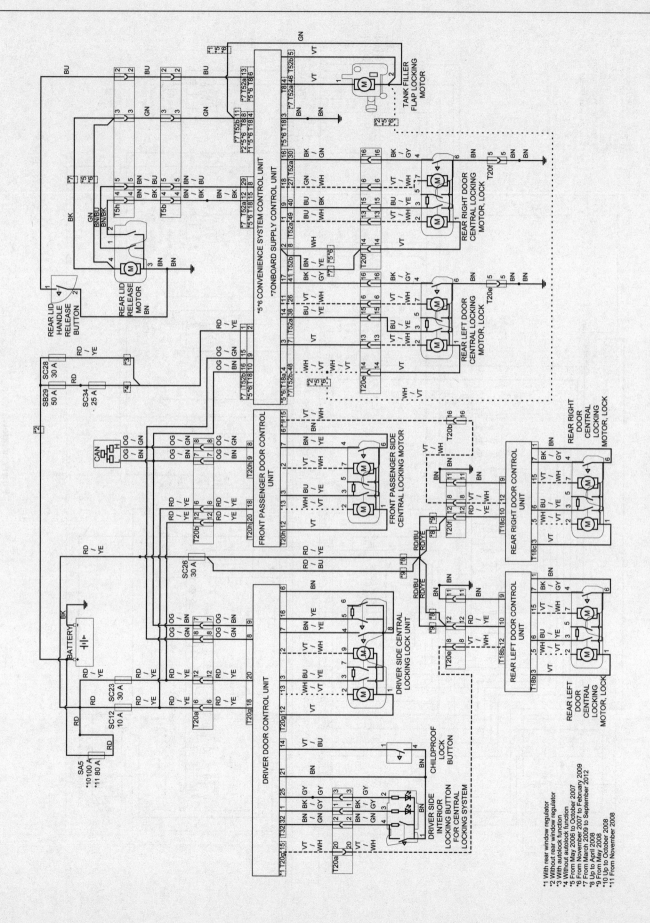

Power door locks – from May 2006 to September 2012

*1 With rear window regulator
*2 Without rear window regulator
*3 With autolock function
*4 Without autolock function
*5 From May 2006 to October 2007
*6 From November 2007 to October 2009
*7 From March 2009 to February 2012
*8 Up to April 2008
*9 From May 2008
*10 Up to October 2008
*11 From November 2008

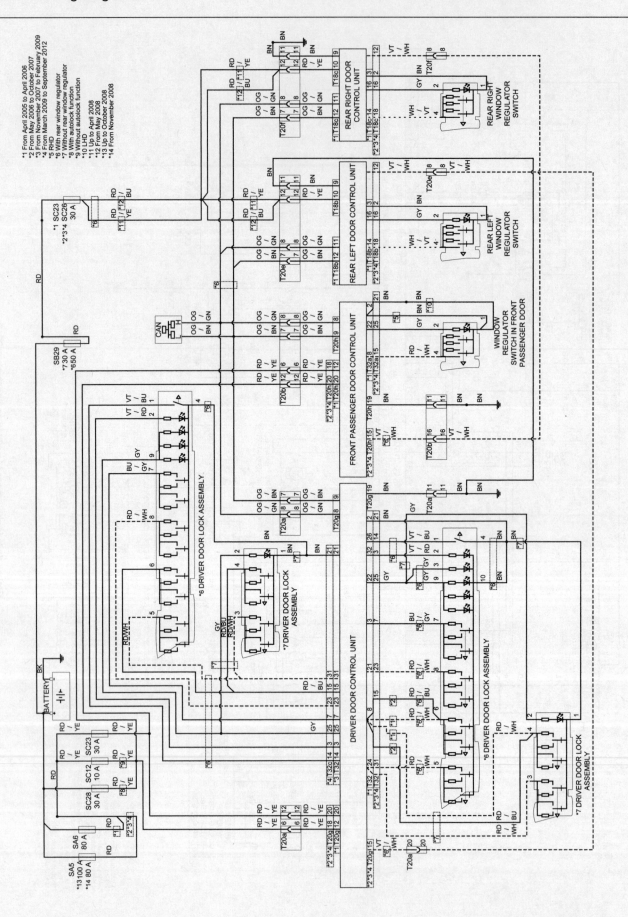

Power windows

*1 From April 2005 to April 2006
*2 From May 2006 to October 2007
*3 From November 2007 to February 2009
*4 From March 2009 to September 2012
*5 RHD
*6 With rear window regulator
*7 Without rear window regulator
*8 With autolock function
*9 Without autolock function
*10 LHD
*11 Up to April 2008
*12 From May 2008
*13 Up to October 2008
*14 From November 2008

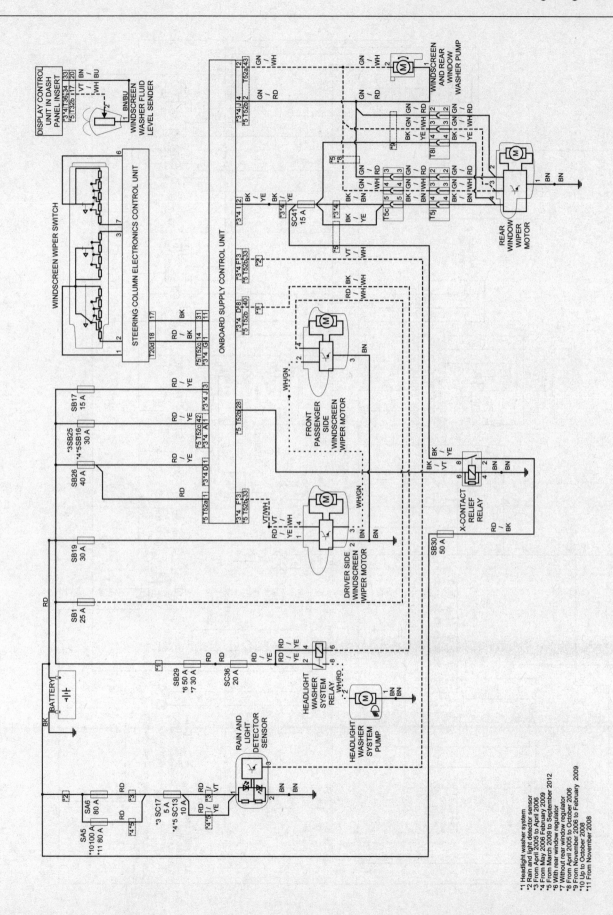

Wiper washer

*1 Headlight washer system
*2 Rain and light detector sensor
*3 From April 2005 to April 2006
*4 From May 2006 February 2009
*5 From March 2009 to September 2012
*6 With rear window regulator
*7 Without rear window regulator
*8 From April 2005 to October 2006
*9 From November 2006 to February 2009
*10 Up to October 2008
*11 From November 2008

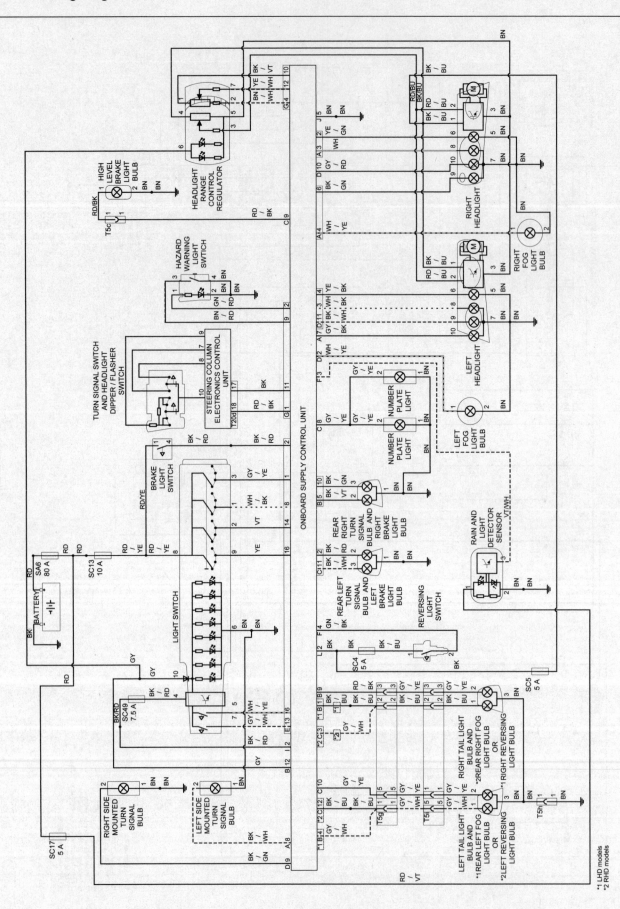

Exterior lights - normal headlights and rear lights - from April 2005 to April 2006

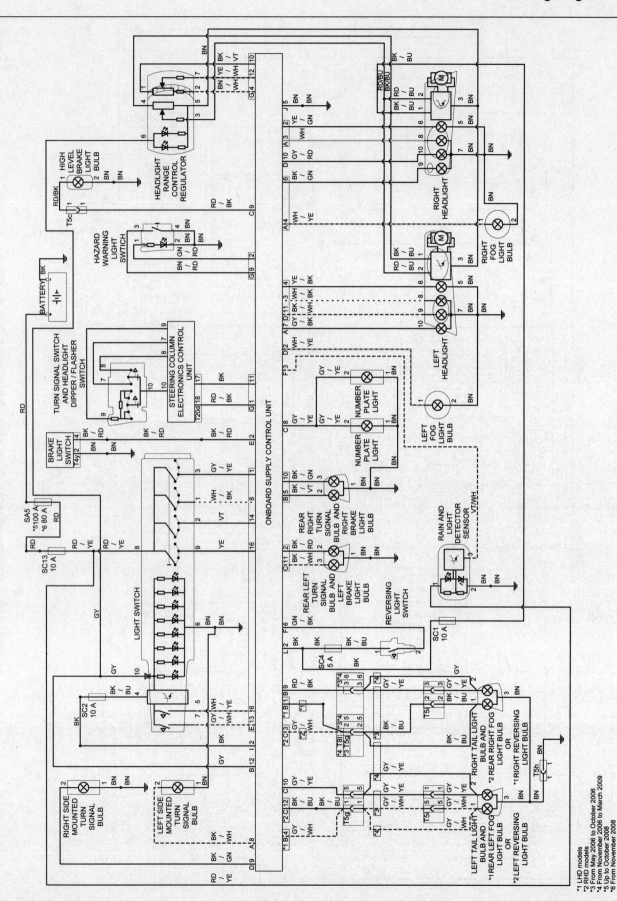

Exterior lights - normal headlights and rear lights - from May 2006 to February 2009

*1 LHD models
*2 RHD models
*3 From May 2006 to October 2006
*4 From November 2006 to March 2009
*5 Up to October 2008
*6 From November 2008

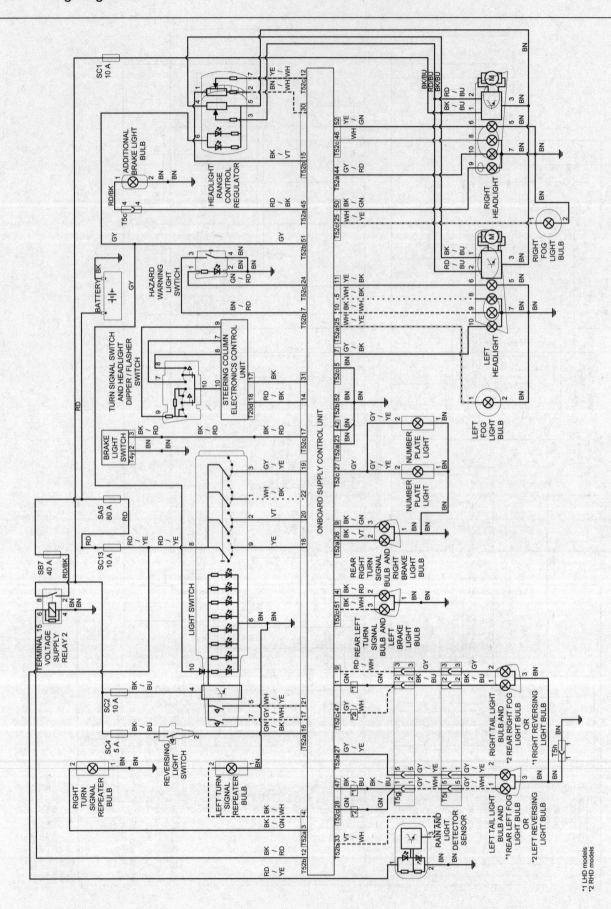

Exterior lights - normal headlights and rear lights - from March 2009 to April 2012

*1 LHD models
*2 RHD models

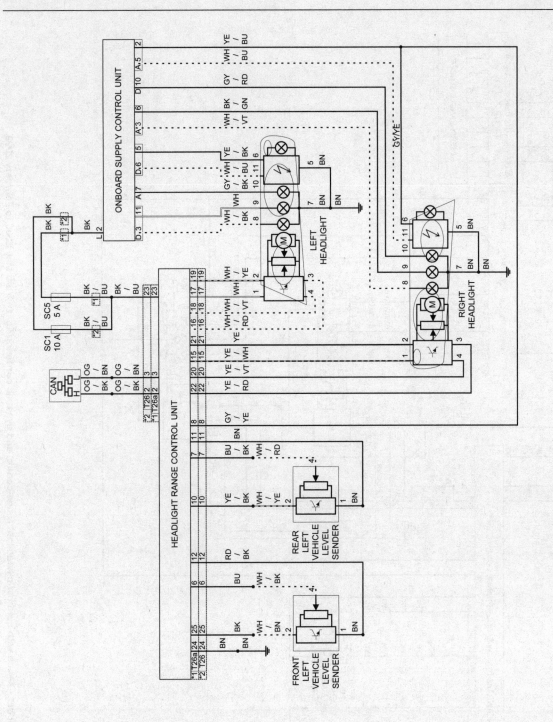

Exterior lights - Xenon with automatic headlight range control – from April 2005 to February 2009

*1 From April 2005 to April 2006
*2 From May 2006 to February 2009

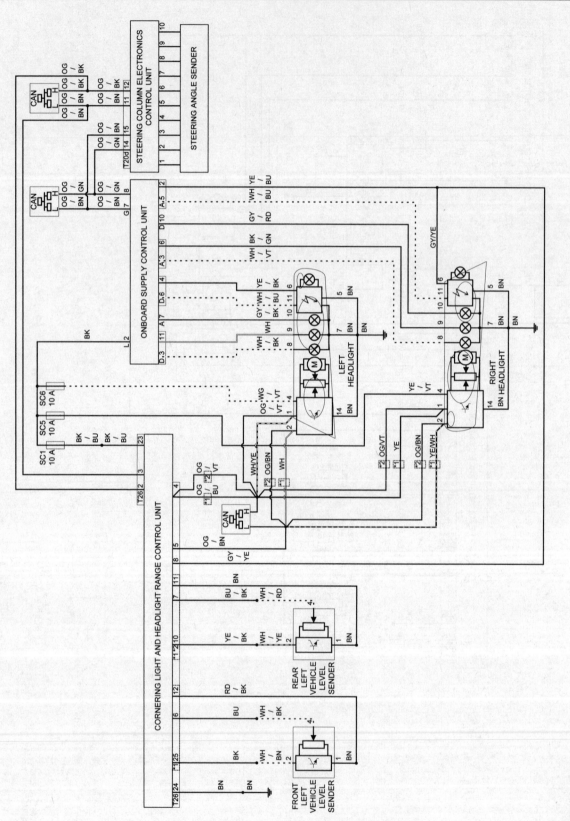

Exterior Lights - Xenon with automatic headlight range control and dynamic cornering light – from May 2006 to February 2009

*1 From May 2006 to April 2008
*2 From May 2008 to February 2009

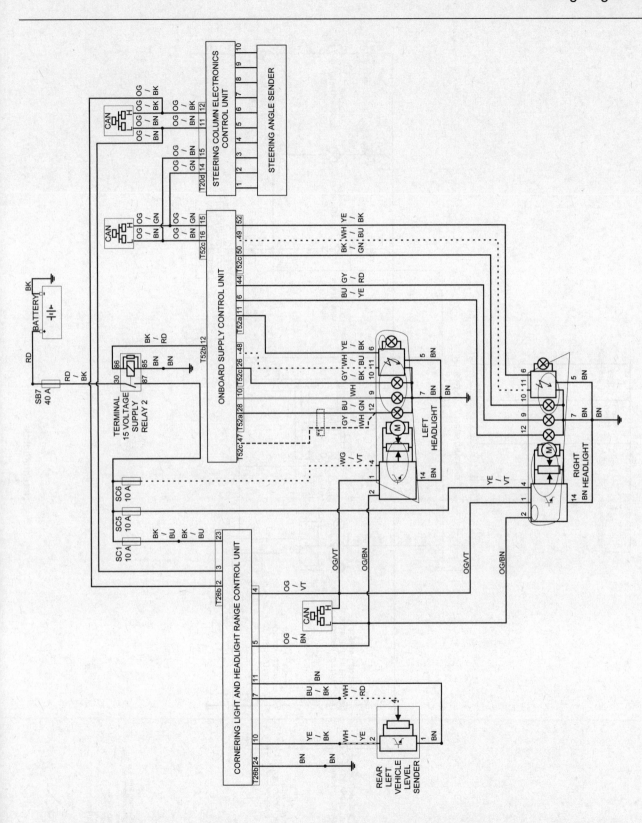

Exterior Lights - Xenon with automatic headlight range control and dynamic cornering light – from March 2009 to September 2012

*1 According to equipment

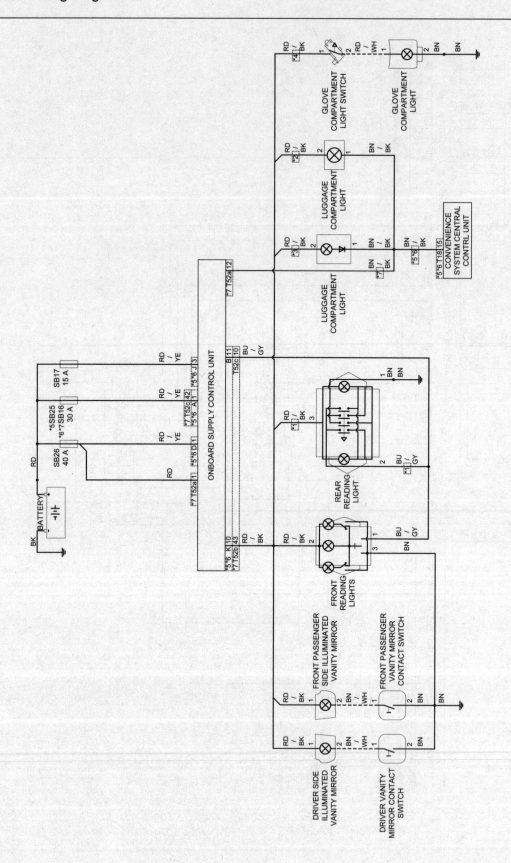

Interior lights

*1 Only models with rear reading light
*2 Only models without alarm horn
*3 Only models with alarm horn only
*4 Glove compartment light switch
*5 From April 2005 to April 2008
*6 From May 2008 February 2009
*7 From March 2009 to September 2012

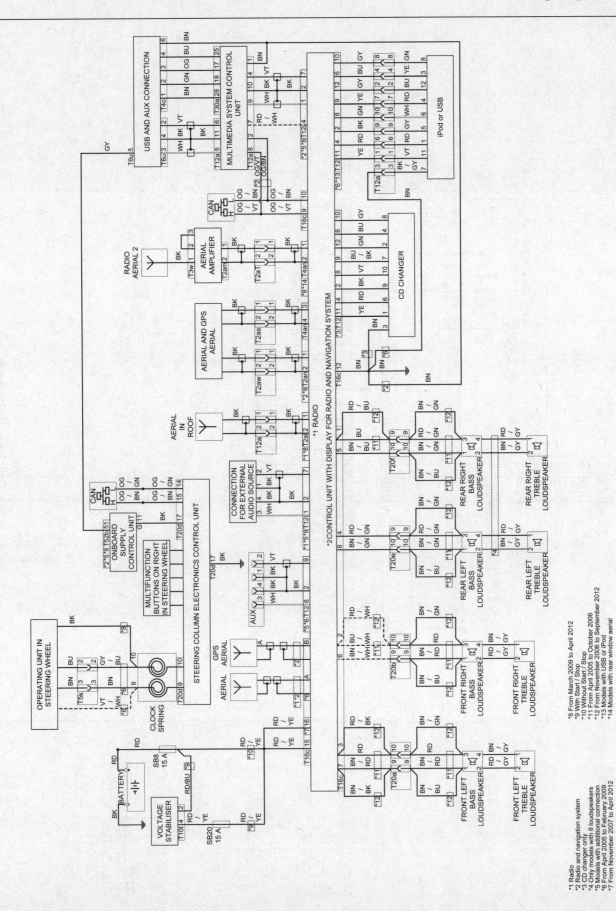

Sound System – radio and navigation system – from April 2005 up to September 2012

*1 Radio
*2 Radio and navigation system
*3 CD changer only
*4 Only models with 8 loudspeakers
*5 Models with additional connection
*6 From April 2005 to February 2009
*7 From November 2007 to April 2012

*8 From March 2009 to April 2012
*9 Without Start / Stop
*10 With Start / Stop
*11 From April 2005 to October 2006
*12 From November 2006 to September 2012
*13 Models with USB or iPod
*14 Models with rear window aerial

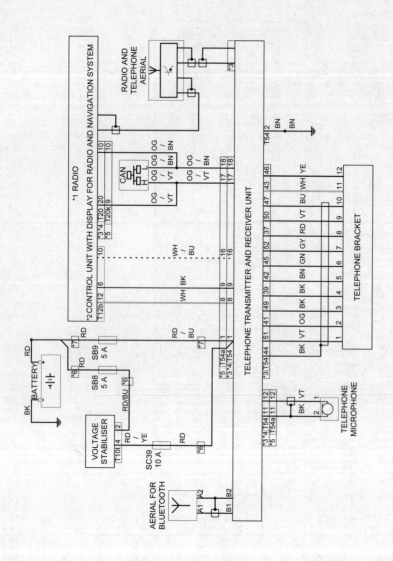

Sound system – telephone and bluetooth

*1 Radio
*2 Radio and navigation system
*3 From April 2005 up to April 2008
*4 From May 2008 up to September 2012
*5 From March 2009 up to September 2012
*6 With Start / Stop
*7 Without Start / Stop

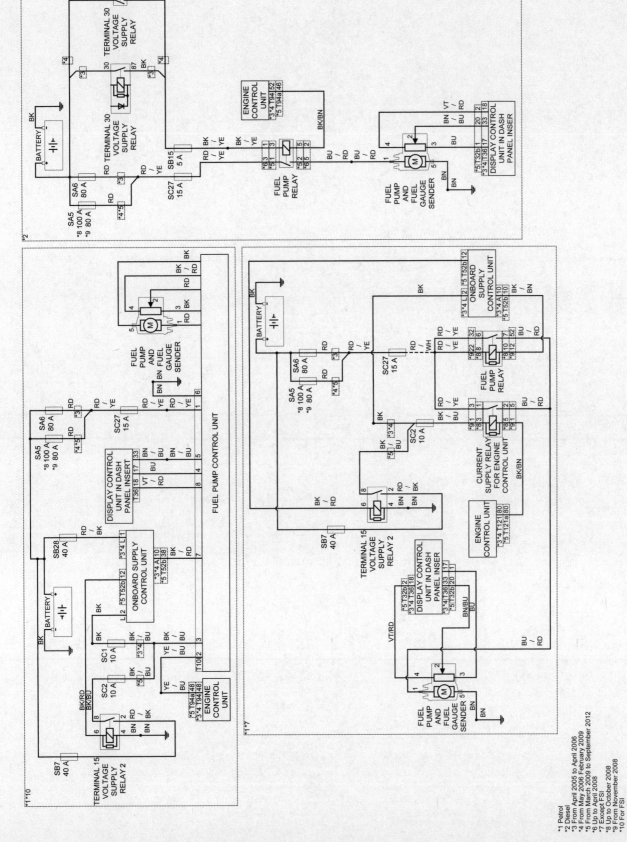

Fuel pump

*1 Petrol
*2 Diesel
*3 From April 2005 to April 2006
*4 From May 2006 February 2009
*5 From March 2009 to September 2012
*6 Up to April 2008
*7 Except FSI
*8 Up to October 2008
*9 From November 2008
*10 For FSI

Dimensions and weights

Note: *All figures and dimensions are approximate and may vary according to model. Refer to manufacturer's data for exact figures.*

Dimensions

Length	4315 mm
Width	1523 mm
Height	1448 mm
Wheelbase	2578 mm
Turning circle	10.9 M

Weights

Kerb weight	1154 to 1521 kg
Maximum gross vehicle weight	1740 to 1940 kg
Maximum roof rack load	75 kg

Towing weights

Maximum towing weights	**Unbraked trailer**	**Braked trailer**
Petrol engines	610 to 690 kg	1000 to 1500 kg
Diesel engines	640 to 690 kg	1000 to 1500 kg

Fuel economy

Although depreciation is still the biggest part of the cost of motoring for most car owners, the cost of fuel is more immediately noticeable. These pages give some tips on how to get the best fuel economy.

Working it out

Manufacturer's figures

Car manufacturers are required by law to provide fuel consumption information on all new vehicles sold. These 'official' figures are obtained by simulating various driving conditions on a rolling road or a test track. Real life conditions are different, so the fuel consumption actually achieved may not bear much resemblance to the quoted figures.

How to calculate it

Many cars now have trip computers which will

display fuel consumption, both instantaneous and average. Refer to the owner's handbook for details of how to use these.

To calculate consumption yourself (and maybe to check that the trip computer is accurate), proceed as follows.

1. Fill up with fuel and note the mileage, or zero the trip recorder.
2. Drive as usual until you need to fill up again.
3. Note the amount of fuel required to refill the tank, and the mileage covered since the previous fill-up.
4. Divide the mileage by the amount of fuel used to obtain the consumption figure.

For example:

Mileage at first fill-up (a) = 27,903
Mileage at second fill-up (b) = 28,346
Mileage covered (b - a) = 443
Fuel required at second fill-up = 48.6 litres

The half-completed changeover to metric units in the UK means that we buy our fuel in litres, measure distances in miles and talk about fuel consumption in miles per gallon. There are two ways round this: the first is to convert the litres to gallons before doing the calculation (by dividing by 4.546, or see Table 1). So in the example:

48.6 litres ÷ 4.546 = 10.69 gallons
443 miles ÷ 10.69 gallons = 41.4 mpg

The second way is to calculate the consumption in miles per litre, then multiply that figure by 4.546 (or see Table 2).

So in the example, fuel consumption is:

443 miles ÷ 48.6 litres = 9.1 mpl
9.1 mpl x 4.546 = 41.4 mpg

The rest of Europe expresses fuel consumption in litres of fuel required to travel 100 km (l/100 km). For interest, the conversions are given in Table 3. In practice it doesn't matter what units you use, provided you know what your normal consumption is and can spot if it's getting better or worse.

Table 1: conversion of litres to Imperial gallons

litres	1	2	3	4	5	10	20	30	40	50	60	70
gallons	0.22	0.44	0.66	0.88	1.10	2.24	4.49	6.73	8.98	11.22	13.47	15.71

Table 2: conversion of miles per litre to miles per gallon

miles per litre	5	6	7	8	9	10	11	12	13	14
miles per gallon	23	27	32	36	41	46	50	55	59	64

Table 3: conversion of litres per 100 km to miles per gallon

litres per 100 km	4	4.5	5	5.5	6	6.5	7	8	9	10
miles per gallon	71	63	56	51	47	43	40	35	31	28

Maintenance

A well-maintained car uses less fuel and creates less pollution. In particular:

Filters

Change air and fuel filters at the specified intervals.

Oil

Use a good quality oil of the lowest viscosity specified by the vehicle manufacturer (see *Lubricants and fluids*). Check the level often and be careful not to overfill.

Spark plugs

When applicable, renew at the specified intervals.

Tyres

Check tyre pressures regularly. Under-inflated tyres have an increased rolling resistance. It is generally safe to use the higher pressures specified for full load conditions even when not fully laden, but keep an eye on the centre band of tread for signs of wear due to over-inflation.

When buying new tyres, consider the 'fuel saving' models which most manufacturers include in their ranges.

Driving style

Acceleration

Acceleration uses more fuel than driving at a steady speed. The best technique with modern cars is to accelerate reasonably briskly to the desired speed, changing up through the gears as soon as possible without making the engine labour.

Air conditioning

Air conditioning absorbs quite a bit of energy from the engine – typically 3 kW (4 hp) or so. The effect on fuel consumption is at its worst in slow traffic. Switch it off when not required.

Anticipation

Drive smoothly and try to read the traffic flow so as to avoid unnecessary acceleration and braking.

Automatic transmission

When accelerating in an automatic, avoid depressing the throttle so far as to make the transmission hold onto lower gears at higher speeds. Don't use the 'Sport' setting, if applicable.

When stationary with the engine running, select 'N' or 'P'. When moving, keep your left foot away from the brake.

Braking

Braking converts the car's energy of motion into heat – essentially, it is wasted. Obviously some braking is always going to be necessary, but with good anticipation it is surprising how much can be avoided, especially on routes that you know well.

Carshare

Consider sharing lifts to work or to the shops. Even once a week will make a difference.

Electrical loads

Electricity is 'fuel' too; the alternator which charges the battery does so by converting some of the engine's energy of motion into electrical energy. The more electrical accessories are in use, the greater the load on the alternator. Switch off big consumers like the heated rear window when not required.

Freewheeling

Freewheeling (coasting) in neutral with the engine switched off is dangerous. The effort required to operate power-assisted brakes and steering increases when the engine is not running, with a potential lack of control in emergency situations.

In any case, modern fuel injection systems automatically cut off the engine's fuel supply on the overrun (moving and in gear, but with the accelerator pedal released).

Gadgets

Bolt-on devices claiming to save fuel have been around for nearly as long as the motor car itself. Those which worked were rapidly adopted as standard equipment by the vehicle manufacturers. Others worked only in certain situations, or saved fuel only at the expense of unacceptable effects on performance, driveability or the life of engine components.

The most effective fuel saving gadget is the driver's right foot.

Journey planning

Combine (eg) a trip to the supermarket with a visit to the recycling centre and the DIY store, rather than making separate journeys.

When possible choose a travelling time outside rush hours.

Load

The more heavily a car is laden, the greater the energy required to accelerate it to a given speed. Remove heavy items which you don't need to carry.

One load which is often overlooked is the contents of the fuel tank. A tankful of fuel (55 litres / 12 gallons) weighs 45 kg (100 lb) or so. Just half filling it may be worthwhile.

Lost?

At the risk of stating the obvious, if you're going somewhere new, have details of the route to hand. There's not much point in achieving record mpg if you also go miles out of your way.

Parking

If possible, carry out any reversing or turning manoeuvres when you arrive at a parking space so that you can drive straight out when you leave. Manoeuvering when the engine is cold uses a lot more fuel.

Driving around looking for free on-street parking may cost more in fuel than buying a car park ticket.

Premium fuel

Most major oil companies (and some supermarkets) have premium grades of fuel which are several pence a litre dearer than the standard grades. Reports vary, but the consensus seems to be that if these fuels improve economy at all, they do not do so by enough to justify their extra cost.

Roof rack

When loading a roof rack, try to produce a wedge shape with the narrow end at the front. Any cover should be securely fastened – if it flaps it's creating turbulence and absorbing energy.

Remove roof racks and boxes when not in use – they increase air resistance and can create a surprising amount of noise.

Short journeys

The engine is at its least efficient, and wear is highest, during the first few miles after a cold start. Consider walking, cycling or using public transport.

Speed

The engine is at its most efficient when running at a steady speed and load at the rpm where it develops maximum torque. (You can find this figure in the car's handbook.) For most cars this corresponds to between 55 and 65 mph in top gear.

Above the optimum cruising speed, fuel consumption starts to rise quite sharply. A car travelling at 80 mph will typically be using 30% more fuel than at 60 mph.

Supermarket fuel

It may be cheap but is it any good? In the UK all supermarket fuel must meet the relevant British Standard. The major oil companies will say that their branded fuels have better additive packages which may stop carbon and other deposits building up. A reasonable compromise might be to use one tank of branded fuel to three or four from the supermarket.

Switch off when stationary

Switch off the engine if you look like being stationary for more than 30 seconds or so. This is good for the environment as well as for your pocket. Be aware though that frequent restarts are hard on the battery and the starter motor.

Windows

Driving with the windows open increases air turbulence around the vehicle. Closing the windows promotes smooth airflow and

reduced resistance. The faster you go, the more significant this is.

And finally . . .

Driving techniques associated with good fuel economy tend to involve moderate acceleration and low top speeds. Be considerate to the needs of other road users who may need to make brisker progress; even if you do not agree with them this is not an excuse to be obstructive.

Safety must always take precedence over economy, whether it is a question of accelerating hard to complete an overtaking manoeuvre, killing your speed when confronted with a potential hazard or switching the lights on when it starts to get dark.

Spare parts are available from many sources, including maker's appointed garages, accessory shops, and motor factors. To be sure of obtaining the correct parts, it will sometimes be necessary to quote the vehicle identification number. If possible, it can also be useful to take the old parts along for positive identification. Items such as starter motors and alternators may be available under a service exchange scheme – any parts returned should be clean.

Our advice regarding spare parts is as follows.

Officially appointed garages

This is the best source of parts which are peculiar to your car, and which are not otherwise generally available (eg, badges, interior trim, certain body panels, etc). It is also the only place at which you should buy parts if the car is still under warranty.

Accessory shops

These are very good places to buy materials and components needed for the maintenance of your car (oil, air and fuel filters, light bulbs, drivebelts, greases, brake pads, touch-up paint, etc). Components of this nature sold by a reputable shop are usually of the same standard as those used by the car manufacturer.

Besides components, these shops also sell tools and general accessories, usually have convenient opening hours, charge lower prices, and can often be found close to home. Some accessory shops have parts counters where components needed for almost any repair job can be purchased or ordered.

Motor factors

Good factors will stock all the more important components which wear out comparatively quickly, and can sometimes supply individual components needed for the overhaul of a larger assembly (eg, brake seals and hydraulic parts, bearing shells, pistons, valves). They may also handle work such as cylinder block reboring, crankshaft regrinding, etc.

Engine reconditioners

These specialise in engine overhaul and can also supply components. It is recommended that the establishment is a member of the Federation of Engine Re-Manufacturers, or a similar society.

Tyre and exhaust specialists

These outlets may be independent, or members of a local or national chain. They frequently offer competitive prices when compared with a main dealer or local garage, but it will pay to obtain several quotes before making a decision. When researching prices, also ask what extras may be added – for instance fitting a new valve, balancing the wheel and tyre disposal all both commonly charged on top of the price of a new tyre.

Other sources

Beware of parts or materials obtained from market stalls, car boot sales, on-line auctions or similar outlets. Such items are not invariably sub-standard, but there is little chance of compensation if they do prove unsatisfactory. In the case of safety-critical components such as brake pads, there is the risk not only of financial loss, but also of an accident causing injury or death.

Second-hand components or assemblies obtained from a car breaker can be a good buy in some circumstances, but this sort of purchase is best made by the experienced DIY mechanic.

Vehicle identification numbers

Modifications are a continuing and unpublicised process in vehicle manufacture, quite apart from major model changes. Spare parts manuals and lists are compiled upon a numerical basis, the individual vehicle identification numbers being essential to correct identification of the component concerned.

When ordering spare parts, always give as much information as possible. Quote the car model, year of manufacture and registration, chassis and engine numbers as appropriate.

The Vehicle Identification Number (VIN) plate is visible from the outside of the vehicle, through the left-hand lower corner of the windscreen, and is also stamped on the top of the right-hand inner wing in the engine compartment (see illustrations).

The Vehicle Data Sticker is located in the spare wheel well accessed through the luggage compartment. It contains the VIN, vehicle type, engine power, transmission type, engine and transmission codes, paint number, interior equipment, optional extras, and PR numbers (for maintenance schedule).

The Type plate and factory plate is located at the bottom of the front, left-hand door A-pillar, and is visible with the door open. It contains the gross vehicle weight, front axle weight and rear axle weight.

The Engine Number is stamped into the left-hand end of the cylinder block and on the right-hand end of the cylinder head on petrol engines. On diesel engines it is stamped into the front of the cylinder block, next to the engine-to-transmission joint. A barcode identification sticker is located on the top of the timing cover or on the right-hand end of the cylinder head (see illustrations).

Engine codes

1.6 litre

SOHC petrol engine	BGU, BSE and BSF

2.0 litre petrol engine

Non-turbo	BLR, BVY and BLY,
Turbo	BWA BVZ, BWJ, CDLA and CDLA

Diesel engine

1.6 litre (CR injection)	CAYC
1.9 litre (PD injection)	BJB, BKC, BLS, BXE and BXF
2.0 litre (PD injection)	BMM, BKD and BMN
2.0 litre (CR injection)	CFJA, CLCB, CEGA and CFHC

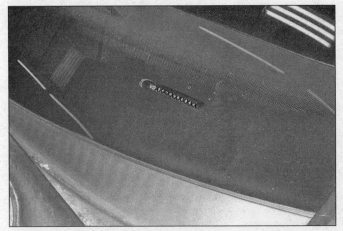

Vehicle Identification Number (VIN) located on the left-hand front edge of the windscreen

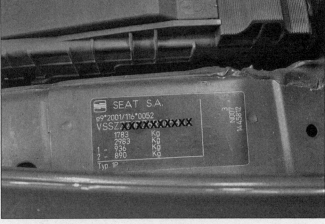

Vehicle Identification Number (VIN) located on the top of the left-hand inner wing

Engine number located on the left-hand end of the block

Engine code on the right-hand end of the cylinder head (2.0 litre engine shown)

Engine code sticker on the top of the timing belt cover

Whenever servicing, repair or overhaul work is carried out on the car or its components, observe the following procedures and instructions. This will assist in carrying out the operation efficiently and to a professional standard of workmanship.

Joint mating faces and gaskets

When separating components at their mating faces, never insert screwdrivers or similar implements into the joint between the faces in order to prise them apart. This can cause severe damage which results in oil leaks, coolant leaks, etc upon reassembly. Separation is usually achieved by tapping along the joint with a soft-faced hammer in order to break the seal. However, note that this method may not be suitable where dowels are used for component location.

Where a gasket is used between the mating faces of two components, a new one must be fitted on reassembly; fit it dry unless otherwise stated in the repair procedure. Make sure that the mating faces are clean and dry, with all traces of old gasket removed. When cleaning a joint face, use a tool which is unlikely to score or damage the face, and remove any burrs or nicks with an oilstone or fine file.

Make sure that tapped holes are cleaned with a pipe cleaner, and keep them free of jointing compound, if this is being used, unless specifically instructed otherwise.

Ensure that all orifices, channels or pipes are clear, and blow through them, preferably using compressed air.

Oil seals

Oil seals can be removed by levering them out with a wide flat-bladed screwdriver or similar implement. Alternatively, a number of self-tapping screws may be screwed into the seal, and these used as a purchase for pliers or some similar device in order to pull the seal free.

Whenever an oil seal is removed from its working location, either individually or as part of an assembly, it should be renewed.

The very fine sealing lip of the seal is easily damaged, and will not seal if the surface it contacts is not completely clean and free from scratches, nicks or grooves. If the original sealing surface of the component cannot be restored, and the manufacturer has not made provision for slight relocation of the seal relative to the sealing surface, the component should be renewed.

Protect the lips of the seal from any surface which may damage them in the course of fitting. Use tape or a conical sleeve where possible. Where indicated, lubricate the seal lips with oil before fitting and, on dual-lipped seals, fill the space between the lips with grease.

Unless otherwise stated, oil seals must be fitted with their sealing lips toward the lubricant to be sealed.

Use a tubular drift or block of wood of the appropriate size to install the seal and, if the seal housing is shouldered, drive the seal down to the shoulder. If the seal housing is unshouldered, the seal should be fitted with its face flush with the housing top face (unless otherwise instructed).

Screw threads and fastenings

Seized nuts, bolts and screws are quite a common occurrence where corrosion has set in, and the use of penetrating oil or releasing fluid will often overcome this problem if the offending item is soaked for a while before attempting to release it. The use of an impact driver may also provide a means of releasing such stubborn fastening devices, when used in conjunction with the appropriate screwdriver bit or socket. If none of these methods works, it may be necessary to resort to the careful application of heat, or the use of a hacksaw or nut splitter device. Before resorting to extreme methods, check that you are not dealing with a left-hand thread!

Studs are usually removed by locking two nuts together on the threaded part, and then using a spanner on the lower nut to unscrew the stud. Studs or bolts which have broken off below the surface of the component in which they are mounted can sometimes be removed using a stud extractor.

Always ensure that a blind tapped hole is completely free from oil, grease, water or other fluid before installing the bolt or stud. Failure to do this could cause the housing to crack due to the hydraulic action of the bolt or stud as it is screwed in.

For some screw fastenings, notably cylinder head bolts or nuts, torque wrench settings are no longer specified for the latter stages of tightening, "angle-tightening" being called up instead. Typically, a fairly low torque wrench setting will be applied to the bolts/nuts in the correct sequence, followed by one or more stages of tightening through specified angles.

When checking or retightening a nut or bolt to a specified torque setting, slacken the nut or bolt by a quarter of a turn, and then retighten to the specified setting. However, this should not be attempted where angular tightening has been used.

Locknuts, locktabs and washers

Any fastening which will rotate against a component or housing during tightening should always have a washer between it and the relevant component or housing.

Spring or split washers should always be renewed when they are used to lock a critical component such as a big-end bearing retaining bolt or nut. Locktabs which are folded over to retain a nut or bolt should always be renewed.

Self-locking nuts can be re-used in non-critical areas, providing resistance can be felt when the locking portion passes over the bolt or stud thread. However, it should be noted that self-locking stiffnuts tend to lose their effectiveness after long periods of use, and should then be renewed as a matter of course.

Split pins must always be replaced with new ones of the correct size for the hole.

When thread-locking compound is found on the threads of a fastener which is to be re-used, it should be cleaned off with a wire brush and solvent, and fresh compound applied on reassembly.

Special tools

Some repair procedures in this manual entail the use of special tools such as a press, two or three-legged pullers, spring compressors, etc. Wherever possible, suitable readily-available alternatives to the manufacturer's special tools are described, and are shown in use. In some instances, where no alternative is possible, it has been necessary to resort to the use of a manufacturer's tool, and this has been done for reasons of safety as well as the efficient completion of the repair operation. Unless you are highly-skilled and have a thorough understanding of the procedures described, never attempt to bypass the use of any special tool when the procedure described specifies its use. Not only is there a very great risk of personal injury, but expensive damage could be caused to the components involved.

Environmental considerations

When disposing of used engine oil, brake fluid, antifreeze, etc, give due consideration to any detrimental environmental effects. Do not, for instance, pour any of the above liquids down drains into the general sewage system, or onto the ground to soak away, as this is likely to pollute your local environment. Many local council refuse tips provide a facility for waste oil disposal, as do some garages. You can find your nearest disposal point by calling the Environment Agency on 03708 506 506 or by visiting www.oilbankline.org.uk.

Note: It is illegal and anti-social to dump oil down the drain. To find the location of your local oil recycling bank, call 03708 506 506 or visit www.oilbankline.org.uk.

The jack supplied with the vehicle tool kit should only be used for changing the roadwheels – see Wheel changing at the front of this book. When carrying out any other kind of work, raise the vehicle using a hydraulic (or 'trolley') jack, and always supplement the jack with axle stands positioned under the vehicle jacking points.

When using a hydraulic jack or axle stands, always position the jack head or axle stand head under one of the relevant jacking points.

To raise the front and/or rear of the vehicle, use the jacking/support points at the front and rear ends of the door sills, indicated by the triangular depressions in the sill panel **(see illustration)**. Position a block of wood with a groove cut in it on the jack head to prevent the vehicle weight resting on the sill edge; align the sill edge with the groove in the wood so that the vehicle weight is spread evenly over the surface of the block. Supplement the jack with axle stands (also with slotted blocks of wood) positioned as close as possible to the jacking points **(see illustrations)**.

Do not jack the vehicle under any other part of the sill, sump, floor pan, or any of the steering or suspension components. With the vehicle raised, an axle stand should be positioned beneath the vehicle jack location point on the sill.

⚠️ *Warning: Never work under, around, or near a raised car, unless it is supported in at least two places.*

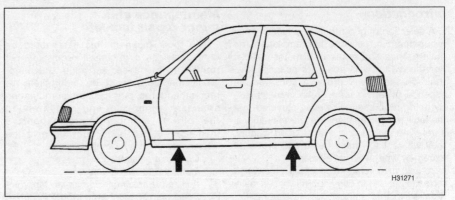

Front and rear jacking points

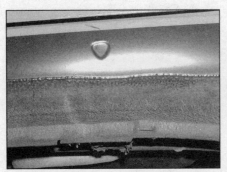

The jacking points are indicated by an arrow on the sill

Use an axle stand with a suitable block of wood

Introduction

A selection of good tools is a fundamental requirement for anyone contemplating the maintenance and repair of a motor vehicle. For the owner who does not possess any, their purchase will prove a considerable expense, offsetting some of the savings made by doing-it-yourself. However, provided that the tools purchased meet the relevant national safety standards and are of good quality, they will last for many years and prove an extremely worthwhile investment.

To help the average owner to decide which tools are needed to carry out the various tasks detailed in this manual, we have compiled three lists of tools under the following headings: *Maintenance and minor repair, Repair and overhaul*, and *Special*. Newcomers to practical mechanics should start off with the *Maintenance and minor repair* tool kit, and confine themselves to the simpler jobs around the vehicle. Then, as confidence and experience grow, more difficult tasks can be undertaken, with extra tools being purchased as, and when, they are needed. In this way, a *Maintenance and minor repair* tool kit can be built up into a *Repair and overhaul* tool kit over a considerable period of time, without any major cash outlays. The experienced do-it-yourselfer will have a tool kit good enough for most repair and overhaul procedures, and will add tools from the *Special* category when it is felt that the expense is justified by the amount of use to which these tools will be put.

Maintenance and minor repair tool kit

The tools given in this list should be considered as a minimum requirement if routine maintenance, servicing and minor repair operations are to be undertaken. We recommend the purchase of combination spanners (ring one end, open-ended the other); although more expensive than open-ended ones, they do give the advantages of both types of spanner.

☐ *Combination spanners:*
 Metric - 8 to 19 mm inclusive
☐ *Adjustable spanner - 35 mm jaw (approx.)*
☐ *Spark plug spanner (with rubber insert) - petrol models*
☐ *Spark plug gap adjustment tool - petrol models*
☐ *Set of feeler gauges*
☐ *Brake bleed nipple spanner*
☐ *Screwdrivers:*
 Flat blade - 100 mm long x 6 mm dia
 Cross blade - 100 mm long x 6 mm dia
 Torx - various sizes (not all vehicles)
☐ *Combination pliers*
☐ *Hacksaw (junior)*
☐ *Tyre pump*
☐ *Tyre pressure gauge*
☐ *Oil can*
☐ *Oil filter removal tool (if applicable)*
☐ *Fine emery cloth*
☐ *Wire brush (small)*
☐ *Funnel (medium size)*
☐ *Sump drain plug key (not all vehicles)*

Repair and overhaul tool kit

These tools are virtually essential for anyone undertaking any major repairs to a motor vehicle, and are additional to those given in the *Maintenance and minor repair* list. Included in this list is a comprehensive set of sockets. Although these are expensive, they will be found invaluable as they are so versatile - particularly if various drives are included in the set. We recommend the half-inch square-drive type, as this can be used with most proprietary torque wrenches.

The tools in this list will sometimes need to be supplemented by tools from the *Special* list:

☐ *Sockets to cover range in previous list (including Torx sockets)*
☐ *Reversible ratchet drive (for use with sockets)*
☐ *Extension piece, 250 mm (for use with sockets)*
☐ *Universal joint (for use with sockets)*
☐ *Flexible handle or sliding T "breaker bar" (for use with sockets)*
☐ *Torque wrench (for use with sockets)*
☐ *Self-locking grips*
☐ *Ball pein hammer*
☐ *Soft-faced mallet (plastic or rubber)*
☐ *Screwdrivers:*
 Flat blade - long & sturdy, short (chubby), and narrow (electrician's) types
 Cross blade – long & sturdy, and short (chubby) types
☐ *Pliers:*
 Long-nosed
 Side cutters (electrician's)
 Circlip (internal and external)
☐ *Cold chisel - 25 mm*
☐ *Scriber*
☐ *Scraper*
☐ *Centre-punch*
☐ *Pin punch*
☐ *Hacksaw*
☐ *Brake hose clamp*
☐ *Brake/clutch bleeding kit*
☐ *Selection of twist drills*
☐ *Steel rule/straight-edge*
☐ *Allen keys (inc. splined/Torx type)*
☐ *Selection of files*
☐ *Wire brush*
☐ *Axle stands*
☐ *Jack (strong trolley or hydraulic type)*
☐ *Light with extension lead*
☐ *Universal electrical multi-meter*

Sockets and reversible ratchet drive

Brake bleeding kit

Torx key, socket and bit

Hose clamp

Angular-tightening gauge

Special tools

The tools in this list are those which are not used regularly, are expensive to buy, or which need to be used in accordance with their manufacturers' instructions. Unless relatively difficult mechanical jobs are undertaken frequently, it will not be economic to buy many of these tools. Where this is the case, you could consider clubbing together with friends (or joining a motorists' club) to make a joint purchase, or borrowing the tools against a deposit from a local garage or tool hire specialist.

The following list contains only those tools and instruments freely available to the public, and not those special tools produced by the vehicle manufacturer specifically for its dealer network. You will find occasional references to these manufacturers' special tools in the text of this manual. Generally, an alternative method of doing the job without the vehicle manufacturers' special tool is given. However, sometimes there is no alternative to using them. Where this is the case and the relevant tool cannot be bought or borrowed, you will have to entrust the work to a dealer.

☐ Angular-tightening gauge
☐ Valve spring compressor
☐ Valve grinding tool
☐ Piston ring compressor
☐ Piston ring removal/installation tool
☐ Cylinder bore hone
☐ Balljoint separator
☐ Coil spring compressors (where applicable)
☐ Two/three-legged hub and bearing puller
☐ Impact screwdriver
☐ Micrometer and/or vernier calipers
☐ Dial gauge
☐ Tachometer
☐ Fault code reader
☐ Cylinder compression gauge
☐ Hand-operated vacuum pump and gauge
☐ Clutch plate alignment set
☐ Brake shoe steady spring cup removal tool
☐ Bush and bearing removal/installation set
☐ Stud extractors
☐ Tap and die set
☐ Lifting tackle

Buying tools

Reputable motor accessory shops and superstores often offer excellent quality tools at discount prices, so it pays to shop around.

Remember, you don't have to buy the most expensive items on the shelf, but it is always advisable to steer clear of the very cheap tools. Beware of 'bargains' offered on market stalls, on-line or at car boot sales. There are plenty of good tools around at reasonable prices, but always aim to purchase items which meet the relevant national safety standards. If in doubt, ask the proprietor or manager of the shop for advice before making a purchase.

Care and maintenance of tools

Having purchased a reasonable tool kit, it is necessary to keep the tools in a clean and serviceable condition. After use, always wipe off any dirt, grease and metal particles using a clean, dry cloth, before putting the tools away. Never leave them lying around after they have been used. A simple tool rack on the garage or workshop wall for items such as screwdrivers and pliers is a good idea. Store all normal spanners and sockets in a metal box. Any measuring instruments, gauges, meters, etc, must be carefully stored where they cannot be damaged or become rusty.

Take a little care when tools are used. Hammer heads inevitably become marked, and screwdrivers lose the keen edge on their blades from time to time. A little timely attention with emery cloth or a file will soon restore items like this to a good finish.

Working facilities

Not to be forgotten when discussing tools is the workshop itself. If anything more than routine maintenance is to be carried out, a suitable working area becomes essential.

It is appreciated that many an owner-mechanic is forced by circumstances to remove an engine or similar item without the benefit of a garage or workshop. Having done this, any repairs should always be done under the cover of a roof.

Wherever possible, any dismantling should be done on a clean, flat workbench or table at a suitable working height.

Any workbench needs a vice; one with a jaw opening of 100 mm is suitable for most jobs. As mentioned previously, some clean dry storage space is also required for tools, as well as for any lubricants, cleaning fluids, touch-up paints etc, which become necessary.

Another item which may be required, and which has a much more general usage, is an electric drill with a chuck capacity of at least 8 mm. This, together with a good range of twist drills, is virtually essential for fitting accessories.

Last, but not least, always keep a supply of old newspapers and clean, lint-free rags available, and try to keep any working area as clean as possible.

Micrometers

Dial test indicator ("dial gauge")

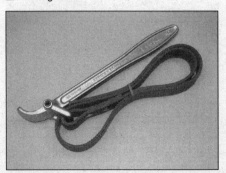
Oil filter removal tool (strap wrench type)

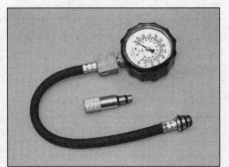
Compression tester

Bearing puller

This is a guide to getting your vehicle through the MOT test. Obviously it will not be possible to examine the vehicle to the same standard as the professional MOT tester. However, working through the following checks will enable you to identify any problem areas before submitting the vehicle for the test.

It has only been possible to summarise the test requirements here, based on the regulations in force at the time of printing. Test standards are becoming increasingly stringent, although there are some exemptions for older vehicles.

An assistant will be needed to help carry out some of these checks.

The checks have been sub-divided into four categories, as follows:

1 Checks carried out **FROM THE DRIVER'S SEAT**

2 Checks carried out **WITH THE VEHICLE ON THE GROUND**

3 Checks carried out **WITH THE VEHICLE RAISED AND THE WHEELS FREE TO TURN**

4 Checks carried out on **YOUR VEHICLE'S EXHAUST EMISSION SYSTEM**

1 Checks carried out **FROM THE DRIVER'S SEAT**

Handbrake (parking brake)

☐ Test the operation of the handbrake. Excessive travel (too many clicks) indicates incorrect brake or cable adjustment.
☐ Check that the handbrake cannot be released by tapping the lever sideways. Check the security of the lever mountings.

☐ If the parking brake is foot-operated, check that the pedal is secure and without excessive travel, and that the release mechanism operates correctly.
☐ Where applicable, test the operation of the electronic handbrake. The brake should engage and disengage without excessive delay. If the warning light does not extinguish when the brake is disengaged, this could indicate a fault which will need further investigation.

Footbrake

☐ Depress the brake pedal and check that it does not creep down to the floor, indicating a master cylinder fault. Release the pedal,

wait a few seconds, then depress it again. If the pedal travels nearly to the floor before firm resistance is felt, brake adjustment or repair is necessary. If the pedal feels spongy, there is air in the hydraulic system which must be removed by bleeding.

☐ Check that the brake pedal is secure and in good condition. Check also for signs of fluid leaks on the pedal, floor or carpets, which would indicate failed seals in the brake master cylinder.
☐ Check the servo unit (when applicable) by operating the brake pedal several times, then keeping the pedal depressed and starting the engine. As the engine starts, the pedal will move down slightly. If not, the vacuum hose or the servo itself may be faulty.

Steering wheel and column

☐ Examine the steering wheel for fractures or looseness of the hub, spokes or rim.
☐ Move the steering wheel from side to side and then up and down. Check that the steering wheel is not loose on the column, indicating wear or a loose retaining nut. Continue moving the steering wheel as before, but also turn it slightly from left to right.

☐ Check that the steering wheel is not loose on the column, and that there is no abnormal movement of the steering wheel, indicating wear in the column support bearings or couplings.
☐ Check that the ignition lock (where fitted) engages and disengages correctly.
☐ Steering column adjustment mechanisms (where fitted) must be able to lock the column securely in place with no play evident.

Windscreen, mirrors and sunvisor

☐ The windscreen must be free of cracks or other significant damage within the driver's field of view. (Small stone chips are acceptable.) Rear view mirrors must be secure, intact, and capable of being adjusted.

☐ The driver's sunvisor must be capable of being stored in the "up" position.

Seat belts and seats

Note: *The following checks are applicable to all seat belts, front and rear.*

☐ Examine the webbing of all the belts (including rear belts if fitted) for cuts, serious fraying or deterioration. Fasten and unfasten each belt to check the buckles. If applicable, check the retracting mechanism. Check the security of all seat belt mountings accessible from inside the vehicle, ensuring any height adjustable mountings lock securely in place.

☐ Seat belts with pre-tensioners, once activated, have a "flag" or similar showing on the seat belt stalk. This, in itself, is not a reason for test failure.

☐ The front seats themselves must be securely attached and the backrests must lock in the upright position.

Doors

☐ Both front doors must be able to be opened and closed from outside and inside, and must latch securely when closed.

Bonnet and boot/tailgate

☐ The bonnet and boot/tailgate must latch securely when closed.

2 Checks carried out WITH THE VEHICLE ON THE GROUND

Vehicle identification

☐ Number plates must be in good condition, secure and legible, with letters and numbers correctly spaced – spacing at (A) should be 33 mm and at (B) 11 mm. At the front, digits must be black on a white background and at the rear black on a yellow background. Other background designs (such as honeycomb) are not permitted.

☐ The VIN plate and/or homologation plate must be permanently displayed and legible.

Electrical equipment

☐ Switch on the ignition and check the operation of the horn.

☐ Check the windscreen washers and wipers, examining the wiper blades; renew damaged or perished blades. Also check the operation of the stop-lights.

☐ Check the operation of the sidelights and number plate lights. The lenses and reflectors must be secure, clean and undamaged.

☐ Check the operation and alignment of the headlights. The headlight reflectors must not be tarnished and the lenses must be undamaged.

☐ Switch on the ignition and check the operation of the direction indicators (including the instrument panel tell-tale) and the hazard warning lights. Operation of the sidelights and stop-lights must not affect the indicators - if it does, the cause is usually a bad earth at the rear light cluster. Indicators should flash at a rate of between 60 and 120 times per minute – faster or slower than this could indicate a fault with the flasher unit or a bad earth at one of the light units.

☐ Check the operation of the rear foglight(s), including the warning light on the instrument panel or in the switch.

☐ The warning lights must illuminate in accordance with the manufacturer's design. For most vehicles, the ABS and other warning lights should illuminate when the ignition is switched on, and (if the system is operating properly) extinguish after a few seconds. Refer to the owner's handbook.

Footbrake

☐ Examine the master cylinder, brake pipes and servo unit for leaks, loose mountings, corrosion or other damage. If ABS is fitted, this unit should also be examined for signs of leaks or corrosion.

☐ The fluid reservoir must be secure and the fluid level must be between the upper (A) and lower (B) markings.

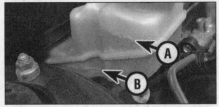

☐ Inspect both front brake flexible hoses for cracks or deterioration of the rubber. Turn the steering from lock to lock, and ensure that the hoses do not contact the wheel, tyre, or any part of the steering or suspension mechanism. With the brake pedal firmly depressed, check the hoses for bulges or leaks under pressure.

Steering and suspension

☐ Have your assistant turn the steering wheel from side to side slightly, up to the point where the steering gear just begins to transmit this movement to the roadwheels. Check for excessive free play between the steering wheel and the steering gear, indicating wear or insecurity of the steering column joints, the column-to-steering gear coupling, or the steering gear itself.

☐ Have your assistant turn the steering wheel more vigorously in each direction, so that the roadwheels just begin to turn. As this is done, examine all the steering joints, linkages, fittings and attachments. Renew any component that shows signs of wear or damage. On vehicles with power steering, check the security and condition of the steering pump, drivebelt and hoses.

☐ Check that the vehicle is standing level, and at approximately the correct ride height.

Shock absorbers

☐ Depress each corner of the vehicle in turn, then release it. The vehicle should rise and then settle in its normal position. If the vehicle continues to rise and fall, the shock absorber is defective. A shock absorber which has seized will also cause the vehicle to fail.

Exhaust system

☐ Start the engine. With your assistant holding a rag over the tailpipe, check the entire system for leaks. Repair or renew leaking sections.

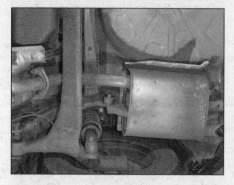

3 Checks carried out
WITH THE VEHICLE RAISED AND THE WHEELS FREE TO TURN

Jack up the front and rear of the vehicle, and securely support it on axle stands. Position the stands clear of the suspension assemblies. Ensure that the wheels are clear of the ground and that the steering can be turned from lock to lock.

Steering mechanism

☐ Have your assistant turn the steering from lock to lock. Check that the steering turns smoothly, and that no part of the steering mechanism, including a wheel or tyre, fouls any brake hose or pipe or any part of the body structure.
☐ Examine the steering rack rubber gaiters for damage or insecurity of the retaining clips. If power steering is fitted, check for signs of damage or leakage of the fluid hoses, pipes or connections. Also check for excessive stiffness or binding of the steering, a missing split pin or locking device, or severe corrosion of the body structure within 30 cm of any steering component attachment point.

Front and rear suspension and wheel bearings

☐ Starting at the front right-hand side, grasp the roadwheel at the 3 o'clock and 9 o'clock positions and rock gently but firmly. Check for free play or insecurity at the wheel bearings, suspension balljoints, or suspension mount-ings, pivots and attachments.
☐ Now grasp the wheel at the 12 o'clock and 6 o'clock positions and repeat the previous inspection. Spin the wheel, and check for roughness or tightness of the front wheel bearing.

☐ If excess free play is suspected at a component pivot point, this can be confirmed by using a large screwdriver or similar tool and levering between the mounting and the component attachment. This will confirm whether the wear is in the pivot bush, its retaining bolt, or in the mounting itself (the bolt holes can often become elongated).

☐ Carry out all the above checks at the other front wheel, and then at both rear wheels.

Springs and shock absorbers

☐ Examine the suspension struts (when applicable) for serious fluid leakage, corrosion, or damage to the casing. Also check the security of the mounting points.
☐ If coil springs are fitted, check that the spring ends locate in their seats, and that the spring is not corroded, cracked or broken.
☐ If leaf springs are fitted, check that all leaves are intact, that the axle is securely attached to each spring, and that there is no deterioration of the spring eye mountings, bushes, and shackles.

☐ The same general checks apply to vehicles fitted with other suspension types, such as torsion bars, hydraulic displacer units, etc. Ensure that all mountings and attachments are secure, that there are no signs of excessive wear, corrosion or damage, and (on hydraulic types) that there are no fluid leaks or damaged pipes.
☐ Inspect the shock absorbers for signs of serious fluid leakage. Check for wear of the mounting bushes or attachments, or damage to the body of the unit.

Driveshafts (fwd vehicles only)

☐ Rotate each front wheel in turn and inspect the constant velocity joint gaiters for splits or damage. Also check that each driveshaft is straight and undamaged.

Braking system

☐ If possible without dismantling, check brake pad wear and disc condition. Ensure that the friction lining material has not worn excessively, (A) and that the discs are not fractured, pitted, scored or badly worn (B).

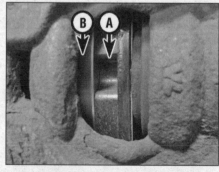

☐ Examine all the rigid brake pipes underneath the vehicle, and the flexible hose(s) at the rear. Look for corrosion, chafing or insecurity of the pipes, and for signs of bulging under pressure, chafing, splits or deterioration of the flexible hoses.
☐ Look for signs of fluid leaks at the brake calipers or on the brake backplates. Repair or renew leaking components.
☐ Slowly spin each wheel, while your assistant depresses and releases the footbrake. Ensure that each brake is operating and does not bind when the pedal is released.

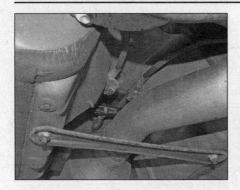

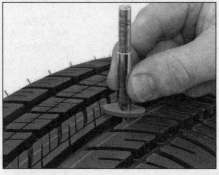

□Examine the handbrake mechanism, checking for frayed or broken cables, excessive corrosion, or wear or insecurity of the linkage. Check that the mechanism works on each relevant wheel, and releases fully, without binding.

□It is not possible to test brake efficiency without special equipment, but a road test can be carried out later to check that the vehicle pulls up in a straight line.

Fuel and exhaust systems

□Inspect the fuel tank (including the filler cap), fuel pipes, hoses and unions. All components must be secure and free from leaks. Locking fuel caps must lock securely and the key must be provided for the MOT test.

□Examine the exhaust system over its entire length, checking for any damaged, broken or missing mountings, security of the retaining clamps and rust or corrosion.

Wheels and tyres

□Examine the sidewalls and tread area of each tyre in turn. Check for cuts, tears, lumps, bulges, separation of the tread, and exposure of the ply or cord due to wear or damage. Check that the tyre bead is correctly seated on the wheel rim, that the valve is sound and properly seated, and that the wheel is not distorted or damaged.

□Check that the tyres are of the correct size for the vehicle, that they are of the same size and type on each axle, and that the pressures are correct.

□Check the tyre tread depth. The legal minimum at the time of writing is 1.6 mm over the central three-quarters of the tread width. Abnormal tread wear may indicate incorrect front wheel alignment or wear in steering or suspension components.

□If the spare wheel is fitted externally or in a separate carrier beneath the vehicle, check that mountings are secure and free of excessive corrosion.

Body corrosion

□Check the condition of the entire vehicle structure for signs of corrosion in load-bearing areas. (These include chassis box sections, side sills, cross-members, pillars, and all suspension, steering, braking system and seat belt mountings and anchorages.) Any corrosion which has seriously reduced the thickness of a load-bearing area (or is within 30 cm of safety-related components such as steering or suspension) is likely to cause the vehicle to fail. In this case professional repairs are likely to be needed.

□Damage or corrosion which causes sharp or otherwise dangerous edges to be exposed will also cause the vehicle to fail.

Towbars

□Check the condition of mounting points (both beneath the vehicle and within boot/hatchback areas) for signs of corrosion, ensuring that all fixings are secure and not worn or damaged. There must be no excessive play in detachable tow ball arms or quick-release mechanisms.

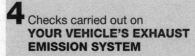

4 Checks carried out on
YOUR VEHICLE'S EXHAUST EMISSION SYSTEM

Petrol models

□The engine should be warmed up, and running well (ignition system in good order, air filter element clean, etc).

□Before testing, run the engine at around 2500 rpm for 20 seconds. Let the engine drop to idle, and watch for smoke from the exhaust. If the idle speed is too high, or if dense blue or black smoke emerges for more than 5 seconds, the vehicle will fail. Typically, blue smoke signifies oil burning (engine wear);

black smoke means unburnt fuel (dirty air cleaner element, or other fuel system fault).

□An exhaust gas analyser for measuring carbon monoxide (CO) and hydrocarbons (HC) is now needed. If one cannot be hired or borrowed, have a local garage perform the check.

CO emissions (mixture)

□The MOT tester has access to the CO limits for all vehicles. The CO level is measured at idle speed, and at 'fast idle' (2500 to 3000 rpm). The following limits are given as a general guide:

At idle speed – Less than 0.5% CO
At 'fast idle' – Less than 0.3% CO
Lambda reading – 0.97 to 1.03

□If the CO level is too high, this may point to poor maintenance, a fuel injection system problem, faulty lambda (oxygen) sensor or catalytic converter. Try an injector cleaning treatment, and check the vehicle's ECU for fault codes.

HC emissions

□The MOT tester has access to HC limits for all vehicles. The HC level is measured at 'fast idle' (2500 to 3000 rpm). The following limits are given as a general guide:

At 'fast idle' – Less then 200 ppm

□Excessive HC emissions are typically caused by oil being burnt (worn engine), or by a blocked crankcase ventilation system ('breather'). If the engine oil is old and thin, an oil change may help. If the engine is running badly, check the vehicle's ECU for fault codes.

Diesel models

□The only emission test for diesel engines is measuring exhaust smoke density, using a calibrated smoke meter. The test involves accelerating the engine at least 3 times to its maximum unloaded speed.

Note: *On engines with a timing belt, it is VITAL that the belt is in good condition before the test is carried out.*

□With the engine warmed up, it is first purged by running at around 2500 rpm for 20 seconds. A governor check is then carried out, by slowly accelerating the engine to its maximum speed. After this, the smoke meter is connected, and the engine is accelerated quickly to maximum speed three times. If the smoke density is less than the limits given below, the vehicle will pass:

Non-turbo vehicles: 2.5m-1
Turbocharged vehicles: 3.0m-1

□If excess smoke is produced, try fitting a new air cleaner element, or using an injector cleaning treatment. If the engine is running badly, where applicable, check the vehicle's ECU for fault codes. Also check the vehicle's EGR system, where applicable. At high mileages, the injectors may require professional attention.

Conversion factors

Length (distance)

Inches (in)	x 25.4	= Millimetres (mm)	x 0.0394	= Inches (in)	
Feet (ft)	x 0.305	= Metres (m)	x 3.281	= Feet (ft)	
Miles	x 1.609	= Kilometres (km)	x 0.621	= Miles	

Volume (capacity)

Cubic inches (cu in; in^3)	x 16.387	= Cubic centimetres (cc; cm^3)	x 0.061	= Cubic inches (cu in; in^3)	
Imperial pints (Imp pt)	x 0.568	= Litres (l)	x 1.76	= Imperial pints (Imp pt)	
Imperial quarts (Imp qt)	x 1.137	= Litres (l)	x 0.88	= Imperial quarts (Imp qt)	
Imperial quarts (Imp qt)	x 1.201	= US quarts (US qt)	x 0.833	= Imperial quarts (Imp qt)	
US quarts (US qt)	x 0.946	= Litres (l)	x 1.057	= US quarts (US qt)	
Imperial gallons (Imp gal)	x 4.546	= Litres (l)	x 0.22	= Imperial gallons (Imp gal)	
Imperial gallons (Imp gal)	x 1.201	= US gallons (US gal)	x 0.833	= Imperial gallons (Imp gal)	
US gallons (US gal)	x 3.785	= Litres (l)	x 0.264	= US gallons (US gal)	

Mass (weight)

Ounces (oz)	x 28.35	= Grams (g)	x 0.035	= Ounces (oz)	
Pounds (lb)	x 0.454	= Kilograms (kg)	x 2.205	= Pounds (lb)	

Force

Ounces-force (ozf; oz)	x 0.278	= Newtons (N)	x 3.6	= Ounces-force (ozf; oz)	
Pounds-force (lbf; lb)	x 4.448	= Newtons (N)	x 0.225	= Pounds-force (lbf; lb)	
Newtons (N)	x 0.1	= Kilograms-force (kgf; kg)	x 9.81	= Newtons (N)	

Pressure

Pounds-force per square inch (psi; lbf/in^2; lb/in^2)	x 0.070	= Kilograms-force per square centimetre (kgf/cm^2; kg/cm^2)	x 14.223	= Pounds-force per square inch (psi; lbf/in^2; lb/in^2)	
Pounds-force per square inch (psi; lbf/in^2; lb/in^2)	x 0.068	= Atmospheres (atm)	x 14.696	= Pounds-force per square inch (psi; lbf/in^2; lb/in^2)	
Pounds-force per square inch (psi; lbf/in^2; lb/in^2)	x 0.069	= Bars	x 14.5	= Pounds-force per square inch (psi; lbf/in^2; lb/in^2)	
Pounds-force per square inch (psi; lbf/in^2; lb/in^2)	x 6.895	= Kilopascals (kPa)	x 0.145	= Pounds-force per square inch (psi; lbf/in^2; lb/in^2)	
Kilopascals (kPa)	x 0.01	= Kilograms-force per square centimetre (kgf/cm^2; kg/cm^2)	x 98.1	= Kilopascals (kPa)	
Millibar (mbar)	x 100	= Pascals (Pa)	x 0.01	= Millibar (mbar)	
Millibar (mbar)	x 0.0145	= Pounds-force per square inch (psi; lbf/in^2; lb/in^2)	x 68.947	= Millibar (mbar)	
Millibar (mbar)	x 0.75	= Millimetres of mercury (mmHg)	x 1.333	= Millibar (mbar)	
Millibar (mbar)	x 0.401	= Inches of water (inH$_2$O)	x 2.491	= Millibar (mbar)	
Millimetres of mercury (mmHg)	x 0.535	= Inches of water (inH$_2$O)	x 1.868	= Millimetres of mercury (mmHg)	
Inches of water (inH$_2$O)	x 0.036	= Pounds-force per square inch (psi; lbf/in^2; lb/in^2)	x 27.68	= Inches of water (inH$_2$O)	

Torque (moment of force)

Pounds-force inches (lbf in; lb in)	x 1.152	= Kilograms-force centimetre (kgf cm; kg cm)	x 0.868	= Pounds-force inches (lbf in; lb in)	
Pounds-force inches (lbf in; lb in)	x 0.113	= Newton metres (Nm)	x 8.85	= Pounds-force inches (lbf in; lb in)	
Pounds-force inches (lbf in; lb in)	x 0.083	= Pounds-force feet (lbf ft; lb ft)	x 12	= Pounds-force inches (lbf in; lb in)	
Pounds-force feet (lbf ft; lb ft)	x 0.138	= Kilograms-force metres (kgf m; kg m)	x 7.233	= Pounds-force feet (lbf ft; lb ft)	
Pounds-force feet (lbf ft; lb ft)	x 1.356	= Newton metres (Nm)	x 0.738	= Pounds-force feet (lbf ft; lb ft)	
Newton metres (Nm)	x 0.102	= Kilograms-force metres (kgf m; kg m)	x 9.804	= Newton metres (Nm)	

Power

Horsepower (hp)	x 745.7	= Watts (W)	x 0.0013	= Horsepower (hp)	

Velocity (speed)

Miles per hour (miles/hr; mph)	x 1.609	= Kilometres per hour (km/hr; kph)	x 0.621	= Miles per hour (miles/hr; mph)	

Fuel consumption*

Miles per gallon, Imperial (mpg)	x 0.354	= Kilometres per litre (km/l)	x 2.825	= Miles per gallon, Imperial (mpg)	
Miles per gallon, US (mpg)	x 0.425	= Kilometres per litre (km/l)	x 2.352	= Miles per gallon, US (mpg)	

Temperature

Degrees Fahrenheit = ($^\circ$C x 1.8) + 32   Degrees Celsius (Degrees Centigrade; $^\circ$C) = ($^\circ$F - 32) x 0.56

It is common practice to convert from miles per gallon (mpg) to litres/100 kilometres (l/100km), where mpg x l/100 km = 282

Engine

- [ ] Engine fails to rotate when attempting to start
- [ ] Engine rotates, but will not start
- [ ] Engine difficult to start when cold
- [ ] Engine difficult to start when hot
- [ ] Starter motor noisy or excessively rough in engagement
- [ ] Engine starts, but stops immediately
- [ ] Engine idles erratically
- [ ] Engine misfires at idle speed
- [ ] Engine misfires throughout the driving speed range
- [ ] Engine hesitates on acceleration
- [ ] Engine stalls
- [ ] Engine lacks power
- [ ] Engine backfires
- [ ] Oil pressure warning light illuminated with engine running
- [ ] Engine runs-on after switching off
- [ ] Engine noises

Cooling system

- [ ] Overheating
- [ ] Overcooling
- [ ] External coolant leakage
- [ ] Internal coolant leakage
- [ ] Corrosion

Fuel and exhaust systems

- [ ] Excessive fuel consumption
- [ ] Fuel leakage and/or fuel odour

Clutch

- [ ] Pedal travels to floor – no pressure or very little resistance
- [ ] Clutch fails to disengage (unable to select gears)
- [ ] Clutch slips (engine speed increases, with no increase in vehicle speed)
- [ ] Judder as clutch is engaged
- [ ] Noise when depressing or releasing clutch pedal

Manual transmission

- [ ] Noisy in neutral with engine running
- [ ] Noisy in one particular gear
- [ ] Difficulty engaging gears
- [ ] Vibration
- [ ] Jumps out of gear
- [ ] Lubricant leaks

DSG transmissions

- [ ] Fluid leakage
- [ ] General gear selection problems

Braking system

- [ ] Vehicle pulls to one side under braking
- [ ] Noise (grinding or high-pitched squeal) when brakes applied)
- [ ] Brakes binding
- [ ] Excessive brake pedal travel
- [ ] Brake pedal feels spongy when depressed
- [ ] Excessive brake pedal effort required to stop vehicle
- [ ] Judder felt through brake pedal or steering wheel when braking
- [ ] Rear wheels locking under normal braking

Driveshafts

- [ ] Clicking or knocking noise on turns (at slow speed on full-lock)

Suspension and steering

- [ ] Vehicle pulls to one side
- [ ] Excessive pitching and/or rolling around corners, or during braking
- [ ] Lack of power assistance
- [ ] Wandering or general instability
- [ ] Excessively stiff steering
- [ ] Excessive play in steering
- [ ] Wheel wobble and vibration
- [ ] Tyre wear excessive

Electrical system

- [ ] Battery will not hold a charge for more than a few days
- [ ] Ignition/no-charge warning light remains illuminated with engine running
- [ ] Ignition/no-charge warning light fails to come on
- [ ] Lights inoperative
- [ ] Instrument readings inaccurate or erratic
- [ ] Horn inoperative, or unsatisfactory in operation
- [ ] Windscreen/tailgate wipers inoperative, or unsatisfactory in operation
- [ ] Windscreen/tailgate washers inoperative, or unsatisfactory in operation
- [ ] Electric windows inoperative, or unsatisfactory in operation
- [ ] Central locking system inoperative, or unsatisfactory in operation

Introduction

The vehicle owner who does his or her own maintenance according to the recommended service schedules should not have to use this section of the manual very often. Modern component reliability is such that, provided those items subject to wear or deterioration are inspected or renewed at the specified intervals, sudden failure is comparatively rare. Faults do not usually just happen as a result of sudden failure, but develop over a period of time. Major mechanical failures in particular are usually preceded by characteristic symptoms over hundreds or even thousands of miles. Those components that do occasionally fail without warning are often small and easily carried in the vehicle.

With any fault-finding, the first step is to decide where to begin investigations. Sometimes this is obvious, but on other occasions, a little detective work will be necessary. The owner who makes half a dozen haphazard adjustments or component renewals may be successful in curing a fault (or its symptoms). However, will be none the wiser if the fault recurs, and ultimately may have spent more time and money than was necessary. A calm and logical approach will be found to be more satisfactory in the long run. Always take into account any warning signs or abnormalities that may have been noticed in the period preceding the fault – power loss, high or low gauge readings, unusual smells, etc – and remember that failure of components such as fuses or spark plugs may only be pointers to some underlying fault.

The pages which follow provide an easy-reference guide to the more common problems which may occur during the operation of the vehicle. These problems and their possible causes are grouped under headings denoting various components or systems, such as Engine, Cooling system, etc. The general Chapter which deals with the problem is also shown in brackets; refer to the relevant part of that Chapter for system-specific information. Whatever the fault, certain basic principles apply. These are as follows:

Verify the fault. This is simply a matter of being sure that you know what the symptoms are before starting work. This is particularly important if you are investigating a fault for someone else, who may not have described it very accurately.

Do not overlook the obvious. For example, if the vehicle will not start, is there petrol in the tank? (Do not take anyone else's word on this particular point, and do not trust the fuel gauge either!) If an electrical fault is indicated, look for loose or broken wires before digging out the test gear.

Cure the disease, not the symptom. Substituting a flat battery with a fully-charged one will get you off the hard shoulder, but if the underlying cause is not attended to, the new battery will go the same way. Similarly, changing oil-fouled spark plugs for a new set will get you moving again, but remember that the reason for the fouling (if it was not simply an incorrect grade of plug) will have to be established and corrected.

Do not take anything for granted. Particularly, do not forget that a new component may itself be defective (especially if it's been rattling around in the boot for months). Also do not leave components out of a fault diagnosis sequence just because they are new or recently fitted. When you do finally diagnose a difficult fault, you will probably realise that all the evidence was there from the start.

Diesel fault diagnosis

The majority of starting problems on small diesel engines are electrical in origin. The mechanic who is familiar with petrol engines but less so with diesel may be inclined to view the diesel's injectors and pump in the same light as the spark plugs and distributor, but this is generally a mistake.

When investigating complaints of difficult starting for someone else, make sure that the correct starting procedure is understood and is being followed. Some drivers are unaware of the significance of the preheating warning light – many modern engines are sufficiently forgiving for this not to matter in mild weather, but with the onset of winter, problems begin.

As a rule of thumb, if the engine is difficult to start but runs well when it has finally got going, the problem is electrical (battery, starter motor or preheating system). If poor performance is combined with difficult starting, the problem is likely to be in the fuel system. The low-pressure (supply) side of the fuel system should be checked before suspecting the injectors and high-pressure pump. The most common fuel supply problem is air getting into the system, and any pipe from the fuel tank forwards must be scrutinised if air leakage is suspected. Normally the pump is the last item to suspect, since unless it has been tampered with, there is no reason for it to be at fault.

Engine

Engine fails to rotate when attempting to start

☐ Battery terminal connections loose or corroded (Weekly checks).
☐ Battery discharged or faulty (Chapter 5A Section 3).
☐ Broken, loose or disconnected wiring in the starting circuit (Chapter 5A Section 7).
☐ Defective starter solenoid or switch (Chapter 5A Section 8).
☐ Defective starter motor (Chapter 5A Section 8).
☐ Starter pinion or flywheel ring gear teeth loose or broken (Chapter 2A Section 15, Chapter 2B Section 13, Chapter 2C Section 14, Chapter 2D Section 14, Chapter 2E Section 16, and Chapter 5A Section 8).
☐ Engine earth strap broken or disconnected (Chapter 12 Section 2).

Engine rotates, but will not start

☐ Fuel tank empty.
☐ Battery discharged (engine rotates slowly) (Chapter 5A Section 2).
☐ Battery terminal connections loose or corroded (*Weekly checks*).
☐ Ignition components damp or damaged – petrol models (Chapter 1A Section 29 and Chapter 5B Section 3).
☐ Broken, loose or disconnected wiring in the ignition circuit – petrol models (Chapter 5B Section 2 and Chapter 12 Section 2).
☐ Worn, faulty or incorrectly gapped spark plugs – petrol models (Chapter 1A Section 29).
☐ Fuel injection system fault (Chapter 4A Section 5 and Chapter 4B Section 4).
☐ Air in fuel system – diesel models ().
☐ Major mechanical failure (eg, timing belt) (Chapter 2A Section 7, Chapter 2B Section 4, Chapter 2C Section 7, Chapter 2D Section 7 or Chapter 2E Section 7).

Engine difficult to start when cold

☐ Battery discharged (Chapter 5A Section 2).
☐ Battery terminal connections loose or corroded (Weekly checks).
☐ Worn, faulty or incorrectly gapped spark plugs – petrol models (Chapter 1A Section 29).
☐ Fuel injection system fault (Chapter 4A Section 5 and Chapter 4B Section 5).
☐ Other ignition system fault – petrol models (Chapter 5B Section 2).
☐ Preheating system fault – diesel models (Chapter 5C Section 2).
☐ Low cylinder compressions (Chapter 2A Section 2, Chapter 2B Section 3, Chapter 2C Section 2, Chapter 2D Section 2, Chapter 2E Section 2).

Engine difficult to start when hot

☐ Air filter element dirty or clogged (Chapter 1A Section 28 or Chapter 1B Section 28).
☐ Fuel injection system fault (Chapter 4A Section 5 and Chapter 4B Section 4).
☐ Low cylinder compressions (Chapter 2A Section 2, Chapter 2B Section 3, Chapter 2C Section 2, Chapter 2D Section 2, or Chapter 2E Section 2).

Starter motor noisy or excessively rough in engagement

☐ Starter pinion or flywheel ring gear teeth loose or broken (Chapter 5A Section 8, Chapter 2A Section 15, Chapter 2A Section 15, Chapter 2C Section 14, Chapter 2D Section 14 and Chapter 2E Section 16).
☐ Starter motor mounting bolts loose or missing (Chapter 5A Section 8).
☐ Starter motor internal components worn or damaged (Chapter 5A Section 8).

Engine starts, but stops immediately

☐ Loose or faulty electrical connections in the ignition circuit – petrol models (Chapter 5B Section 2 and Chapter 12 Section 2).
☐ Vacuum leak at the throttle body or inlet manifold – petrol models (Chapter 4A Section 10).
☐ Blocked injector/fuel injection system fault – petrol models (Chapter 4A Section 5 or Chapter 12 Section 2).
☐ Faulty injector(s) – diesel models (Chapter 4B Section 5).
☐ Air in fuel system – diesel models (Chapter 4B Section 11).

Engine idles erratically

☐ Air filter element clogged (Chapter 1A Section 28 or Chapter 1B Section 28).
☐ Vacuum leak at the throttle body, inlet manifold or associated hoses – petrol models (Chapter 4A).
☐ Worn, faulty or incorrectly gapped spark plugs – petrol models (Chapter 1A Section 29).
☐ Uneven or low cylinder compressions (Chapter 2A Section 2, Chapter 2B Section 3, Chapter 2C Section 2, Chapter 2D Section 2 or Chapter 2E Section 2.
☐ Camshaft lobes worn (Chapter 2A Section 9, Chapter 2B Section 11, Chapter 2C Section 9, Chapter 2D Section 9, or Chapter 2E Section 10).
☐ Timing belt incorrectly tensioned (Chapter 2A Section 7, Chapter 2B Section 4, Chapter 2C Section 7,Chapter 2D Section 7 or Chapter 2E Section 7).
☐ Blocked injector/fuel injection system fault (Chapter 4A Section 5.)
☐ Faulty injector(s) – diesel models (Chapter 4B Section 5).

Engine misfires at idle speed

☐ Worn, faulty or incorrectly gapped spark plugs – petrol models (Chapter 1A Section 29).
☐ Faulty spark plug HT leads – petrol models (Chapter 5B Section 3).
☐ Vacuum leak at the throttle body, inlet manifold or associated hoses (Chapter 4A).
☐ Blocked injector/fuel injection system fault (Chapter 4A Section 5 or Chapter 4B Section 5).
☐ Faulty injector(s) – diesel models (Chapter 4B Section 5).
☐ Uneven or low cylinder compressions (Chapter 2A Section 2, Chapter 2B Section 3, Chapter 2C Section 2, Chapter 2D Section 2 or Chapter 2E Section 2).
☐ Disconnected, leaking, or perished crankcase ventilation hoses (Chapter 4C or Chapter 4D)

Engine (continued)

Engine misfires throughout the driving speed range

- ☐ Fuel filter choked (Chapter 4A Section 6 or Chapter 1B Section 29).
- ☐ Fuel pump faulty, or delivery pressure low (Chapter 4A Section 7).
- ☐ Fuel tank vent blocked, or fuel pipes restricted (Chapter 4A Section 8).
- ☐ Vacuum leak at the throttle body, inlet manifold or associated hoses – petrol models (Chapter 4A).
- ☐ Worn, faulty or incorrectly gapped spark plugs – petrol models (Chapter 1A Section 29).
- ☐ Faulty spark plug HT leads (Chapter 5B Section 3).
- ☐ Faulty injector(s) – diesel models (Chapter 4B Section 5).
- ☐ Faulty ignition coil – petrol models (Chapter 5B Section 3).
- ☐ Uneven or low cylinder compressions (Chapter 2A Section 2, Chapter 2B Section 3, Chapter 2C Section 2, Chapter 2D Section 2 or Chapter 2E Section 2).
- ☐ Blocked injector/fuel injection system fault (Chapter 4A Section 5 or Chapter 4B Section 5).

Engine hesitates on acceleration

- ☐ Worn, faulty or incorrectly gapped spark plugs – petrol models (Chapter 1A Section 29).
- ☐ Vacuum leak at the throttle body, inlet manifold or associated hoses – petrol models (Chapter 4A Section 10).
- ☐ Blocked injector/fuel injection system fault (Chapter 4A Section 5 or Chapter 4B Section 4).
- ☐ Faulty injector(s) – diesel models (Chapter 4B Section 5).

Engine stalls

- ☐ Vacuum leak at the throttle body, inlet manifold or associated hoses – petrol models (Chapter 4A Section 10).
- ☐ Fuel filter choked (Chapter Chapter 4A Section 6 or Chapter 1B Section 29).
- ☐ Fuel pump faulty, or delivery pressure low – petrol models (Chapter 4A Section 7).
- ☐ Fuel tank vent blocked, or fuel pipes restricted (Chapter 4A Section 8 or Chapter 4B Section 9).
- ☐ Blocked injector/fuel injection system fault (Chapter 4A Section 5 or Chapter 4B Section 4).
- ☐ Faulty injector(s) – diesel models (Chapter 4B Section 5).
- ☐ Air in fuel system – diesel models (Chapter 4B Section 11).

Engine lacks power

- ☐ Timing belt incorrectly fitted or tensioned (Chapter 2A Section 7, Chapter 2B Section 4, Chapter 2C Section 7, Chapter 2D Section 7 or Chapter 2E Section 7).
- ☐ Fuel filter choked (Chapter 4A Section 6 or Chapter 1B Section 29).
- ☐ Fuel pump faulty, or delivery pressure low – petrol models (Chapter 4A Section 7).
- ☐ Uneven or low cylinder compressions (Chapter 2A Section 2, Chapter 2B Section 3, Chapter 2C Section 2, Chapter 2D Section 2 or Chapter 2E Section 2).
- ☐ Worn, faulty or incorrectly gapped spark plugs – petrol models (Chapter 1A Section 29).
- ☐ Vacuum leak at the throttle body, inlet manifold or associated hoses – petrol models (Chapter 4A Section 10).
- ☐ Blocked injector/fuel injection system fault (Chapter 4A Section 5 or Chapter 4B Section 5).
- ☐ Turbocharger fault (Chapter 4C Section 6 or Chapter 4D Section 5).
- ☐ Brakes binding (Chapter 9 Section 5 or Chapter 9 Section 9).
- ☐ Clutch slipping (Chapter 6 Section 6).

Engine backfires

- ☐ Timing belt incorrectly fitted or tensioned (Chapter 2A Section 7, Chapter 2B Section 4, Chapter 2C Section 7, Chapter 2D Section 7 or Chapter 2E Section 7).
- ☐ Vacuum leak at the throttle body, inlet manifold or associated hoses – petrol models (Chapter 4A Section 10).

- ☐ Blocked injector/fuel injection system fault (Chapter 4A Section 5 or Chapter 4B Section 4).

Oil pressure warning light illuminated with engine running

- ☐ Low oil level, or incorrect oil grade (Weekly checks Section 5).
- ☐ Faulty oil pressure warning light switch (Chapter 2A Section 20, Chapter 2B Section 16, Chapter 2C Section 18, Chapter 2D Section 18 or Chapter 2E Section 20).
- ☐ Worn engine bearings and/or oil pump (Chapter 2A Section 14, Chapter 2B Section 15, Chapter 2C Section 13, Chapter 2D Section 13 or Chapter 2E Section 15).
- ☐ Oil pressure relief valve defective (Chapter 2A Section 19, Chapter 2B Section 15, Chapter 2C Section 13, Chapter 2D Section 13 or Chapter 2E Section 15).
- ☐ Oil pick-up strainer clogged (Chapter 2A Section 14, Chapter 2B Section 15, Chapter 2C Section 13, Chapter 2D Section 13 or Chapter 2E Section 15).

Engine runs-on after switching off

- ☐ Excessive carbon build-up in engine (Chapter 2A Section 12, Chapter 2B Section 10, Chapter 2C Section 11, Chapter 2D Section 11 or Chapter 2E Section 13).
- ☐ High engine operating temperature (Chapter 3).
- ☐ Fuel injection system fault – petrol models (Chapter 4A Section 5).
- ☐ Fuel injection system fault – diesel models (Chapter 4B Section 4).

Engine noises

Pre-ignition (pinking) or knocking during acceleration or under load

- ☐ Ignition system fault – petrol models (Chapter 1A Section 29 and Chapter 5B Section 2).
- ☐ Incorrect grade of spark plug – petrol models (Chapter 1A Section 29).
- ☐ Incorrect grade of fuel (Chapter 4A Section 1).
- ☐ Vacuum leak at the throttle body, inlet manifold or associated hoses – petrol models (Chapter 4A Section 10).
- ☐ Excessive carbon build-up in engine (Chapter 2A Section 12, Chapter 2B Section 10, Chapter 2C Section 11, Chapter 2D Section 11 or Chapter 2E Section 13).
- ☐ Blocked injector/fuel injection system fault – petrol models (Chapter 4A Section 5).

Whistling or wheezing noises

- ☐ Leaking inlet manifold or throttle body gasket – petrol models (Chapter 4A Section 10).
- ☐ Leaking exhaust manifold gasket or pipe-to-manifold joint (Chapter 4C Section 9 or Chapter 4D Section 9).
- ☐ Leaking turbocharger air ducts (Chapter 4C Section 7 or Chapter 4D Section 7).
- ☐ Leaking vacuum hose (Chapter 9 Section 11 or Chapter 9 Section 21).
- ☐ Blowing cylinder head gasket (Chapter 2A Section 12, Chapter 2B Section 10, Chapter 2C Section 11, Chapter 2D Section 11 or Chapter 2E Section 13).

Tapping or rattling noises

- ☐ Worn valve gear or camshaft (Chapter 2A Section 9, Chapter 2B Section 11, Chapter 2C Section 9, Chapter 2D Section 9 or Chapter 2E Section 10).
- ☐ Ancillary component fault (coolant pump, alternator, etc) (Chapter 3 Section 7 and Chapter 5A Section 5).

Knocking or thumping noises

- ☐ Worn big-end bearings (regular heavy knocking, perhaps worsening under load) (Chapter 2F Section 14).
- ☐ Worn main bearings (rumbling and knocking, perhaps less under load) (Chapter 2F Section 14).
- ☐ Piston slap (most noticeable when cold) (Chapter 2F Section 12).
- ☐ Ancillary component fault (coolant pump, alternator, etc) (Chapter 3 Section 7 or Chapter 5A Section 5) etc.

Cooling system

Overheating

☐ Insufficient coolant in system (Weekly checks Section 5).
☐ Thermostat faulty (Chapter 3 Section 4).
☐ Radiator core blocked, or grille restricted (Chapter 3 Section 3).
☐ Electric cooling fan or thermoswitch faulty (Chapter 3 Section 5).
☐ Pressure cap faulty (Chapter 3 Section 1).
☐ Ignition system fault – petrol engines (Chapter 1A Section 29 or Chapter 5B Section 2).
☐ Inaccurate temperature gauge sender unit (Chapter 3 Section 6).
☐ Airlock in cooling system (Chapter 1A Section 33 or Chapter 1B Section 33).

Overcooling

☐ Thermostat faulty (Chapter 3 Section 4).
☐ Inaccurate temperature gauge sender unit (Chapter 3 Section 6).

External coolant leakage

☐ Deteriorated or damaged hoses or hose clips (Chapter 3 Section 2).

☐ Radiator core or heater matrix leaking (Chapter 3 Section 3 or Chapter 3 Section 9).
☐ Pressure cap faulty (Chapter 3 Section 1).
☐ Water pump seal leaking (Chapter 3 Section 7).
☐ Boiling due to overheating (Weekly checks).
☐ Core plug leaking (Chapter 2F Section 7).

Internal coolant leakage

☐ Leaking cylinder head gasket (Chapter 2A Section 12, Chapter 2B Section 10, Chapter 2C Section 11, Chapter 2D Section 11 or Chapter 2E Section 11).
☐ Cracked cylinder head or cylinder bore (Chapter 2F Section 7).

Corrosion

☐ Infrequent draining and flushing (Chapter 1A Section 33 or Chapter 1B Section 33).
☐ Incorrect coolant mixture or inappropriate coolant type (Chapter 1A Section 33 or Chapter 1B Section 33).

Fuel and exhaust systems

Excessive fuel consumption

☐ Air filter element dirty or clogged (Chapter 1A Section 28 or Chapter 1B Section 28).
☐ Fuel injection system fault (Chapter 4A Section 11 or Chapter 4B Section 4).
☐ Ignition system fault – petrol models (Chapter 1A Section 29 and Chapter 5B Section 3).
☐ Faulty injector(s) – diesel models (Chapter 4B Section 5).
☐ Tyres under-inflated (Weekly checks Section 5).

Fuel leakage and/or fuel odour

☐ Damaged or corroded fuel tank, pipes or connections (Chapter 4A Section 8 or Chapter 4B Section 9)Excessive noise or fumes from exhaust system (Chapter 4C Section 9 or Chapter 4D Section 9).
☐ Leaking exhaust system or manifold joints (Chapter 4C Section 8, Chapter 4C Section 9, Chapter 4D Section 5 or Chapter 4D Section 9).
☐ Leaking, corroded or damaged silencers or pipe (Chapter 4C Section 9 or Chapter 4D Section 9).
☐ Broken mountings causing body or suspension contact (Chapter 4C Section 9 or Chapter 4D Section 9).

Clutch

Pedal travels to floor – no pressure or very little resistance

- [ ] Hydraulic fluid level low/air in hydraulic system (*Weekly checks*).
- [ ] Broken clutch release bearing or fork (Chapter 6 Section 7).
- [ ] Broken diaphragm spring in clutch pressure plate (Chapter 6 Section 6).

Clutch fails to disengage (unable to select gears)

- [ ] Clutch disc sticking on transmission input shaft splines (Chapter 6 Section 6).
- [ ] Clutch disc sticking to flywheel or pressure plate (Chapter 6 Section 6).
- [ ] Faulty pressure plate assembly (Chapter 6 Section 6).
- [ ] Clutch release mechanism worn or incorrectly assembled (Chapter 6 Section 7).

Clutch slips (engine speed increases, with no increase in vehicle speed)

- [ ] Clutch disc linings excessively worn (Chapter 6 Section 6).
- [ ] Clutch disc linings contaminated with oil or grease (Chapter 6 Section 6).

- [ ] Faulty pressure plate or weak diaphragm spring (Chapter 6 Section 6).

Judder as clutch is engaged

- [ ] Clutch disc linings contaminated with oil or grease (Chapter 6 Section 6).
- [ ] Clutch disc linings excessively worn (Chapter 6 Section 6).
- [ ] Faulty or distorted pressure plate or diaphragm spring (Chapter 6 Section 6).
- [ ] Worn or loose engine or transmission mountings (Chapter 2A Section 17, Chapter 2B Section 17, Chapter 2C Section 16, Chapter 2D Section 16 or Chapter 2E Section 18).
- [ ] Clutch disc hub or transmission input shaft splines worn (Chapter 6 Section 6).

Noise when depressing or releasing clutch pedal

- [ ] Worn clutch release bearing (Chapter 6 Section 7).
- [ ] Worn or dry clutch pedal bushes (Chapter 6 Section 3).
- [ ] Faulty pressure plate assembly (Chapter 6 Section 6).
- [ ] Pressure plate diaphragm spring broken (Chapter 6 Section 6).
- [ ] Broken clutch disc cushioning springs (Chapter 6 Section 6).

Manual transmission

Noisy in neutral with engine running

- [ ] Input shaft bearings worn (noise apparent with clutch pedal released, but not when depressed) (Chapter 7A Section 5).*
- [ ] Clutch release bearing worn (noise apparent with clutch pedal depressed, possibly less when released) (Chapter 6 Section 7).

Noisy in one particular gear

- [ ] Worn, damaged or chipped gear teeth (Chapter 7A Section 5).*

Difficulty engaging gears

- [ ] Clutch fault (Chapter 6 Section 6).
- [ ] Worn or damaged gear linkage (Chapter 7A Section 2).
- [ ] Incorrectly adjusted gear linkage (Chapter 7A Section 2).
- [ ] Worn synchroniser units (Chapter 7A Section 5).*

Vibration

- [ ] Lack of oil (Chapter 7A Section 4).
- [ ] Worn bearings (Chapter 7A Section 5).*

Jumps out of gear

- [ ] Worn or damaged gear linkage (Chapter 7A Section 2).
- [ ] Incorrectly adjusted gear linkage (Chapter 7A Section 2).
- [ ] Worn synchroniser units (Chapter 7A Section 5).*
- [ ] Worn selector forks (Chapter 7A Section 5).*

Lubricant leaks

- [ ] Leaking differential output oil seal (Chapter 8 Section 2).
- [ ] Leaking housing joint (Chapter 7A Section 5).*
- [ ] Leaking input shaft oil seal (Chapter 7A Section 5).*

* Although the corrective action necessary to remedy the symptoms described is beyond the scope of the home mechanic, the above information should be helpful in isolating the cause of the condition. This should enable the owner can communicate clearly with a professional mechanic.

DSG transmissions

Fluid leakage

Note: *Due to the complexity of the DSG transmissions, it is difficult for the home mechanic to properly diagnose and service this unit. For problems other than the following, the vehicle should be taken to a dealer service department or automatic transmission specialist. Do not be too hasty in removing the transmission if a fault is suspected, as most of the testing is carried out with the unit still fitted.*

To determine the source of a leak, first remove all built-up dirt and grime from the transmission housing and surrounding areas using a degreasing agent, or by steam-cleaning. Drive the vehicle at low speed, so airflow will not blow the leak far from its source. Raise and support the vehicle, and determine where the leak is coming from.

General gear selection problems

☐ Chapter 7B Section 4 deals with checking and adjusting the selector cable on transmissions.

Braking system

Vehicle pulls to one side under braking

Note: *Before assuming that a brake problem exists, make sure that the tyres are in good condition and correctly inflated, that the front wheel alignment is correct, and that the vehicle is not loaded with weight in an unequal manner. Apart from checking the condition of all pipe and hose connections, any faults occurring on the anti-lock braking system should be referred to a Seat dealer or suitably equipped garage for diagnosis.*

☐ Worn, defective, damaged or contaminated brake pads on one side (Chapter 9 Section 4 and Chapter 9 Section 8).
☐ Seized or partially seized brake caliper piston (Chapter 9 Section 5 or Chapter 9 Section 9).
☐ A mixture of brake pad lining materials fitted between sides (Chapter 9 Section 4 and Chapter 9 Section 8).
☐ Brake caliper mounting bolts loose (Chapter 9 Section 5 or Chapter 9 Section 9).
☐ Worn or damaged steering or suspension components (Chapter 10).

Noise (grinding or high-pitched squeal) when brakes applied)

☐ Brake pad friction lining material worn down to metal backing (Chapter 9 Section 4 or Chapter 9 Section 8).
☐ Excessive corrosion of brake disc. This may be apparent after the vehicle has been standing for some time (Chapter 9 Section 6).
☐ Foreign object (stone chipping, etc.) trapped between brake disc and shield (Chapter 9 Section 6).

Brakes binding

☐ Seized brake caliper piston(s) (Chapter 9 Section 5 or Chapter 9 Section 9).
☐ Incorrectly adjusted handbrake mechanism (Chapter 9 Section 14).
☐ Faulty master cylinder (Chapter 9 Section 13).

Excessive brake pedal travel

☐ Faulty master cylinder (Chapter 9 Section 13).
☐ Air in hydraulic system (Chapter 9 Section 2).
☐ Faulty vacuum servo unit (Chapter 9 Section 11).

Brake pedal feels spongy when depressed

☐ Air in hydraulic system (Chapter 9 Section 2).
☐ Deteriorated flexible rubber brake hoses (Chapter 9 Section 3).
☐ Master cylinder mounting nuts loose (Chapter 9 Section 13).
☐ Faulty master cylinder (Chapter 9 Section 13).

Excessive brake pedal effort required to stop vehicle

☐ Faulty vacuum servo unit (Chapter 9 Section 11).
☐ Faulty brake vacuum pump – diesel models (Chapter 9 Section 21).
☐ Disconnected, damaged or insecure brake servo vacuum hose (Chapter 9 Section 11).
☐ Primary or secondary hydraulic circuit failure (Chapter 9 Section 1).
☐ Seized brake caliper piston(s) (Chapter 9 Section 5 or Chapter 9 Section 9).
☐ Brake pads incorrectly fitted (Chapter 9 Section 4 or Chapter 9 Section 8).
☐ Incorrect grade of brake pads fitted (Chapter 9 Section 4 or Chapter 9 Section 8).
☐ Brake pads contaminated (Chapter 9 Section 4 or Chapter 9 Section 8).

Judder felt through brake pedal or steering wheel when braking

☐ Excessive run-out or distortion of discs (Chapter 9 Section 6).
☐ Brake pads worn (Chapter 9 Section 4 or Chapter 9 Section 8).
☐ Brake caliper mounting bolts loose (Chapter 9 Section 5 or Chapter 9 Section 9).
☐ Wear in suspension or steering components or mountings (Chapter 10).

Rear wheels locking under normal braking

☐ Rear brake pads contaminated (Chapter 9 Section 8).
☐ ABS system fault (Chapter 9 Section 19).

Driveshafts

Clicking or knocking noise on turns (at slow speed on full-lock)

☐ Lack of constant velocity joint lubricant, possibly due to damaged gaiter (Chapter 8 Section 3).

☐ Worn outer constant velocity joint (Chapter 8 Section 5).
☐ Vibration when accelerating or decelerating (Chapter 8 Section 5).
☐ Worn inner constant velocity joint (Chapter 8 Section 5).
☐ Bent or distorted driveshaft (Chapter 8 Section 5).

Suspension and steering

Vehicle pulls to one side

Note: *Before diagnosing suspension or steering faults, be sure that the trouble is not due to incorrect tyre pressures, mixtures of tyre types, or binding brakes.*

☐ Defective tyre (Weekly checks).
☐ Excessive wear in suspension or steering components (Chapter 10).
☐ Incorrect front wheel alignment (Chapter 10 Section 25).
☐ Accident damage to steering or suspension components (Chapter 10).

Excessive pitching and/or rolling around corners, or during braking

☐ Defective shock absorbers (Chapter 10 Section 4 or Chapter 10 Section 15).
☐ Broken or weak spring and/or suspension component (Chapter 10).
☐ Worn or damaged anti-roll bar or mountings – where applicable (Chapter 10 Section 7 or Chapter 10 Section 16).

Lack of power assistance

☐ Faulty rack-and-pinion steering gear (Chapter 10 Section 22).

Wandering or general instability

☐ Incorrect front wheel alignment (Chapter 10 Section 25).
☐ Worn steering or suspension joints, bushes or components (Chapter 10).
☐ Roadwheels out of balance (*Weekly checks*).
☐ Faulty or damaged tyre (Weekly checks Section 5).
☐ Wheel bolts loose (refer to Chapter 10 Section 0 for correct torque).
☐ Defective shock absorbers (Chapter 10 Section 4 or Chapter 10 Section 15).

Excessively stiff steering

☐ Lack of steering gear lubricant (Chapter 10 Section 22).
☐ Seized track rod end balljoint or suspension balljoint Chapter 10 Section 24 or Chapter 10 Section 6.
☐ Incorrect front wheel alignment (Chapter 10 Section 25).
☐ Steering rack or column bent or damaged (Chapter 10 Section 22 or Chapter 10 Section 20).

Excessive play in steering

☐ Worn steering column intermediate shaft universal joint (Chapter 10 Section 20).

☐ Worn steering track rod end balljoints (Chapter 10 Section 23).
☐ Worn rack-and-pinion steering gear (Chapter 10 Section 22).
☐ Worn steering or suspension joints, bushes or components (Chapter 10).

Wheel wobble and vibration

☐ Front roadwheels out of balance (vibration felt mainly through the steering wheel) (Chapter 1A Section 20, Chapter 1B Section 19 and Chapter 10 Section 1).
☐ Rear roadwheels out of balance (vibration felt throughout the vehicle) (Chapter 1A Section 20, Chapter 1B Section 19, and Chapter 10).
☐ Roadwheels damaged or distorted (Chapter 1A Section 27, Chapter 1B Section 26 and Chapter 10).
☐ Faulty or damaged tyre (Weekly checks).
☐ Worn steering or suspension joints, bushes or components (Chapter 1A Section 20, Chapter 1B Section 19 and Chapter 10).
☐ Wheel bolts loose (*Weekly checks*).

Tyre wear excessive

Tyres worn on inside or outside edges

☐ Tyres under-inflated (wear on both edges) (Weekly checks).
☐ Incorrect camber or castor angles (wear on one edge only) (*Weekly checks*).
☐ Worn steering or suspension joints, bushes or components (Chapter 1A Section 20, Chapter 1B Section 19 and Chapter 10).
☐ Excessively hard cornering.
☐ Accident damage.

Tyre treads exhibit feathered edges

☐ Incorrect toe setting (Chapter 10 Section 25).

Tyres worn in centre of tread

☐ Tyres over-inflated (Weekly checks).

Tyres worn on inside and outside edges

☐ Tyres under-inflated (Weekly checks).

Tyres worn unevenly

☐ Tyres/wheels out of balance (Chapter 1A Section 20 or Chapter 1B Section 19).
☐ Excessive wheel or tyre run-out (Chapter 1A Section 20, Chapter 1B Section 19 and Chapter 10).
☐ Worn shock absorbers (Chapter 10 Section 4 and Chapter 10 Section 15).
☐ Faulty tyre (Weekly checks).

Electrical system

Battery will not hold a charge for more than a few days

Note: *For problems associated with the starting system, refer to the faults listed under 'Engine' earlier in this Section.*

- ☐ Battery defective internally (Chapter 5A Section 2).
- ☐ Battery terminal connections loose or corroded (Weekly checks).
- ☐ Auxiliary drivebelt worn or incorrectly adjusted (Chapter 1A Section 30 or Chapter 1B Section 30).
- ☐ Alternator not charging at correct output (Chapter 5A Section 4).
- ☐ Alternator or voltage regulator faulty (Chapter 5A Section 4).
- ☐ Short-circuit causing continual battery drain (Chapter 12 Section 2)

Ignition/no-charge warning light remains illuminated with engine running

- ☐ Auxiliary drivebelt broken, worn, or incorrectly adjusted (Chapter 1A Section 30 or Chapter 1B Section 30).
- ☐ Alternator brushes worn, sticking, or dirty (Chapter 5A Section 6).
- ☐ Alternator brush springs weak or broken (Chapter 5A Section 6).
- ☐ Internal fault in alternator or voltage regulator (Chapter 5A Section 6).
- ☐ Broken, disconnected, or loose wiring in charging circuit (Chapter 12 Section 2).

Ignition/no-charge warning light fails to come on

- ☐ Warning light LED defective (Chapter 12 Section 11).
- ☐ Broken, disconnected, or loose wiring in warning light circuit (Chapter 12 Section 2).
- ☐ Alternator faulty (Chapter 5A Section 4).

Lights inoperative

- ☐ Bulb blown or LED unit faulty (Chapter 12 Section 6 and Chapter 12 Section 7).
- ☐ Corrosion of bulb or bulbholder contacts (Chapter 12 Section 6).
- ☐ Blown fuse (Chapter 12 Section 3).
- ☐ Faulty relay (Chapter 12 Section 3).
- ☐ Broken, loose, or disconnected wiring (Chapter 12 Section 2).
- ☐ Faulty switch (Chapter 12 Section 5).

Instrument readings inaccurate or erratic

Fuel or temperature gauges give no reading

- ☐ Faulty gauge sender unit (Chapter 4A Section 7 or Chapter 4B Section 8).
- ☐ Wiring open-circuit (Chapter 12 Section 2).
- ☐ Faulty gauge (Chapter 12 Section 11).

Fuel or temperature gauges give continuous maximum reading

- ☐ Faulty gauge sender unit (Chapter 4A Section 7 or Chapter 4B Section 8).
- ☐ Wiring short-circuit (Chapter 12 Section 2).
- ☐ Faulty gauge (Chapter 12 Section 11).

Horn inoperative, or unsatisfactory in operation

Horn operates all the time

- ☐ Horn push either earthed or stuck down (Chapter 12 Section 5).
- ☐ Horn cable-to-horn push earthed (Chapter 12 Section 2).

Horn fails to operate

- ☐ Blown fuse (Chapter 12 Section 3).
- ☐ Cable or cable connections loose, broken or disconnected (Chapter 12 Section 2).
- ☐ Faulty horn (Chapter 12 Section 16).

Horn emits intermittent or unsatisfactory sound

- ☐ Cable connections loose (Chapter 12 Section 2).
- ☐ Horn mountings loose (Chapter 12 Section 16).
- ☐ Faulty horn (Chapter 12 Section 16).

Windscreen/tailgate wipers inoperative, or unsatisfactory in operation

Wipers fail to operate, or operate very slowly

- ☐ Bonnet not closed (built-into the electronic control system) (Chapter 11 Section 8).
- ☐ Wiper blades stuck to screen, or linkage seized or binding (Weekly checks and Chapter 12 Section 18).
- ☐ Blown fuse (Chapter 12 Section 3).
- ☐ Cable or cable connections loose, broken or disconnected (Chapter 12 Section 2).
- ☐ Faulty relay (Chapter 12 Section 3).
- ☐ Faulty wiper motor (Chapter 12 Section 19).

Wiper blades sweep over too large or too small an area of the glass

- ☐ Wiper arms incorrectly positioned on spindles (Chapter 12 Section 18).
- ☐ Excessive wear of wiper linkage (Chapter 12 Section 19).
- ☐ Wiper motor or linkage mountings loose or insecure (Chapter 12 Section 19).

Electrical system (continued

Windscreen/tailgate wipers inoperative, or unsatisfactory in operation (continued)

Wiper blades fail to clean the glass effectively

☐ Wiper blade rubbers worn or perished (Weekly checks).
☐ Wiper arm tension springs broken, or arm pivots seized (Chapter 12 Section 18).
☐ Insufficient windscreen washer additive to adequately remove road film (Weekly checks).

Windscreen/tailgate washers inoperative, or unsatisfactory in operation

One or more washer jets inoperative

☐ Blocked washer jet (Chapter Chapter 1A Section 24 or Chapter 1B Section 23).
☐ Disconnected, kinked or restricted fluid hose (Chapter 12 Section 21).
☐ Insufficient fluid in washer reservoir (Chapter 1A Section 24 or Chapter 1B Section 23).

Washer pump fails to operate

☐ Broken or disconnected wiring or connections (Chapter 12 Section 2).
☐ Blown fuse (Chapter 12 Section 3).
☐ Faulty washer switch (Chapter 12 Section 5).
☐ Faulty washer pump (Chapter 12 Section 21).

Electric windows inoperative, or unsatisfactory in operation

Window glass will only move in one direction

☐ Faulty switch (Chapter 12 Section 5).

Window glass slow to move

☐ Regulator seized or damaged, or in need of lubrication (Chapter 11 Section 14).
☐ Door internal components or trim fouling regulator (Chapter 11 Section 12).
☐ Faulty motor (Chapter 11 Section 14).

Window glass fails to move

☐ Blown fuse (Chapter 12 Section 3).
☐ Faulty relay (Chapter 12 Section 3).
☐ Broken or disconnected wiring or connections (Chapter 12 Section 2).
☐ Faulty motor (Chapter 11 Section 14).

Central locking system inoperative, or unsatisfactory in operation

Complete system failure

☐ Blown fuse (Chapter 12 Section 3).
☐ Faulty relay (Chapter 12 Section 3).
☐ Broken or disconnected wiring or connections (Chapter 12 Section 2).
☐ Faulty control unit (Chapter 11 Section 17).

Latch locks but will not unlock, or unlocks but will not lock

☐ Faulty control unit (Chapter 11 Section 17).
☐ Broken or disconnected latch operating rods or levers (Chapter 11 Section 13).
☐ Faulty relay (Chapter 12 Section 3).

One actuator fails to operate

☐ Broken or disconnected wiring or connections (Chapter 12 Section 2).
☐ Faulty actuator (Chapter 11 Section 13).
☐ Broken, binding or disconnected latch operating rods or levers (Chapter 11 Section 13).
☐ Fault in door lock (Chapter 11 Section 13).

A

ABS (Anti-lock brake system) A system, usually electronically controlled, that senses incipient wheel lockup during braking and relieves hydraulic pressure at wheels that are about to skid.

Air bag An inflatable bag hidden in the steering wheel (driver's side) or the dash or glovebox (passenger side). In a head-on collision, the bags inflate, preventing the driver and front passenger from being thrown forward into the steering wheel or windscreen.

Air cleaner A metal or plastic housing, containing a filter element, which removes dust and dirt from the air being drawn into the engine.

Air filter element The actual filter in an air cleaner system, usually manufactured from pleated paper and requiring renewal at regular intervals.

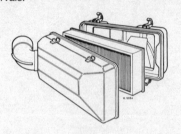

Air filter

Allen key A hexagonal wrench which fits into a recessed hexagonal hole.

Alligator clip A long-nosed spring-loaded metal clip with meshing teeth. Used to make temporary electrical connections.

Alternator A component in the electrical system which converts mechanical energy from a drivebelt into electrical energy to charge the battery and to operate the starting system, ignition system and electrical accessories.

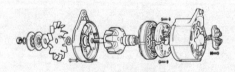

Alternator (exploded view)

Ampere (amp) A unit of measurement for the flow of electric current. One amp is the amount of current produced by one volt acting through a resistance of one ohm.

Anaerobic sealer A substance used to prevent bolts and screws from loosening. Anaerobic means that it does not require oxygen for activation. The Loctite brand is widely used.

Antifreeze A substance (usually ethylene glycol) mixed with water, and added to a vehicle's cooling system, to prevent freezing of the coolant in winter. Antifreeze also contains chemicals to inhibit corrosion and the formation of rust and other deposits that

would tend to clog the radiator and coolant passages and reduce cooling efficiency.

Anti-seize compound A coating that reduces the risk of seizing on fasteners that are subjected to high temperatures, such as exhaust manifold bolts and nuts.

Anti-seize compound

Asbestos A natural fibrous mineral with great heat resistance, commonly used in the composition of brake friction materials. Asbestos is a health hazard and the dust created by brake systems should never be inhaled or ingested.

Axle A shaft on which a wheel revolves, or which revolves with a wheel. Also, a solid beam that connects the two wheels at one end of the vehicle. An axle which also transmits power to the wheels is known as a live axle.

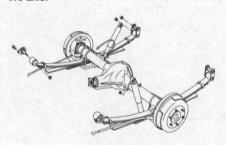

Axle assembly

Axleshaft A single rotating shaft, on either side of the differential, which delivers power from the final drive assembly to the drive wheels. Also called a driveshaft or a halfshaft.

B

Ball bearing An anti-friction bearing consisting of a hardened inner and outer race with hardened steel balls between two races.

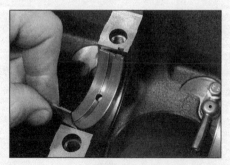

Bearing

Bearing The curved surface on a shaft or in a bore, or the part assembled into either, that permits relative motion between them with minimum wear and friction.

Big-end bearing The bearing in the end of the connecting rod that's attached to the crankshaft.

Bleed nipple A valve on a brake wheel cylinder, caliper or other hydraulic component that is opened to purge the hydraulic system of air. Also called a bleed screw.

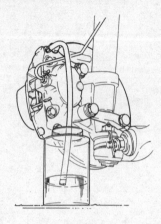

Brake bleeding

Brake bleeding Procedure for removing air from lines of a hydraulic brake system.

Brake disc The component of a disc brake that rotates with the wheels.

Brake drum The component of a drum brake that rotates with the wheels.

Brake linings The friction material which contacts the brake disc or drum to retard the vehicle's speed. The linings are bonded or riveted to the brake pads or shoes.

Brake pads The replaceable friction pads that pinch the brake disc when the brakes are applied. Brake pads consist of a friction material bonded or riveted to a rigid backing plate.

Brake shoe The crescent-shaped carrier to which the brake linings are mounted and which forces the lining against the rotating drum during braking.

Braking systems For more information on braking systems, consult the *Haynes Automotive Brake Manual*.

Breaker bar A long socket wrench handle providing greater leverage.

Bulkhead The insulated partition between the engine and the passenger compartment.

C

Caliper The non-rotating part of a disc-brake assembly that straddles the disc and carries the brake pads. The caliper also contains the hydraulic components that cause the pads to pinch the disc when the brakes are applied. A caliper is also a measuring tool that can be set to measure inside or outside dimensions of an object.

Camshaft A rotating shaft on which a series of cam lobes operate the valve mechanisms. The camshaft may be driven by gears, by sprockets and chain or by sprockets and a belt.

Canister A container in an evaporative emission control system; contains activated charcoal granules to trap vapours from the fuel system.

Canister

Carburettor A device which mixes fuel with air in the proper proportions to provide a desired power output from a spark ignition internal combustion engine.

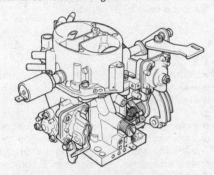

Carburettor

Castellated Resembling the parapets along the top of a castle wall. For example, a castellated balljoint stud nut.

Castellated nut

Castor In wheel alignment, the backward or forward tilt of the steering axis. Castor is positive when the steering axis is inclined rearward at the top.

Catalytic converter A silencer-like device in the exhaust system which converts certain pollutants in the exhaust gases into less harmful substances.

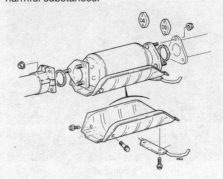

Catalytic converter

Circlip A ring-shaped clip used to prevent endwise movement of cylindrical parts and shafts. An internal circlip is installed in a groove in a housing; an external circlip fits into a groove on the outside of a cylindrical piece such as a shaft.

Clearance The amount of space between two parts. For example, between a piston and a cylinder, between a bearing and a journal, etc.

Coil spring A spiral of elastic steel found in various sizes throughout a vehicle, for example as a springing medium in the suspension and in the valve train.

Compression Reduction in volume, and increase in pressure and temperature, of a gas, caused by squeezing it into a smaller space.

Compression ratio The relationship between cylinder volume when the piston is at top dead centre and cylinder volume when the piston is at bottom dead centre.

Constant velocity (CV) joint A type of universal joint that cancels out vibrations caused by driving power being transmitted through an angle.

Core plug A disc or cup-shaped metal device inserted in a hole in a casting through which core was removed when the casting was formed. Also known as a freeze plug or expansion plug.

Crankcase The lower part of the engine block in which the crankshaft rotates.

Crankshaft The main rotating member, or shaft, running the length of the crankcase, with offset "throws" to which the connecting rods are attached.

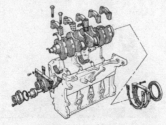

Crankshaft assembly

Crocodile clip See Alligator clip

D

Diagnostic code Code numbers obtained by accessing the diagnostic mode of an engine management computer. This code can be used to determine the area in the system where a malfunction may be located.

Disc brake A brake design incorporating a rotating disc onto which brake pads are squeezed. The resulting friction converts the energy of a moving vehicle into heat.

Double-overhead cam (DOHC) An engine that uses two overhead camshafts, usually one for the intake valves and one for the exhaust valves.

Drivebelt(s) The belt(s) used to drive accessories such as the alternator, water pump, power steering pump, air conditioning compressor, etc. off the crankshaft pulley.

Accessory drivebelts

Driveshaft Any shaft used to transmit motion. Commonly used when referring to the axleshafts on a front wheel drive vehicle.

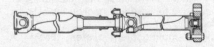

Driveshaft

Drum brake A type of brake using a drum-shaped metal cylinder attached to the inner surface of the wheel. When the brake pedal is pressed, curved brake shoes with friction linings press against the inside of the drum to slow or stop the vehicle.

Drum brake assembly

E

EGR valve A valve used to introduce exhaust gases into the intake air stream.

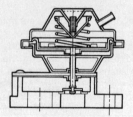

EGR valve

Electronic control unit (ECU) A computer which controls (for instance) ignition and fuel injection systems, or an anti-lock braking system. For more information refer to the *Haynes Automotive Electrical and Electronic Systems Manual.*

Electronic Fuel Injection (EFI) A computer controlled fuel system that distributes fuel through an injector located in each intake port of the engine.

Emergency brake A braking system, independent of the main hydraulic system, that can be used to slow or stop the vehicle if the primary brakes fail, or to hold the vehicle stationary even though the brake pedal isn't depressed. It usually consists of a hand lever that actuates either front or rear brakes mechanically through a series of cables and linkages. Also known as a handbrake or parking brake.

Endfloat The amount of lengthwise movement between two parts. As applied to a crankshaft, the distance that the crankshaft can move forward and back in the cylinder block.

Engine management system (EMS) A computer controlled system which manages the fuel injection and the ignition systems in an integrated fashion.

Exhaust manifold A part with several passages through which exhaust gases leave the engine combustion chambers and enter the exhaust pipe.

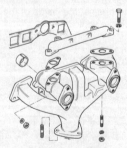

Exhaust manifold

F

Fan clutch A viscous (fluid) drive coupling device which permits variable engine fan speeds in relation to engine speeds.

Feeler blade A thin strip or blade of hardened steel, ground to an exact thickness, used to check or measure clearances between parts.

Feeler blade

Firing order The order in which the engine cylinders fire, or deliver their power strokes, beginning with the number one cylinder.

Flywheel A heavy spinning wheel in which energy is absorbed and stored by means of momentum. On cars, the flywheel is attached to the crankshaft to smooth out firing impulses.

Free play The amount of travel before any action takes place. The "looseness" in a linkage, or an assembly of parts, between the initial application of force and actual movement. For example, the distance the brake pedal moves before the pistons in the master cylinder are actuated.

Fuse An electrical device which protects a circuit against accidental overload. The typical fuse contains a soft piece of metal which is calibrated to melt at a predetermined current flow (expressed as amps) and break the circuit.

Fusible link A circuit protection device consisting of a conductor surrounded by heat-resistant insulation. The conductor is smaller than the wire it protects, so it acts as the weakest link in the circuit. Unlike a blown fuse, a failed fusible link must frequently be cut from the wire for replacement.

G

Gap The distance the spark must travel in jumping from the centre electrode to the side

Adjusting spark plug gap

electrode in a spark plug. Also refers to the spacing between the points in a contact breaker assembly in a conventional points-type ignition, or to the distance between the reluctor or rotor and the pickup coil in an electronic ignition.

Gasket Any thin, soft material - usually cork, cardboard, asbestos or soft metal - installed between two metal surfaces to ensure a good seal. For instance, the cylinder head gasket seals the joint between the block and the cylinder head.

Gasket

Gauge An instrument panel display used to monitor engine conditions. A gauge with a movable pointer on a dial or a fixed scale is an analogue gauge. A gauge with a numerical readout is called a digital gauge.

H

Halfshaft A rotating shaft that transmits power from the final drive unit to a drive wheel, usually when referring to a live rear axle.

Harmonic balancer A device designed to reduce torsion or twisting vibration in the crankshaft. May be incorporated in the crankshaft pulley. Also known as a vibration damper.

Hone An abrasive tool for correcting small irregularities or differences in diameter in an engine cylinder, brake cylinder, etc.

Hydraulic tappet A tappet that utilises hydraulic pressure from the engine's lubrication system to maintain zero clearance (constant contact with both camshaft and valve stem). Automatically adjusts to variation in valve stem length. Hydraulic tappets also reduce valve noise.

I

Ignition timing The moment at which the spark plug fires, usually expressed in the number of crankshaft degrees before the piston reaches the top of its stroke.

Inlet manifold A tube or housing with passages through which flows the air-fuel mixture (carburettor vehicles and vehicles with throttle body injection) or air only (port fuel-injected vehicles) to the port openings in the cylinder head.

J

Jump start Starting the engine of a vehicle with a discharged or weak battery by attaching jump leads from the weak battery to a charged or helper battery.

L

Load Sensing Proportioning Valve (LSPV) A brake hydraulic system control valve that works like a proportioning valve, but also takes into consideration the amount of weight carried by the rear axle.

Locknut A nut used to lock an adjustment nut, or other threaded component, in place. For example, a locknut is employed to keep the adjusting nut on the rocker arm in position.

Lockwasher A form of washer designed to prevent an attaching nut from working loose.

M

MacPherson strut A type of front suspension system devised by Earle MacPherson at Ford of England. In its original form, a simple lateral link with the anti-roll bar creates the lower control arm. A long strut - an integral coil spring and shock absorber - is mounted between the body and the steering knuckle. Many modern so-called MacPherson strut systems use a conventional lower A-arm and don't rely on the anti-roll bar for location.

Multimeter An electrical test instrument with the capability to measure voltage, current and resistance.

N

NOx Oxides of Nitrogen. A common toxic pollutant emitted by petrol and diesel engines at higher temperatures.

O

Ohm The unit of electrical resistance. One volt applied to a resistance of one ohm will produce a current of one amp.

Ohmmeter An instrument for measuring electrical resistance.

O-ring A type of sealing ring made of a special rubber-like material; in use, the O-ring is compressed into a groove to provide the sealing action.

O-ring

Overhead cam (ohc) engine An engine with the camshaft(s) located on top of the cylinder head(s).

Overhead valve (ohv) engine An engine with the valves located in the cylinder head, but with the camshaft located in the engine block.

Oxygen sensor A device installed in the engine exhaust manifold, which senses the oxygen content in the exhaust and converts this information into an electric current. Also called a Lambda sensor.

P

Phillips screw A type of screw head having a cross instead of a slot for a corresponding type of screwdriver.

Plastigage A thin strip of plastic thread, available in different sizes, used for measuring clearances. For example, a strip of Plastigage is laid across a bearing journal. The parts are assembled and dismantled; the width of the crushed strip indicates the clearance between journal and bearing.

Plastigage

Propeller shaft The long hollow tube with universal joints at both ends that carries power from the transmission to the differential on front-engined rear wheel drive vehicles.

Proportioning valve A hydraulic control valve which limits the amount of pressure to the rear brakes during panic stops to prevent wheel lock-up.

R

Rack-and-pinion steering A steering system with a pinion gear on the end of the steering shaft that mates with a rack (think of a geared wheel opened up and laid flat). When the steering wheel is turned, the pinion turns, moving the rack to the left or right. This movement is transmitted through the track rods to the steering arms at the wheels.

Radiator A liquid-to-air heat transfer device designed to reduce the temperature of the coolant in an internal combustion engine cooling system.

Refrigerant Any substance used as a heat transfer agent in an air-conditioning system. R-12 has been the principle refrigerant for many years; recently, however, manufacturers have begun using R-134a, a non-CFC substance that is considered less harmful to the ozone in the upper atmosphere.

Rocker arm A lever arm that rocks on a shaft or pivots on a stud. In an overhead valve engine, the rocker arm converts the upward movement of the pushrod into a downward movement to open a valve.

Rotor In a distributor, the rotating device inside the cap that connects the centre electrode and the outer terminals as it turns, distributing the high voltage from the coil secondary winding to the proper spark plug. Also, that part of an alternator which rotates inside the stator. Also, the rotating assembly of a turbocharger, including the compressor wheel, shaft and turbine wheel.

Runout The amount of wobble (in-and-out movement) of a gear or wheel as it's rotated. The amount a shaft rotates "out-of-true." The out-of-round condition of a rotating part.

S

Sealant A liquid or paste used to prevent leakage at a joint. Sometimes used in conjunction with a gasket.

Sealed beam lamp An older headlight design which integrates the reflector, lens and filaments into a hermetically-sealed one-piece unit. When a filament burns out or the lens cracks, the entire unit is simply replaced.

Serpentine drivebelt A single, long, wide accessory drivebelt that's used on some newer vehicles to drive all the accessories, instead of a series of smaller, shorter belts. Serpentine drivebelts are usually tensioned by an automatic tensioner.

Serpentine drivebelt

Shim Thin spacer, commonly used to adjust the clearance or relative positions between two parts. For example, shims inserted into or under bucket tappets control valve clearances. Clearance is adjusted by changing the thickness of the shim.

Slide hammer A special puller that screws into or hooks onto a component such as a shaft or bearing; a heavy sliding handle on the shaft bottoms against the end of the shaft to knock the component free.

Sprocket A tooth or projection on the periphery of a wheel, shaped to engage with a chain or drivebelt. Commonly used to refer to the sprocket wheel itself.

Starter inhibitor switch On vehicles with an automatic transmission, a switch that prevents starting if the vehicle is not in Neutral or Park.

Strut See MacPherson strut.

T

Tappet A cylindrical component which transmits motion from the cam to the valve stem, either directly or via a pushrod and rocker arm. Also called a cam follower.

Thermostat A heat-controlled valve that regulates the flow of coolant between the cylinder block and the radiator, so maintaining optimum engine operating temperature. A thermostat is also used in some air cleaners in which the temperature is regulated.

Thrust bearing The bearing in the clutch assembly that is moved in to the release levers by clutch pedal action to disengage the clutch. Also referred to as a release bearing.

Timing belt A toothed belt which drives the camshaft. Serious engine damage may result if it breaks in service.

Timing chain A chain which drives the camshaft.

Toe-in The amount the front wheels are closer together at the front than at the rear. On rear wheel drive vehicles, a slight amount of toe-in is usually specified to keep the front wheels running parallel on the road by offsetting other forces that tend to spread the wheels apart.

Toe-out The amount the front wheels are closer together at the rear than at the front. On front wheel drive vehicles, a slight amount of toe-out is usually specified.

Tools For full information on choosing and using tools, refer to the *Haynes Automotive Tools Manual*.

Tracer A stripe of a second colour applied to a wire insulator to distinguish that wire from another one with the same colour insulator.

Tune-up A process of accurate and careful adjustments and parts replacement to obtain the best possible engine performance.

Turbocharger A centrifugal device, driven by exhaust gases, that pressurises the intake air. Normally used to increase the power output from a given engine displacement, but can also be used primarily to reduce exhaust emissions (as on VW's "Umwelt" Diesel engine).

U

Universal joint or U-joint A double-pivoted connection for transmitting power from a driving to a driven shaft through an angle. A U-joint consists of two Y-shaped yokes and a cross-shaped member called the spider.

V

Valve A device through which the flow of liquid, gas, vacuum, or loose material in bulk may be started, stopped, or regulated by a movable part that opens, shuts, or partially obstructs one or more ports or passageways. A valve is also the movable part of such a device.

Valve clearance The clearance between the valve tip (the end of the valve stem) and the rocker arm or tappet. The valve clearance is measured when the valve is closed.

Vernier caliper A precision measuring instrument that measures inside and outside dimensions. Not quite as accurate as a micrometer, but more convenient.

Viscosity The thickness of a liquid or its resistance to flow.

Volt A unit for expressing electrical "pressure" in a circuit. One volt that will produce a current of one ampere through a resistance of one ohm.

W

Welding Various processes used to join metal items by heating the areas to be joined to a molten state and fusing them together. For more information refer to the *Haynes Automotive Welding Manual*.

Wiring diagram A drawing portraying the components and wires in a vehicle's electrical system, using standardised symbols. For more information refer to the *Haynes Automotive Electrical and Electronic Systems Manual*.

Note: *References throughout this index are in the form* **"Chapter number"** • **"Page number"**. *So, for example, 2C•15 refers to page 15 of Chapter 2C.*

Note: *References throughout this index are in the form "**Chapter number**" • "**Page number**". So, for example, 2C•15 refers to page 15 of Chapter 2C.*

Preserving Our Motoring Heritage

< The Model J Duesenberg Derham Tourster. Only eight of these magnificent cars were ever built – this is the only example to be found outside the United States of America

Almost every car you've ever loved, loathed or desired is gathered under one roof at the Haynes Motor Museum. Over 300 immaculately presented cars and motorbikes represent every aspect of our motoring heritage, from elegant reminders of bygone days, such as the superb Model J Duesenberg to curiosities like the bug-eyed BMW Isetta. There are also many old friends and flames. Perhaps you remember the 1959 Ford Popular that you did your courting in? The magnificent 'Red Collection' is a spectacle of classic sports cars including AC, Alfa Romeo, Austin Healey, Ferrari, Lamborghini, Maserati, MG, Riley, Porsche and Triumph.

A Perfect Day Out

Each and every vehicle at the Haynes Motor Museum has played its part in the history and culture of Motoring. Today, they make a wonderful spectacle and a great day out for all the family. Bring the kids, bring Mum and Dad, but above all bring your camera to capture those golden memories for ever. You will also find an impressive array of motoring memorabilia, a comfortable 70 seat video cinema and one of the most extensive transport book shops in Britain. The Pit Stop Cafe serves everything from a cup of tea to wholesome, home-made meals or, if you prefer, you can enjoy the large picnic area nestled in the beautiful rural surroundings of Somerset.

> John Haynes O.B.E., Founder and Chairman of the museum at the wheel of a Haynes Light 12.

< Graham Hill's Lola Cosworth Formula 1 car next to a 1934 Riley Sports.

The Museum is situated on the A359 Yeovil to Frome road at Sparkford, just off the A303 in Somerset. It is about 40 miles south of Bristol, and 25 minutes drive from the M5 intersection at Taunton.
Open 9.30am - 5.30pm (10.00am - 4.00pm Winter) 7 days a week, *except Christmas Day, Boxing Day and New Years Day*
Special rates available for schools, coach parties and outings Charitable Trust No. 292048